Artificial Intelligence and Competition Policy

Foreword by
Cani Fernández

Editors
Alden Abbott
Thibault Schrepel

Concurrences
Paris - New York - London

All rights reserved. No photocopying: copyright licences do not apply. The information provided in this publication is general and may not apply in a specific situation. The publisher accepts no responsibility for any acts or omissions contained herein. Enquiries concerning reproduction should be sent to the Institute of Competition Law, at the address below.

Copyright © 2024 by Institute of Competition Law
106 West 32nd Street, Suite 144 New York, NY, 10001, USA
www.concurrences.com
book@concurrences.com

First Printing, September 2024
978-1-954750-42-5 (Hardcover)
Library of Congress Control Number: 2024945284

Cover Design: Yves Buliard, www.yvesbuliard.fr
Book Design and Layout implementation: Nord Compo

Concurrences Books

Tributes

Harry First Liber Amicorum, *S. Waller, A. Gavil, D. Bush (eds.), Forthcoming 2025*
James F. Rill – A Life in Antitrust, *J. Taladay, P. Lugard, J. Antonio (eds.), 2024*
Eleanor M. Fox – Antitrust Ambassador to the World, *2021*
Herbert Hovenkamp – The Dean of American Antitrust Law, *2021*
Frédéric Jenny – Standing Up for Convergence and Relevance in Antitrust, (Vol. I & II), *2019 & 2021*
Albert Foer – A Consumer Voice in the Antitrust Arena, *2020*
Richard Whish – Taking Competition Law Outside the Box, *2020*
Douglas H. Ginsburg – An Antitrust Professor on the Bench (Vol. I & II), *2018 & 2020*
Wang Xiaoye – The Pioneer of Competition Law in China, *A. Emch, W. Ng (eds.), 2019*
Ian S. Forrester – A Scot without Borders (Vol. I & II), *A. Komninos (eds.), 2015*
William E. Kovacic – An Antitrust Tribute (Vol. I & II), *2013 & 2014*

Practical Books

The DMA and more – Future application and margin of manoeuvre for national jurisdictions, *G. Muscolo & A. Massolo (eds.), Forthcoming 2025*
State Aid & National Enforcement, *J. Derenne, D. Jouve, C. Lemaire, F. Martucci (eds.), Forthcoming 2024*
The 2023 U.S. Merger Guidelines: A Review, *Sean Sullivan (ed.), 2024*
Competition Inspections in 25 Jurisdictions 2nd Edition – A Practitioner's Guide, *N. Jalabert-Doury (ed.), 2024*
EU Antitrust Enforcement – Law, Economics, History, Policy & Practice, *Wouter P. J. Wils, 2024*
The EU Foreign Subsidies Regulation, *Andreas Reindl, Isabelle Van Damme, 2024*
Compendium of Antitrust Damages Actions – ICC – 2nd edition, *J.W.H. Denton AO, F. Brunet, S. Williams, C. Inthavisay (eds.), 2023*
Pharmaceutical Antitrust: An Analysis of US and EU Law, *M. Thill-Tayara, G. Gordon (eds.), 2023*
Innovation Paradox in Merger Control, *G. Gurkaynak, 2023*
Antitrust and the Digital Economy, *Y. Katsoulacos (ed.), 2023 (in collaboration with CRESSE)*
Judicial Review of Competition Cases, *D. Ginsburg, T. Eicke (eds.), 2023*
Competition Law Treatment of Joint Ventures, *B. Bleicher, N. Campbell, A. Hamilton, N. Hukkinen, A. Khan, A. Mordaunt (eds.), 2022 (in collaboration with the IBA)*
Information Exchange & Related Risks, *Z. Marosi, M. Soares (eds.), 2022 (in collaboration with the IBA)*
Rulemaking Authority of the US Federal Trade Commission, *D. Crane (ed.), 2022*
The International Competition Network at Twenty, *D. Anderson & P. Lugard (eds.), 2022*
Competition Case Law Digest – 5th Edition, *F. Jenny, N. Charbit (eds.), 2022*
Competition Inspections in 21 Jurisdictions – A Practitioner's Guide, *N. Jalabert-Doury (ed.), 2022*
Perspectives on Antitrust Compliance, *A. Riley, A. Stephan, A. Tubbs (eds.), 2022 (in collaboration with the ICC)*
Turkish Competition Law, *G. Gürkaynak, 2021*
Competition Law – Climate Change & Environmental Sustainability, *S. Holmes, D. Middelschulte, M. Snoep (eds.), 2021*
Merger Control in Latin America – A Jurisdictional Guide, *P. Burnier da Silveira, P. Sittenfeld (eds.), 2020*
Competition Inspections under EU Law – A Practitioner's Guide, *N. Jalabert-Doury, 2020*
Gun Jumping in Merger Control – A Jurisdictional Guide, *C. Hatton, Y. Comtois, A. Hamilton (eds.), 2019 (in collaboration with the IBA)*
Choice – A New Standard for Competition Analysis? *P. Nihoul (ed.), 2016*

PhD Theses

The Economics of Digital Markets – Essays in Theoretical and Empirical Industrial Organization, *E. Arnaud-Joufray, 2024*
Abuse of Platform Power, *F. Bostoen, 2023*
Reform of Chinese State-Owned Enterprises, *X. Bai, 2023*
Competition & Regulation in Network Industries – Essays in Industrial Organization, *J-M. Zogheib, 2021*
The Role of Media Pluralism in the Enforcement of EU Competition Law, *K. Bania, 2019*
Buyer Power, *I. Herrera Anchustegui, 2017*

General Interest

Why Competition? Voices from the Antitrust Community and Beyond, *D. Crane, D. Gerard, R. Tritell (eds.), Forthcoming 2024*
Competition Law Dictionary, *D. Healey, R. Whish, W. E. Kovacic, P. Trevisán (eds.), 2024*
The 5 Labours of Europe, *P.-E. Partsch, 2024*
Great Antitrust Enforcers, *W. E. Kovacic, 2023*
Competition – How to Speak Like an Expert, *E. Combe, 2023*
Women and Antitrust – Voices from the Field (Vol I & II), *E. Kurgonaite & K. Nordlander, 2020*

e-Book versions available for **Concurrences+** subscribers

Contributors

Cora Allen
Wilson Sonsini Goodrich & Rosati LLP

Jonathan M. Barnett
University of Southern California

Christian Bergqvist
University of Copenhagen

Oliver Budzinski
Ilmenau University of Technology

Yo Sop Choi
Graduate School of International and Area Studies, Hankuk University of Foreign Studies

Diane Coyle
University of Cambridge

Daniel Crane
University of Michigan Paul, Weiss, Rifkind, Wharton & Garrison LLP

Hayane Dahmen
University of Toronto

Stephen Dnes
Royal Holloway, University of London

Fausto Gernone
University of California, Berkeley

Teodora Groza
Sciences Po Law School

Aaron Hoag
Antitrust Division, United States Department of Justice

William Lehr
Massachusetts Institute of Technology

Mariateresa Maggiolino
Bocconi University

Godefroy de Boiscuillé
Côte d'Azur University Paris Panthéon Assas University

Tejas N. Narechania
University of California, Berkeley

Mark Niefer
Antonin Scalia Law School, George Mason University

Victoriia Noskova
Ilmenau University of Technology

Maureen Ohlhausen
Wilson Sonsini Goodrich & Rosati LLP

Taylor Owings
Wilson Sonsini Goodrich & Rosati LLP

Jennifer Pullen
University of St. Gallen

Lazar Radic
International Center for Law & Economics IE University

Camila Ringeling
The George Washington University Hausfeld LLP

Quentin B. Schäfer
University of Strathclyde

Ganesh Sitaraman
Vanderbilt University Law School

Daniel F. Spulber
Northwestern University

Volker Stocker
Weizenbaum-Institute

Kristian Stout
International Center for Law & Economics

David J. Teece
University of California, Berkeley

Aleksandra Wierzbicka
Cleary Gottlieb Steen & Hamilton

John M. Yun
Antonin Scalia Law School, George Mason University

Laura Zoboli
University of Brescia

Overview

Contributors	I
Foreword	V
Cani Fernández	
Introduction	VII
Alden Abbott and Thibault Schrepel	
Biographies	XIII
Table of Contents	XXI

PART I
Market Dynamics, Mergers, and Partnerships in AI

Open-Source Generative AI from a Competition Policy Perspective	3
Diane Coyle and Hayane Dahmen	
Competing in the Age of AI: Firm Capabilities and Antitrust Considerations ...	17
Fausto Gernone and David J. Teece	
Antitrust and Innovation Competition: Investments and Partnerships in Artificial Intelligence	35
Daniel F. Spulber	
Mergers by Other Means? AI Partnerships and the Frontiers of (Post-)Industrial Organization	47
Teodora Groza and Aleksandra Wierzbicka	
Preserving Competition in Generative AI: Addressing the Merger Conundrum ...	65
Mariateresa Maggiolino and Laura Zoboli	
Artificial Intelligence, Uncertainty, and Merger Review	83
Mark J. Niefer and Aaron D. Hoag	

PART II
AI Challenges for Competition Law

What is the Relevant Product Market in AI?	107
Lazar Radic and Kristian Stout	
Defining AI Markets: Who is Afraid of Digital Ghosts?	133
Stephen Dnes	
Finding the *Ghost in the Shell*: EU and US Antitrust Enforcement of AI Collusion	151
Christian Bergqvist and Camila Ringeling	

What About Bob? Revisiting the Intersection of Antitrust Law
and Algorithmic Pricing in 2024 187
    Maureen Ohlhausen, Taylor Owings and Cora Allen

Korean Competition Rules on Algorithmic Discrimination 201
    Yo Sop Choi

The Recoupment Conundrum: Rethinking Predatory Pricing
in the Age of Algorithms 215
    Jennifer Pullen

Computational Methods in the Evaluation of Mergers
and Acquisitions 231
    Victoriia Noskova and Oliver Budzinski

PART III
Policy Responses to the AI Boom

The Folly of AI Regulation 247
    John M. Yun

The Case Against Preemptive Antitrust in the Generative
Artificial Intelligence Ecosystem 261
    Jonathan M. Barnett

Antimonopoly Tools for Regulating Artificial Intelligence 287
    Tejas N. Narechania and Ganesh Sitaraman

Competition Policy after the Coming Wave of General Purpose
Technologies 303
    Daniel A. Crane

The Tortuous Path to AI Act Compliance: A Competitive Burden
for Companies 321
    Godefroy de Boiscuillé

Competition Policy over the Generative AI Waterfall 335
    William Lehr and Volker Stocker

AI, IP, and Competition Policy: Adjusting Policy Levers to a New GPT 359
    Quentin B. Schäfer

Foreword

CANI FERNÁNDEZ

President of the Spanish National Markets
and Competition Commission (CNMC)
Spain

When I was asked to write a few words to present the book Artificial Intelligence and Competition Policy, I felt very excited considering the strong implications it has in relation with competition assessment and enforcement, which are part of the critical duties of the Spanish National Markets and Competition Commission, CNMC.

The long road taken by Artificial Intelligence to reach almost every corner of advanced societies, from its tentative beginnings in the 60's to the present day, when extremely powerful computing facilities have made it a practical and ubiquitous reality, is one of no-return, and it would be wise to be prepared to exploit its enormous potentialities, while managing its inherent risks.

This is precisely the goal of this book, to help public officials, practitioners, firms, and academia to have a better understanding of what it means and its implications.

In my opinion, the present work offers a comprehensive analysis of the implications of AI on competition policy and policy makers. It features a large number of chapters by distinguished economists and lawyers, discussing various aspects of AI's impact on market competition, regulatory frameworks, and legal challenges.

In fact, the key topics covered by this remarkable work should be in the current agenda of any Competition Authority, including among others:

- *AI-Driven Collusion*: can AI really facilitate collusion among competitors? Quite probably the answer is yes, posing new challenges for competition policy.
- *Addressing Algorithmic Discrimination*: a closely related issue is the adequacy of current competition rules to address algorithmic discrimination, the book highlights the need for more nuanced rules.
- *Predatory Pricing Strategies*: the authors analyze predatory pricing strategies in the context of AI, exploring how these strategies can distort market competition.
- *Potential for Monopolistic Megafirms*: another issue scarcely explored to date is the potential for AI to give rise to monopolistic megafirms, which would carry important implications for competition policy.

- *Antitrust Interventions*: the need for a nuanced approach to antitrust interventions in the age of AI is discussed.
- *Merger Review Involving AI*: another point of view explores the challenges of merger review involving AI, which could force competition authorities to use computational methods in such merger evaluations.

The list of the remaining topics covered in the book is extensive, including the enforcement gaps in US and EU competition law, the role of digital infrastructure in AI regulation, and the undeniable importance of human-AI hybrids in strategic decision-making.

Each of these topics would certainly warrant its own dedicated work, but the level of objective, academic discussion achieved here provides a solid foundation for those who wish to delve into these emerging subjects of interest for competition enforcement and regulation.

Accordingly, the CNMC has among its top priorities to adapt its competition enforcement policy to the current wave of digitization and use of artificial intelligence. The initiatives already in hand, applying new techniques of data analysis and using artificial intelligence to detect traditional and new forms of anticompetitive conduct, have experienced an evolution and, in some respects, a relevant breakthrough.

From my personal point of view, a book like this is welcomed, appreciated and is a much-needed tool. As the work states, competition policy frameworks must evolve to address the complexities of competition in the age of AI. As such, it provides valuable insights for legislators, policymakers, and enforcers navigating the intersection of AI and competition policy.

Introduction

ALDEN ABBOTT AND THIBAULT SCHREPEL

Setting the stage

It's Friday night, 7:00 pm. You have two tickets to the New York Metropolitan Opera House. As you make your way to your front-row seats, you take in the elegant crowd and prepare for a magical evening. To your surprise, a speaker comes to announce that tonight's conductor will be a robot powered by artificial intelligence (AI). Curious and slightly nervous, you are intrigued. The orchestra begins with the repertoire of Verdi and Léo Delibes, producing harmonious music. But it soon transitions into playing dissonant free jazz pieces by Sun Ra and Ornette Coleman.

This book in your hands explores the wide-ranging effects of AI on competition law. Several contributions assign AI the role of an orchestra conductor fine-tuning the performance of each player. They detail how AI reinforces the positions of large technology companies in the intricate symphony of the global economy. Or is that so? AI can also act as an agent of chaos. Other contributions explore how AI can accelerate disruption, challenging and breaking free from the harmonious patterns of the structure-conduct-performance paradigm. They illustrate how AI dismantles barriers to entry and empowers startups to deviate from traditional frameworks. Like a free jazz musician, AI fosters unpredictable (market) outcomes.

The positioning of competition agencies within these musical ensembles is yet to be defined. Artificial Intelligence and Competition Policy sheds some light on the stage, providing insights for anyone interested in the transformative impact it may foster on the melody of digital markets.

Roadmap

AI encompasses a variety of innovative technologies that imitate or automate human cognitive functions through mechanisms like machine learning. AI applications range from mimicking of human reasoning to automation of specific complex tasks. Generative AI, which involves the creation of text and media, has experienced rapid adoption for business and personal use in recent years.

The far-reaching implications of this "AI revolution" have caught the attention of government regulators worldwide. In March 2024, the European Parliament adopted the AI Act, the first comprehensive regulatory framework for dealing with AI. And in October 2023, the Biden Administration issued a Presidential Executive Order on Artificial Intelligence. Unsurprisingly, public enforcers are beginning to investigate the effects on market competition stemming from the introduction of AI in business settings.

Introduction

In the light of these developments, Artificial Intelligence and Competition Policy is particularly timely. The volume features 20 chapters by distinguished economists and lawyers exploring a wide range of questions bearing on the implications of AI for the competitive process.

These chapters are organized into three sections that reflect the depth of the subject. The first section, "Market Dynamics, Mergers, and Partnerships in AI," delves into the intricate ways AI is reshaping market structures, driving mergers and acquisitions, and fostering strategic partnerships. The second section, "AI Challenges for Competition Law," addresses the novel challenges AI poses to existing competition frameworks, including market definitions, algorithmic collusion, and predatory pricing. The final section, "Policy Responses to the AI Boom," examines the various policy mechanisms and regulatory strategies designed to harness the benefits of AI while mitigating potential risks.

Market Dynamics, Mergers, and Partnerships in AI

Diane Coyle of the University of Cambridge and Hayane Dahmen of the University of Toronto provide a surprising insight on the competitive effects of "openness" in AI. As they explain, "open" has been presumed to be a procompetitive attribute for generative artificial intelligence (genAI). Yet each genAI model's "openness" can vary significantly, and this variation affects firms' incentives. The degree of openness may also vary at different levels of the technology stack. Also, a number of open models have partnerships with incumbent large technology companies, on terms that are not public. The authors conclude that competition agencies should employ a graduated concept of "open" in genAI markets, taking into account not only the specific "open" features of the model and any partnerships involved, but also the specific character of the ecosystems involving related upstream models and downstream AI products.

Fausto Gernone and David J. Teece of the Haas School of Business, University of California at Berkeley, evaluate the integration of AI within routine business operations and its potential to reconfigure markets, as well as its effects on dynamic capabilities and strategic management. After reviewing the role of AI in shaping organizations, the authors analyze some potential areas of concern for competition policy. They stress that AI is poised to increase the complexity of economic interactions further, adding to the urgency to update the competition policy framework. It is essential to move beyond a static view of markets as isolated pools of substitutable products and build a framework that acknowledges the interconnected and ever-evolving nature of competition in the age of AI.

Daniel Spulber of Northwestern University evaluates US FTC orders to Alphabet (Google), Amazon, Microsoft, OpenAI, and Anthropic, requesting information regarding investments and partnerships in AI. The FTC expressed concerns about the potential harm to innovation and competition. Spulber points out that investments and partnerships are an alternative to M&A and vertical integration. He argues that these investments and partnerships are not likely to diminish innovation or competition in either cloud computing or AI. Antitrust policymakers should not discourage investments and partnerships based on hypothetical effects on future innovation. He concludes that investments and partnerships can make important contributions to innovation, competition and the market for technology.

Teodora Groza of Sciences Po Law School and Aleksandra Wierzbicka of the Cleary Gottlieb Steen & Hamilton law firm (Brussels office) explain that AI has ushered in a new era of collaboration between tech giants and AI startups, reshaping competition dynamics. AI partnerships pose a challenge for the traditional EU competition law instruments. Using the Microsoft–OpenAI partnership as a case study, the authors discuss the need to balance innovation incentives with competition concerns.

Mariateresa Maggiolino of Bocconi University and Laura Zoboli of the University of Brescia explore a merger control regulatory framework tailored to scrutinize Big Tech's generative AI investments. They also delve into the theories of harm that could justify authorities intervening to block such transactions. Ultimately, their chapter seeks to offer nuanced reflections on the practicality and efficacy of deploying antitrust authorities' interventions within this rapidly evolving landscape.

Attorneys Mark J. Niefer of the Scalia Law School and Aaron D. Hoag of the US Department of Justice Antitrust Division discuss AI, uncertainty, and merger review. They note that rapid innovation can take firms, industries, and economies in unexpected directions, with unpredictable implications for competition. They draw on the existing literature on innovation, uncertainty, and mergers, as well as our own experience as competition law enforcers, to discuss the ways in which a competition authority might encounter and address uncertainty about the competitive effect of a merger involving AI.

AI Challenges for Competition Law

Lazar Radic and Kristian Stout of the International Center for Law and Economics highlight the problem of AI market definition. They argue that the lack of a proper understanding of the outward and inward boundaries of AI markets has practical implications for antitrust policy and regulation because it may lead to inaccurate assessments of market concentration and market power, resulting in both under and over-enforcement of competition law compared to the social optimum. For example, it is likely – or at the very least plausible – that as soon as one accounts for the substitutability of AI and non-AI products, the concentration in some of those markets that hitherto appeared to be monopolistic withers away. Conversely, a failure to account for the internal heterogeneity of AI could lead to an under-estimation of market concentration and market power, by artificially expanding the universe of products that comprise the same relevant market. Confounding the boundaries of product markets obfuscates the real competitive dynamics in those markets, rendering the enforcer myopic to incentives, the direction of competitive threats, the potential for procompetitive benefits, and the possibility and likelihood of entry and expansion.

Introduction

Stephen Dnes of Royal Holloway, University of London, assesses the current state of AI merger evaluation in major jurisdictions. He notes that "digital ghosts" from perceived missteps with Web 2.0 technologies pervade antitrust analysis of artificial intelligence. Chief amongst these is a concern that orthodox requirements to define markets led to undue precaution in relation to network effects. As such, Dnes considers the underlying debate as to the interaction of effects and relevant markets and considers recent developments relating to AI based on this history. He identifies the boundary between demand and supply side analysis as a key to understanding future dynamics. Changes will be evolutionary, not revolutionary, as the orthodoxy requiring a statement of market definition remains even in AI merger cases.

Christian Bergqvist of the University of Copenhagen and Camilla Ringeling of the George Washington University Competition and Innovation Lab examine the assessment of AI collusion under US and EU competition law. Their focus is on whether the proliferation of AI-driven decision-making could enable entities to coordinate in anticompetitive ways that are beyond the reach of enforcers. According to the authors, there are indeed enforcement gaps. In particular, they explain that US enforcers face greater challenges than their EU counterparts, as American case law has developed a narrow notion of "agreement" or "understanding" under high and inconsistent evidentiary burdens. They argue that US courts should adopt a more welcoming approach to considering "plus factors" in determining whether illegal anticompetitive agreements may be inferred.

Wilson Sonsini Goodrich & Rosati attorneys Maureen Ohlhausen (a former Acting FTC Chair), Taylor Owings, and Cora Allen, focus on the use by competitors of the same price setting algorithm to facilitate hard core collusion over price. The authors review recent US case law and evaluate the legal implications of various pricing scenarios. One scenario involves situations when an algorithm may "teach itself" to initiate collusion without human intervention

Yo Sop Choi of the Graduate School of International and Area Studies, Hankuk University of Foreign Studies, Korea, provides an overview of Korean competition rules on algorithmic discrimination. The author explains that the Korean competition authority has applied the competition provisions on abuses of market dominance and on unfair trade practices to self-preferencing. Moreover, personalized pricing is becoming one of the most important competition issues in Korea. The author argues against a new digital law in Korea on the basis that the existing competition rules are sufficient to tackle conduct involving algorithmic discrimination.

Jennifer Pullen of the University of St. Gallen examines the evolving challenge of predatory pricing in the context of artificial intelligence. The chapter explores the contrasting approaches to assessing recoupment, generally regarded as the second phase of predatory pricing strategies after predation, under both US and EU law. The author argues that sophisticated AI tools make recoupment easier to achieve, making a potential convergence between US and EU approaches to recoupment more likely. She concludes, however, that both jurisdictions may need to adapt their frameworks to effectively address the nuances of predatory pricing in the era of algorithmic pricing strategies.

Economists Victoriia Noskova and Oliver Budzinski of the Institute of Economics, Ilmenau University of Technology, Germany, delve into computational methods in the evaluation of mergers and acquisitions. They conclude that each step of the merger review process offers space for computational tools, which, despite having considerable imperfections, can enhance merger control proceedings. However, some elements of the institutional framework of merger control require adaptations to specific characteristics of computational tools, which should contribute to their wider application.

Policy Responses to the AI Boom

John M. Yun of Scalia Law School argues against "the folly of AI regulation." While AI offers great promise, the technology is powered by people. Since people are involved, AI can be used to generate immense value but also facilitate harms including perpetuating false information and deception. Further, AI could itself hallucinate. Consequently, a call for restraint on AI regulation is not a call to ignore the dangers that AI generated output can pose. Rather it is a recognition that regulation itself creates costs on society including the potential to dampen the rate of innovation, crowd out market-based solutions, and entrench incumbents and larger technology companies.

Jonathan M. Barnett of the University of Southern California argues against preemptive application of antitrust in the generative AI (genAI) ecosystem. As he explains, available evidence indicates that technically competitive entrants can generally secure access to the inputs required to achieve entry, including funding, semiconductors, cloud-computing services, datasets, and foundation models. Consistent with this view, entry into the models and applications segments of the genAI is especially robust. Investments and alliances involving large technology platforms, venture-capital, and institutional investors, and model developers, which have elicited regulatory concern, currently appear to be efficient arrangements to aggregate the complementary assets required to produce genAI models and applications and face competition from a variety of other business models.

Tejas Narechania of the University of California, Berkely, and Ganesh Sitaraman of Vanderbilt University, make the case for an antimonopoly approach to governing AI. They show that AI's industrial organization, rooted in AI's technological structure, suffers from market concentration at and across a number of layers. They argue that an unregulated AI oligopoly has a range of undesirable consequences. In light of these conclusions, they show how antimonopoly market-shaping tools – the law of networks, platforms, and utilities, industrial policy, public options, and cooperative governance – can help facilitate competition.

Daniel Crane of the University of Michigan postulates that a coming technological wave, consisting of a variety of overlapping and mutually reinforcing general-purpose technologies including AI, robotics, quantum computing, synthetic biology, energy expansion, and nanotechnology, is likely to fundamentally reshape the economic order, with profound consequences for competition policy. The author explains that the continuing shift in economic value from atoms to bits will likely generate an inevitable and perhaps unstoppable tendency toward monopolistic megafirms. According to Crane, antitrust's technologies for controlling anticompetitive behavior will run into a wall as far more powerful technologies for engaging in anticompetitive behavior emerge.

Godefroy de Boiscuillé of the University of Nice Côte d'Azur argues that EU regulation of AI will have a negative effect on competition. He notes that the EU AI Act imposes new rules "without prejudice to the application of Union competition law." Despite this legitimate goal, he posits that the AI Act will impact EU competition law for two main reasons. First, the AI Act both extends and limits the investigative powers of competition agencies in Europe. Accordingly, it will impact competition authorities. Second, the AI Act could create new regulatory barriers to market entry. It will therefore impact EU based companies in the internal market, permanently preventing firms from entering a market or delaying the arrival of new companies.

Economists William Lehr of the Massachusetts Institute of Technology and Volcker Stocker of the Weizman-Institute assert that the emergence of generative AI signals has brought us to a global "waterfall moment," an inflection point that will result in profound and wide-ranging changes for all aspects of society and the economy, including competition policy. They note the essential role digital infrastructure plays in making (gen)AI possible, in establishing the effective bounds of what may be accomplished with it, and in shaping our digital future. This infrastructure is the ship that is carrying us over the falls, and its steerage capabilities will prove essential in enabling any hope for regulating AI, including enabling effective competition policy in the future digital economy. The authors focus in particular on the importance of evidence-based decision-making (EBDM) enabled through a digital infrastructure platform.

Quentin B. Schäfer of the University of Strathclyde, Glasgow, examines the suitability of different policy levers to improve and maintain incentives to invent and invest in AI. He argues that the fundamental technological and economic uncertainty renders anticipatory doctrinal adjustments in IP law, such as AI inventorship, undesirable. Rather, scholarship should focus on the design of policy levers capable of flexibly accommodating a variety of different technological and market outcomes. He also proposes a research agenda for future issues on the IP-Competition Interface in relation to AI, focusing on obligations surrounding obligations to share access to closed systems and IP rights.

In sum, these diverse chapters offer a provocative look at key areas of competition-related AI scholarship. They should help inform legislators, policymakers, and enforcers, as they evaluate how competition law can be adapted to confront the challenges of AI.

PART I
Market Dynamics, Mergers, and Partnerships in AI

Open-Source Generative AI from a Competition Policy Perspective

DIANE COYLE[*] AND HAYANE DAHMEN[**]

University of Cambridge

University of Toronto

Abstract

This chapter explores the implications of open-source generative artificial intelligence (genAI) for competition policy, examining the conflicting ways in which open-source characteristics could be interpreted in competition analysis. In general, "open" has been presumed to be positive for competition in genAI. Yet each model's "openness" can vary significantly, and this variation affects firms' incentives. The degree of openness may also vary at different levels of the technology stack, so vertical or ecosystem relationships between upstream and downstream markets need to be considered. Finally, a number of open models have partnerships with incumbent large technology companies, on terms that are not public. Competition agencies will therefore need to employ a graduated concept of "open" in genAI markets, and to take into account not only the specific "open" features of the model and any partnerships involved, but also the specific character of the ecosystems involving related upstream models and downstream AI products. They should avoid a presumption that open-source genAI models are automatically pro-competitive.

---

[*] Diane Coyle is Bennett Professor of Public Policy at the University of Cambridge, and an academic advisor to the Competition and Markets Authority and Office for National Statistics.

[**] Hayane Dahmen is a Postdoctoral Fellow at the University of Toronto.

I. Introduction

1. The launch of generative AI (genAI) models has prompted new questions for competition authorities, further amplifying existing concerns about the extent of concentration and the power of Big Tech companies in digital markets.[1] On the one hand, Big Tech companies are among the leaders in the latest frontier AI models and may be able to cement their dominant position further with the new advances. On the other hand, there are several new competitors and, importantly, open-source models are assumed to offer hope for a more competitive landscape.

2. This chapter examines the implications of open-source genAI for competition policy. We focus on foundation models (FMs), which include Large Language Models (LLMs) as well as image generative models trained with image data.[2] Several factors introduce elements of uncertainty for competition analysis in this area. The field is marked by rapidly advancing technology and use-cases as yet largely undefined. Moreover, market structures and relationships of new players such as OpenAI and Mistral with large tech incumbents are in flux, as are their commercial models. While open-source models are seen as a potential constraint on the market power of incumbent "closed" models developed by a small number of already large tech companies, some of these models are being integrated with other services offered by Big Tech, especially the "hyperscalers" – Amazon, Google, and Microsoft. These factors pose a challenge in terms of how competition in and for the market should be understood in the already concentrated landscapes of many digital markets, where gatekeepers exist at several levels of the technology stack.

3. In what follows, we argue that competition agencies will need to employ a graduated concept of "open" in genAI markets, and take into account not only the specific features of the model and any partnerships involved, but also the specific character of the ecosystems involving related upstream models and downstream AI products. First, we begin by discussing the open-source tradition and the need for categorizing the degrees of openness in genAI. This classification paves the way for a discussion of their likely implications from a traditional competition perspective. In general, "open" has been presumed to be positive for genAI competition. Yet the level of "openness" can vary significantly; it is important to appreciate that "open" in this context has a fluid

---

[1] The terminology is fluid, so we use "genAI" here as an umbrella term. *Cf. generative AI*, CAMBRIDGE DICTIONARY, https://dictionary.cambridge.org/dictionary/english/generative-ai (last visited 22 July 2024).

[2] *See e.g.*, COMPETITION & MARKETS AUTHORITY, AI FOUNDATION MODELS: INITIAL REPORT 2-2–2-4 (2023) [hereinafter CMA REPORT], https://assets.publishing.service.gov.uk/media/64528e622f62220013a6a491/AI_Foundation_Models_-_Initial_review_.pdf. "FMs are a type of AI technology that are trained on vast amounts of data that can be adapted to a wide range of tasks and operations... The type of data that is used to train a FM determines its 'mode'." See also COMPETITION AND MARKETS AUTHORITY, AI FOUNDATION MODELS: UPDATE PAPER (Apr. 11, 2024), https://assets.publishing.service.gov.uk/media/661941a6c1d297c6ad1dfeed/Update_Paper__1_.pdf.

meaning. Second, we analyze potential arguments regarding the competitive dynamics of open-source models in genAI. Finally, we consider how the antitrust toolkit needs to expand by taking an ecosystem-wide approach in the case of markets involving genAI.

II. The Meaning of Open Source

4. The concept of open-source software has generated significant contention in computer science literature. In its early iterations, the free-software movement was concerned with pushing back against the expansion of proprietary mechanisms in the tech industry.[3] The concept of freely available software is often attributed to Richard Stallman's GNU Project, first announced in 1983.[4] Stallman had argued that anyone with a software copy should have the freedom to cooperate with others in using it.[5] This movement laid the groundwork for the concept of open-source software to develop, which emphasizes collaboration and accessibility.[6]

5. Open source has become an important feature of corporate software engineering even as proprietary software became the norm. For instance, the Linux Foundation noted in 2022 that 70–90 percent of any given piece of modern software was estimated to involve open-source components.[7] A 2023 analysis estimated that 96 percent of all code bases use some kind of open-source software.[8] This suggests the widespread prevalence of open-source components.

6. More recently, open source has become a key focus of the wider AI safety debate over who should have control over genAI. In general, more open models involve increasing amounts of community research, auditability, and broader perspectives. But they also involve lower risk controls. Conversely, more closed models reflect only internal research and more narrow perspectives, lower auditability, and higher risk control.[9] Critics have focused on the argument that open models increase the possibility of harms such as the production of bioweapons or prevalence of cyberattacks

---

[3] See e.g., Rebecca Ackermann, *The Future of Open-Source is Still Very Much in Flux*, MIT TECH. REV. (Aug. 17, 2023), https://www.technologyreview.com/2023/08/17/1077498/future-open-source/.

[4] Richard Stallman, Initial Announcement, *GNU stands for GNU's not Unix* GNU (Sept. 27, 1983), https://www.gnu.org/gnu/initial-announcement.html.

[5] *Id.*

[6] Christine Peterson coined the term in 1998. See Christine Peterson, *How I Coined the Term "Open Source"*, OPENSOURCE (Feb. 1, 2018), https://opensource.com/article/18/2/coining-term-open-source-software.

[7] THE LINUX FOUNDATION AND THE LABORATORY FOR INNOVATION SCIENCE AT HARVARD, CENSUS II OF FREE AND OPEN SOURCE SOFTWARE – APPLICATION LIBRARIES (Mar. 2022), https://www.linuxfoundation.org/research/census-ii-of-free-and-open-source-software-application-libraries.

[8] Ackermann, *supra* note 3.

[9] Irene Solaiman, The Gradient of Generative AI Release: Methods and Considerations, Proceedings of the 2023 ACM Conference on Fairness, Accountability, and Transparency 4 (Feb. 5, 2023), https://arxiv.org/abs/2302.04844.

7. Proponents of open source have focused on the argument that open models bring widely ranging economic and non-economic benefits. An important strand of economic analysis of the open-source movement has focused on individual incentives, emphasizing issues such as reputational enhancement.[12] Other key arguments have centered around increasing the size of innovation communities, changing the balance between competition and collaboration, and affecting monetization strategies.[13] Some scholars have cited the potential for open-source models to mitigate business model monoculture and market concentration,[14] as well as antitrust concerns.[15] Open-sourcing has also been correlated to greater entrepreneurship.[16]

8. On the non-economic side, proponents have argued open-source AI models can accelerate scientific research by enhancing the reproducibility of research, whereas closed models tend to get retired.[17] They have also argued that increased access of open-source FMs can contribute to a democratization of AI, enhancing the ability for actors across a greater economic range to participate in the market.[18] By allowing others to inspect, modify, and distribute the source code, open source can contribute to an environment of trust and accountability, enabling the public and regulators to have a greater understanding of how results were derived.

---

[10] *See e.g.*, Elizabeth Seger et al., *Open-Sourcing Highly Capable Foundation Models: An Evaluation of Risks, Benefits, and Alternative Methods for Pursuing Open-Source Objectives*, CTR. FOR THE GOVERNANCE OF AI (Sept. 29, 2023), https://arxiv.org/pdf/2311.09227. "While open-sourcing has historically provided substantial net benefits for most software and AI development processes, we argue that for some highly capable foundation models likely to be developed in the near future, open-sourcing may pose sufficiently extreme risks to outweigh the benefits. In such a case, highly capable foundation models should not be open-sourced, at least not initially," at 1.

[11] *See e.g.*, John Thornhill, *How to Keep the Lid on the Pandora's Box of Open AI*, FIN. TIMES (Nov. 16, 2023), https://www.ft.com/content/b62a32c9-a068-40c4-8612-9da8cffb396c. "The danger of open source is that it enables more crazies to do crazy things," Geoffrey Hinton, one of the pioneers of modern AI, has warned. See also Anita Balakrishnan & Catherine McIntyre, *AI Can't Be Slowed Down, Hinton Says in Call to Ban Open-Source Models*, THE LOGIC (Feb. 9, 2024), at https://thelogic.co/news/ai-cant-be-slowed-down-hinton-says-in-call-to-ban-open-source-models/, where Hinton is quoted as saying that: "Open-sourcing big models is like being able to buy nuclear weapons at Radio Shack."

[12] Josh Lerner & Jean Tirole, *The Economics of Technology Sharing: Open-source and Beyond*, 19 J.E.P. 99–120 (2005).

[13] MIRKO BÖHM, *Economics of Open Source*, in OPEN SOURCE LAW, POLICY AND PRACTICE, (Amanda Brock ed., 2nd ed. 2022).

[14] Sayash Kapoor & Rishi Bommasani et al., *On the Societal Impact of Open Foundation Models*, STAN. (Feb. 27, 2024), https://crfm.stanford.edu/open-fms/paper.pdf.

[15] Thibault Schrepel & Jason Potts, *Measuring the Openness of AI Foundation Models: Competition and Policy Implications* (SCI. PO DIGIT., Working Paper, 2024), https://papers.ssrn.com/sol3/papers.cfm?abstract_id=4827358.

[16] Nataliya Wright, Frank Nagle & Shane Greenstein, *Open source software and global entrepreneurship*, 52 RSCH. POL'Y 104846 (2023).

[17] *Id.* See also Sayash Kapoor & A. Rishi Narayanan, *OpenAI's Policies Hinder Reproducible Research on Language Models*, AISNAKEOIL (Mar. 2023), https://www.aisnakeoil.com/p/openais-policies-hinder-reproducible.

[18] *See e.g.*, Bruce Schneier, *Big Tech Isn't Prepared for A.I.'s Next Chapter*, BERKMAN KLEIN CTR (May 30, 2023), https://cyber.harvard.edu/story/2023-05/big-tech-isnt-prepared-ais-next-chapter.

9. At its core, this debate is normative in nature. It pits goals such as security against goals such as competition, transparency, and innovation. The difficult task of weighing these objectives, opportunities, and risks is ultimately a broader question of AI regulation. For its part, competition policy continues to be focused on consumer welfare, which is the standard adopted in our analysis here.

10. The main arguments in favor of open-source genAI models from a traditional consumer welfare perspective can be summarized as:
    – Reducing barriers to entry: By lowering the cost of entry for start-ups and smaller firms, open-source AI may foster a more competitive market environment.
    – Increased innovation: By promoting contestable markets, open-source AI can lead to overall increased innovation. Encouraging interoperability, supporting collaborative models, and recognizing the value of diverse contributions can benefit consumers.
    – Increased product choice: Open-source models can increase product choice by allowing developers to iterate more easily on existing technologies, leading to a greater variety of products and services for consumers.

11. But these arguments require greater nuance in terms of what constitutes "open" genAI and how the choice of degree of openness interacts with business strategies. For the most part, the debate has treated "open" and "closed" as binary terms; however, these labels are not binary, and this dichotomous nomenclature can be misleading.[19]

12. In the context of genAI today, providers have a range of development and release strategies for AI models.[20] The models may be completely proprietary and therefore accessible only to the firm that owns them, such as DeepMind's Chinchilla. They may be restricted, such as OpenAI's GPT-4 or Anthropic's Claude, which are available to other users on subscription but in a "black box" approach where the weights used and other technical information – particularly training data – is not made available. Models can be open in some or all respects, and allow for some scrutiny and downstream modifications, such as Meta's LLaMA 2.[21] Few models are fully open; for instance, xAI's Grok has closed elements, and many models described as open are not actually so.[22] A few models are maximally open,

---

19 See e.g., David Gray Widder, Sarah West & Meredith Whittaker, *Open (For Business): Big Tech, Concentrated Power, and the Political Economy of Open AI*, SSRN (Aug. 16, 2023), https://papers.ssrn.com/sol3/papers.cfm?abstract_id=4543807.

20 *Cf.* Solaiman, *supra* note 9.

21 *Cf.* Rishi Bommasani et al., *Considerations for Governing Open Foundation Models*, 1 (Issue Brief for HAI Pol'y & Soc'y, Dec. 2023), https://hai.stanford.edu/sites/default/files/2023-12/Governing-Open-Foundation-Models.pdf.

22 See e.g., Pascale Davies, *Sorry Elon, Grok Is Not Open Source AI. Here's Why, According to the Creator of the Definition*, Euro News (Mar. 28, 2024), https://www.euronews.com/next/2024/03/28/sorry-elon-grok-is-not-open-source-ai-heres-why-according-to-the-creator-of-the-definition. "AI companies that claim they are open source, such as Musk's latest AI venture, open source the weights – the numerical parameters that influence how an AI model performs – but not the data it is trained on nor the training process. Maffulli said this means

but many others are portrayed as being open while retaining many closed features, a phenomenon dubbed "openwashing".[23] Some authors have argued that the discussion around "open" AI is in fact being leveraged by some of the incumbent companies in order to bolster their interests in light of increasing calls for AI regulation.[24]

13. "Open" systems can combine to different degrees the attributes of transparency (access to data, model weights and source code), reusability (licensing enabling access to code and perhaps data), and extensibility (ability to tune or build onto existing models). Minimal and maximal versions of open will have different attributes in terms of the capabilities or complementary assets needed to use them, and their impact on market dynamics. Although the terms "open" and "closed" have often been treated in the literature as dichotomous, the spectrum below (Table 1) makes clear that they are more nuanced:

Table 1. The closed-open spectrum for genAI models

Fully closed, internal use only	Hosted access, freemium	Cloud-based/API access, freemium or subscription	Partly open, downloadable	Partly open, e.g., license needed, some weights/ code	Partly open, e.g., model weights/ code/documentation	Fully open & replicable including training data
Gopher (DeepMind)	Midjourney	GPT4 (OpenAI)	Opt (Meta)	Llama2 (Meta)	Grok (xAI)	OLMo

14. Given the wide spectrum of "openness" in FMs, competition enforcers should approach each model with a gradient rather than a binary understanding.

III. Competitive Effects of Open-Source Models in GenAI

15. Competition authorities have, however, often started from the assumption that open models would clearly enhance competition. For example, the European Parliament has adopted the position that open-source models

---

it cannot be open source because it is not transparent in what data is used to train the weight which can cause copyright issues and ethical questions around if the data is biased." See also Schrepel & Potts, *supra* note 15.

23 The term was first coined by Audrey Watters in 2014 where she defines it as "having an appearance of open-source and open-licensing for marketing purposes, while continuing proprietary practices." Audrey Watters, *From "Open" to Justice #OpenCon2014*, HACKEDUCATION (Nov. 16, 2014), http://hackeducation.com/2014/11/16/from-open-to-justice.

24 Widder et al, *supra* note 19, at 1. "While we recognize that there is a vibrant community of earnest contributors building and contributing to 'open' AI efforts in the name of expanding access and insight, we also find that marketing around openness and investment in (somewhat) open AI systems is being leveraged by powerful companies to bolster their positions in the face of growing interest in AI regulation. And that some companies have moved to embrace 'open' AI as a mechanism to entrench dominance, using the rhetoric of 'open' AI to expand market power while investing in 'open' AI efforts in ways that allow them to set standards of development while benefiting from the free labor of open-source contributors."

contribute to research, innovation, and growth.²⁵ In the US, the Federal Trade Commission (FTC) has noted that "the open-source ecosystem may help open up the playing field once the base models become available to the public, if it can reach parity with the quality of proprietary models."²⁶

16. Likewise, the UK's Competition and Markets Authority (CMA) has suggested that the existence of open source is an indicator of competition in the market. In its September 2023 report and subsequent April 2024 update, the CMA sets out a list of proposed overarching principles that could guide the development of FMs – access, diversity, choice, flexibility, fair dealing, and transparency.²⁷ In the context of open source, the CMA notes that a positive market outcome would arise if multiple competing firms produce leading FM models. Firms would be able to experiment with various business models – such as open and closed-source – and market entrants would have access to the inputs they need in order to keep markets contestable.²⁸ In contrast, market power would allow firms to provide models on a closed basis only, the weights and algorithms in the model a trade secret, and these could have unfair prices and terms.²⁹ In this view, open source would be a presumptive indicator of competition.

17. We argue against any assumptions of pro-competitiveness regarding open-source models. First – and enforcement agencies have mentioned this issue – incumbent firms can use open-source components to attract users in order to cement their positions.³⁰ "Open" now does not necessarily mean open later, and agencies will be watchful of possible "bait and switch" strategies.³¹

18. Second, the pro-competitive aspects of open source apply mainly downstream from FMs, which are recognized to be very costly to create and maintain. The scale of the open-source movement today has far outgrown its idealistic origins, with products – including open-source genAI

---

25 *European Parliament legislative resolution of 13 March 2024 on the proposal for a regulation of the European Parliament and of the Council on laying down harmonised rules on Artificial Intelligence (Artificial Intelligence Act) and amending certain Union Legislative Acts*, COM (2021) 0206 final, at para. 102 (Mar. 13, 2024), https://www.europarl.europa.eu/doceo/document/TA-9-2024-0138_EN.html. "Software and data, including models, released under a free and open-source licence that allows them to be openly shared and where users can freely access, use, modify and redistribute them or modified versions thereof, can contribute to research and innovation in the market and can provide significant growth opportunities for the Union economy." It also notes at para. 103 that "AI components that are provided against a price or otherwise monetised… should not benefit from the exceptions provided to free and open source AI components."

26 Fed. Trade Comm'n, *Generative AI Raises Competition Concerns*, FTC BLOG (June 29, 2023), https://www.ftc.gov/policy/advocacy-research/tech-at-ftc/2023/06/generative-ai-raises-competition-concerns. It also cautioned that firms can use "open first, closed later" tactics that can lock-in customers and lock-out competition.

27 CMA REPORT, *supra* note 2, at para. 7.2.

28 *Id.* at para. 3.60.

29 *Id.* at para. 3.61.

30 *See e.g.*, Fed. Trade Comm'n, *supra* note 26, cautioning that firms can use "open first, closed later" tactics that can lock-in customers and lock-out competition.

31 The terms and conditions of a license could help guarantee continuing access to an existing model.

models – provided by for-profit corporations to garner varying business advantages. These include strengthening a firm's position against its market competitors by curbing their dominance, using open-source tools to create or extend a platform (e.g., Google and Android), or creating demand for tools that integrate with proprietary products and strengthen an ecosystem (as Meta has with PyTorch).

19. All FMs require significant access to compute, data, and human talent as inputs. The first step in building an FM consists of pretraining data in order to create the model. The second step requires finetuning the model for downstream tasks.[32] Any upstream model developer needs access to specialist chips and to compute, where there is embedded market power due to massive scale.[33] Unlike previous types of open-source software, developing and operating any AI model relies on scarce and costly resources in the hands of a small number of large companies in order to operate.[34] The resources required to build open AI models remain closed; some authors have gone as far as to suggest that FMs tend toward natural monopoly.[35]

20. Moreover, human talent remains a significant input constraint, even in the open-source context. In its 2022 Census II of open-source software, the Harvard Laboratory for Innovation Science (LISH) and the Open Source Security Foundation (OpenSSF) noted a prevalence of "super-coders" in the open-source community who are responsible for the majority of open-source contributions. The Census II found that much of the most widely used free and open-source software (FOSS) is developed by only a handful of contributors: the project found that 94 percent of the top 49 Census II projects had fewer than ten developers accounting for more than 90 percent of the LOC added.[36] These findings are counter to the typically held belief that open source is substantially more distributed or democratic than closed.

21. Project atrophy and contributor abandonment are also known issues: the number of developer contributors who work on projects to ensure updates for feature improvements, security, and stability decreases over time as they prioritize other software development work in their professional lives or decide to leave the project for any number of reasons. Therefore, it is much more likely that these communities may face challenges without sufficient developers to act as maintainers.[37]

---

32 Jai Vipra & Anton Korinek, *Market Concentration Implications of Foundation Models: The Invisible Hand of ChatGPT*, Brookings 5 (Sept. 7, 2023), https://www.brookings.edu/articles/market-concentration-implications-of-foundation-models-the-invisible-hand-of-chatgpt/.

33 *See e.g.*, Jai Vipra & Sarah Myers West, *Computational Power and AI*, AI Now Inst. (Sept. 27, 2023), https://ainowinstitute.org/publication/policy/compute-and-ai.

34 *Id.* at *NVIDIA's Market Dominance and Competitors*.

35 Vipra & Korinek, *supra* note 32, at 9.

36 The Linux Foundation and The Laboratory for Innovation Science at Harvard, *supra* note 7, at 4.

37 *Id.*

22. Given the significant costs to build and maintain an FM, open source tends to have an impact primarily on downstream competition. FMs provide the infrastructure on which downstream applications can be built. This creates the risk of algorithmic monoculture, where many applications depend on the same FMs.[38] Downstream competition can help reduce some of the severity of algorithmic homogeneity by yielding more diverse downstream model behavior.[39] But this also creates the potential for some downstream players to have greater access to essential inputs than others.[40]

23. Finally, FM ecosystems are particularly affected by adjacent market considerations. In this respect, the antitrust toolkit will need to adapt by expanding its traditional scope of analysis. Competition analysis restricted to a particular market will likely fail to capture key insights into the ability of dominant firms to yield market power along the FM value chain.

IV. The AI Ecosystem: Business Models and Incentives

24. FMs will require the traditional antitrust toolkit to expand by going beyond traditional market definitions and by taking into account the specific character of the ecosystem under analysis. This includes investments and partnerships, as well as relevant upstream and downstream AI products. Some economists previously argued that the acquisition of nascent, potentially vertically adjacent, or complementary competitors carries enough competitive concerns to merit a reconsideration of permissive safe harbors in non-horizontal mergers.[41] We agree and argue further that enforcers should take an ecosystem-wide view in any type of competition analysis involving FMs. This is because of the particularly interconnected nature of firms across the FM value chain.

25. A number of features are relevant to the assessment of competition in this complex ecosystem. First, there is a question about which dimensions of "open" on the genAI spectrum provide a competitive constraint and at which level of the stack. If key information about a model is available

---

38 Kapoor and Bommasani et al., *supra* note 14, at 5.

39 *Id.* at 8. Kapoor and Bommasani et al., also note that this downstream competition can help "reduce market concentration at the foundation model level from vertical cascading," at 5. But see Massimo Motta, *Self-preferencing and Foreclosure in Digital Markets: Theories of Harm for Abuse Cases*, 90 INT. J. IND. ORGAN. (Sept. 2023).

40 Vipra & Korinek, *supra* note 32, at 30.

41 *See e.g.*, Steven Salop, *Potential Competition and Antitrust Analysis*, (submission to the OECD, June 10, 2021), https://one.oecd.org/document/DAF/COMP/WD(2021)37/en/pdf, where he argues that the safe harbours in the EU's Non-Horizontal Merger Guidelines (Guidelines on The Assessment of Non-Horizontal Mergers under The Council Regulation on The Control of Concentrations between Undertakings, 2008 O.J. (C 265) 6) should be removed. "Indeed, my own view is that such safe harbors also are inappropriate for evaluating acquisitions of established firms in vertically adjacent or complementary product mergers, except perhaps when both firms compete in unconcentrated markets." *See also* Jonathan B. Baker, Nancy L. Rose, Steven C. Salop & Fiona Scott Morton, *Five Principles for Vertical Merger Enforcement*, 33 ANTITRUST 12 (2019). *But see* Carl Shapiro, *Evolution of the Merger Guidelines: Is This Fox Too Clever by Half?*, REV. IND. ORGAN. (2024), https://link.springer.com/article/10.1007/s11151-024-09956-y.

and no license is required, the likelihood of competition will be higher. But openness in one level might lead to foreclosures on another.[42]

26. Firms have significant incentives to integrate vertically, in order to control various levels of the genAI ecosystem.[43] The CMA in 2024 observed that incumbent firms are rapidly engaging in the vertical integration of their FMs across the genAI ecosystem.[44] It has also noted that the most established firms are becoming active across several levels of the FM value chain through partnerships, strategic investments, and agreements, which often include access to key inputs like data and compute, as well as distribution agreements.[45] Indeed, the already complex relationships between large companies and start-ups in genAI models (including open-source models) indicate significant and at times indirect vertical integration.[46]

27. In general, the terms of these arrangements are not public. What are the implications for "open" model incentives – can they provide a competitive constraint on the proprietary models of those who have invested in them? It is not obvious that they can, and competition authorities would need to consider the arrangements in detail on a case-by-case basis. It was assumed for a long time that downstream competition would constrain upstream market power in a vertical relationship, but this presumption has been eroded in the digital context.[47]

28. In the notorious leaked Google "moat" memo, which prompted much excitement about the prospect for a competitive environment in genAI, the author's conclusion was that a moat could be built by "owning the ecosystem." Arguing that future innovation did not need more large, newly trained FMs, and that the availability of open models with weights available had destroyed Google's "moat," the author wrote:

> *The value of owning the ecosystem cannot be overstated.* Google itself has successfully used this paradigm in its open-source offerings, like Chrome and Android. By owning the platform where innovation happens, Google cements itself as a thought leader and direction-setter, earning the ability to shape the narrative on ideas that are larger than itself.[48]

---

42 For instance, if openness at one level leads to greater market power at another level of the stack.

43 Vipra & Korinek, *supra* note 32, at 6.

44 CMA Report, *supra* note 2, at 16.

45 *Id.* at 8.

46 *Cf. id.* at Figures 2 and 5. "[I]t is notable that not all such partnerships and investments will fall within the scope of merger control rules and some may have been structured to seek to avoid them."

47 Motta, *supra* note 39.

48 Dylan Patel & Afzal Ahmad, *Google "We Have No Moat, And Neither Does OpenAI": Leaked Internal Google Document Claims Open Source AI Will Outcompete Google and OpenAI*, Semianalysis (May 4, 2023), https://www.semianalysis.com/p/google-we-have-no-moat-and-neither. They also noted that "Paradoxically, the one clear winner in all of this is Meta. Because the leaked model was theirs, they have effectively garnered an entire planet's worth of free labor. Since most open-source innovation is happening on top of their architecture, there is nothing stopping them from directly incorporating it into their products."

29. Theories of harm concerning digital ecosystems must move away from traditional analysis in terms of product markets and the incentives and abilities of individual firms, toward a systemic viewpoint. This would gauge firms as actors across a totality of markets – as an ecosystem.[49] In line with the importance of ecosystem leadership[50] and the role played by capabilities in structuring ecosystems,[51] competition analysis would entail a consideration of firms' assets and capabilities rather than existing products, and the ability to augment or entrench market power by restricting access of competitors to key assets, as well as by raising entry and expansion barriers.[52] Emerging ecosystem theories of harm have some similarities to vertical theories of harm in that there may be an argument to be made for efficiencies, such as gains from indirect network effects or scale in this case, but the incentives to realize and pass on the efficiencies need to be considered. As ever, there will also be context-specific considerations.

30. There is some evidence that agencies are moving toward this wider-lens approach. The CMA hinted at this type of analysis in its decision on *Facebook/Giphy*, although the stated theory of harm was a standard vertical foreclosure possibility, and also in its later (subsequently reversed) preliminary ruling on *Microsoft/Activision*.[53] The EU blocked Booking's acquisition of e-Traveli on a similar basis.[54] The ACCC in Australia is currently considering the analysis of digital ecosystems.[55] The US Merger

---

[49] *See e.g.*, Thibault Schrepel & Alex Sandy Pentland, *Competition between AI Foundation Models: Dynamics and Policy Recommendations* 8–9 (AMSTERDAM L. & TECH. INST., Working Paper No. 1-2023, 2023), https://papers.ssrn.com/sol3/papers.cfm?abstract_id=4493900. They argue that although the state of competition between FMs as of mid-2023 was very dynamic, regulation must ensure an open, public FM ecosystem as a whole.

[50] *See* Nicolai J. Foss, Jens Schmidt & David J. Teece, *Ecosystem Leadership as a Dynamic Capability*, 56 L.R. PLAN. 102270 (2023), https://www.sciencedirect.com/science/article/pii/S0024630122000899?via%3Dihub.

[51] See DAVID J. TEECE, *Dynamic Capabilities and (Digital) Platform Lifecycles*, *in* 37 ENTREPRENEURSHIP, INNOVATION, AND PLATFORMS ADVANCES IN STRATEGIC MANAGEMENT 211–25 (Emerald Publ'g Ltd. 2017).

[52] Michael G. Jacobides, Carmelo Cennamo & Annabelle Gawer, *Towards a Theory of Ecosystems* 39 STRATEGIC MGMT. J. 2255, 2276 (2018). See also Cristina Caffarra, Matthew Elliott & Andrea Galeotti, *"Ecosystem" Theories of Harm in Digital Mergers: New Insights from Network Economics, Part 1*, CEPR (June 5, 2023), https://cepr.org/voxeu/columns/ecosystem-theories-harm-digital-mergers-new-insights-network-economics-part-1.

[53] CMA, *Facebook, Inc (now Meta Platforms, Inc) / Giphy, Inc Merger Inquiry*, GOV.UK (June 12, 2020), https://www.gov.uk/cma-cases/facebook-inc-giphy-inc-merger-inquiry; CMA, *Microsoft / Activision Blizzard Merger Inquiry*, GOV.UK (July 6, 2022), https://www.gov.uk/cma-cases/microsoft-slash-activision-blizzard-merger-inquiry, *see also* Sara Warner, *The UK Competition Authority clears a landmark modified bid of a Big Tech company to acquire a video game producer (Microsoft / Activision Blizzard)*, E-COMPETITIONS Oct. 2023, art. No. 116377.

[54] Eliana Garces, Olga Kozlova & Devin Reilly, *Ecosystem Theories of Harm in Merger Enforcement: Current Direction and Open Questions*, (Feb. 29, 2024), https://papers.ssrn.com/sol3/papers.cfm?abstract_id=4742444.

[55] ACCC, DIGITAL PLATFORM SERVICES INQUIRY – SEPTEMBER 2023 REPORT ON THE EXPANDING ECOSYSTEMS OF DIGITAL PLATFORM SERVICE PROVIDERS (Mar. 2023), https://www.accc.gov.au/system/files/Digital%20platform%20services%20inquiry%20-%20September%202023%20report%20-%20Issues%20paper_0.pdf. "The Report will closely examine the extensive web of interconnected products and services of digital platform service providers and the extent to which this may have increased the risk of competition issues and consumer harms."

Guidelines make specific reference to ecosystem competition.[56] And the FTC has noted that incumbents may use their power in one market to affect adjacent markets in the ecosystem.[57]

31. Competition authorities should take into account the dynamics of the ecosystem across vertical layers of the technology stack, building on the growing literature on digital ecosystems to formalize theories of harm. Alongside potential scale efficiencies, these would also use potential economies of scale and scope in supply, access to scarce technological skills and/or the use of easier use of the technology in integrated software, and potential network effects in demand to point to harms such as lessened interoperability with competitors' products or using the ecosystem to entrench dominance or weaken competition by foreclosure and raising entry and expansion barriers.[58]

32. The technological landscape remains fluid, and future innovations might yet reduce the importance of some of the current input constraints and market power concerns in genAI. As matters stand, however, scale remains immensely important. The stakes are high, and the costs of participating are becoming increasingly prohibitive. We have previously discussed the access constraints that compute, data, and human talent pose for entrants. As incumbent firms reach new peaks, the costs of competing with them become excessive. It is widely reported that Microsoft and OpenAI are currently working on a $100 billion supercomputer called Stargate.[59] In contrast, Databricks spent $10 million to build an LLM that roughly approximates ChatGPT 3.5. Jack Clark, co-founder of Anthropic, has argued that the capital expenditures required to participate in FMs going forward will be so high that genAI will become intertwined with the industrial policy of governments.[60] It is nearly inconceivable that no monetary recoupment will be expected at this level of investment. Unfortunately, this suggests that open-source models are unlikely to reach the kind of parity with proprietary models that would be required for them to provide sufficient competitive pressure in the context of the market position of the large incumbents.

---

56 Fed. Trade Comm'n. and U.S. Dep't of Justice, *Merger Guidelines (2023)*, (Dec. 18, 2023), https://www.ftc.gov/system/files/ftc_gov/pdf/2023_merger_guidelines_final_12.18.2023.pdf. "[E]cosystem competition refers to a situation where an incumbent firm that offers a wide array of products and services may be partially constrained by other combinations of products and services from one or more providers, even if the business model of those competing services is different," at 20.

57 Fed. Trade Comm'n, *supra* note 26. "Incumbents that control key inputs or adjacent markets, including the cloud computing market, may be able to use unfair methods of competition to entrench their current power or use that power to gain control over a new generative AI market."

58 Zhijun Chen & Patrick Rey, *A Theory of Conglomerate Mergers* (TSE, Working Paper No. 23-1447, 2023), https://ideas.repec.org/p/tse/wpaper/128159.html.

59 Anissa Gardizy & Amir Efrati, *Microsoft and OpenAI Plot $100 Billion Stargate AI Supercomputer*, THE INFO. (Mar. 29, 2024, 10:05 AM PDT), https://www.theinformation.com/articles/microsoft-and-openai-plot-100-billion-stargate-ai-supercomputer.

60 Jack Clark, *Import AI 367*, (Apr. 1, 2024), https://importai.substack.com/p/import-ai-367-googles-world-spanning. "AI is going to look more like a vast CapEx intensive industry like oil extraction, mining, heavy industry, and so on. These industries all end up being heavily regulated, having a tiny number of participants, and also become intertwined with the industrial policy of governments. It's worth bearing this in mind when we look at things like openly accessible models being released that cost $10m to train (see: Databricks). Is anyone going to openly release a model that costs $100 billion? $10 billion? $1 billion? All seems doubtful to me!"

V. Conclusions

33. Open-source genAI models can bring significant benefits to the competitive landscape, but their treatment in competition policy will require greater nuance. First, the meaning of "open" can vary considerably between models. Second, there can be no presumption that open-source models will constrain the largest providers. Firms may have strategic motives for providing open-source AI models, such as the "bait and switch" approach of establishing de facto standards, and subsequently switching to proprietary offerings and services, or arguing strategically that open-source models imply no need for competition scrutiny of genAI. The more open the model, the less likely it is to have the scale of the most powerful models, and there is as yet no sign that scale (in terms of data or compute) is becoming less important. Finally, given the proliferation of partnerships within the industry, including providers of "open" models, competition authorities will need to expand the traditional antitrust toolkit and look to ecosystem theories of harm in order to assess the ability of dominant firms to yield market power along the FM value chain.

Competing in the Age of AI: Firm Capabilities and Antitrust Considerations

FAUSTO GERNONE[*] AND DAVID J. TEECE[**]
University of California, Berkeley

Abstract

In the 1960s, Nobel Laureate Herbert Simon posited that intelligent systems exhibit their intelligence by achieving their objectives in the face of diverse and evolving circumstances, within certain physiological or computational limits. He argued that both human organizations and computer systems are inherently "artificial" in that they dynamically update in response to changing environments.[1] Artificial Intelligence (AI) is a fundamental enabling technology that has captured the imagination for over half a century and is now becoming a tangible reality. It stands out both quantitatively and qualitatively from other technologies due to its capacity to assimilate tacit knowledge and rival human intelligence. AI's human-like adaptability makes it an incredibly powerful tool for a variety of applications, promising not only to automate routine tasks, but to revolutionize how businesses operate and compete. This chapter reviews AI's role as an enabling technology, exploring its impact on organizational capabilities and strategic management. Lastly, it examines some potential issues in competition policy.

---

[*] Fausto Gernone is a competition economist currently visiting Haas School of Business, University of California Berkeley, and a Ph.D candidate at the UCL Institute for Innovation and Public Purpose. His research explores competition dynamics in industrial ecosystem, focusing on the interplay of industrial organization, information theory and firm strategy.

[**] David J. Teece is Professor of the Graduate School at the University of California, Berkeley. He is the author of over thirty books and two hundred scholarly papers and co-editor of the Palgrave Encyclopedia of Strategic Management. Dr. Teece has received nine honorary doctorates and has been recognized by Royal Honors.

1 Herbert A. Simon, *Cognitive Science: The Newest Science of the Artificial*, 4 COGNITIVE SCI. 33, 46 (1980).

I. The Economics of AI

1. What is AI?

34. Artificial Intelligence (AI) refers to the capability of a machine to imitate intelligent human behavior. It is a broad field of computer science dedicated to building smart machines capable of performing tasks that typically require human intelligence. One remarkable capability provided by AI lies in its ability to grasp tacit knowledge – uncodified information that is often unspoken, which humans typically acquire through experience rather than through written instructions. This is why we speak of "machine *learning*."[2] The ability to learn from data – whether structured or unstructured, explicit or implicit – is achieved through algorithms to identify patterns and make inferences, which may involve calculating probabilities, detecting correlations, or applying predetermined rules. This enables AI to adapt to new circumstances without needing to be reprogrammed.

35. Despite the recent surge in interest, the concept of AI has a rather long history. The notion dates back to ancient myths and stories of artificial beings endowed with intelligence or consciousness by master craftsmen; however, the formal foundation of AI as a scientific discipline was laid in the 1950s. Pioneers like Alan Turing, Herbert Simon and John McCarthy – who coined the term "artificial intelligence" in 1956[3] – explored the possibilities of machines performing functions associated with human intelligence, such as reasoning, knowledge representation, planning, learning, natural language processing, perception, and the ability to move and manipulate objects.

2. Constituents of an AI Model

36. This section summarizes the key steps typically involved in developing and deploying an AI model.
    - *Data Collection.* AI systems require data to learn and make decisions. This data can be diverse: text, images, numbers, and more, depending on the domain and the problem at hand. Access to data is frequently identified as the primary bottleneck and a significant source of competitive advantage. The relative merits of this view are discussed in Section III.1 below.
    - *Data Preparation.* The collected data is then prepared and cleaned. This step involves handling missing data, encoding, and splitting the data into training sets and test sets.

---

[2] Not all AI systems use machine learning, but machine learning is currently one of the most effective approaches to building AI applications. For example, unlike machine learning, rule-based systems make decisions based on hardcoded rules. They do not learn from data but follow pre-programmed instructions.

[3] John McCarthy, Dartmouth Summer Research Project on Artificial Intelligence (Dartmouth College 1956), https://home.dartmouth.edu/about/artificial-intelligence-ai-coined-dartmouth.

- *Model Building.* Machine learning models are selected according to the problem. This might involve supervised learning, where the model is trained on a labeled dataset; unsupervised learning, in which the system tries to learn the patterns without a specific target variable; or reinforcement learning, in which an artificial agent learns to behave in an environment by performing certain actions and receiving rewards in return. For a problem like email spam detection, where the goal is to classify emails as spam or not spam, a supervised learning model such as logistic regression or a support vector machine might be used. On the other hand, for customer segmentation in marketing, where there might not be a clear label, an unsupervised learning technique could be appropriate to group customers based on purchasing behaviors.
- *Training.* Algorithms are trained on the data to create a model, using techniques such as neural networks, decision trees, or support vector machines. During the training phase, the model learns from the data by adjusting its parameters to minimize error. This phase requires considerable computational power, especially for deep learning models, which process vast amounts of data through complex network structures. The computational requirements of training are typically met by means of integration with cloud computing. Cloud platforms provide the necessary computational power and data storage solutions that AI systems require, often reducing the cost barrier for smaller enterprises and startups to develop AI solutions. This synergy enables rapid scalability and accessibility of AI technologies, reducing both time to market and operational costs. As discussed in Section III.3, the market for computing power is currently populated by a handful of cloud providers, incentivizing smaller players to engage in partnerships and strategic alliances.
- *Evaluation.* Once trained, the model is tested against a set of data it hasn't seen before (test set) to evaluate its accuracy and effectiveness.
- *Deployment and Monitoring.* After evaluation, the AI system is deployed into a real-world environment where it needs to make decisions based on new data. Continuous monitoring and adjusting are necessary as the model may degrade over time or need to be updated with new data or outcomes.

37. Most AI models are characterized by high setup and training costs, and negligible costs for deploying them, leading to large economies of scale. At the same time, online developers' communities are perfecting ways to set up bespoke AI models locally, running on the user's machine using open-source software such as Meta's Llama. While this development hints at the possibility of a future where AI resources are affordable and easily accessible, the scale and reliability required from AI by organizations currently necessitate considerable investment.

38. Based on the process to build an AI system, the three critical inputs are data, skilled labor and computational power. Access to each of these can have important implications for the ability of companies to build AI systems. These issues are discussed in Section III.

3. AI as a Platform

39. There are numerous foundational models and more specialized application models within the AI landscape. General-purpose AI models, often referred to as foundation models, are designed for a broad output generality, making them suitable for a wide array of applications. As Berkeley computer scientist Stuart Russell points out, "research on tool AI... often leads to progress on general purpose techniques that are applicable to a wide range of other problems."[4] These models can function as standalone systems or serve as the foundational "building blocks" for narrow-purpose AI systems designed to perform a growing array of specific tasks.

40. Typically, as an enabling technology is adaptable to a larger number of markets, higher applicability might come at the cost of it being tuned less perfectly to each individual market. Yet, general purpose AI models have shown remarkable effectiveness in narrow applications across various fields. For instance, foundation models like OpenAI's GPT have been fine-tuned to perform tasks that range from predicting protein structures to improving drug development, which has significantly accelerated advancements in healthcare.[5] The versatility of general models in handling specialized functions allows them to overcome the usual tradeoff between their potential to create value and their range of applications.[6]

41. However, the width and variety of its realms of applicability make the value proposition of foundational AI particularly complex. In the context of enabling technologies, traditional methods for securing profits, such as strong appropriability regimes, are less effective because of the difficulty of monitoring their application across a wide variety of domains.

42. To navigate this complexity, companies in the foundation AI sector are increasingly turning to partnerships and alliances to secure access to essential complementary assets. This includes both upstream resources like computing power and chips, and downstream applications that incorporate AI-powered programs. The pursuit of these downstream components results in a platform-based approach, where general-purpose models act as a base upon which other services and applications can be developed

---

[4] STUART RUSSELL, HUMAN COMPATIBLE: ARTIFICIAL INTELLIGENCE AND THE PROBLEM OF CONTROL (Viking Press 2019).

[5] For a comprehensive overview of foundation models, see: Elliot Jones, *Explainer: What is A Foundation Model?*, ADA LOVELACE INST. (July 17, 2023), https://www.adalovelaceinstitute.org/resource/foundation-models-explainer/.

[6] A discussion on value capture in the context of enabling technology is provided by, Alfonso Gambardella, Sohvi Heaton, Elena Novelli & David J. Teece, *Profiting from Enabling Technologies?*, 6 STRATEGY SCI. 1 75–90 (2021).

through Application Programming Interfaces (APIs). This platform-based approach enables multiple user interactions, creating network effects that can enhance value with increased usage.

43. Platforms like Google Cloud and Amazon Web Services (AWS) offer both the tools for developing AI applications and the infrastructure to run them, fostering a rich ecosystem of innovation. OpenAI allows complementary innovations to develop AI-based apps for specific geographies and use cases. Meanwhile, Apple is taking a distinct approach by planning to deploy generative AI directly on its mobile devices, allowing AI chatbots and applications to operate using the phone's own hardware and software, instead of relying on cloud services hosted in data centers.[7] By doing so, Apple might be able to build upon its existing ecosystem rather than creating a new external AI platform.

4. Competitive Landscape

44. The market for AI is characterized by its remarkable dynamism, with frequent disruptions and an increasing array of participants. This section outlines the main types of firms offering AI systems and their distinguishing characteristics.

45. The first category comprises of AI-specialized companies that have recently captured public attention with technological breakthroughs, such as those achieved by ChatGPT or Dall-E. These are typically young companies focused on general-purpose AI, often partnering with established firms to secure access to computational power and downstream applications. Notable examples include OpenAI and Anthropic, which are, respectively, affiliated with Microsoft and Amazon by means of special collaboration agreements.[8,9] The success of these models has spurred industry interest, leading to a rapid influx of resources for developing new AI products and competition has intensified as a consequence. While some AI systems in this category are leading in performance, the gap is narrowing, and competition is increasing.

46. Second, there is a growing trend for Big Tech companies developing AI models internally to aim at integrating AI technologies into their platforms. These models benefit from large volumes of high-quality user data generated from other sides of the business, as well as guaranteed access to

---

[7] Emilia David, *Apple's New AI Model Hints at How AI Could Come to the iPhone*, THE VERGE (Apr. 24, 2024, 7:49 PM GMT+1), https://www.theverge.com/2024/4/24/24139266/apple-ai-model-openelm-iphone-laptops-strategy.

[8] The collaboration between Microsoft and OpenAI is unique and complex. Since 2019, Microsoft has invested approximately $13 billion in OpenAI, which remains independently operated, although Microsoft holds an observer's seat on its board. Microsoft is OpenAI's exclusive provider of cloud services and it enjoys preferential access to some of OpenAI's APIs. Moreover, Microsoft has the right to 49 percent of OpenAI profits until it recoups its investments. However, the agreement does not prevent either company from competing for customers, nor from doing business with the partner's competitors, as demonstrated by Microsoft's investment in other AI startups such as Mistral and OpenAI's recent commercial agreement with Apple.

[9] As part of their partnership, Amazon holds a minority stake in Anthropic amounting to a cumulative $4 billion investment. Anthropic exclusively uses AWS's clouding services and chips, and its products are available on the AWS generative AI platform, Amazon Bedrock.

computational power through the company's cloud services. Often, these models compete with the AI systems offered by the company's own partners. For example, Microsoft's Copilot and Bing AI initiatives directly compete with products from OpenAI and Mistral, despite Microsoft's investments in these companies. Among Big Tech AI models, special mention is deserved by Meta's Llama. As an open-source model, it leverages a large community of developers and has arguably slowed down the adoption of competing products. Although Meta has yet to find a way to monetize it, it is easy to see how it can benefit Meta, as Llama 3-70B is integrated across its platforms i.e., Facebook, Instagram, and WhatsApp. While it does not have yet the capacity of other language models – it is well positioned to be adopted by mobile device users, especially in emerging markets.[10]

47. The availability of public AI software modules, including data and code, has facilitated the emergence of a third category of models: the narrow-purpose AI systems tailored by companies across various sectors for internal applications or specialized uses. Notable examples include PepsiCo's Ada and McKinsey's Lilli, though many other models remain lesser known or within the private domain. Like the AI companies in the first category, these narrow AI models typically depend on third-party services to varying extents for essential functions like model maintenance and computing services.

48. In summary, the market for AI is expanding rapidly, with new actors innovating both in product designs and efficiency. An example is DeepSeek, a China-based open-source model that outperforms Meta's Llama in both cost and performance thanks to an innovative architecture Multi-Head Latent Attention mechanism. Many AI products are not yet profitable due to the substantial costs of training – they are developed in hopes of becoming dominant platforms or capturing lucrative downstream applications in the future. While there is little doubt that AI will profoundly affect, even revolutionize, the way companies are run, the market for the technology is currently marked by unforgiving levels of competition and uncertainty.

II. Organizational Benefits of AI

49. AI holds a promise of great benefits. The relative success of companies in mastering this new technology is likely to be a key factor shaping the competitive landscape in the coming years. Rather than documenting the benefits of its applications in various fields, such as medical research and health and safety, in this chapter we focus on the impact on the corporation and competition. We examine the potential impacts on the ordinary and dynamic capabilities of the business enterprise, and then we discuss potential areas of competition policy concern.

---

10 Dylan Patel & Daniel Nishball, *OpenAI Is Doomed? – Et tu, Microsoft?*, SEMIANALYSIS (May 7, 2024), https://www.semianalysis.com/p/openai-is-doomed-et-tu-microsoft?utm_source=substack&utm_medium=email&utm_campaign=email-restack-comment.

50. As AI technologies are integrated into the day-to-day running of a company, the implications might be far reaching. While the effect is likely to vary from industry to industry, integrating AI could influence the optimal size and structure of a company by reducing internal coordination costs. Traditionally, larger firms have benefited from economies of scale in administrative functions, spreading the fixed costs of management over a larger volume of production or services. With AI reducing the marginal cost of adding additional administrative complexity, smaller firms might be able to compete more effectively without necessarily increasing in size. External transaction costs might also be reduced, as the technology facilitates the coordination of economic interactions and potentially allows enterprises to have a better grip on their external environment.[11] Overall, the implication could be a more nuanced relationship between market share and market power.

1. Ordinary Capabilities

51. Most activities within a business are driven by what we call ordinary capabilities. These capabilities are fundamental for performing routine tasks efficiently, enabling an organization to produce and deliver a prolific lineup of products and services. Having strong ordinary capabilities means that a business can maintain efficiency in its products and services, even if the current approach may not be the most innovative or future-proof path. Ordinary capabilities are grouped under administration, operations, and governance. This foundation is crucial for sustaining standard business operations efficiently.[12]

52. AI models stand out as a powerful tool to rationalize and coordinate economic activities, which can greatly assist in running a company. As explained in a report by PepsiCo on AI:

> At PepsiCo, AI has become a vital tool for business and employees alike. It's a fundamental part of PepsiCo's digital transformation as the company builds on its digital resources: Starting with the best time to plant potatoes all the way to predicting how many bags of Lay's should go on store shelves, AI is reshaping how the company plans, makes, moves, sells and delivers products.[13]

53. This type of automation can greatly reduce the internal transaction costs of a company, as it allows for better coordination, communication, and monitoring within the firm. Crucially, it is likely to greatly reduce the

---

11 Nobel laurate Olivier Williamson notes: "Changes in information processing technology may occur which alter the degree to which bounded rationality limits apply, with the result that a different assignment of activities between markets and hierarchies than was selected initially becomes appropriate later." *See* OLIVER E. WILLIAMSON, MARKETS AND HIERARCHIES: ANALYSIS AND ANTITRUST IMPLICATIONS 10. (The Free Press 1975).

12 David J. Teece, *The Foundations of Enterprise Performance: Dynamic and Ordinary Capabilities in an (Economic) Theory of Firms*, 28 ACAD. MGMT. PERSP. 328 (2014).

13 *Artificial Intelligence at PepsiCo*, PEPSICO (Oct. 21, 2021), https://www.pepsico.com/our-stories/story/artificial-intelligence-at-pepsico.

amount of workload on staff, pointing to a high degree of substitutability between human labor and AI in the context of ordinary capabilities.

54. In *administration*, AI can automate routine tasks such as scheduling, email sorting, and document management. For example, AI-powered virtual assistants can handle appointment bookings and manage calendars, freeing up human employees. These technologies reduce the likelihood of human error, leading to smoother administrative processes.

55. In the realm of *operations*, AI promises to reduce risk by improving forecasting and allowing for better management of supply chains, logistics and inventories. An example is the use of AI in warehouse management systems, where algorithms predict stock levels, optimize routes for delivery vehicles, and manage inventory through automated robots, as seen in Amazon's fulfillment centers.

56. Lastly, within *governance*, AI can play a crucial role in risk management and compliance by monitoring and analyzing vast amounts of data to identify potential risks or non-compliance issues. Financial institutions, for instance, use AI to detect patterns indicative of fraudulent activities, significantly enhancing their ability to adhere to regulatory requirements and protect against financial crimes.

2. Superordinary Capabilities

57. Certain skills can be described as "super-ordinary." These skills arise from the complementarity between a foundation of specialized knowledge, tailored to specific applications or markets, and generic expertise, such as industry best-practices, and they often develop into what are known as "signature" processes.[14] Since super-ordinary capabilities are idiosyncratic and part of a firm's organizational heritage and values, competitors have difficulty replicating them.

58. AI can be instrumental in developing or strengthening super-ordinary capabilities. As AI learns from data to simulate human behavior, it can be integrated into a company's assets to systematize its processes and help recognize successful, company-specific patterns. This can allow companies to leverage their specific knowledge to distil super-ordinary capabilities and take industry best practices beyond best practices. However, the effective deployment of AI in this context requires the AI models to be well-integrated with the firm's unique systems. For example, AI could be trained on the firm's internal data to encapsulate the firm's specific operational nuances and build upon its idiosyncratic elements to enhance its competitive advantage. This is the case with Ada, PepsiCo's AI platform, built on a combination of external data – such as household and consumption statistics – and internal resources on marketing and production. For several years the company has relied on the platform to

---

[14] Lynda Gratton & Sumantra Ghoshal, *Beyond Best Practice*, MIT SLOAN MGMT. REV. (Apr. 15, 2005), https://sloanreview.mit.edu/article/beyond-best-practice/.

optimize product design and marketing campaigns. In this context, Ada represents a good example of how AI can be deeply integrated into a company's operations, enhancing distinctive capabilities that are closely tied to the organization's heritage and strategic objectives. This integration supports PepsiCo's ability to leverage its unique strengths in a manner that enhances its market position.[15]

59. The use of AI to boost a company's signature processes, however, comes with its risks. First, the integration of AI into these processes often requires a significant upfront investment in technology and expertise to develop and maintain such bespoke AI systems. Second, and more concerningly, the systematization of a companies' signature practices might inadvertently lead to their codification and standardization. Once super-ordinary processes are well understood and standardized, they are at a higher risk of imitation by competitors and ultimately can turn into an industry best practice. In other words, the idiosyncrasy of super-ordinary capabilities often lies in an element of tacit knowledge, which AI systems could intuit and make easier to adopt elsewhere.

3. Dynamic Capabilities

60. Dynamic capabilities hinge on the entrepreneurial activities that steer a company's resources so as to better align with evolving customer needs and the fresh opportunities created by new technologies and innovations introduced by suppliers, partners and the enterprise itself. Entrepreneurial management enables a business to overcome current challenges and launch itself forward. At the core of dynamic capabilities are three key entrepreneurial management activities: sensing, seizing, and transforming.

61. AI has profound implications on the way dynamic capabilities are implemented. Strong dynamic capabilities require the ability to collect, analyze, and process very large amounts of data regarding change in technologies and in the business environment. The use of AI reduces the cycle time and enlarges the opportunity set for monitoring, analyzing, and adjusting a business. In this sense, AI can be regarded as the latest manifestation of what Beninger called "the control revolution,"[16] which views progress in information processing technologies as a response to the fundamental need to manage increasingly complex productive systems. However, the decisions required for robust dynamic capabilities are highly complex, as they require management to engage with an ever-changing set of factors whose identification and understanding are riddled with ambiguities, and

---

15 McKinsey's AI tool, Lilli, is trained on its internal company archives, documents and reports, and it represents another case of a company attempting to develop its super-ordinary capabilities through AI. For more information, *see* MCKINSEY & CO. (2024), https://www.mckinsey.com/about-us/new-at-mckinsey-blog/meet-lilli-our-generative-ai-tool.

16 JAMES R. BENINGER, THE CONTROL REVOLUTION: TECHNOLOGICAL AND ECONOMIC ORIGINS OF THE INFORMATION SOCIETY (Harv. Univ. Press 1986).

for which past data points are of little use. This makes it unlikely for AI to completely replace human entrepreneurial insight and judgement. Whereas ordinary capabilities can be displaced by AI, dynamic capabilities and AI are better regarded as complementary.

62. AI has the potential to reduce risks through enhanced analytics, yet it currently falls short in scenarios characterized by deep uncertainty. Risk management involves evaluating possible outcomes and their likelihood of occurrence. However, deep uncertainty presents a situation where little is known about potential outcomes, their probabilities, and the criteria for assessing them.[17] While managing risk typically involves optimization – a task well-suited to AI's capabilities – dealing effectively with deep uncertainty requires dynamic capabilities or making insightful decisions based on insufficient information. Consequently, such scenarios cannot yet be effectively handled by algorithms.

63. *Sensing capabilities* are those that can potentially be assisted the most. These have to do with the environmental scanning, hypothesis generation and testing of the market, technological and regulatory dynamics influencing a company, its customers and its suppliers. This requires gathering and interpreting data from both internal and external sources to monitor the business environment effectively, prioritizing strategies and issues, and pinpointing emerging threats and opportunities, such as untapped markets or new revenue streams. Sensing capabilities require an efficient flow of information to the top management from the external environment and the rest of the organization: a task whose challenges were already highlighted by Hayek as the key problem of centralized organizations.[18] AI significantly enhances this process by automating the collection and synthesis of vast amounts of data, allowing for real-time insights and faster responsiveness. AI can also support the rigorous assessment of this information by applying advanced analytics to detect patterns and trends that may not be visible to the human eye. This capability allows companies to generate more accurate hypotheses and predictions about market forces, customer behavior, and potential disruptions.

64. Once a firm senses opportunities and threats, the ability to respond to them is what we call *seizing capabilities*. Although AI will undoubtedly provide assistance, the activities for "seizing" require a nuanced approach that blends technology with human oversight. AI can assist "seizing" capabilities by providing advanced analytics and decision-support tools to allow firms to make more informed and quicker decisions. AI-based tools can support the timely development of business strategies, quickly assessing their impact under different scenarios and adjust them if circumstances change. This capability is especially beneficial in uncertain and dynamically changing environments. For example, better forecasting

---

17   David J. Teece, Innovation and Dynamic Capabilities: Strategic Management Under Deep Uncertainty in Innovation: Flexibility, Organization, and Strategy (L. Trigeorgis ed., MIT Press 2020).

18   Friedrich A. Hayek, *The Use of Knowledge in Society*, 35 Am. Econ. Rev. 519 (1945).

techniques can allow decision-makers to craft more robust business models. However, while AI can propose actions based on data-driven insights, human intervention remains crucial. Over-reliance on AI might bias management judgement in favor of quantifiable elements, leading to poor seizing. Strategic decisions often involve complex considerations that go beyond quantitative analysis, including ethical implications, long-term brand impact, and stakeholder relationships. Humans bring context, experience, and judgment to the decision-making process, elements that are currently beyond the reach of AI.

65. *Transforming capabilities* have to do with the ability to reconfigure an organization in line with the company's strategy. A changing business landscape might require organizational change in the form of new policies, procedures and incentive schemes. While AI can play a role in engendering organizational transformation, the reconfiguration of assets integral to dynamic capabilities still heavily depends on human involvement. AI's promise is also in helping to decentralize authority by empowering employees at any level with access to insights and analytics that support their decision-making, thus promoting a more agile and responsive organizational structure. Encouraging a collaborative organizational culture is further enhanced by AI-driven platforms that facilitate communication and information sharing across diverse teams and departments. A stronger system of control and knowledge sharing is a necessary condition for any organizational change. However, an organization can be viewed as a web of human relationships, where the implementation of change is fundamentally a human endeavor. As such, despite the significant potential of AI tools, the reconfiguration of an organization will continue to require substantial human engagement for the foreseeable future.

4. Outsourced versus In-House AI

66. Integrating AI into an organization necessitates a strategic decision: should AI development be managed in-house, or outsourced to external providers? Outsourcing AI development can be more cost-effective, allowing companies to quickly tap into advanced technology and expertise without the overhead of developing these capabilities internally. These AI services can be general-purpose or tailored to a company's specific needs. For instance, a small e-commerce business might utilize an outsourced AI service for customer support chatbots, providing 24/7 customer service without the need to hire additional staff or develop the technology in-house.

67. However, outsourcing can introduce vulnerabilities, especially when the AI model is specialized and deeply integrated into the company's operations, or, more critically, when it supports the company's core business. In such scenarios, dependency on an external provider for essential, specialized AI inputs can financially disadvantage the company in future contract negotiations, as well as creating risks of disruption and security concerns.

68. While developing AI capabilities in-house can be costly, especially in the short term, and demands hard-to-find technical resources, it provides a company with better resources to capture value. Moreover, it would allow the company to better exploit the benefits of asset specialization and integration. For example, Ant Financial Services operates a range of services – from consumer lending to health insurance – powered by an AI system that leverages data from its primary mobile payments platform, Alipay. The business core of Ant Financial is AI-powered, with no human involvement. AI autonomously handles tasks from loan approval to financial advising and medical expense authorizations.[19] For Ant Financial Services, outsourcing AI would create insurmountable hold-up problems, as it would give the provider disproportionate bargaining power. Moreover, giving up the control of sensitive information creates risks in terms of data breaches and business disruption.

69. It should be noted that the choice of whether to rely on another firm for the deployment of AI is not strictly binary; in practice, most firms opt for a hybrid approach that combines elements of both in-house and outsourced AI solutions. In fact, even firms that choose to develop their own systems typically need to rely on third parties not only for storage and computational power, but also for software services that support the development of AI. For example, both Coca-Cola, which uses OpenAI's foundation model, and PepsiCo, which built an in-house AI platform, rely on Microsoft Azure some AI services.[20,21] The importance of access to computational power to develop AI is explored more in depth in Section III.3.

70. In summary, the decision to develop AI systems in-house or to outsource them should consider the extent to which a company's core business relies on AI and the specificity required of the model. For example, while ordinary capabilities might be adequately supported by off-the-shelf, general-purpose AI tools due to their generic nature, super-ordinary capabilities, which are unique to a company's operations, may necessitate bespoke algorithms, making in-house development a more attractive option.

III. Potential Competition Issues

71. As AI technologies promise to integrate organizational capabilities, firms' ability to adopt AI tools will affect their competitive edge. In this section we discuss the potential barriers that some firms could face in rolling out their own AI systems. These considerations are valid not only for companies developing and selling generative AI but for any enterprise aiming

---

[19] MARCO IANSITI & KARIM R. LAKHANI, COMPETING IN THE AGE OF AI: STRATEGY AND LEADERSHIP WHEN ALGORITHMS AND NETWORKS RUN THE WORLD (Harvard Bus. Rev. Press 2020).

[20] *PepsiCo uses Azure Machine Learning to Identify Consumer Shopping Trends and Produce Store-Level Actionable Insights*, PEPSICO (Jan. 28, 2021), https://ms-f1-sites-03-ea.azurewebsites.net/es-es/story/754571-pepsico-consumer-goods-azure-machine-learning.

[21] Christopher Doering, *Coca-Cola Turns to Microsoft's AI Services for Its Supply Chain*, SUPPLY CHAIN DIVE (May 2, 2024), https://www.supplychaindive.com/news/coca-cola-ai-artificial-intelligence-microsoft/714889/.

to restructure its capabilities in light of the new technology. As AI compels businesses to update ordinary capabilities and complement dynamic capabilities, this change is likely to be cross-sectoral and widespread.

72. Each of the resources necessary to build and integrate AI models can potentially lead to the emergence of new opportunities for some actors and bottlenecks for others. While training data is commonly perceived to be the key input for AI, the availability of skilled labor currently represents another limiting factor. The bottleneck, however, is likely to remain computing infrastructure, especially if technological specialization prevents its commodification. Investing in multihoming and cloud interoperability might be a sensible strategy for businesses wishing to reduce their reliance on compute providers.

1. Access to Data

73. Data is a necessary input for AI models. Its quality is crucial, as there appears to be a tradeoff between data quality and quantity. On the one hand, a large quantity of data can average out bad data. On the other hand, for the same amount of data and computing power, high-quality data can make a model stand out.[22] Many AI models have been trained using vast amounts of data freely available on the web. As the availability of computing power increases, both established firms and startups will be able to utilize the majority of the data available online. As more models leverage this free data, access to exclusive data will become increasingly important, including peer-reviewed academic journals, reputable news articles, and meticulously curated resources like Wikipedia. In fact, internet data is becoming more exclusive, as resources that were once free are being enclosed and marketed – Reddit and X (formerly Twitter) being prominent examples. Specialized data for companies wishing to build their own tailored AI solutions might be even harder to come by. It is likely that those in control of good corpuses of data will be able to exert some degree of gatekeeping.

74. However, while access to data is a *sine qua non* condition for an AI model, the value derived from accumulating larger datasets does not always increase indefinitely. Marco Iansiti notes that

> firms typically face diminishing returns to data value, ... additional data may not necessarily confer learning across other users of the produce. For example, Netflix discovered that there were rapidly diminishing returns to data in driving improvements in their recommendations algorithm, beyond a modest threshold.[23]

---

22 Google's PaLM 2 technical report highlights that smaller training datasets can achieve better performances if the quality is higher: "A smaller but higher quality model significantly improves inference efficiency, reduces serving cost, and enables the model's downstream application for more applications and users." *See* Rohan Anil et al., *PaLM 2 Technical Report*, Computer Sci. (Sept. 13, 2023), https://doi.org/10.48550/ARXIV.2305.10403

23 Marco Iansiti, *The Value of Data and Its Impact on Competition* (Harv. Bus. Sch., NOM Unit Working Paper No. 22-002, July 20, 2021).

75. Moreover, new techniques allow companies to achieve better performances with the data they have.[24] Where data is scarce or cannot be used for privacy reasons, algorithmically generated *synthetic data* can be used to validate models or to deepen machine learning.[25] This points to the fact that, while access to high quality data does confer an advantage, this is unlikely to be the main bottleneck for the development of AI models.

2. Access to Skilled Labor

76. As AI affects the capability structure of businesses across industries, its adoption is likely to lead to a shift in the composition of skills. The ability of firms to navigate this set of challenges will be in part the result of their dynamic capabilities. Enterprises with strong *seizing* and *transforming* capabilities are more likely to successfully and swiftly reorient their composition of skills in line with the new technological paradigm. Laggards, on the other hand, are more likely to be disrupted and eventually outcompeted.

77. The first challenge concerns the access to the supply of skilled professionals required to build and train AI, including expertise in programming, machine learning and data interpretation. Given the recent rush to develop AI models, the demand for such personnel is currently extremely high. A recent article by the *Financial Times* revealed that Apple has been poaching dozens of AI experts from Google, as part of the company's efforts to reorient its strategy towards AI.[26] Competition for talent might intensify even further as the US Federal Trade Commission (FTC) recently imposed a ban on noncompete agreements,[27] which are contractual restrictions that prevent workers from taking new jobs in related fields or starting new businesses within a specified time frame.[28] Another common strategy to acquire capabilities is the acquisition of another company: a phenomenon which became known as "acqui-hiring,"

78. However, the shift in the composition of demand for skills does not concern only Big Tech or AI firms, but also those businesses wishing to integrate AI technologies into their business processes. This task requires

---

24 In their review of the types of foundation models, Schrepel and Pentland describe a number of technical advancements that allow models with smaller datasets to compete with larger ones – not necessarily at the expense of higher computational requirements. *See* Thibault Schrepel & Alex Sandy Pentland, *Competition between AI Foundation Models: Dynamics and Policy Recommendations* 8–9 (Amsterdam L. & Tech. Inst., Working Paper No. 1-2023, 2023).

25 174 Sergey I. Nikolenko, Synthetic Data for Deep Learning (Springer Nature 2021).

26 Michael Acton, *Apple Targets Google Staff to Build Artificial Intelligence Team*, Fin. Times (Apr. 30, 2024), https://www.ft.com/content/87054a60-dc4d-4238-a4b9-93ab48f22f56.

27 Fed. Trade Comm'n, *FTC Announces Rule Banning Noncompetes*, FTC (2024), https://www.ftc.gov/news-events/news/press-releases/2024/04/ftc-announces-rule-banning-noncompetes.

28 The news has sparked mixed reactions. The US Chamber of Commerce along with other business groups filed a lawsuit against the FTC, arguing that the move could negatively impact businesses' ability to safeguard their proprietary information and diminish their incentives to train employees. Conversely, labor advocates have welcomed the ban, stating that it will grant Americans greater liberty to change jobs, start new businesses, or market new innovations. They contend that by lowering the hurdles to acquiring skills, the ban is advantageous to new market entrants – as AI companies tend to be.

large amounts of specialized skills not only for the initial development of AI models but also for their ongoing management, refinement, integration in existing routines, and alignment with business goals. Companies aiming to integrate AI solutions into their operations face a dual challenge for their human resources. They must secure talent that possesses the necessary technical skills to develop and curate AI models, and they also need managers with a deep understanding of the specific industry and company context. In order to successfully integrate AI, they ultimately need employees with both types of knowledge. To address this shortage, some multinational companies are building AI teams in geographies with more accessible skilled labor. PepsiCo, for example, is planning on expanding its presence in India and build a third AI laboratory.[29]

79. Finally, the substitutability between ordinary capabilities and generative AI might lead to the displacement of some jobs. AI company Cognition has recently released Devin, an "AI employee" that can be hired for engineering tasks, such as coding and project management. Instead of hiring a software engineer, companies can now hire Devin.[30] This example also demonstrates that ordinary capabilities are by no means synonymous with what are traditionally regarded as "low-skill jobs", and the displacement might affect several types of profession.

3. Access to Compute

80. Computational power, often referred to as "compute," is a fundamental requirement for companies wishing to develop their own AI model or upgrade their capabilities by incorporating the technology into their business. As AI models become more complex, the need to access computational resources increases exponentially. For instance, while OpenAI's ChatGPT 3.5 utilized 175 billion parameters, its successor, ChatGPT 4, employs 1.75 trillion parameters, with an estimated training cost of around $100 million.[31] As AI makers scramble to secure more compute, backed by the support of massive capital investments, its supply struggles to keep up with demand. The scale of computation required to manage these tasks is thus a limiting factor in the AI industry.[32]

81. Computing power encompasses a stack composed of three types of products: infrastructure hardware (including data centers, servers, cables and

---

29 Beena Parmar, *PepsiCo Leans on Indian Tech Talent to Leverage GenAI*, THE ECON. TIMES (Mar. 15, 2024, 06:01 AM IST), https://economictimes.indiatimes.com/tech/technology/pepsico-leans-on-indian-tech-talent-to-leverage-genai/articleshow/108502739.cms.

30 Scott Wu, *Introducing Devin, the First AI Software Engineer*, COGNITION BLOG (Mar. 12, 2024), https://www.cognition.ai/blog/introducing-devin.

31 Will Knight, *OpenAI's CEO Says the Age of Giant AI Models Is Already Over*, WIRED (Apr. 17, 2023), https://www.wired.com/story/openai-ceo-sam-altman-the-age-of-giant-ai-models-is-already-over/.

32 The computational demands of AI models have raised widespread concerns about their environmental sustainability, particularly in relation to the resource-intensive cooling systems used in data centers. While these considerations are crucial for the long-term development of the industry, they fall beyond the scope of this chapter.

cooling systems), specialized chips (both design and manufacturing), and software applications. It represents both a fixed cost and a recurring expense, as it is necessary for both training and running AI models.

Table 1. Top AI startups by funding raised and respective cloud providers[33]

AI startup	Cloud partner
AI21 Labs	Google Cloud
Anthropic	AWS, Google Cloud
Anyscale	AWS
Cohere.ai	Google Cloud, Oracle Cloud Infrastructure
Hugging Face	AWS, Google Cloud, Microsoft Azure
Inflection AI	Microsoft Azure
Instadeep	Google Cloud
Mistral	Microsoft Azure, Google Cloud, AWS
OctoML	AWS
OpenAI	Microsoft Azure
Scale AI	AWS, Microsoft Azure
Weights & Biases	AWS

82. Key providers of this infrastructure, who orchestrate the integration of hardware and software, offer compute as a comprehensive cloud service. Leading players in the Infrastructure as a Service (IaaS) market include Amazon Web Services (AWS), Google Cloud, and Microsoft Azure.[34] These companies play a significant role across all layers of compute production but are particularly influential in the realm of physical infrastructure. They either purchase specialized chips from manufacturers, principally Nvidia, or design their own, such as AWS's Graviton4 or Google's Tensor chips, which are manufactured by a chip foundry such as Taiwan Semiconductor Manufacturing Company (TSMC). Additionally, they develop specialized software to enhance hardware capabilities, while allowing third parties to create further software layers, such as various applications and programs.

83. Both AI and traditional firms adopting the new technology are reliant on a handful of companies for cloud services. Partnerships are the best

---

33 See ANSWERIQ (June 19, 2024), https://www.answeriq.com/openai-statistics/ (last visited Aug. 6, 2024), and authors' own research.

34 In the first quarter of 2024, the market shares of AWS, Microsoft Azure and Google Cloud were respectively 31, 25, and 11 percent. The next biggest competitors were Alibaba Cloud (4 percent), Salesforce (3 percent), IBM Cloud (2 percent), Oracle (2 percent) and Tencent Cloud (2 percent). However, these were aggregated across public and private cloud services, while computing power is mostly obtained through public cloud. Since some companies such as Salesforce and IBM Cloud mostly provide private cloud infrastructure, the actual market shares for the supply of compute to AI firms are even more concentrated in favor of the top three actors. For more information *see Cloud Market Gets Its Mojo Back; AI Helps Push Q4 Increase in Cloud Spending to New Highs*, SYNERGY RSCH. GRP. (Feb. 1, 2024), https://www.srgresearch.com/articles/cloud-market-gets-its-mojo-back-q4-increase-in-cloud-spending-reaches-new-highs.

way a smaller player can hedge against the risk of chip supply shortages and ensure scalability. As a result, AI companies, including OpenAI, Anthropic and Mistral, typically pair with one of the leading cloud providers. Alphabet's CEO Sundar Pitchai in 2023 declared that over 70 percent of generative AI startups rely on Google's cloud infrastructure.[35] Table 1 above, shows the affiliations of the main emerging AI firms with Big Tech. By co-designing all the layers of cloud computing, Big Tech firms create immense value and efficiency. At the same time, they might act as bottlenecks and may create vendor switching costs, as migrating from one provider to another could require redesigning a product if the application is integrated with the provider's specific hardware/software environment and services. In order to compete for new customers, some providers, including Google Cloud and AWS, mitigate these switching costs by eliminating data transfer fees for migrations to their services. However, as Google's Head of Platform Amit Zavery pointed out, egress fees represent only about 2 percent of the total cost involved in switching providers.[36]

84. Cloud providers' relations with AI startups have attracted antitrust scrutiny, with the FTC launching an inquiry on the partnerships and investments involving Alphabet (Google Cloud), Amazon (AWS), Anthropic, Microsoft (Azure), and OpenAI.[37] The FTC has requested information on the nature of the existing arrangements, influence over the partner product releases and governance, and competitive impact. On the one hand, such partnerships are essential to the development of the industry, as they unlock essential complementarities and without which startups and small businesses would struggle to secure the computing power required to develop an AI system. On the other hand, relying on a specific complement can create vulnerabilities with respect to the provider. Current agreements are often opaque and, as Google pointed out, the investigation might "shine a bright light on companies that have a "history of locking-in customers."[38]

85. To minimize complementarity risks, some AI startups employ a strategy of multihoming, where their products are supported on multiple cloud platforms. This approach offers several benefits, such as competitive pricing

---

35 Johan Moreno, *70% Of Generative AI Startups Rely on Google Cloud, AI Capabilities*, FORBES (July 25, 2023, 10:39 PM EDT), https://www.forbes.com/sites/johanmoreno/2023/07/25/70-of-generative-ai-startups-rely-on-google-cloud-ai-capabilities-says-alphabet-ceo-sundar-pichai/.

36 Dina Bass, *Google Ends Cloud Switching Fees, Pressuring Amazon and Microsoft*, BLOOMBERG (Jan. 11, 2024), https://www.bloomberg.com/news/articles/2024-01-11/google-googl-ends-switching-fees-for-cloud-data-pressuring-amazon-microsoft.

37 Fed. Trade Comm'n, *FTC Launches Inquiry into Generative AI Investments and Partnerships*, FTC (Jan. 25, 2024), https://www.ftc.gov/news-events/news/press-releases/2024/01/ftc-launches-inquiry-generative-ai-investments-partnerships.

38 Zachary Folk, *FTC Investigating AI Deals and Investments with Big Tech Firms, including Microsoft, Amazon, and Alphabet*, FORBES (Jan. 25, 2024, 01:44 PM EST). Retrieved from: https://www.forbes.com/sites/zacharyfolk/2024/01/25/ftc-investigating-ai-deals-and-investments-with-big-tech-firms-including-microsoft-amazon-and-alphabet/.

per service and reduced dependency risks. Yet, developing applications that are compatible across different cloud environments can be expensive due to the lack of standardized cloud provision. Furthermore, multihoming often leads to increased overall cloud consumption. Companies may find themselves procuring additional services to ensure interoperability or to facilitate data transfers between cloud providers, consequently inflating their overall cloud expenditure. This approach typically results in higher costs compared to using a single cloud provider.

86. Cloud providers are currently competing fiercely. Whether the current competitive environment will undergo a process of commodification remains unclear. Some economic forces might prevent that. In particular, as computing power requirements grow alongside the demand for AI services, efficiency gains might accrue from vertical integration and specialization. While traditional chips were general purpose, in-house chips tend to be co-designed with software, enhancing purpose-specific performance. The specialization of hardware and software also creates switching costs for cloud compute services, which may prevent commodification.

IV. Conclusions

87. This chapter has discussed the integration of AI within routine business operations and its potential to reconfigure markets, as well as its effects on dynamic capabilities and strategic management. After reviewing the role of AI in shaping organizations, we analyzed some potential areas of concern for competition policy. AI is poised to increase the complexity of economic interactions further, adding to the urgency to update the competition policy framework. It is essential to move beyond a static view of markets as isolated pools of substitutable products and build a framework that acknowledges the interconnected and ever-evolving nature of competition in the age of AI.

88. AI is without a doubt a disruptive technology. It learns like us, grasping and making use of tacit knowledge. As it integrates into organizations, it promises to reshape the very economics of industries and market structures in profound and perhaps unpredictable ways. AI is a harbinger of transformation, offering unprecedented opportunities for those who can harness its potential and presenting challenges to those who stand still. As a new technological paradigm emerges, it may dissolve existing barriers while introducing new bottlenecks, underscoring the fluid nature of innovation and competition. The pace of change is relentless – a clear call for maintaining a dynamic perspective and looking ahead.

Antitrust and Innovation Competition: Investments and Partnerships in Artificial Intelligence

DANIEL F. SPULBER[*]

Northwestern University

Abstract

The Federal Trade Commission (FTC) issued orders to Alphabet (Google), Amazon, Microsoft, OpenAI, and Anthropic requesting information regarding investments and partnerships in artificial intelligence (AI). The FTC expressed concerns about potential harm to innovation and competition. I point out that investments and partnerships are an alternative to Mergers and Acquisitions (M&A) and vertical integration. I argue that these investments and partnerships are not likely to diminish innovation or competition in either cloud computing or AI. Antitrust policymakers should not discourage investments and partnerships based on hypothetical effects on future innovation. I conclude that investments and partnerships can make important contributions to innovation competition and the market for technology.

---

[*] Elinor Hobbs Distinguished Professor of International Business and Professor of Strategy, Strategy Department, Kellogg School of Management, Northwestern University; Professor of Law (Courtesy), Pritzker School of Law, Northwestern University. Email: jems@kellogg.northwestern.edu.

Antitrust and Innovation Competition:
Investments and Partnerships in Artificial Intelligence

I. Introduction

89. The Federal Trade Commission (FTC) issued orders to Alphabet (Google), Amazon, Microsoft, OpenAI, and Anthropic for information regarding some of their investments and partnerships.[1] The FTC's study, according to Chair Lina M. Khan, "will shed light on whether investments and partnerships pursued by dominant companies risk distorting innovation and undermining fair competition."[2] Khan noted that the FTC was concerned about foreclosure of competition to "develop and monetize" artificial intelligence (AI).[3] In this chapter, I consider investments and partnerships in the context of innovation competition. I argue that these investments and partnerships are not likely to diminish innovation or competition in either cloud computing or AI.

90. I apply a Coasian perspective to examine these investments and partnerships. I point out that investments and partnerships among firms are market contracts that allow the exchange of technology and complementary inputs. Investments and partnerships provide an alternative to increased vertical integration, whether through mergers and acquisitions (M&A) or expansion of internal research and development (R&D). I find that investments and partnerships can contribute to competition and the market for technology.

91. The potential for widespread applications of AI motivated antitrust policy interest. AI contributes to what I have termed "The Business Revolution", which I defined as "the augmentation and replacement of human effort in business transactions by computers, communications systems, and the Internet."[4] As I explained, "The Business Revolution is changing the office, the store, and the market, just as the Industrial Revolution earlier changed the factory."[5] Rapid diffusion of AI technology also motivated antitrust scrutiny. For example, it was reported that OpenAI's ChatGPT reached over 180 million monthly users with 1.6 billion visits to the website.[6]

92. Changes in the characteristics of competition have drawn antitrust attention. Firms are competing through innovation, resulting in creative destruction.[7] Consequently, antitrust policy must adapt to an economy

---

[1] The FTC's orders were under the FTC Act, 15 U.S.C. § 6(b). Fed. Trade Comm'n, *FTC Launches Inquiry into Generative AI Investments and Partnerships*, FTC (Jan. 25, 2024), https://www.ftc.gov/news-events/news/press-releases/2024/01/ftc-launches-inquiry-generative-ai-investments-partnerships.

[2] *Id.*

[3] Khan stated: "As companies race to develop and monetize AI, we must guard against tactics that foreclose this opportunity." *Id.*

[4] Daniel F. Spulber, *Should Business Method Inventions be Patentable?*, J. Leg. Anal. Spring no. 1, 2011, 265-340, at 265.

[5] *Id.* at 269.

[6] Oskar Mortensen, *How Many Users on ChatGPT?*, SEO.AI (Apr. 24, 2024), https://seo.ai/blog/how-many-users-does-chatgpt-have#:~:text=How%20Many%20Users%20on%20ChatGPT,users%20on%20a%20weekly%20basis.

[7] Joseph A. Schumpeter, The Theory of Economic Development (Harvard Univ. Press, new ed., Routlege 1980) (1934); Joseph A. Schumpeter, Capitalism, Socialism and Democracy (Harper Perennial 1976) (1942).

93. with substantial technological change. Antitrust policymakers need to evaluate competitive conduct and industry performance with economic tools that go beyond price competition with static technologies.

93. Firms are shifting toward greater innovation competition, as I have observed elsewhere.[8] Innovation competition is a form of rivalry in which firms introduce new products, production processes, and transaction methods. Incumbent firms and entrepreneurs invest in R&D to create inventions and produce innovations, often by combining inventions.[9] Also incumbent firms and entrants invest in promoting adoption of innovations, which generates technological change. This contrasts with traditional competition in which technologies do not change and firms emphasize alternative instruments such as prices, output levels, productive capacity, marketing, and sales.

94. This does not mean that competition causes innovation or that innovation causes competition. Rather, the extent of innovation and the intensity of competition are the result of strategic interaction among firms. Firms make strategic decisions about the direction of innovation and competitive entry. This means that innovation decisions and other competitive decisions are interconnected. This explains why it is typically not possible to identify any causation between competition and innovation.

95. Economic analysis showed that the direction of causation in the traditional structure-conduct-performance (SCP) framework was inconsistent with strategic interaction among firms. The SCP framework used market structure observations to make predictions about competitive conduct. Even without technological change, however, strategic interaction among firms affects competitive strategies and market structure. With technological change, strategic interaction among firms affects both innovation and other aspects of competition.

96. The FTC has recognized the significance of innovation as a means of competition: "Innovation is a central aspect of rivalries among technology firms, and the markets are dynamic: new ideas topple formerly dominant technologies and consumers line up to buy products that are smaller, faster, and better."[10] The FTC argues that competition encourages innovation. The FTC stated that it "promotes competition in technology industries (like computers, software, communications, and biotechnology) as the best way to reduce costs, encourage innovation, and expand choices for consumers."[11]

97. The FTC, through its orders, sought to determine how investments and partnerships affect innovation competition in AI. This raises the general

---

8 Daniel F. Spulber, *Antitrust and Innovation Competition*, 11 J. ANTITRUST ENFORC. 5-50, (Mar. 2023), https://academic.oup.com/antitrust/article/11/1/5/6593929.

9 DANIEL F. SPULBER, THE INNOVATIVE ENTREPRENEUR (Cambridge Univ. Press 2014).

10 Fed. Trade Comm'n, *Competition in the Technology Marketplace*, FTC (2024), https://www.ftc.gov/advice-guidance/competition-guidance/industry-guidance/competition-technology-marketplace.

11 *Id.*

question of whether antitrust scrutiny of investments and partnerships strengthens or weakens innovation competition. A related question is how antitrust policies targeting investments and partnerships related to technological change affect consumer welfare and economic efficiency.

98. The FTC focused on three relationships: Microsoft and OpenAI, Amazon and Anthropic, and Google and Anthropic. Among the FTC requests for information were: "the strategic rationale of an investment/partnership," "decisions around new product releases," and "[c]ompetition for AI inputs and resources."[12] The FTC also sought information on "the transactions' competitive impact, including information related to market share, competition, competitors, markets, potential for sales growth, or expansion into product or geographic markets."[13]

99. The chapter is organized as follows. In Section II, I consider the three sets of investments and partnerships that are the subject of the FTC orders. In Section III, I examine some of the implications of a Coasian perspective for antitrust policy toward innovation competition. Section IV concludes the discussion.

II. Investments and Partnerships in Artificial Intelligence

100. The three sets of investments and partnerships considered by the FTC involve invention and innovation. These company relationships have two-way vertical aspects. Microsoft, Amazon, and Google provide cloud computing, which is an input to the development of Large Language Models (LLMs). Conversely, OpenAI and Anthropic provide AI LLM products, which are inputs in the development of consumer and corporate products offered by Microsoft, Amazon, and Google.

1. Microsoft and OpenAI

101. Microsoft entered a multiyear partnership with OpenAI that involved several phases of investment in 2019, 2021, and 2023.[14] By May 1, 2024, Microsoft invested $13 billion in OpenAI.[15] It was reported that the investments and the partnership initially at least were driven by innovation competition with Google. According to internal emails, Microsoft was "'very worried' that Google was years ahead in scaling up its AI efforts."[16]

---

12 Fed. Trade Comm'n, *supra* note 1.

13 *Id.*

14 *Microsoft and OpenAI Extend Partnership*, MICROSOFT CORP. BLOGS (Jan. 23, 2023), https://blogs.microsoft.com/blog/2023/01/23/microsoftandopenaiextendpartnership/.

15 Tom Warren, *Microsoft's OpenAI Investment was triggered by Google Fears, Emails Reveal*, THE VERGE (May 1, 2024), https://www.theverge.com/2024/5/1/24146302/microsoft-openai-investment-google-worries-internal-emails.

16 *Id.*

102. Describing the partnership, OpenAI stated that it used Microsoft's cloud platform to train its AI: "We've worked together to build multiple supercomputing systems powered by Azure, which we use to train all of our models. Azure's unique architecture design has been crucial in delivering best-in-class performance and scale for our AI training and inference workloads."[17] OpenAI emphasized that Microsoft's investment would grow: "Microsoft will increase their investment in these systems to accelerate our independent research and Azure will remain the exclusive cloud provider for all OpenAI workloads across our research, API and products."[18]

103. For its part, Microsoft would use OpenAI's models: "Microsoft will deploy OpenAI's models across our consumer and enterprise products and introduce new categories of digital experiences built on OpenAI's technology."[19] Microsoft's Azure OpenAI Service would offer developers "direct access to OpenAI models backed by Azure's trusted, enterprise-grade capabilities and AI-optimized infrastructure and tools."[20]

104. Leading executives from both companies endorsed the partnership. According to Sam Altman, CEO of OpenAI, "Microsoft shares our values and we are excited to continue our independent research and work toward creating advanced AI that benefits everyone."[21] In turn, Satya Nadella, Chairman and CEO, of Microsoft stated: "In this next phase of our partnership, developers and organizations across industries will have access to the best AI infrastructure, models, and toolchain with Azure to build and run their applications."[22]

105. The Microsoft and OpenAI partnership involved both invention and innovation. According to the companies, they "pushed the frontier of cloud supercomputing technology, announcing our first top-5 supercomputer in 2020, and subsequently constructing multiple AI supercomputing systems at massive scale."[23] In addition, the partnership was involved in invention and product development. Microsoft stated: "OpenAI has used this infrastructure to train its breakthrough models, which are now deployed in Azure to power category-defining AI products like GitHub Copilot, DALL·E 2 and ChatGPT."[24] According to Microsoft: "These innovations have captured imaginations and introduced large-scale AI as a powerful, general-purpose technology platform

---

17 *OpenAI and Microsoft Extend Partnership*, OPENAI (Jan. 23, 2023), https://openai.com/index/openai-and-microsoft-extend-partnership/.

18 *Id.*

19 Microsoft, *supra* note 14.

20 *Id.*

21 *Id.*

22 *Id.*

23 *Id.*

24 *Id.*

that we believe will create transformative impact at the magnitude of the personal computer, the internet, mobile devices and the cloud."[25]

2. Amazon and Anthropic

106. Amazon and Anthropic also formed a partnership that involved investment in invention and innovation. Amazon invested about $4 billion in Anthropic and took a minority ownership share.[26] Anthropic has developed an LLM named Claude. Anthropic states that it is "an AI safety and research company. We build reliable, interpretable, and steerable AI systems."[27] Anthropic describes its R&D as follows: "We search for simple relations among data, compute, parameters, and performance of large-scale networks. Then we leverage these relations to train networks more efficiently and predictably, and to evaluate our own progress. We're also investigating what scaling laws for the safety of AI systems might look like, and this will inform our future research."[28]

107. Amazon would provide cloud computing that Anthropic would use to train its AI models, "Anthropic selects AWS [Amazon Web Services] as its primary cloud provider and will train and deploy its future foundation models on AWS Trainium and Inferentia chips, taking advantage of AWS's high-performance, low-cost machine learning accelerators."[29] Anthropic would provide its AI models to Amazon: "Anthropic makes a long-term commitment to provide AWS customers around the world with access to future generations of its foundation models via Amazon Bedrock, AWS's fully managed service that provides secure access to the industry's top foundation models. In addition, Anthropic will provide AWS customers with early access to unique features for model customization and fine-tuning capabilities."[30]

108. Senior executives from both companies promoted the partnership. Andy Jassy, Amazon CEO stated: "Customers are quite excited about Amazon Bedrock, AWS's new managed service that enables companies to use various foundation models to build generative AI applications on top of, as well as AWS Trainium, AWS's AI training chip, and our collaboration with Anthropic should help customers get even more value from these two capabilities."[31] Dario Amodei, co-founder and CEO of Anthropic said: "Since announcing our support of Amazon Bedrock

---

25 *Id.*

26 Amazon Press Release, *Amazon and Anthropic Announce Strategic Collaboration to Advance Generative AI*, AMAZON (Sept. 25, 2023), https://press.aboutamazon.com/2023/9/amazon-and-anthropic-announce-strategic-collaboration-to-advance-generative-ai.

27 ANTHROPIC, https://www.anthropic.com/company (last visited July 17, 2024).

28 ANTHROPIC, https://www.anthropic.com/research#entry:8@1:url.

29 Amazon, *supra* note 26.

30 *Id.*

31 *Id.*

in April, Claude has seen significant organic adoption from AWS customers. By significantly expanding our partnership, we can unlock new possibilities for organizations of all sizes, as they deploy Anthropic's safe, state-of-the-art AI systems together with AWS's leading cloud technology."[32]

109. The collaboration between Amazon and Anthropic involved invention and innovation in the layers of the generative AI stack. First, "[a]t the bottom layer, AWS continues to offer compute instances from NVIDIA as well as AWS's own custom silicon chips, AWS Trainium for AI training and AWS Inferentia for AI inference."[33] Second, "[a]t the middle layer, AWS... customers will have early access to features for customizing Anthropic models, using their own proprietary data to create their own private models, and will be able to utilize fine-tuning capabilities via a self-service feature within Amazon Bedrock."[34] Third, "[a]t the top layer, AWS offers generative AI applications and services for customers like Amazon CodeWhisperer, a powerful AI-powered coding companion, which recommends code snippets directly in the code editor, accelerating developer productivity as they code."

3. Google and Anthropic

110. Google in 2023 stated its intention to invest $2 billion in Anthropic.[35] In 2024, Google emphasized the innovation aspect of the agreement: "This is just the beginning of our partnership with Anthropic, and we're excited to enable customer innovation with the newest models."[36] Google highlighted its use of Anthropic's products, "[t]hrough our partnership, we will bring Anthropic's latest models to our customers via Vertex AI, the comprehensive AI development platform. The Claude 3 family joins over 130 models already available on Vertex AI Model Garden, further expanding customer choice and flexibility as gen AI use cases continue to rapidly evolve."[37]

111. Senior executives from both companies announced the partnership. Thomas Kurian, CEO of Google Cloud, said: "This expanded partnership with Anthropic, built on years of working together, will bring AI to more people safely and securely, and provides another example of how the most innovative and fastest growing AI startups are building

---

32 *Id.*

33 *Id.*

34 *Id.*

35 Krystal Hu, *Google Agrees to Invest up to $2 billion in OpenAI Rival Anthropic*, REUTERS (Oct. 27, 2023), https://www.reuters.com/technology/google-agrees-invest-up-2-bln-openai-rival-anthropic-wsj-2023-10-27/.

36 Warren Barkley, *Announcing Anthropic's Claude 3 models on Google Cloud Vertex AI*, GOOGLE CLOUD (Mar. 4, 2024), https://cloud.google.com/blog/products/ai-machine-learning/announcing-anthropics-claude-3-models-in-google-cloud-vertex-ai.

37 *Id.*

on Google Cloud."³⁸ Dario Amodei, co-founder and CEO of Anthropic, stated: "Our longstanding partnership with Google is founded on a shared commitment to develop AI responsibly and deploy it in a way that benefits society."³⁹

III. Implications for Antitrust Policy Toward Innovation Competition

112. The FTC expressed concern about AI and competition: "The rising importance of AI to the economy may further lock in the market dominance of large incumbent technology firms."⁴⁰ Additionally, the FTC worried about R&D in AI and the potential for vertical foreclosure: "These powerful, vertically integrated incumbents control many of the inputs necessary for the effective development and deployment of AI tools, including cloud-based or local computing power and access to large stores of training data. These dominant technology companies may have the incentive to use their control over these inputs to unlawfully entrench their market positions in AI and related markets, including digital content markets."⁴¹

113. Antitrust policies motivated by innovation competition should not be based on speculative theories of harm in the distant future. As I have noted elsewhere, it often is feasible to measure dynamic efficiency by observing innovation, rather than theorizing about how competitive conduct might affect the capacity to innovate.⁴² Economic analysis can measure dynamic efficiency by considering the effects of observed innovation in comparison to pre-existing technologies.

114. Antitrust concerns should not be directed at hypothetical future harms to innovation. Rather, antitrust policymakers should focus on whether there is evidence of anticompetitive conduct. Some FTC scrutiny may be justified if joint arrangements provide cover for coordination and collusive information sharing. Such FTC questions, however, can create litigation risk that might discourage joint technology arrangements motivated by efficiency. Antitrust policymakers should evaluate whether the harm from discouraging innovation outweighs the chance of finding hidden collusion.

---

38 *Id.*

39 *Id.*

40 FED. TRADE COMM'N, GENERATIVE ARTIFICIAL INTELLIGENCE AND THE CREATIVE ECONOMY STAFF REPORT: PERSPECTIVES AND TAKEAWAYS (Dec. 2023), https://www.ftc.gov/system/files/ftc_gov/pdf/12-15-2023AICEStaffReport.pdf.

41 *Id. See also* Fed. Trade Comm'n, *Staff in the Bureau of Competition & Office of Technology, Generative AI Raises Competition Concerns*, FTC (June 29, 2023), https://www.ftc.gov/policy/advocacy-research/tech-at-ftc/2023/06/generative-ai-raises-competition-concerns.

42 Daniel F. Spulber, *Antitrust Policy Toward Innovation Competition: Measuring Dynamic Efficiency,* CPI (Sept. 2023), https://www.pymnts.com/cpi-posts/antitrust-policy-toward-innovation-competition-measuring-dynamic-efficiency/.

115. Speculative concerns about future innovation should not be used to prevent economic arrangements that can lead to improved technologies and efficiencies in R&D. Such restrictive antitrust policies would damage the innovation competition they seek to protect.

116. As Ronald Coase first pointed out, firms choose between carrying out activities within the organization and relying on transactions with other firms. If production costs within the organization and in other organizations were similar, the firm's decision would depend on whether there were significant costs of using the market. The firm would undertake the activity if transaction costs were relatively high, and the firm would outsource the activity if transaction costs were relatively low.[43] In this way, transaction costs fundamentally affect the boundaries of the firm.[44]

117. Coase understood that economic activities can be achieved through organizational governance or through market contracts. This also applies to invention and innovation. Firms can bring invention and innovation in-house or they can obtain technologies outside the firm. Firms increasingly combine internal R&D and with technology obtained from other firms. This has accompanied rapid growth of the technology marketplace.

118. Antitrust policy should recognize that firms face a trade-off between production within the organization and obtaining goods and services in the marketplace. Robert Bork observed, citing Coase, "[o]nce the category of visibly naked restraints is set aside as per se illegal, the category of ancillary agreements is seen to be the same economic phenomenon as the category of mergers or close-knit combinations. Their difference is merely one of legal form: the difference between integration accomplished by contract and integration accomplished by ownership."[45]

119. This Coasian insight extends to invention and innovation. Firms also face a trade-off between conducting R&D and obtaining technologies in the marketplace. The FTC orders suggest that antitrust enforcers are concerned about the effects of investments and partnerships on competition in AI.

120. The investments and partnerships are an alternative to M&A. Antitrust scrutiny of investments and partnerships in AI could provide incentives for firms to pursue M&A or to grow internal R&D. Limiting investments and partnerships could influence firms to pursue greater vertical integration. The result could be less efficient organizations and increases in the size of firms.

---

43 Ronald H. Coase, *The Nature of the Firm*, 4 ECONOMICA, 386-405 (1937); Ronald H. Coase, *The Institutional Structure of Production*, in RONALD H. COASE, ESSAYS ON ECONOMICS AND ECONOMISTS (Chicago Univ. Press 1994); Ronald H. Coase, *The Firm, the Market, and the Law* (Chicago Univ. Press 1990). *See also* Ronald H. Coase, *The Problem of Social Cost*, 3 J. LAW ECON. 1–44 (1960).

44 Francine Lafontaine & Margaret Slade, *Vertical Integration and Firm Boundaries: The Evidence*, J. ECON. LIT. 45, 629 (2007).

45 See Robert H. Bork, *The Rule of Reason and the Per Se Concept: Price Fixing and Market Division*, 75 YALE L.J. 373-475, at 384 (1966). See also Robert H. Bork, *Contrasts in Antitrust Theory: I*, 65 COLUM. L. REV. 401 1965.

121. At the same time, antitrust policymakers have taken on M&A in the name of protecting innovation. The revised merger guidelines have increased scrutiny of M&A based on concerns about potential effects on R&D.[46] For example, the FTC highlighted innovation effects in their complaint against the vertical merger between Illumina and Grail.[47] The issue in *Illumina* was whether the vertical merger would create incentives for foreclosure and decrease innovation. There was little if any evidence that the merger would diminish innovation.[48]

122. Investments and partnerships provide inputs to R&D in AI. Cloud computing is a useful input for the development and so-called training of AI models. Cloud computing is an important technological change that has generated efficiencies in computing and storage through economics of scale.

123. Conversely, AI products help in developing apps related to cloud computing. Google, Microsoft, and Amazon can commercialize advances in AI through various types of products for consumers and firms.

124. Innovation competition in AI has resulted in rapid technological change with a quick succession of new LLMs. According to a report: "There are dozens of major LLMs, and hundreds that are arguably significant for some reason or other. Listing them all would be nearly impossible, and in any case, it would be out of date within days because of how quickly LLMs are being developed."[49] Among the many LLMs are GPT, Gemini, Gemma, Llama 3, Vicuna, Claude 3, Stable Beluga, StableLM 2, Coral, Falcon, DBRX, Mixtral, Xgen-7B, and Grock.[50]

125. There also are many competing cloud providers, including Amazon Web Services (AWS), Microsoft Azure, Google Cloud Platform, Alibaba Cloud, Salesforce, IBM, Digital Ocean, and Dell.[51] There are over 800 cloud providers worldwide.[52] According to a report, the global cloud industry was over $626 billion in 2023 and is expected to grow to more than $626 billion by 2028.[53]

---

46 Alden Abbott & Daniel F. Spulber, *Antitrust Merger Policy and Innovation Competition*, 19 J. Bus. & Tech. L. (forthcoming Spring 2024).

47 Illumina, Inc., and Grail, Inc. (hereafter *Illumina*), Fed. Reg. 201 0144, Docket Number 9401, (Fed. Trade Comm'n Oct. 14, 2022), https://www.ftc.gov/legal-library/browse/cases-proceedings/201-0144-illumina-inc-grail-inc-matter.

48 Daniel F. Spulber, *How Do Vertical Mergers Affect Innovation? Learning from Illumina*, Network L. Rev. (Nov. 3, 2022) https://www.networklawreview.org/spulber-mergers/.

49 Harry Guinness, *The Best Large Language Models (LLMs) in 2024*, Zapier (May 2, 2024), https://zapier.com/blog/best-llm/#mixtral.

50 *Id.*

51 *Top 10 Cloud Providers in 2024*, Allcode, https://allcode.com/cloud-providers/ (last visited July 17, 2024).

52 *Id.*

53 Cloud Computing Market Report: Global Forecast to 2028, Markets and Markets (Dec. 2023), https://www.marketsandmarkets.com/Market-Reports/cloud-computing-market-234.html. See also Belle Lin, *Technology Chiefs Seek Help Wrangling Cloud Costs*, Wall St. J. (Mar. 3, 2023), https://www.wsj.com/articles/technology-chiefs-seek-help-wrangling-cloud-costs-61ba0b50. *Gartner Says Cloud Will Become a Business Necessity by 2028*, Gartner (Nov. 29, 2023), https://www.gartner.com/en/newsroom/press-releases/2023-11-29-gartner-says-cloud-will-become-a-business-necessity-by-2028.

126. A key question for antitrust policy is whether AI investments and partnerships will have vertical foreclosure effects. It does not seem likely that these investments and partnerships will diminish competition either among LLM providers or among cloud providers. The investments and partnerships considered here do not appear to foreclose LLM providers. There are many LLM providers, and these providers can continue to obtain cloud computing from the various cloud computing companies. The investments and partnerships considered here do not appear to foreclose cloud computing providers. There are many cloud computing providers, and these providers have many corporate customers, with AI projects being only a part of the large cloud computing business.

127. Antitrust actions that limit contracts for technology transfers and provision of complementary inputs could create inefficiencies that discourage innovation competition. Antitrust policy that narrowly defines competition based on market structure or contractual relationships could discourage innovation competition.

IV. Conclusion

128. FTC scrutiny of investments and partnerships in AI raises the general question of whether investments and partnerships diminish innovation competition. The characteristics of investments and partnerships in AI help address this general question. Looking at investments and partnerships from a Coasian perspective suggests that these contractual relationships may be beneficial in supporting innovation competition and fostering transactions in the market for technology.

129. First, it does not appear that investments and partnerships at issue here foreclose entry in either LLMs or cloud computing. These investments and partnerships do not seem to have created entry barriers in either the supply of cloud computing or the development of LLMs. These types of contractual relationships are more likely to encourage entrepreneurship and transactions in the market for technology.

130. Second, investments and partnerships in AI allow the companies involved to avoid the financial and organizational costs of M&A. Investments and partnerships allow firms greater flexibility in comparison to M&A. Companies can reduce commitments to specific products and technologies in contrast to M&A. Investments and partnerships allow firms to engage in contracts with many firms. This allows firms access to more technologies than could be attained through either M&A or internal R&D.

131. Third, investments and partnerships involve contracts for providing cloud computing services and developing AI. Observing the existence of these business relationships suggests that the benefits of the contracts outweigh their transaction costs. Investments and partnerships allow firms to share the costs of R&D and exchange inventions and innovations. Limiting the exchange of ideas would constrain the market for technology.

132. Fourth, contracts for the exchange of R&D and complementary inputs allow companies engaged in R&D to obtain the benefits of specialization of function and division of labor. By specialization, companies can improve their expertise in cloud computing and development of LLM models. Specialization and market transactions help firms achieve economies of scale, which creates cost efficiencies in technology development.

133. To protect competition, antitrust policymakers need to focus on anticompetitive conduct of firms instead of hypothetical future innovation. I have emphasized the need for antitrust policy toward Big Tech to consider competitive conduct rather than simply the size of firms or platform technology.[54]

134. Antitrust policy should rely more on observed technological developments rather than focusing solely on predictions of future technologies. Speculation about potential effects of investments and partnerships on future technologies should not limit transactions in the market for ideas. Antitrust policymakers can support the market for technology by recognizing the benefits of investments and partnerships for innovation competition.

---

54 Daniel F. Spulber, *Antitrust Policy toward Intermediaries: Digital Platforms and "Big Tech"*, CPI (June 29, 2022), https://www.pymnts.com/cpi-posts/antitrust-policy-toward-intermediaries-digital-platforms-and-big-tech/.

Mergers by Other Means? AI Partnerships and the Frontiers of (Post-)Industrial Organization

TEODORA GROZA[*]
Sciences Po Law School, Paris
AND
ALEKSANDRA WIERZBICKA[**]
Cleary Gottlieb Steen & Hamilton, Brussels

Abstract

The surge of artificial intelligence (AI) has ushered in a new era of collaboration between tech giants and AI startups, reshaping competition dynamics. AI partnerships, i.e., deals between Big Tech firms and AI startups, pose a challenge for competition law. Existing competition law instruments, from Articles 101 and 102 of the Treaty on the Functioning of the European Union (TFEU) to the EU Merger Control Regulation, struggle to address the nuances of these new modes of organizing economic activity. Using the Microsoft–OpenAI partnership as a case study, this chapter highlights the need to balance innovation incentives with competition concerns.

---

[*] Teodora Groza is a Ph. D student at Sciences Po Law School, Paris.
[**] Aleksandra Wierzbicka is a law clerk at Cleary Gottlieb Steen & Hamilton, Brussels.

Mergers by Other Means?
AI Partnerships and the Frontiers of (Post-)Industrial Organization

I. Introduction

135. Artificial intelligence (AI) is the talk of the town. AI systems, from self-driving vehicles to medical imaging analysis tools to attention-grabbing generative AI models, are already transfiguring our daily lives.[1] Their impact on how we acquire information and perform routine tasks – from writing emails to self-diagnosing minor injuries – is palpable.[2] Public-facing AI systems – i.e., systems directly available to the general public – have generated an unprecedented hype: it took the AI-powered ChatGPT chatbot only five days to reach the milestone of one million users.[3] By comparison, Instagram needed 76 days, and Netflix almost 1,300.[4] In two months, ChatGPT's user base rose to 100 million.[5]

136. The hype did not go unnoticed. In disciplines ranging from ethics to computer science, significant academic efforts have been channeled towards understanding, explaining, and trying to tame AI systems.[6] Legal scholars and regulators alike have also done their job. The EU made headlines last December pursuant to the adoption of the AI Act, dubbed "the world's first rules on AI"; ever since, much was written on how (not) to regulate AI systems.[7] Nonetheless, a persistent gap remains in the legal literature. The emphasis has so far been almost exclusively on the technological advances enabled by AI and the risks and benefits associated with them. Contrastingly, discussions on how companies active in the AI space are structured and governed are scarce.

137. There is a fundamental reason why we should address this question: AI firms are pushing not just the frontiers of technology, but also the boundaries of the organization of industry, with important consequences for market competition and innovation outcomes. Two examples are noteworthy.[8]

---

[1] Jamie Berryhill et al., *Hello, World: Artificial Intelligence and Its Use in the Public Sector* (OECD PUBL'G, Working Papers on Public Governance No. 36, 2019); Mickael Brossard et al., *Deep learning in product design*, McKINSEY (Dec. 14, 2022), https://www.mckinsey.com/capabilities/operations/our-insights/deep-learning-in-product-design.

[2] Delphine Strauss, *The Algorithms by Hilke Schellmann – Why AI Really Is Coming for Your Job*, FIN. TIMES (Apr. 28, 2024), https://www.ft.com/content/e27ee51f-ea02-4489-b223-51fed88fd6a8; Sam Sabin, *ChatGPT-written Phishing Emails Are Already Scary Good*, AXIOS (Oct. 24, 2023), https://www.axios.com/2023/10/24/chatgpt-written-phishing-emails; Muhammad Sufyan et al., *Artificial intelligence in Cancer Diagnosis and Therapy: Current Status and Future Perspectives*, 165 COMPUT. BIOL. MED. 107356 (2023).

[3] *Adoption Rate for Major Milestone Internet-Of-Things Services and Technology in 2022, in Days*, STATISTA (Dec. 2022), https://www.statista.com/statistics/1360613/adoption-rate-of-major-iot-tech/.

[4] *Id.*

[5] Michael Chui et al., *What Every CEO Should Know about Generative AI*, McKINSEY (May 12, 2023), https://www.mckinsey.com/capabilities/mckinsey-digital/our-insights/what-every-ceo-should-know-about-generative-ai.

[6] Stanford AI Alignment, *AI Might Change the World as We Know It*, SAIA (2024), https://stanfordaialignment.org/ (laying down the key stakes for the development of AI systems).

[7] European Parliament, *EU AI Act: First Regulation on Artificial Intelligence*, (updated June 18, 2024), https://www.europarl.europa.eu/topics/en/article/20230601STO93804/eu-ai-act-first-regulation-on-artificial-intelligence (referring to the AI Act as "the world's first comprehensive AI law").

[8] There are, however, other developments worth mentioning. A notable example is the emergence of open-source foundation models. *See* Thibault Schrepel & Jason Potts, *Measuring the Openness of AI Foundation Models: Competition Policy and Implications*, (SCIS PO CHAIR DIGIT., GOVERNANCE & SOVEREIGNTY, Working Paper, 2024), https://papers.ssrn.com/sol3/papers.cfm?abstract_id=4827358.

138. First, when OpenAI's board of directors decided to increase the entity's ability to raise capital, they opted for the creation of a subsidiary which took an unprecedented form – a "capped-profit" company which enables investors and employees to get returns but capped by a fixed rate.[9] The company's refusal to adopt a standard corporate structure seems to be underlined by a belief that AI systems are unlike any other technologies. Indeed, there is a widespread conviction that AI can "create unprecedentedly large externalities, ranging from national security risks, to large-scale economic disruption, to fundamental threats to humanity, to enormous benefits to human safety and health."[10] As a consequence of these features of the industry, a similar non-standard entity has been crafted for Anthropic, another prominent AI firm, which opted for a "Long-Term Benefit Trust".[11] These legal innovations testify to the need for bespoke and unusual legal structures to meet the governance requirements of firm active at in the AI space, calling into question the fitness of legal institutions, designed for the economy of the past century, for the dynamics of contemporary markets.

139. Second, AI firms do not operate in isolation: they are often involved in close-knit partnerships with larger companies with an established foothold in digital markets. Microsoft has partnerships with OpenAI and Mistral which involve multi-billion dollar investments, close collaboration in research and development (R&D), and certain exclusivities.[12] Amazon poured several billion dollars into Anthropic as part of a renewed agreement under which Anthropic will use Amazon Web Services (AWS) as its main cloud provider, as well as Amazon's chips.[13] The list can go on: the UK Competition and Markets Authority (CMA) counted, as of April 2024, more than 90 partnerships, each of which involved one of the Big Tech firms (i.e., Google, Amazon, Meta, Microsoft, Apple and Nvidia) and smaller actors developing AI capabilities.[14]

140. These partnerships have already caught the attention of competition authorities on both sides of the Atlantic. The UK's CMA and the German Bundeskartellamt (BKA) have assessed the applicability of the merger control

---

[9] *OpenAI LP*, OpenAI (Mar. 11, 2019), https://openai.com/index/openai-lp/ (explaining the functioning and role of the OpenAI LP as a "capped-profit" company).

[10] *The Long-Term Benefit Trust*, Anthropic (Sept. 19, 2023), https://www.anthropic.com/news/the-long-term-benefit-trust.

[11] *Id.*

[12] *OpenAI and Microsoft Extend Partnership*, OpenAI (Jan. 23, 2023), https://openai.com/index/openai-and-microsoft-extend-partnership/; *Microsoft and Mistral AI Announce New Partnership to Accelerate AI innovation and Introduce Mistral Large First on Azure*, Microsoft Corp. Blogs (Feb. 26, 2024), https://azure.microsoft.com/en-us/blog/microsoft-and-mistral-ai-announce-new-partnership-to-accelerate-ai-innovation-and-introduce-mistral-large-first-on-azure/.

[13] *Amazon and Anthropic Deepen Their Shared Commitment to Advancing Generative AI*, Amazon (Mar. 27, 2024), https://www.aboutamazon.com/news/company-news/amazon-anthropic-ai-investment.

[14] Competition & Markets Authority, AI Foundation Models: Technical update report (Apr. 16, 2024), at [hereinafter CMA Technical Update] (note that the paper does not mention any partnerships not involving big tech firms and thus paints an incomplete picture of the industry).

framework to the Microsoft–OpenAI partnership.[15] The latter agency has concluded that the merger control regime remains inapplicable for now, whereas the former's investigation is pending.[16] The European Commission has also decided not to proceed with a merger review, but is looking into whether the partnership breaches the rest of the antitrust toolkit.[17]

141. Although these partnerships are not mergers in the traditional sense of uniting two firms into one,[18] the question whether they should be assessed as such is legitimate. Since antitrust laws aim to deal with economic realities, bypassing legal formalities,[19] it comes as no surprise that competition agencies are concerned about the extent to which an up-and-coming entity reliant on funding and critical infrastructure provided by a big player can retain its agency and operate independently of the demands of its sponsor and/or supplier. Furthermore, the European Merger Control Regulation (EUMR) targets "concentrations" which arise pursuant to "a change of control on a lasting basis,"[20] appearing at first sight malleable enough to capture these unusual arrangements. Nonetheless, as we show later, a deeper look at the notion of "control" and the relevant thresholds to be met complicates matters. So does the fact that merger control is neither the only tool in the box nor necessarily the sharpest. As these partnerships are *collaborations* between separate entities, wouldn't Article 101(1) TFEU be a better framework for the competitive assessment, offering the possibility of redemption under Article 101(3)?

142. The novelty of these partnerships that define the AI market environment makes it impossible to reach any definitive conclusions as of now, and regulators are right to take their time before declaring them anti-competitive from the outset. However, because competition authorities will have to deal with this question imminently, we aim to provide an overview of the challenges that AI partnerships pose to the EU competition law framework. The goal is not to provide any definitive answers, but rather to map out the questions that should guide the competitive assessment.

---

15   Competition and Markets Authority, *Microsoft/OpenAI Partnership Merger Inquiry*, (Dec. 8, 2023), https://www.gov.uk/cma-cases/microsoft-slash-openai-partnership-merger-inquiry; Bundeskartellamt Press Release, Cooperation between Microsoft and OpenAI Currently Not Subject to Merger Control, (Nov. 15, 2023), https://www.bundeskartellamt.de/SharedDocs/Meldung/EN/Pressemitteilungen/2023/15_11_2023_Microsoft_OpenAI.html.

16   According to industry news, Microsoft expects the CMA to launch a Phase I investigation of its relationship with OpenAI. Bethan John, *Microsoft Prepares for Formal CMA Probe into OpenAI Partnership*, Glob. Competition Rev. (May 23, 2024), https://globalcompetitionreview.com/article/microsoft-prepares-formal-cma-probe-openai-partnership.

17   Javier Espinoza & Tim Bradshaw, *Brussels Explores Antitrust Probe into Microsoft's Partnership with OpenAI*, Fin. Times (June 28, 2024), https://www.ft.com/content/cdb1ab92-9148-47c4-add1-a079a7652ddb.

18   Marshall Hargrave, *Merger*, Investopedia (June 12, 2024), https://www.investopedia.com/terms/m/merger.asp.

19   Mariana Pargendler et al., *Family Ties and the Boundaries of the Firm in Antitrust Law*, *in* Research Handbook on Competition and Corporate Law 2 (Florence Thépot & Anna Tzanaki eds., forthcoming Oct. 2024); Herbert Hovenkamp, *American Needle and the Boundaries of the Firm in Antitrust Law*, 14 (Aug. 15, 2010), https://papers.ssrn.com/sol3/papers.cfm?abstract_id=1616625.

20   Council Regulation (EC) 139/2004 on the control of concentrations between undertakings (EC Merger Regulation), 2004, O.J. (L 24) 1, art. 3(1).

143. The chapter is structured as follows. Section two provides an overview of existing AI partnerships, zooming in on the Microsoft–OpenAI collaboration. The third section leverages the industrial organization (IO) literature on inter-firm cooperation, which points in the direction that AI partnerships are neither mergers nor contracts between fully independent market actors, but rather third-type hybrid entities that are currently not recognized by antitrust laws. The fourth section provides an overview of the European competition law instruments that could catch AI partnerships, exploring their relative fitness. The conclusion summarizes the findings of the chapter, highlighting what we know, and most importantly, what we don't know, about AI partnerships.

144. Jurisdictionally, we focus exclusively on the EU regime. Nonetheless, it is worth noting that the US antitrust rules are more malleable than their European counterparts. Section 7 of the Clayton Act has a broad reach – it catches any acquisitions of stocks or assets which may substantially lessen competition or tend to create a monopoly, even when these acquisitions do not involve the conferral of control.[21] This may mean that even the attribution of certain governance rights could potentially be scrutinized.[22] Furthermore, Section 1 of the Sherman Act prohibits "any *contract* [...] in restraint of trade," and could thus be activated against these contractual arrangements.[23] Despite the fact that a comparative analysis falls beyond the scope of this chapter, it is interesting to observe that US antitrust laws seem to be better equipped to capture AI partnership. Whether or not this will translate into more aggressive enforcement is an open question.

II. If You Can't Acquire Them, Team Up with Them: AI Partnerships

145. The cemented market power of Big Tech firms has long been a focal point of scrutiny for antitrust watch dogs around the world.[24] These players have been exerting unparalleled influence over digital markets by controlling access to online platforms and digital services, and allegedly impeding the entry of new players.[25] Then, all of a sudden, the release of natural language processing chatbots driven by generative AI seemed to give rise to a

---

21 Denver & Rio Grande W. RR. Co. v. U.S., 387 U.S. 485 (1967); U.S. v. E.I. du Pont de Nemours & Co., 353 U.S. 586 (1957).

22 For an analysis, *see* Ana Tzanaki, *Minority Shareholdings, Global Dictionary of Competition Law*, CONCURRENCES, art. No. 89176.

23 Sherman Act 15 U.S.C. §§ 1.

24 The epitome of the backlash against Big Tech is the EU's adoption of the Digital Markets Act. *See* Commission Regulation 2022/1925 of the European Parliament and of the Council of Sept. 14, 2022, on Contestable and Fair Markets in the Digital Sector and Amending Directives (EU) 2019/1937 and (EU) 2020/1828 (Digital Markets Act, hereinafter DMA), 2022, O.J. (L 265) 1.

25 Jorge Guzman & Scott Stern, *The State of American Entrepreneurship*, 12 AM. ECON. J. 212 (2020) (finding that the increasing power of incumbents is one of the reasons why less startups succeed).

textbook example of Schumpeterian creative destruction. Some voices proclaimed the advent of ChatGPT as the end of Google Search.[26] Others went as far as to declare the emergence of public-facing AI systems as a broader shift from the dominant pattern of Web2 business in which "applications are developed, delivered, and monetized in a proprietary way" to a new economy defined by "open standards and protocols," "control no longer centralized in large platforms and aggregators," and "governance [taking place] in the community rather than behind closed doors."[27]

146. Fearing these developments, Big Tech firms came up with unique ways of mitigating the competitive threat. While the Big Tech companies dominate the Web2 space, they have been lagging behind small, up-and-coming startups which seem to have a competitive advantage in the AI space. Embracing the wisdom of the adage "if you can't beat them, join them," the Big Tech firms started to look for backdoor entry in the emerging AI space. Instead of competing, the chosen option was collaboration: the mammoths of Web2 were quick to enmesh themselves in close partnerships with promising AI startups. These partnerships are more than just financing agreements: on top of billions of dollars, AI firms receive access to cloud computing, accelerator chips, sizable databases, and, perhaps most importantly – a customer base.[28] These are critical inputs which are all but impossible to replicate by small firms. It is unclear, though, what the Big Tech firms receive in exchange. Some partnerships come with exclusive licenses of IP rights, while others are limited to non-exclusive commitments to make any resulting AI systems compatible with the existing infrastructures of Web2 companies. For example, the partnership between Apple and OpenAI is a non-exclusive deal involving the integration ChatGPT capabilities into the iOS, iPadOS and MacOS ecosystems as a "plug-in," meaning that "any AI models developed in the future which will surpass ChatGPT will be easy to add."[29]

147. Understanding the nuts and bolts of these agreements and the exact purpose they serve is a hard task, for two reasons. First, the Big Tech firms have systematically downplayed their importance, highlighting that they do not hold any ownership stakes and that all collaborating parties remain free to commercially exploit the resulting technologies. Second, even if we had access to the granularity of the agreements, there arguably is much

---

26 Bernard Marr, *Is Google's Reign Over? ChatGPT Emerges as a Serious Competitor*, FORBES (Feb. 20, 2023, 03:36 AM EST), https://www.forbes.com/sites/bernardmarr/2023/02/20/is-googles-reign-over-chatgpt-emerges-as-a-serious-competitor/.

27 Anutosh Banerjee et al., *Web3 Beyond the Hype*, MCKINSEY (Sept. 26, 2022), https://www.mckinsey.com/industries/financial-services/our-insights/web3-beyond-the-hype.

28 CMA TECHNICAL UPDATE, *supra* note 14.

29 John Koetsier, *Apple, ChatGPT, iOS 18: Here's How It Will Work*, FORBES (June 10, 2024, 05:18 PM EDT), https://www.forbes.com/sites/johnkoetsier/2024/06/10/apple-chatgpt-ios-18-heres-how-it-will-work/; *OpenAI and Apple Announce Partnership to Integrate ChatGPT into Apple Experiences*, OPENAI (June 10, 2024), https://openai.com/index/openai-and-apple-announce-partnership/; *Introducing Apple Intelligence, the Personal Intelligence System That Puts Powerful Generative Models at the Core of iPhone, iPad, and Mac*, APPLE (June 10, 2024), https://www.apple.com/newsroom/2024/06/introducing-apple-intelligence-for-iphone-ipad-and-mac/.

more to these collaborations than can be captured by written contractual provisions. More often than not, such agreements remain deliberately vague, enabling collaborating parties to renegotiate and reconfigure the design of their collaborations.[30] This vagueness, however, represents a feature, not a bug: the contracts at stake are meant to represent "framework agreements describe the cooperative behavior in which parties agree to engage" and not exhaustive to-do lists which bind the parties to concrete courses of action.[31]

148. The CMA identified an "interconnected web" of over 90 partnerships and "strategic investments" involving Big Tech firms and AI startups.[32] The epitome of this phenomenon is the Microsoft–OpenAI deal: a closer look at its intricacies illustrates the dynamics of these collaborations. Since 2019, Microsoft has made several multi-billion-dollar investments in OpenAI and its flagship generative AI chatbot, ChatGPT. In January 2023, the alliance entered its third phase: on top of an additional investment of $10 billion, Microsoft would deploy OpenAI's models across its products and its cloud-computing platform, and Azure would become the exclusive cloud provider for OpenAI.[33] Despite these significant inter-links, Microsoft denied any form of control over its partner. The gatekeeper has emphasized OpenAI's complete discretion over its R&D strategy and the absence of any formal governance rights.[34]

149. The reality, however, is more complicated. Since March 2019, the original OpenAI non-profit entity has a for-profit subsidiary, OpenAI LP. Most recently, the company's CEO Sam Altman has declared the company could morph into a for-profit entity.[35] In its current structure, the LP takes an unusual form: it is a "capped-profit" company which places a limit on the returns investors can claim, i.e., when the profits earned by the LP are higher than 100 times the initial investment, the outstanding sums are automatically redirected to the OpenAI non-profit.[36] The operations of the LP are foreseen by the OpenAI non-profit's board, where Microsoft does not have

---

30 Matthew Jennejohn, *The Private Order for Innovation Networks*, 68 STAN. L. REV. 334 (2016) (explaining that in the context of collaborations for highly innovative products or services, the parties involved rely on flexible contractual terms rather than restrictive clauses).

31 Alan Schwartz & Simone Sepe, *Contract Remedies for New Economy Collaborations*, 101 TEX. L. REV. 5–6 (forthcoming).

32 CMA TECHNICAL UPDATE, *supra* note 14.

33 *Microsoft and OpenAI extend partnership*, MICROSOFT CORP. BLOGS (Jan. 23, 2023), https://blogs.microsoft.com/blog/2023/01/23/microsoftandopenaiextendpartnership/ (detailing the specifics of the renewed agreement).

34 Nicholas Hirst, *Microsoft-OpenAI Partnership Drives Competition and Innovation, Senior Company Lawyer Says*, MLEX (Apr. 23, 2024, 11:20 AM GMT), https://mlexmarketinsight.com/news/insight/microsoft-openai-partnership-drives-competition-innovation-senior-company-lawyer-says.

35 *OpenAI CEO Says Company Could Become For-Profit Corporation, The Information Reports*, REUTERS (June 15, 2024, 06:37 AM GMT+1), https://www.reuters.com/technology/artificial-intelligence/openai-ceo-says-company-could-become-benefit-corporation-information-2024-06-15/#:~:text=June%2014%20 (Reuters)%20%2D%20OpenAI,The%20Information%20reported%20on%20Friday.

36 Tim Bradshaw et al., *How Microsoft's Multibillion-Dollar Alliance with OpenAI Really Works*, FIN. TIMES (Dec. 15, 2023), https://www.ft.com/content/458b162d-c97a-4464-8afc-72d65afb28ed.

any seats. But the lack of formal governance rights does not translate into a lack of involvement of Microsoft in the management of the entity. During the November 2023 crisis, when the board decided to fire Sam Altman from his role as CEO, Microsoft pushed for his return and subsequently obtained a non-voting observer role on the board. The reality of Microsoft's influence over the management of OpenAI seems much divorced from the Web2 giant's claims that it is "simply entitled to a share of profits." Some commentators have pointed out that the structure of the partnership seems to have been deliberately engineered with the strategic aim to circumvent regulatory oversight. Others have gone as far as to declare that "while billed as a partnership, the deal looks more like a killer acquisition."[37]

150. The Microsoft–OpenAI deal is not one of a kind. In February 2024, Microsoft announced that it had entered into a similar arrangement with Mistral AI, a French startup, commercializing AI products.[38] Likewise, Amazon funded Anthropic – one of OpenAI's strongest competitors – a total of $4 billion as part of a partnership which also included agreements for purchasing computing capacity and non-exclusive commitments to make Anthropic's models available on Amazon's existing infrastructure. The sweeping presence of the largest technology firms in the booming AI space raises legitimate concerns about the ability of gatekeepers to undermine competitive dynamics to the detriment of consumer welfare. This ability stems not only from the risk of leveraging their established market power in these neighboring markets,[39] but also – and *especially* – from their control of key assets for the development of foundation models (FM), i.e., chips, computing power, and customer bases.[40] Commentators were quick to point out that "with vanishingly few exceptions, every startup, new entrant, and even AI research lab is dependent on these firms."[41] This is because they "all rely on the computing infrastructure of Microsoft, Amazon, and Google to train their systems, and on those same firms' vast consumer market reach to deploy and sell their AI products."[42] These skeptical voices argue that these partnerships give Big Tech profound influence over the trajectory of AI,[43] and that the promise that AI technologies will "give back control of data and its benefits to individuals and communities" is illusory.[44]

---

37 Courtney Radsch, *The Real Story of the OpenAI Debacle is the Tyranny of Big Tech*, THE GUARDIAN (Nov. 27, 2023, 11:02 AM GMT), https://www.theguardian.com/commentisfree/2023/nov/27/openai-microsoft-big-tech-monopoly.

38 MICROSOFT CORP. BLOGS, *supra* note 12.

39 FRISO BOSTOEN, ABUSE OF PLATFORM POWER – LEVERAGING CONDUCT IN DIGITAL MARKETS UNDER EU COMPETITION LAW AND BEYOND (Concurrences 2023); *see also* Radsch, *supra* note 37 (arguing that the emerging AI space is already controlled by existing big tech players).

40 CMA TECHNICAL UPDATE, *supra* note 14.

41 Amba Kak et al., *Make No Mistake – AI is Owned by Big Tech*, MIT TECH. REV. (Dec. 5, 2023), https://www.technologyreview.com/2023/12/05/1084393/make-no-mistake-ai-is-owned-by-big-tech/.

42 *Id.*

43 *Id.*

44 Alex Pentland, *Building a New Economy: Data, AI, and Web3*, 65 COMMC'NS ACM 27 (2022).

151. It is too early for any doomsday scenarios. Before reaching any sweeping conclusion as to the negative impact of partnerships between Big Tech firms and AI startups, a required – yet often overlooked – step is to assess the counterfactual scenario. Would the public-facing AI products that are already part and parcel of our everyday lives have reached the market as quickly as they have without these partnerships? OpenAI claims that the multi-billion dollar investment from Microsoft "allow[s] [them] to continue [their] independent research and develop AI that is increasingly safe, useful, and powerful."[45] Even anti-Big-Tech voices agree that "for companies hoping to build base models, there is little alternative to working with either Microsoft, Google, or Amazon."[46] The reality is that building FM requires access to the aforementioned critical inputs – accelerator chips, computing power via the cloud, and data for training models – which are as of now held almost exclusively by Big Tech firms.[47] The question, then, is whether we prefer a world in which AI startups fail due to their inability to access critical inputs, or the current one in which they succeed through partnerships with Big Tech players. In its research paper on FM, the CMA noted that "such partnerships [...] may be an essential ingredient for the success of independent developers."[48] A prominent technology think tank has noted that such "partnerships [are] the catalyst for the current wave of innovation and competition in generative AI," and that rather than killing competition, they nourish it, by enabling the AI startups to grow into viable companies.[49]

152. There is no correct answer to the question whether these partnerships are a positive development. Until we have more data as to how AI startups will evolve and what impact the Big Tech partners have on their governance, any answer is bound to be a byproduct of political economy preferences. Pro-innovation voices will favor these collaborations, whereas anti-bigness sentiments will advocate for nascent AI players to develop their own capabilities in-house, even at the cost of delaying product development, slowing innovation, and potentially translating into consumer welfare losses. We are only at the beginning of this conversation. The CMA's investigation into the evolution of FM is ongoing – an updated report is expected to be published in Autumn 2024. The European Commission is also expected to look closely into the patterns of AI arrangements.[50]

---

45 OPENAI, *supra* note 12.
46 Kak et al., *supra* note 41.
47 CMA TECHNICAL UPDATE, *supra* note 14.
48 *Id.* at § 44.
49 Daniel Castro & Aswin Prabhakar, Comments to the UK's Competition and Markets Authority on Microsoft's Partnership with OpenAI, ITIF (Dec. 14, 2023), https://itif.org/publications/2023/12/14/comments-to-uks-competition-markets-authority-on-microsofts-partnership-with-openai/.
50 Lewis Croft, *Pattern of AI Deals Dodging Merger Review May Draw EU Scrutiny, Guersent Says*, MLEX (Apr. 11, 2024, 22:57 PM GMT), https://mlexmarketinsight.com/news/insight/pattern-of-ai-deals-dodging-merger-review-may-draw-eu-scrutiny-guersent-says#:~:text=Mergers%20%26%20Acquisitions-,Pattern%20of%20AI%20deals%20dodging%20merger,draw%20EU%20scrutiny%2C%20Guersent%20says&text=Microsoft%2C%20Amazon%2C%20Google%20and%20others,that%20have%20avoided%20merger%20scrutiny.

Nonetheless, before more data becomes available, a task that can be performed is to survey what the economic literature can teach us about inter-firm partnerships and how they can be approached by regulators. The next section delves into this subject.

III. Post-Industrial Organization

153. When competition lawyers think about how interactions between firms should be structured, the vision that comes to mind is highly influenced by the textbook model of perfect competition. In this ideal scenario, there are many firms without market power and, most importantly, they operate in isolation, as independent atoms. The AI partnerships introduced in the previous section are difficult to fit into this framework, as they involve close-knit collaborations between market actors that remain nominally independent but have significant joint operations. To theorize them, we need to rely on a different economic toolkit.

154. IO economics has made significant advances in understanding the functioning of real-life markets and firms. The crux of this literature is to make sense of how economic activities are structured and to assess why they take different forms, ranging from firms to inter-firm collaborations to market transactions between separate actors. Consequently, this literature is best suited to explain the emergence of inter-firm partnerships as a recurrent mode of doing business in the AI space.

155. The starting point for this strand of literature was Ronald Coase's seminal piece *The Nature of the Firm,* which is credited for the insight that there are two different modes of organizing economic activity – firms and markets – and each come with their own costs and benefits.[51] Nonetheless, a lesser-known insight of the Coasean piece is that the distinction between the two types of structures is not that sharp. In Coase's view, firms are bundles of transactions that are placed under a centralized authority, i.e., a management structure.[52] They emerge because reliance on management helps save certain costs, the most important of which is the cost of discovering and negotiating the price of each input that is being transacted. These costs are referred to in the literature as "transaction costs," or, in layman terms, "costs within markets."[53] On the flip side, firms come with their own "management costs": acquiring information about different sub-units, transmitting orders, overseeing their implementation, etc. Since both firms and markets have their own costs and benefits, there is no point in talking about one or the other as being more efficient. Centralized authority is a better fit for some transactions, while others are more efficiently carried out in markets as arms-length negotiations between independent actors.

---

51 Ronald Coase, *The Nature of the Firm*, 4 ECONOMICA 286 (1937).

52 *Id.*

53 Harold Demstez, *The Theory of the Firm Revisited*, IV J.L. ECON. & ORG. 142 (1988).

156. The Coasean insight that firms are bundled transactions was taken further by Oliver Williamson, who argued that instead of thinking of the organization of economic activity as bifurcated between firms and markets, we need to analyze each individual transaction as a relevant unit and assess what governance framework is best suited for it.[54] Building on this, Williamson argued that some governance frameworks are "hybrid," i.e., are neither firms, nor markets, but intermediary organizations which need to be theorized in their own right. These hybrid modes include "various forms of long-term contracting, reciprocal trading, [...] and the like" and are based on "elastic" contracts which leave collaborating parties with plenty of room for maneuver in case of changing circumstances or expectations.[55] IO economists have labeled these contracts as "relational" because they represent the backbone of close inter-firm relations that are characterized by repetitive, long-term collaborations which often involve the joint execution of economic functions traditionally associated with firms, such as research and development activities.[56] To enable such close collaborations, these relational contractual arrangements are not confined to written contracts, but are made out of "a rich braiding of explicit – legally enforceable – and implicit – legally unenforceable obligations."[57] Think of the dynamics of the Microsoft–OpenAI partnership, particularly the events of November 2023: while legally Microsoft did not have any governance rights, the company was heavily involved in the reinstatement of Sam Altman as CEO of OpenAI.[58] Even if Microsoft's sizeable financial investment did not come with any written expectations of a specific governance configuration, in practice the sponsor could enforce its unwritten demands that "things at OpenAI be done by the book."[59]

157. The existence of long-term, close-knit partnerships between firms is nothing new. Williamson documents a decades-long coal supply agreement between a coal company and a fur trading company dating back to the 1960s, which was based on a contract mandating high levels of information disclosure and adaptation in case of unforeseen circumstances. This is not an isolated agreement. Business historians have argued that "long-term relationships [...] between otherwise independent economic actors in which the parties voluntarily choose to continue dealing with each other for significant periods of time" represent "an intermediate form [that] is distinctive and common enough" to be recognized as a

---

54 Oliver E. Williamson, *Comparative Economic Organization: The Analysis of Discrete Structural Alternatives*, 36 ADMIN. SCI. Q. 269 (1991); OLIVER E. WILLIAMSON, MARKETS AND HIERARCHIES: ANALYSIS AND ANTITRUST IMPLICATIONS (The Free Press 1975).

55 Williamson (1991), *supra* note 54.

56 George Baker, Robert Gibbons & Kevin J. Murphy, *Relational Contracts and the Theory of the Firm*, 117 Q. J. ECON. 40 (2002).

57 Ronald J. Gilson, Charles F. Sabel & Robert E. Scott, *Contracting for Innovation: Vertical Disintegration and Interfirm Collaboration*, 109 COLUM. L. REV. 431, 435 (2009).

58 Bradshaw et al., *supra* note 36.

59 *Id.*

form of economic organization *sui generis*.[60] Tracing the evolution of American business, Lamoreaux notes that in the late twentieth century hybrid forms became increasingly frequent, highlighting that this phenomenon was likely to be heightened by the advent of the internet which functions as a "coordination infrastructure" which will further decrease coordination costs and consequently the need for vertical integration.[61]

158. The dynamics of the AI partnerships surveyed in the previous section confirm this hypothesis. What is novel about these partnerships is not their existence but rather their prevalence: they represent one of – if not *the* – main way(s) of doing business in the AI space. As Margarethe Vestager put it, these partnerships represent a "trend" which is "becoming a feature of the industry."[62]

159. Yet the increasing reliance on partnerships and inter-firm collaborations is the byproduct of a broader reconfiguration of the economy at large. IO economists have been arguing since the turn of the twenty-first century that society is shifting towards a "post-industrial" phase where the center of gravity of economic activity is knowledge creation rather than industrial production.[63] This new stage of economic development is characterized by an increasingly rapid pace of innovation and a corresponding need for heavy investments in R&D. As producers recognize that they cannot themselves maintain cutting-edge technology in every field, they are increasingly involved in lateral partnerships with small players who dedicate their efforts to developing targeted technologies.[64] A look at the Microsoft–OpenAI partnership proves this point: the dominant narrative is that since Microsoft could not develop AI technologies by itself that could rival those of other market players, the company chose instead to partner with OpenAI.[65]

160. The turn to a post-industrial society requires a corresponding turn to "post-industrial organization" economics.[66] We need to "brush aside questions of absolutes" and open up more and more space to theorize hybrid and intermediate economic forms. Unfortunately, the legal framework of competition law is still rooted in the economic realities of the previous century, in which inter-firm collaborations were the exception, and market

---

60 Naomi R. Lamoreaux, Daniel M.G. Raff & Peter Temin, *Beyond Markets and Hierarchies: Toward a New Synthesis of American Business* 5 (Nat'l Bureau of Econ. Rsch., Working Paper No. 9029, 2002) 5.

61 *Id.*

62 Margarethe Vestager, Speech, Competition in Virtual Worlds and Generative AI (Brussels, June 28, 2024), https://ec.europa.eu/commission/presscorner/detail/en/SPEECH_24_3550.

63 Daniel F. Spulber, *Antitrust and Innovation Competition*, 11 J. Antitrust Enf't 5, 11 (2023) (introducing the term "post-industrial organization" economics); Daniel Bell, The Coming of Post-Industrial Society: A Venture in Social Forecasting (Basic Books 1973) (1999) (claiming that a "post-industrial society rests on a knowledge theory of value. Knowledge is the source of invention and innovation.").

64 Gilson et al., *supra* note 57, at 431.

65 OpenAI, *supra* note 12.

66 Spulber, *supra* note 63.

players wanting to access the assets of other economic actors would choose the route of full mergers rather than looser partnerships. It is high time to question whether our existing legal institutions can adequately capture the dynamics of AI partnerships. The next section performs this task.

IV. Catch Them If You Can: Competition Law

161. Markets evolve faster than the law, requiring regulators to constantly synchronize existing tools with the latest business practices.[67] Trying to defy this inherent gap, EU competition policy has been designed with the aim of providing enough flexibility to address new forms of anticompetitive behavior. The dominant narrative is that Articles 101 and 102 TFEU allow the Commission to respond to emerging market conditions by repurposing these tools without adopting new instruments.[68] A case in point is that before the adoption of the EUMR, the Commission asserted its jurisdiction over mergers by applying Article 101 and 102 TFEU.[69] More recently, the Commission has substantially broadened its powers – without any regulatory change – by reviving the dormant Article 22 of the EUMR, which empowers it to review mergers falling below the EU and national merger thresholds and hence scrutinize the so-called "killer acquisitions."[70]

162. The above considerations beg the question whether the competition law instruments that the Commission has at its disposal are fit to catch AI partnerships. A straightforward and non-satisfactory answer to this question would be that the Commission has three avenues for scrutinizing such behavior: Articles 101, and 102 TFEU, or the EU Merger Control Regulation. Before assessing which legal pigeonhole would be most suitable the first question to address is whether the EU competition law regime is applicable at all. To be caught by any of the competition law provisions, a market actor needs to be considered as an "undertaking", i.e., an entity engaging in an economic activity. Given that many AI companies are non-profit organizations, the answer is not straightforward. For example, in the US, the Microsoft–OpenAI collaboration was exempted from the

---

67 Justice Brandeis, *The Living Law*, X ILL. L. REV. 464 (1916) (claiming that "the law has everywhere a tendency to lag behind the facts of life").

68 European Parliament, Hearing of Margrethe Vestager (Brussels, Oct. 8, 2019), https://www.europarl.europa.eu/resources/library/media/20191009RES63801/20191009RES63801.pdf; European Commission, Margrethe Vestager Keynote Speech, Keystone Conference: A Triple Shift for Competition Policy (Mar. 2, 2023), https://ec.europa.eu/commission/presscorner/detail/en/SPEECH_23_1342.

69 In 1971, the Commission used Article 102 TFEU to prohibit a merger in the Continental Can case (Commission Decision of Dec. 9, 1971 relating to a procedure in the application of Article 86 of the EEC Treaty Case IV/26811 (Europemballage Corp.), 1972, O.J. (L 7)), *see* Georges Vallindas, *The EU Commission determines that a can manufacturer abused its dominance by refusing to license its technology to competitors thereby denying consumers lower prices (Continental Can)*, E-COMPETITIONS Dec. 1971, art. No. 117223. In 1984, the Commission used Article 101 TFEU for merger control purposes in the Philip Morris case (Commission Decision of Mar. 22, 1984) relating to an infringement of Articles 85 and 86 of the EEC Treaty Joined Cases IV/30.342 and IV/30.926 (BAT & Reynolds v. Eur. Comm'n).

70 Anne Looijestijn-Clearie, Catalin S. Rusu & Marc J.M. Veenbrink, *In Search of the Holy Grail? The EU Commission's New Approach to Article 22 of the EU Merger Regulation*, 29 MAASTRICHT J. EUR. & COMPAR. L. 550 (2022).

pre-merger notification requirement under the Hart-Scott-Rodino (HSR) Act because it involved a non-profit organization.[71] This workaround would not be effective in the EU, though: the catch-all definition of the undertaking allows the Commission to consider a non-profit organization as an undertaking for the purpose of competition rules.[72]

163. The next question is which of the three regimes is best suited. As we demonstrate below, none of them adequately captures the granularity of AI partnerships.

164. Article 101 TFEU applies to collusive conduct between independent market actors. Yet, given the deep entanglement between AI startups and their Big Tech partners, it is not obvious that they remain distinct market actors. To assess whether two entities are separate economic organizations, the test employed under EU competition law is that of control. So far, the test has been deployed only in the context of relations between parent companies and their subsidiaries. However, its reach may very well be stretched to cover different configurations of corporate inter-links. To establish that a parent and a subsidiary are not separate market actors but one unitary entity – an undertaking – case law requires that "the subsidiary does not decide independently upon its own conduct on the market, but carries out, in all material respects, the instructions given to it by the parent company,"[73] Besides full ownership, recent case law has expanded the notion to cover situations where the parent company has *de facto* control because it has all voting rights. The Court declared in the 2021 *Goldman Sachs* case that "it is not the mere holding of all or virtually all the capital of the subsidiary in itself that gives rise to the presumption of the actual exercise of decisive influence, but the degree of control of the parent company over its subsidiary that this holding implies."[74] In this case, the Court established the existence of decisive influence due to several factors, including notably that Goldman Sachs had the power to appoint members of the various boards of directors of the subsidiary, to call shareholder meetings and propose revocations of directors or entire boards, as well as to receive regular updates and monthly reports from the subsidiary.[75]

165. Within this understanding of the notion of control, the entities engaged in AI partnerships seem far from being considered to form a single undertaking. The existing partnerships fall well short of the threshold of decisive influence. Going back to the architecture of the Microsoft–OpenAI

---

[71] 15 U.S.C. § 18a. On the other hand, § 1 and § 2 of the Sherman Act do apply to non-profits, and so does § 7 of the Clayton Act.

[72] *See* Case C-475/99, Firma Ambulanz Glöckner v. Landkreis Südwestpfalz, E.C.R. I-8089, ¶¶ 18–22 (2001).

[73] Case 48/69, Imperial Chem. Indus. (ICI) Ltd. v. Comm'n Eur. Comtys, ECLI:EU:C:1972:70, ¶ 133; Case C-625/13 P, Villeroy & Boch AG v. Eur. Comm'n., ECLI:EU:C:2017:52, ¶ 146.

[74] Case C-595/18 P, Goldman Sachs Grp. v. Eur. Comm'n., ECLI:EU:C:2021:73, ¶ 35, *see* Kyriakos Fountoukakos, Daniel Vowden & Agathe Célarié, *The EU Court of Justice confirms the liability of a parent company for the conduct of its subsidiaries involved in a cartel (Goldman Sachs)*, e-Competitions Jan. 2021, art. No. 99132.

[75] *Id.* at para. 18.

partnership, Microsoft has a "minority economic interest" *only* in the capped-profit branch of the company, and this does not translate into any governance rights on the entity's board.[76] Microsoft even dropped its non-voting observer member on the board that it had secured in the aftermath of the November 2023 governance crisis, allegedly in order to avoid antitrust scrutiny.[77] In this configuration, Microsoft and OpenAI cannot be considered to form a single economic entity, as the former does not have decisive influence over the latter. The consequence is that the partnership risks running afoul of Article 101 TFEU. This is more than a theoretical possibility. When the Bundeskartellamt concluded that the partnership did not represent a concentration within the meaning of the domestic merger control regime, it declared that its compliance with Article 101 TFEU still needed to be scrutinized.[78] Similarly, Margarethe Vestager declared in a June 2024 speech that the exclusivity clauses involved in the partnership could prove anti-competitive.[79]

166. Given that Article 101 has an open-ended language, prohibiting "agreements" which prevent, restrict or distort competition, it would seem to be the default choice for scrutinizing AI partnerships.[80] Contrastingly, European competition authorities have chosen to analyze them through the lens of merger control. Until recently, nascent competitors, such as AI startups, were too small to be captured by the thresholds.[81] With the reinterpretation of Article 22 of the EUMR, this is no longer where the shoe pinches: now that firm size is no longer an issue, the new question is whether the partnerships confer "control" to the Big Tech players. Under the EUMR, a concentration must entail a change of control on a lasting basis, i.e., a transaction which gives an undertaking "decisive influence" over another one.[82] Mirroring the discussion in the previous paragraph, this influence can be either *de jure* or *de facto*. *De jure* control is traditionally associated with shareholdings conferring the majority of voting rights, whereas *de facto* control stems from factual circumstances enabling a minority shareholder to acquire a majority of voting rights at the general meetings.[83] Nonetheless, the control threshold required for

---

76 *Our Structure*, OPENAI (2024), https://openai.com/our-structure/.

77 Mauro Orru & Christian Moess Laursen, *Microsoft Quits OpenAI's Board Amid Antitrust Scrutiny*, WALL ST. J. (July 10, 2024), https://www.wsj.com/tech/ai/microsoft-withdraws-from-openais-board-amid-antitrust-scrutiny-aab6ff1e.

78 Bundeskartellamt, *supra* note 15.

79 Vestager, *supra* note 62.

80 Consolidated Version of the Treaty on the Functioning of the European Union, Oct. 26, 2008, art. 101, 2008 O.J. (C 326) 1.

81 Robert Ryan, James Rutt & Mike Walker, *How to Address Under-Enforcement in Digital Markets?*, in RESEARCH HANDBOOK ON GLOBAL MERGER CONTROL 150 (Ioannis Kokkoris & Nicholas Levy eds., Edward Elgar 2023).

82 EC Merger Regulation 2004 O.J. (L 24), 1, art. 3(2).

83 Alexandre Rouhette & Pierre Garenne, *De Facto Control in EU Merger Control Law*, 4 CONCURRENCES COMPETITION L. REV. 1 (2020).

triggering a merger investigation is not the same as that which is applicable for establishing the existence of an undertaking. Acquisitions of shareholdings well below 50 percent have led to merger investigations when the shareholdings at stake were coupled with seats on the board, information rights entailing access to sensitive information, and the right to block certain resolutions.[84] National merger control regimes, particularly in Germany, have caught even acquisitions of shares amounting to less than 15 percent.[85]

167. Regardless of how low the Commission would be willing to push the threshold, two aspects need to be taken into consideration. First, pursuant to the definition of concentrations under the EUMR, when the acquisition of a minority shareholding is unrelated to acquisition of control, the Commission cannot investigate or intervene against it.[86] This means that only minority shareholdings conferring control can be scrutinized. Second, given the unusual corporate structures that AI firms take, the concept of "shareholdings" is not applicable – in the Microsoft–OpenAI, the former secured only the right to profits in the capped-profit subsidiary of the non-profit entity. Nonetheless, the unsuitability of the notion of shareholding does not render the whole merger control inapplicable. In the Consolidated Jurisdictional Notice, the Commission clarifies that "control can [...] be acquired on a contractual basis" where the contract leads to "control of the management and the resources of the other undertaking as in the case of acquisition of shares and assets."[87] Additionally, the Notice specifies that "a situation of economic dependence may lead to control on a *de facto* basis where [...] very important long-term supply agreements [...] coupled with structural links confer decisive influence."[88]

168. In the case of the Microsoft–OpenAI agreement, Microsoft denies owning any assets of OpenAI or any voting rights.[89] In the absence of any legal hooks hinting at the existence of control, even a bold interpretation of

---

84 *See*, for instance, Commission Decision of Jan. 22, 1997 declaring a concentration to be compatible with the common market and the functioning of the EEA Agreement Case IV/M.794 (Coca-Cola/Amalgamated Beverages GB), 1997, O.J. (218) 15; Commission Decision of July 28, 1993 Case IV/ECSC.1031 (US/Sollac/Bamesa); Commission Decision of Aug. 23, 1995 Case IV/M.625 (Nordic Capital/Transpool).

85 Bundeskartellamt Decision of Jan. 31 2012 Case B8-116-11-2 (Gazprom/VNG), *see* Wolfgang Bosch, *The Federal Cartel Office considers the acquisition of 1.88% of the shares in a competitor resulting in a total shareholding of 10.52% as a notifiable concentration under German law as it enabled to exercise competitively significant influence jointly with a third party (Gazprom / VNG)*, E-COMPETITIONS Jan. 2012, art. No. 48616.

86 European Commission, Staff Working Document Accompanying the White Paper Towards More Effective Merger Control, SWD (2014) 221 final, at § 45.

87 European Commission, Consolidated Jurisdictional Notice under Council Regulation 139/2004 on the control of concentrations between undertakings, 2008, O.J. (C 95), 1, at § 18. *See also* Commission Decision of Dec. 5, 2003 Case COMP/M.3136 (GE/AGFA NDT) 2003, O.J. (C 125), *see* Paul Gorecki, *The EU Commission clears at phase I, subject to a structural remedy, the acquisition whereby a wholly-owned subsidiary would acquire a non-destructive testing business in the electric sector (GE / GEAE)*, E-COMPETITIONS Dec. 2003, art. No. 81738.

88 *Id.* at § 19.

89 *Microsoft Says It Does Not Own Any Portion of OpenAI*, REUTERS (Dec. 8, 2023, 09:57 PM GMT), https://www.reuters.com/technology/microsoft-says-it-does-not-own-any-portion-openai-2023-1208/#:~:text=WASHINGTON%2C%20Dec%208%20(Reuters),OpenAI%2C%20an%20artificial%20intelligence%20powerhouse.

the notion of decisive influence seems to be a stretch. At the same time, though, if we were to follow the voices attributing Sam Altman's reinstatement as CEO to Microsoft, then this gesture could be reason enough to qualify the Big Tech player's influence as "decisive". In the absence of more information about OpenAI's concrete governance architecture and the actual reach of Microsoft's involvement, the applicability of the merger control framework remains an open question.

169. What is clear, though, is that the AI partnerships have been finely and legally engineered to attempt evasion of merger review. This leaves us one last question – whether such arrangements could require consideration under Article 102 TFEU. The prerequisite for scrutiny under Article 102 is the existence of a dominant position. The concept of market dominance is closely related to market power, indicating the capacity to act independently, without being restricted by consumer or supplier choices, thereby deviating from competitive outcomes.[90] Article 102 TFEU goes beyond abuses occurring in the market where the undertaking is dominant, reaching abuses occurring in distinct, yet closely related markets. In the context of AI partnerships, this implies that while Big Tech players may not exhibit dominance in the adjacent AI market, their established stronghold in other markets enables them to potentially leverage this position into the nascent AI sector. However, demonstrating such leverage poses a considerable challenge for the Commission, requiring extensive time and resources to assess the nuanced dynamics of these evolving markets. For this reason, reaching for Article 102 TFEU is likely to be a last resort.

V. Conclusion

170. This chapter has navigated the complex terrain of AI partnerships, shedding light on the challenges they pose to the competition law framework. The novelty of these partnerships renders it difficult to draw definitive conclusions at this stage, warranting cautious antitrust scrutiny. However, as AI markets continue to evolve, the regulatory frameworks must adapt accordingly to ensure that competition thrives and innovation flourishes in this rapidly changing environment. AI partnerships could be seen as a strategic response to prevent Schumpeterian creative destruction, integrating emerging technologies into the existing frameworks established by the current Big Tech players rather than allowing them to disrupt the status quo. Alternatively, such partnerships can be viewed as a springboard for AI startups requiring access to key inputs in order to fulfill their mission. The jury is still out, giving us much needed time for reflection.

---

90 OECD, Abuse of Dominance in Digital Markets (2020), https://web-archive.oecd.org/2021-10-31/566602-abuse-of-dominance-in-digital-markets-2020.pdf.

Preserving Competition in Generative AI: Addressing the Merger Conundrum

MARIATERESA MAGGIOLINO[*] AND LAURA ZOBOLI[**]

Bocconi University, Milan

University of Brescia

Abstract

The recent surge of attention from antitrust authorities signals a deep-seated concern over the substantial investments made by Big Tech companies in the generative artificial intelligence (genAI) market. This attention reflects a broader recognition of the potentially transformative impact of genAI and a firm resolve to maintain competitive and innovative market dynamics. At the heart of this trend lies a fundamental question: why are major antitrust authorities investing considerable effort in gathering information about Big Tech's ventures into genAI markets, and what specific competitive issues are motivating such an interest? From these fundamental questions, the chapter will explore the application of merger control frameworks to scrutinize Big Tech's genAI investments, through the merger control tool. Additionally, it will delve into the theories of harm that could justify authorities intervening to block such transactions. Ultimately, the chapter seeks to offer nuanced reflections on the practicality and efficacy of deploying antitrust authorities' interventions within this rapidly evolving landscape.

---

[*] Full Professor in Antitrust law and Market Regulation, Bocconi University.

[**] Researcher in Business Law, University of Brescia.

I. Introduction

171. In just a few months, the US Federal Trade Commission (FTC), the UK Competition and Markets Authority (CMA), and the European Commission (EC) among many other agencies have collectively directed their attention to a significant development: the substantial investments by Big Tech companies in the generative artificial intelligence (genAI) market. Specifically, in December 2023, the CMA launched an investigation to assess whether Microsoft's partnership with OpenAI, or any modifications thereof, had triggered a relevant merger situation under the Enterprise Act 2002.[1] The focus was to determine if such a situation might result in a substantial lessening of competition within any market or markets in the UK for goods or services. As of today, the CMA has commenced this investigation, inviting comments from December 8, 2023, to January 3, 2024.

172. At the onset of January 2024, the EU Commission announced its examination of whether Microsoft's investment in OpenAI might be subject to EU Merger Regulation review.[2] This action is a component of a broader strategic plan which focuses on evaluating competition within Virtual Reality (VR) and genAI markets. In this context, the Commission not only highlighted the Microsoft–OpenAI investment, but also sent requests for information to several major digital players and shared its inquiry into agreements between large digital market players and genAI developers and providers.

173. Towards the end of January 2024, the FTC also embarked on an inquiry into genAI investments and partnerships.[3] It issued Section 6(b) orders,[4] granting the FTC authority to compel entities – in the case at hand, Alphabet, Amazon.com, Anthropic, Microsoft, and OpenAI – to provide detailed reports or answers regarding their organizational structure, business practices, and affiliations. More specifically, these orders targeted recent investments and partnerships concerning genAI companies and major cloud service providers and the focus included three separate multibillion-dollar investments: Microsoft–OpenAI, Amazon–Anthropic, and Google–Anthropic. As companies employ various strategies in AI development, including partnerships and direct investments with

---

[1] CMA Press Release, Microsoft / OpenAI partnership Merger Inquiry (Dec. 8, 2023), https://www.gov.uk/cma-cases/microsoft-slash-openai-partnership-merger-inquiry.

[2] European Commission Press Release, Commission Launches Calls for Contributions on Competition in Virtual Worlds and Generative AI (Jan. 9, 2024), https://ec.europa.eu/commission/presscorner/detail/en/ip_24_85.

[3] Fed. Trade Comm'n Press Release, *FTC Launches Inquiry into Generative AI Investments and Partnerships*, FTC Blog (Jan. 25, 2024), https://www.ftc.gov/news-events/news/press-releases/2024/01/ftc-launches-inquiry-generative-ai-investments-partnerships.

[4] FTC Act, 15 U.S.C. § 6. Section 46(b), provides for another investigative tool and, specifically, § 6(b) empowers the Commission to require an entity to file "annual or special (…) reports or answers in writing to specific questions" to provide information about the entity's "organization, business, conduct, practices, management, and relation to other corporations, partnerships, and individuals."

AI developers to get access to key technologies and inputs needed for AI development, the FTC seeks insights into the details, implications, and competitive dynamics of these transactions. Specifically, the FTC seeks information regarding: (i) details of specific investments or partnerships, including agreements and the strategic rationale behind them, (ii) practical implications of these partnerships or investments, such as decisions regarding new product releases, governance or oversight rights, and the frequency of regular meetings, (iii) analysis of the transactions' competitive impact, including market share, competitors, potential for sales growth, or expansion into product or geographic markets, (iv) competition for AI inputs and resources, including the competitive dynamics concerning key products and services required for genAI, and (v) information provided to any other government entity, including foreign government entities' requests for information, or other inquiries related to these topics.

174. This recent flurry of attention from antitrust authorities[5] underscores a growing concern and a desire to have a say regarding the significant investments made by Big Tech companies in the genAI market, suggesting a recognition of the transformative potential of genAI and a determination to ensure that market dynamics remain conducive to competition and innovation.

175. Against this background, this chapter aims to delve into the motivations driving major antitrust authorities to closely monitor the entry of Big Tech companies into genAI markets, Section II. In light of these considerations, it will then explore the regulatory frameworks that antitrust authorities could potentially employ to scrutinize Big Tech investments in these markets, Section III, alongside the theories of harm that might prompt authorities to consider blocking such transactions, Section IV. Finally, the chapter will conclude with reflections on the viability and efficacy of wielding the antitrust authorities' "arsenal" in the genAI context.

II. What Are the FTC, the CMA, and the EC Afraid Of?

176. In recent years, Big Tech companies have been aggressively expanding into new markets, driven by several factors, well-documented in the literature.[6] First, they excel at developing new products and services

---

5 While it will not be examined as a separate system, being part of the EU, it is worth noting that the German competition authority (Bundeskartellamt) launched a formal investigation into whether Microsoft was required to notify its investment in OpenAI under merger control rules and, in its decision dated Sept. 25, 2023, concluded that although Microsoft did not formally breach German merger control notification requirements, this does not dismiss the potential for significant concerns regarding the investment. Bundeskartellamt Press Release, Bundeskartellamt Examined Whether Partnership between Microsoft and OpenAI was Subject to Notification Obligation under Merger Control (Sept. 25, 2023). In more general terms, all the main competition authorities are addressing the genAI-related challenges. For an updated database of antitrust initiatives targeting generative AI, *see* Thibault Schrepel, *A Database of Antitrust Initiatives Targeting Generative AI*, NETWORK L. REV. (Jan. 23. 2024), https://www.networklawreview.org/antitrust-generative-ai/.

6 JACQUES CRÉMER, YVES-ALEXANDER DE MONTJOYE & HEIKE SCHWEITZER, COMPETITION POLICY FOR THE DIGITAL ERA, REPORT TO THE EUROPEAN COMMISSION 110 (2019). JASON FURMAN, DIANE COYLE, AMELIA FLETCHER, DEREK MCAULEY & PHILIP MARSDEN, UNLOCKING DIGITAL COMPETITION, REPORT OF THE DIGITAL COMPETITION EXPERT PANEL (HM Treasury 2019) (hereinafter FURMAN REPORT). *See also* FIONA MORTON ET AL., COMMITTEE

at lower costs than their competitors,[7] leveraging economies of scale and scope as well as the wealth of knowledge they obtain from the data they collect and analyze.[8] Second, entering new markets offers them three distinct advantages: access to new data,[9] the potential to quickly dominate these markets due to their "winner takes all" and "tipping point" dynamics,[10] and the strategic goal of pre-empting rivals in developing new technologies that could turn out to be disruptive.[11] Lastly, they constantly enter new markets to utilize resources that would otherwise go unused and invest capital that would otherwise remain idle.[12]

177. This rush to penetrate new markets has several consequences. First – as is widely acknowledged – it has led Alphabet, Amazon, Meta, Apple, and Microsoft to emerge as the central figures of business communities structured as digital ecosystems, capable of offering profiled users a wide array of interoperable goods and services.[13] Second, the

---

FOR THE STUDY OF DIGITAL PLATFORMS: MARKET STRUCTURE AND ANTITRUST SUBCOMMITTEE REPORT (2019); Steven C. Salop, *Modifying Merger Consent Decrees: An Economist Plot to Improve Merger Enforcement Policy*, 31 ANTITRUST 15–20 (2016); Steven C. Salop & Carl Shapiro, *Whither Antitrust Enforcement in the Trump Administration?*, ANTITRUST SOURCE (2017); Herbert Hovenkamp & Carl Shapiro, *Horizontal Mergers, Market Structure, and Burdens of Proof*, 127 YALE L.J. 1996 (2017).

[7] Konstantinos Stylianou, *Exclusion in Digital Markets*, 24 MICH. TELECOMM. & TECH. L. REV. 181 (2018); Georgios Petropoulos, *Competition Economics of Digital Ecosystems*, OECD 91 (Dec. 3, 2020), https://one.oecd.org/document/DAF/COMP/WD(2020)91/en/pdf; Marc Bourreau, *Some Economics of Digital Ecosystems*, OECD 89 (Dec. 3, 2020), https://one.oecd.org/document/DAF/COMP/M(2020)2/ANN5/FINAL/en/pdf; Amelia Fletcher, *Digital Competition Policy: Are Ecosystems Different?*, OECD 96 (Dec. 3, 2020), https://one.oecd.org/document/DAF/COMP/WD(2020)96/en/pdf.

[8] Geoffrey Parker, Georgios Petropoulos & Marshall Van Alstyne, *Platform Mergers and Antitrust*, 30 INDUS. & CORP. CHANGE 1308–1309 (2021); Anne C. Witt, *Who's Afraid of Conglomerate Mergers?*, 67 ANTITRUST BULL. 208 (2022); Jasper van den Boom & Peerawat Samranchit, *Assessing the Long Run Competitive Effects of Digital Ecosystem Mergers* 3 (UNIV. TILBURG TILEC, Discussion Paper, 2020); Marc Bourreau & Alexandre de Streel, *Digital Conglomerates and EU Competition Policy* (Working Paper, 2019).

[9] *See*, for example Commission Decision of Dec. 17, 2020 declaring a concentration compatible with the internal market and the functioning of the EEA Agreement Case M.9660 (*Google/Fitbit*), 2021, O.J. (C 194), at 7; *see also* Simon Vande Walle, *Clearance: The European Commission clears the acquisition of a maker of fitness trackers and smartwatches by a major online platform, subject to long-lasting behavioural remedies (Fitbit / Google)*, CONCURRENCES No. 3-2021, art. No. 101736.

[10] FURMAN ET AL., FURMAN REPORT, *supra* note 6, at 4; Antonio Capobianco & Anita Nyeso, Challenges for Competition Law Enforcement and Policy in the Digital Economy, 9 J. EUR. COMPETITION L. & PRAC. 19 (2018); Özlem Bedre-Defolie & Rainer Nitsche, *When Do Markets Tip? An Overview and Some Insights for Policy*, 11 J. EUR. COMPETITION L. & PRAC. 610 (2020); Nicolas Petit & Natalia Moreno Belloso *A Simple Way to Measure Tipping in Digital Markets*, PROMARKET (Apr. 6, 2021), https://www.promarket.org/2021/04/06/measure-test-tipping-point-digital-markets/; *Tipping: Should Regulators Intervene Before or After? A Policy Dilemma*, OXERA (2021), https://www.oxera.com/insights/agenda/articles/tipping-should-regulators-intervene-before-or-after-a-policy-dilemma/; MAURICE STUCKE & ALLEN GRUNES, BIG DATA AND COMPETITION POLICY 107 (2016); Michael L. Katz & Carl Shapiro, *Systems Competition and Network Effects*, 8 J. ECON. PERS. 106 (1994).

[11] Y Lim, *Tech Wars: Return of the Conglomerate – Throwback or Dawn of a New Series for Competition in the Digital Era?*, 19 J. KOR. LAW 47–55 (2020), who invokes the "fear of displacement" to explain the propensity of digital ecosystems to expand their reach into ever new and different markets through specific concentrations.

[12] *Id*.

[13] Within these communities, there is typically a founding company that not only provides some of the core products and services but also shares technology, financial resources, human capital, and intellectual property needed to establish and grow the ecosystem. This founding company, often identified as the central platform, also has a degree of control over the products or services in the ecosystem, setting technical and quality standards they must meet. Other companies in the ecosystem contribute goods and

expansion of Big Tech companies into new markets suggests that their already substantial size, in terms of both revenue and market presence, is poised to grow even further, suggesting that those digital ecosystems will be present in an increasing number of markets, almost becoming ubiquitous. Third, their pursuit of entry into new markets widens the competitive divide between them and their actual and potential competitors. Indeed, their actual competitors increasingly struggle to match digital ecosystems' integrated and interoperable offerings, missing out on the many efficiencies and advantages listed above. To be sure, the development of new technologies might raise hopes for potential new rivals capable of challenging digital ecosystems' dominance. However, the possibility of these companies seizing the most promising emerging technologies dampens this hope, leaving us seemingly destined to continually face Alphabet, Amazon, Meta, Apple, and Microsoft, for the foreseeable future.

178. We believe that it is mainly this last concern that has driven the FTC, the CMA, and the EC to launch the investigations mentioned earlier. With the rapid advancement of genAI and its potential to revolutionize industries, these prominent antitrust authorities seek to understand how Big Tech firms could be leveraging genAI to solidify their market power and prevent competitors from challenging it, all within a context where these digital ecosystems *already* exert various forms of power beyond mere market power, significantly widening the competitive gap with their rivals, while potentially generating efficiencies. For instance – whether it is an efficient way of managing the many businesses cooperating in the platform or not – they possess the authority to establish the rules governing their ecosystems, obligating other companies to comply, and they dictate the allocation of resources such as interoperability codes and critical datasets crucial for other businesses.[14] Furthermore, leveraging their data analytics capabilities, they may outperform competitors in predicting market trends and subtly shape consumer choices by manipulating the presentation of information.[15] And – if that were not enough – consider that digital ecosystems not only allocate resources and affect

---

services to both the leading company and the users. In more detail, digital ecosystems can be described in three different ways, each of which integrates and overlaps with the others without aiming for exhaustiveness. From a business perspective, ecosystems can be seen as conglomerate organizations that, to ensure users the most comprehensive experience possible, operate in many markets, offering a wide range of complementary products and services within the platform and its interfaces. Not surprisingly, from an engineering standpoint, digital ecosystems also represent a network structure with hardware and software components that are compatible and interoperable. Finally, in today's data economy, digital ecosystems can also be thought of as huge data collectors that are then analyzed not only to develop and test new technologies, but also to infer terabytes of information about users, competitors, and future market dynamics.

14 These are respectively the power of disposition and the power of coercion. They are the oldest forms of power, already known to Machiavelli and Marx.

15 These are the so-called technical power and the power of manipulation, which are forms of power that have emerged in contemporary society, studied by Bachrach, Blau, and Foucault.

income distribution,[16] they also amass significant capital for investment,[17] employ hundreds of thousands of talented individuals,[18] and typically lead innovation by deciding when and how to introduce new products, services, and technologies. In a sentence, they hold significant economic power.[19] Hence, it comes as no surprise that the CMA, the EC, and the FTC have taken steps to comprehend the activities of Alphabet, Amazon, and Microsoft within the genAI market.

179. One aspect, however, remains to be understood: despite the significant concerns raised thus far, like any other legal framework, antitrust law operates on certain principles.

180. In the interest of fostering innovation, antitrust law aims not to impede internal growth processes through which a company develops new products and services. Therefore, when considering the application of antitrust law to a Big Tech company entering a new market, such as genAI, it is essential to first ascertain whether this entry is the result of an external growth process, such as acquiring an equity stake in a company already engaged in genAI development. Indeed, the actions of the FTC, CMA, and EC have been triggered by some investments.

181. Second – and perhaps more importantly – in accordance with the principle of the rule of law, antitrust law's prohibitions have defined boundaries that must be adhered to. Consequently, when dealing with a Big Tech company entering a new market through an acquisition, two questions must be answered: what procedural and substantial rules could antitrust authorities apply to scrutinize and possibly forbid the transaction? And, what theories of harm might authorities contemplate in blocking such a transaction?

182. In each of the following paragraphs we will try to answer both of these questions.

III. The Possible Legal Rules

183. As is well known, while firms in dominant positions always face scrutiny for their behavior, different jurisdictions have a variety of specific rules for determining which mergers merit *ex-ante* examination. These criteria may include firm revenues, transaction value, or other

---

16 Indeed, they hold market power, which is in a *species-genus* relationship with the economic power.

17 It suffices to note that in 2023, Microsoft, Google, and Amazon orchestrated a sequence of investments that, as per recent findings from private market researchers PitchBook, accounted for two-thirds of the total $27 billion raised by emerging genAI enterprises. *See* George Hammond, *Big Tech Outspends Venture Capital Firms in AI Investment Frenzy*, FIN. TIMES (2023), https://www.ft.com/content/c6b47d24-b435-4f41-b197-2d826cce9532. *See also* Gerrit De Vynck & Naomi Nix, *Big Tech Keeps Spending Billions on AI. There's No End in Sight*, WASH. POST (2024), https://www.washingtonpost.com/technology/2024/04/25/microsoft-google-ai-investment-profit-facebook-meta/.

18 Especially the most skilled and hard-to-find workers, again recall the well-known competition among big tech in attracting the most skilled individuals in the JHA context; on this point see Aaron Mok, *Inside Big Tech's Nasty Battle for Coveted AI Talent*, BUS. INSIDER (2024), https://www.businessinsider.com/inside-the-ai-hiring-frenzy-big-tech-talent-war-2024-3.

19 ADOLF A. BERLE, POWER – EPILOGUE IN AMERICA 199–216 (Harcourt, Brace & World 1968).

relevant parameters like the value of goods traded in a particular region. Consequently, many mergers involving Big Techs bypass *ex-ante* antitrust oversight.[20] However, our focus here is not on the tools used to enlarge scrutiny – from Article 14 of the Digital Markets Act (DMA) to the examination of sub-threshold mergers,[21] we aim to explore more specifically the potential qualification for review of Big Techs' investments in start-ups developing genAI. Therefore, our analysis transcends mere concerns about operational visibility and delves into the intricate legal classification of such transactions under the antitrust laws that the FTC, CMA, and EC – the triad upon which this study is centered – may apply.

184. This prompts us to divide the actions whereby Big Tech companies enter the genAI market into two main categories: (i) the acquisition of a minority stake in the target company, and (ii) the acquisition of a majority stake in the target company.

185. In general terms, under the EU Merger Regulation, a merger occurs when a firm gains control over another firm, meaning it obtains the ability to exert "decisive influence" on the target company and define its commercial policy.[22] Therefore, an investment that involves acquiring a minority stake in a company, without joint or *de facto* control, does not constitute a merger, under the assumption that it will not alter the commercial strategies of the target company to the extent of raising antitrust concerns.[23] For what is most relevant here, under EU merger control regulation two types of *de facto* control can be detected, based on their origin: (i) when one or several shareholders obtain a majority of the voting rights in general

---

20   Suffice it to say that the European Commission did not have to investigate the merger between Facebook and Instagram.

21   Commission Regulation 2022/1925 of the European Parliament and of the Council of Sept. 14, 2022, on Contestable and Fair Markets in the Digital Sector and Amending Directives (EU) 2019/1937 and (EU) 2020/1828 (Digital Markets Act, hereinafter DMA), 2022, O.J. (L 265), at 1. The DMA entered into force on Nov. 1, 2022 and requires all companies designated as gatekeepers to notify the Commission of acquisitions in the digital sector or those facilitating data collection, regardless of whether they meet national or EU notification thresholds. Additionally, the DMA grants the European Commission the authority to temporarily ban mergers involving a gatekeeper found to consistently violate its DMA obligations. In a similar direction can be commented the referral process under Article 22 of the European Union Merger Regulation (activated for example in Commission Decision of Nov. 10, 2023 declaring a concentration compatible with the internal market and the functioning of the EEA Agreement Case M.10262 (*Meta (Formerly Facebook)/Kustomer*), https://competition-cases.ec.europa.eu/cases/M.10262.

22   *See* Council Regulation 139/2004 of 20 January 2004 on the control of concentrations between undertakings (the EC Merger Regulation, art. 3(2). The most common form of control is the so-called *de jure* control which recurs when a shareholder, with the majority of the voting rights in an undertaking can define the commercial policy of the undertaking through its governing bodies. On the other hand, in very specific cases, the shareholder with the majority of voting rights cannot define, on its own, the commercial policy of the undertaking because, for example, other shareholders have veto rights. *See*, in this regard, European Commission, Consolidated Jurisdictional Notice under Council Regulation 139/2004 on the control of concentrations between undertakings, 2008, O.J. (C 95), at 1, ¶¶ 56–57).

23   This assumption is questionable. Indeed, the Commission has previously explored this matter and pondered proposing amendments to the EUMR to enable the assessment of minority shareholding acquisitions. Concrete proposals have not yet been developed by the Commission, seemingly allowing Member States to take the lead in experimenting with this issue for the time being. However, there could be a resurgence of interest in this topic at the EU level once Member States provide feedback on the efficacy of such measures.

shareholders' meetings due to factual circumstances,[24] and (ii) in exceptional instances, when an individual or a business entity, through contractual arrangements with a company, can exert its influence over the governing bodies of that company, imposing its own vision.[25] Thus, some seem to be willing to spot *de facto* control of start-ups operating in the genAI market by Big Techs, as, for example, was observed by several organizations in response to the CMA's request for comments regarding the Microsoft–OpenAI partnership.[26]

186. Under US antitrust law, the concept of a merger appears to be defined less rigidly than in the EU,[27] and the 2023 Guidelines explicitly acknowledge that a partial acquisition can still impact competition by granting the partial owner the ability to influence the competitive behavior of the target company, by diminishing the acquiring firm's incentive to compete, or by providing access to confidential, competitively sensitive information held by the target company.[28] Furthermore, and for the specific purposes of this analysis, it is worth noting that in the US the acquisition of voting securities exceeding 10 percent of the outstanding voting securities of the issuer may prompt antitrust intervention, even if the operation is made solely for investment purposes.[29]

---

24 The first form of *de facto* control, which is the most prevalent, occurs when a minority shareholder stands a high chance of securing a majority position during shareholders' meetings, mainly due to the expected low attendance rate at these meetings (European Commission, Consolidated Jurisdictional Notice, at ¶ 59). The Commission undertakes a prospective assessment based on past attendance rates at shareholders' meetings and other factors indicating the potential for future control by the minority shareholder. Additionally, in paragraph 76 of the Consolidated Jurisdictional Notice, the Commission discusses a form of joint *de facto* control associated with the exercise of voting rights in a joint venture: when multiple minority shareholders share significant common interests, suggesting they won't oppose each other in exercising their rights within the joint venture, the Commission may conclude that they hold *de facto* control. However, such *de facto* control is uncommon as it necessitates the Commission's conviction, in the specific factual setting, that the shared interests among minority shareholders are robust and enduring enough to confer collective control. *See* Alexandre Rouhette & Pierre Garenne, *De Facto Control in EU Merger Control Law*, 4 Concurrences Competition L. Rev. 1 (2020).

25 *De facto* control based on contractual relationships can be either relinquished, or effectively imposed upon, the *de jure* control holder. In the first case, the decision-making ability has been voluntarily transferred to a minority shareholder or a third party. In the second case, an undertaking is economically dependent on a minority shareholder, which can have a decisive influence on this undertaking's activity through its contractual relationships. *See* Garenne & Rouhette, *supra* note 24, at 6.

26 According to Article 19, "UK: Submission to the Competition and Markets Authority on Microsoft-OpenAI merger" Article 19 (Jan. 9, 2024), https://www.article19.org/resources/uk-submission-to-the-competition-and-markets-authority-on-microsoft-openai-merger/. Microsoft's status as OpenAI's exclusive cloud provider allows it to hinder competitors' access to OpenAI while profiting consistently. The submission asserts that Microsoft seemingly benefits from privileged access to OpenAI's technology and can license it on its behalf. The recent upheaval at OpenAI highlighted Microsoft's significant influence, with reports of Microsoft pressuring the board to reverse decisions and ultimately controlling critical strategic and business choices. Furthermore, Microsoft's role across the AI stack, from developing AI models to providing infrastructure like Azure, reinforces OpenAI's dependence and further solidifies Microsoft's dominance in various markets. This consolidation of power allows Microsoft unprecedented oversight and influence over OpenAI's operations and decision-making processes.

27 Herbert Hovenkamp, Federal Antitrust Policy, The Law of Competition and its Practice 664–665 (2015).

28 *See* Fed. Trade Comm'n and U.S. Dep't of Justice, *Merger Guidelines (2023)*, (Dec. 18, 2023), at ¶ 2.11, "Guideline 11: When an Acquisition Involves Partial Ownership or Minority Interests, the Agencies Examine Its Impact on Competition" (2023), https://www.ftc.gov/system/files/ftc_gov/pdf/P234000-NEW-MERGER-GUIDELINES.pdf.

29 Indeed, Under the Hart-Scott-Rodino Antitrust Improvements Act (HSR), an acquisition of voting securities that meets the monetary filing threshold is reportable even if the acquiring entity does not obtain control of the target. However, the HSR Act exempts certain acquisitions of voting securities if made "solely for the purpose of investment." This "investment-only" exemption is available if, as a result of the acquisition, an acquiring entity holds 10 percent or less of the voting securities of the target issuer and has only passive

187. In the UK, an acquisition can affect competition if it grants, *de jure* or *de facto*, the ability to exert "material influence" over the competitive conduct of the target company – a fact that can occur even when the EU threshold of the "decisive influence" is not met.[30] Moreover, the CMA can activate a sector inquiry and impose commitments to remedy, mitigate, or prevent adverse effects on competition, and/or to address any detrimental effects on customers so far as these effects have resulted from, or may be expected to result from, the adverse effect on competition.[31]

188. In summary, it appears that prosecuting the minority investments that Big Tech companies are making in the genAI market by using the rules on merger control is easier in the US and the UK than in the EU.

189. This does not exclude the fact that, in all the three systems, minorities investments could still be deemed illicit under laws governing anti-competitive agreements, if they were tools for collusion among more companies investing in the same target entity. Alternatively, albeit less likely, minority investments by the same companies in identical targets might be seen as instigating collusive behaviors among the targets, aiming to align with the economic interests of their investors.

190. Finally, if our hypothetical Big Tech company through its investment acquires a majority stake in the target operating within the genAI market (as per scenario (ii)), the transaction would be qualified as a merger in all three legal systems. And, as said previously, this classification remains consistent regardless of the transaction's visibility, which is often compromised by the fact that start-ups operating in the development market of genAI often have zero or very low turnover.

IV. The Possible Theories of Harm

191. In antitrust lawsuits and proceedings, judges or commissioners do not actually measure the potentially negative effects of a practice against its possible benefits. Instead, as in other legal cases with opposing parties, they decide which of the two groups of effects is prevailing based on the most convincing evidence presented by each side. Therefore, plaintiffs need to prove their "theory of harm," explaining why the practices at hand break antitrust rules, while defendants must present their "defenses," explaining why they do not.

---

investment intent (i.e., has "no intention of participating in the formulation, determination, or direction of the basic business decisions of the issuer" under HSR Rule 801.1(i)(1)). However, ¶ 7A(c)(9), 15 U.S.C. ¶ 18a(c)(9), and Rule 802.9, 16 C.F.R. ¶ 802.9 read: "acquisitions, solely for the purpose of investment, of voting securities, if, as a result of such acquisition, the securities acquired or held do not exceed 10 per centum of the outstanding voting securities of the issuer."

30 As a general rule, indeed, in the UK a shareholder of more than 25 percent is likely to be viewed as giving rise to material influence, and shareholdings of as low as 10–15 percent (with no board representation or other governance rights) might be viewed as conferring material influence, depending on the circumstances.

31 Pursuant to the Enterprise Act 2002 s. 5, "The (CMA) has the function of obtaining, compiling and keeping under review information about matters relating to the carrying out of its functions", "That function is to be carried out with a view to (among other things) ensuring that the (CMA) has sufficient information to take informed decisions and to carry out its other functions effectively." The duty to remedy adverse effects is provided for by s. 138 of the Enterprise Act 2002.

192. In the context of a digital ecosystem entering the genAI market, plaintiffs might explore a broad range of theories of harm. These could include claims related to anticompetitive agreements or mergers, some of which may be unconventional, while others adhere to more traditional legal perspectives.

193. For example, if the aforementioned investigations reveal a trend where all digital ecosystems consistently invest in the genAI industry by acquiring minority stakes in key companies developing this technology, plaintiffs could express concerns about *common ownership*. As already noted, while admittedly a very unusual and controversial scenario (!), the underlying concern would be that companies financed by major tech firms may recognize that limiting competition in developing genAI and, thus, controlling its advances, aligns with the interests of their funders, even without receiving explicit directives to collude in this direction.

194. In the more realistic scenario where two digital ecosystems acquire minority stakes in the same company developing genAI – consider, indeed, the cases of Google and Amazon both investing in Anthropic – plaintiffs could invoke antitrust laws pertaining to *cooperative agreements*. Namely, they could argue that the transaction may: diminish competition between the two investing companies, hinder innovation, resulting in fewer or inferior products reaching the market, or causing delays in the introduction of new products, foster collusion between the involved parties in markets beyond the scope of the cooperation agreement, and/or, if one or more parties wields a certain amount of market power, result in anticompetitive exclusion of third parties.

195. Alternatively, if the transaction under scrutiny could be characterized as a merger, apparently plaintiffs could not resort to the theories of harm specific to horizontal and vertical mergers, because the genAI market is collaterally related to the markets in which digital ecosystems operate. The available theories of harm would still be many, spanning again from the most conventional to the most unconventional ones.

196. For example, it is possible that the entry of a Big Tech player into a new market through consolidation is the result of *a series of acquisitions*, each of which is individually minor, but collectively leads companies like Big Techs to a significant, though not dominant, market position. In this scenario, one might question whether, under the theory of the incipient monopoly, this progressive increase in market share should be somehow controlled and limited, even though each individual acquisition does not substantially reduce competition or, at least, does not meet the specific parameters, such as the HHI index, that usually help to justify prohibiting the merger.[32]

197. In a more orthodox way, plaintiffs may argue that the acquisition is meant to eliminate a *potential competitor* of the ecosystem, that is, a company that could threaten the market position of the ecosystem by

---

[32] Robert H. Lande, John M. Newman & Rebecca K. Slaughter, *The Forgotten Anti-Monopoly Law: The Second Half of Clayton Act § 7*, 103 TEX. L. REV. (2024), https://papers.ssrn.com/sol3/papers.cfm?abstract_id=4769563.

entering the markets in which it already operates. Depending on the jurisdiction, in order to be successful, the plaintiffs should meet several conditions. For example, in the EU they must prove that the potential competitor already exerts (or that there is a significant likelihood that it would exert) a significant constraining influence over the digital ecosystem and that there are not enough other potential competitors that, after the merger, could maintain sufficient competitive pressure on the digital ecosystem.[33]

198. In order to address these challenges inherent in the theories of harm based on potential competition, some plaintiffs may draw on recent insights from economists[34] and categorize certain transactions as "killer acquisitions" or "kill-zone acquisitions" – still two defensive strategies,[35] but that do not posit as a legal requirement that the target company be characterized as a potential competitor.[36] Specifically, plaintiffs could argue that dominant digital ecosystem firms employ acquisitions to safeguard their core technology and business interests, by either stifling and absorbing a company and its evolving technology or by signaling that any new entrant would be acquired. Taking a less aggressive stance, plaintiffs could argue that when Big Tech companies acquire firms developing emerging technologies, they are signaling defensive intentions. Consequently, other innovative start-ups and small companies learn that they will face acquisition, if they enter the market of the Big Tech company. Hence, the ultimate effect of these strategies is to eliminate competitive pressure from rivals, deter new entries into the market, and reduce overall innovation within the market.[37] And, as hinted above, the advantage of these theories lies

---

33 *Guidelines on the assessment of horizontal mergers under the Council Regulation on the control of concentrations between undertakings*, 2004, O.J. (C 31), at 5–18, ¶60.

34 Massimo Motta & Martin Peitz, *Removal of Potential Competitors – A Blind Spot of Merger Policy*, 6 COMPETITION L. & POL'Y DEBATE 19–25 (2020); OECD, *Start-ups, Killer Acquisitions and Merger Control – Background Note; Materials for the Meeting of the Competition Committee*, (2020); Colleen Cunningham, Floria Ederer & Song Ma, *Killer Acquisitions*, 129 J. POL. ECON. 649–702 (2021), who identify as the distinctive element of a killer acquisition the fact that the firm's strategy is "solely to discontinue the target's innovation projects and preempt future competition". *See also* Nicolas Levy, Henry Mostyn & Bianca Buzatu, *Reforming EU Merger Control to capture "Killer Acquisitions" – the Case for Caution*, 19 COMPETITION L.J. 51–67 (2020); and Sai Krishna Kamepalli, Raghuram G. Rajan & Luigi Zingales, *Kill Zone*, (BECKER FRIEDMAN INST., Working Paper No. 2020-19, 2020), https://bfi.uchicago.edu/wp-content/uploads/BFI_WP_202019.pdf.

35 JONATHAN BAKER, THE ANTITRUST PARADIGM: RESTORING A COMPETITIVE ECONOMY 160–161 (Harv. Univ. Press 2019)

36 The CMA embraced the theory of harm discussed here and stated that it could be established that a "substantial lessening of competition" may result from an acquisition that eliminates "a dynamic competitor that is making efforts toward entry or expansion," "even though entry by such a participant is unlikely and may ultimately be unsuccessful.". *See* CMA, *Merger Assessment Guidelines 2021*, (Mar. 18, 2021), https://assets.publishing.service.gov.uk/media/61f952dd8fa8f5388690df76/MAGs_for_publication_2021_--_.pdf, paragraph 5.23 reads: "The likelihood of successful entry by a dynamic competitor and the expected closeness of competition between a dynamic competitor and other firms are both relevant to the constraint exerted by a dynamic competitor on other firms and the CMA will take this into account. The elimination of a dynamic competitor that is making efforts towards entry or expansion may lead to an SLC even where entry by that entrant is unlikely and may ultimately be unsuccessful."

37 *See* Hanna Halaburda, Bruno Jullien & Jaron Yehezkel, *Dynamic Competition with Network Externalities: How History Matters*, 51 RAND J. ECON. 3–31 (2020); Gary Biglaiser & Jacques Crémer, *The Value of Incumbency When Platforms Face Heterogeneous Customers*, 12 AM. ECON. J. MICROECON. 229–269 (2020). Moreover, digital ecosystems would have a strong incentive to proceed with pre-emptive buyouts, by which is meant the acquisition of new entrants with the aim of reducing potential future competition (*see*

in their potential application even when, at present, the target company cannot be classified as an actual or potential rival of the digital ecosystem in question, as it may eventually emerge as a competitor in the long run due to the dynamics of digital markets.[38]

199. For the sake of completeness and to avoid any misuse of the above theories that could lead to false positives, it is worth noting that there are scholars who harbor skepticism toward these theories of harm, offering four distinct arguments.[39] First, they underscore the ambiguity and sector-specific limitation of empirical data buttressing the notion of such killer and kill-zone acquisitions, notably confined to the biotechnological domain rather than encompassing the broader purview of digital technologies herein under scrutiny.[40] Second, they point out that when an incumbent acquires a start-up or small company without suppressing its technology, it opens up a new user base that would not have been reached otherwise. Plus, the incumbent offers the chance for further development of the acquired technology, leveraging its financial and technological resources. Third, this scholarship contends that for start-ups and small-scale enterprises, the prospect of acquisition by a corporate behemoth may paradoxically constitute a strategic avenue for market exit and investor remuneration, especially when they are in dire financial straits, and even in the worst case, in which the technology of the target company perishes with the acquisition.[41] Finally, these scholars observe that the prevailing pattern whereby pioneering firms are subsumed by their more established counterparts is indicative of a nuanced form of market specialization inherent within realms of heightened innovation.[42]

Eric Rasmusen, *Entry for Buyout*, 36 J. INDUS. ECON. 281–299 (1988).

[38] Anne C. Witt, *Big Tech Acquisitions: The Return of Conglomerate Merger Control*, AVAILABLE AT SSRN, 2020, who observes: "(...) while these reports use slightly different terminology – some express the danger in terms of loss of potential competition, others use the term "conglomerate effects" – all seem to be in agreement that a dominant platform's acquisition of innovative upstarts in adjacent markets can be detrimental to competition in the long term, and that enforcement agencies should not presume that such transactions are benign. On the contrary, they recommend scrutinising such acquisitions closely, and revising the existing standards and burdens of proof."

[39] Jonathan Barnett, *"Killer Acquisitions" Reexamined: Economic Hyperbole in the Age of Populist Antitrust* (U. S. CALIF. GOULD SCH. L., Research Paper Series No. CLASS23-1, 2023).

[40] The study by Cunningham, Ederer & Ma, *supra* note 34 offers empirical evidence with respect to killer acquisitions in the pharmaceutical sector-which although has significant differences from the digital sector, for example, due to the presence of strong network effects in the latter sector. For a careful examination related to the digital sector, *see* Axel Gautier & Joe Lamesch, *Mergers in the Digital Economy* (CESIFO, Working Paper No.8056, 2020) who analyze as many as 175 acquisitions made by large technology companies (Google, Amazon, Facebook, and Microsoft) in the period 2015–2017; while most of the acquired products are closed, the underlying technology is integrated into the companies' ecosystems. In this regard, *see also* Bourreau & de Streel, *supra* note 8, at 21–23; Luís Cabral, *Merger Policy in Digital Industries*, 54 INFO. ECON. & POL'Y 100866 (2021). OECD, *Background Note*, *supra* note 34, at 5, 12. In this regard, also VIKTORIA H.S.E. ROBERTSON, MERGER REVIEW IN DIGITAL AND TECHNOLOGY MARKETS FINAL REPORT ¶ 127 (2022).

[41] Carl Shapiro, *Antitrust in a Time of Populism*, INT'L. J. INDUS. ORG. 714 (2018).

[42] Barnett, *supra* note 39, at 24–27. With regard to the relationship between innovation and acquired enterprises, *see* Carlsson et al., *Knowledge Creation, Entrepreneurship and Economic Growth: A Historical Review*, 18 INDUS. & CORP. CHANGE 1222–1223 (2009); Todd R. Zenger & Sergio G. Lazzarini, *Compensating for Innovation: Do Small Firms Offer High- Powered Incentives that Lure Talent and Motivate Effort?*, 25 MANAGERIAL & DECISION ECON. 329 (2004), that observe "(e)mpirical research on innovation and firm size confirms that despite large firms' apparent advantages in scale and access to complementary assets and capabilities (...) small firms are more efficient at innovation, particular radical forms of innovation."

200. However, there is no need to adopt a final theoretical stance on these opposing views. To reconcile them in practice – even if such an operation can prove challenging – antitrust authorities can meticulously analyze the specific circumstances of the case at hand to determine whether it is indeed true that, without the acquisition, the target company would exit the market or enable its technology to flourish, and whether the acquired technology will indeed be disseminated through the acquisition rather than being suppressed by it.[43] For example, agencies could consider factual elements such as whether the acquired company has already achieved the breakeven point, and the business plans of the companies involved or any other internal documents that suggest the economic rationale behind the deal.

201. Finally, if plaintiff cannot resort to the above theories of harm, still they could maintain that the merger executes a defensive strategy aimed at *raising the costs of the ecosystem's rivals*, by drawing on the argumentations commonly applied in cases of abuse of dominance.[44] Namely, plaintiffs may contend that the merger would grant the ecosystem control over crucial assets, including inputs,[45] intellectual property rights,[46] data,[47] and interoperability codes. Similarly, plaintiffs may assert that the merger could discourage the entry of new players by reducing market profitability or could lead consumers to become accustomed to a vast array of products and services that the ecosystem's rivals cannot match.[48] And, in support of these arguments, plaintiffs could certainly argue that dominant ecosystem firms' financial strength relative to their competitors would enable such defensive maneuvers.[49]

---

43 See Federico Ghezzi & Mariateresa Maggiolino, *La nuova disciplina di controllo delle concentrazioni in Italia: alla ricerca di una convergenza con il diritto europeo*, 1 Riv. Soc. 38–39 (2023); Massimo Motta, Competition Policy, Theory and Pratice 320 (Cambridge Univ. Press 2004); Michael L. Katz, *Big Tech Mergers: Innovation, Competition for the Market, and the Acquisition of Emerging Competitors*, 54 Info. Econ. & Pol'y 100883 (2021).

44 Steven C. Salop & David T. Scheffman, *Raising Rivals' Costs*, 73 Am. Econ. Rev. 267–271 (1983); Richard Schmalensee, *Entry Deterrence in the Ready-to-Eat Breakfast Cereal Industry*, 9 Bell J. Econ. 305 (1978); and B. Curtis Eaton & Richard G. Lipsey, *The Theory of Market Pre-Emption: The Persistence of Excess Capacity and Monopoly in Growing Spatial Markets*, 46 Economica 149 (1979); *see also* Mariateresa Maggiolino, Intellectual Property and Antitrust: A Comparative Economic Analysis of US and EU Law (Edward Elgar 2011).

45 *See, e.g.*, Case T-221/95, Endemol Ent. Holding BV v. Comm'n of the Eur. Cmtys., ECLI:EU:T:1999:85, ¶ 167.

46 *See, e.g.*, Commission Decision of May 3, 2000 declaring a concentration to be compatible with the common market and the functioning of the EEA Agreement Case COMP/M.1671 (*Dow Chemical/Union Carbide*), 2001, O.J. (L 245), at 1, ¶¶107–114.

47 Commission Decision of Sept. 6, 2018 declaring a concentration compatible with the internal market and the functioning of the EEA Agreement Case M.8788 (*Apple/Shazam*), 2018, O.J. (C 417), at 4, ¶ 211. The Commission, though, could not prove that after the merger, Apple would have had the legal ability and economic incentives to use Shazam user data to outdo Shazam's competitors in price and quality. *See also* Thomas K. Cheng, *Sherman vs. Goliath?: Tackling the Conglomerate Dominance Problem in Emerging and Small Economies*, 37 Nw. J. Int'l L. & Bus. 45 (2017); Tristan Lécuyer, *Digital conglomerates and killer acquisitions: A discussion of the competitive effects of start-up acquisitions by digital platforms*, Concurrences No. 1-2020, art. No. 92964; *see also* Ianis Girgenson, *User data: The European Commission approves the acquisition of a music recognition app by a provider of digital music streaming services (Apple / Shazam)*, Concurrences No. 1-2019, art. No. 88970.

48 Yong Lim, *Tech Wars: Return of the Conglomerate – Throwback or Dawn of a New Series for Competition in the Digital Era?*, 19 J. Kor. Law 55 (2017); Cheng, *supra* note 47, at 45; Lécuyer, *supra* note 47.

49 Case T-156/98, RJB Mining Plc. v. Comm'n of the Eur. Cmtys., ECLI:EU:T:2001:29.

202. In conclusion, when a digital ecosystem ventures into a new market, such as genAI, plaintiffs can invoke theories of harm that suggest collusion hypotheses, as well as theories that frame consolidation as a tool to defend the digital ecosystem against competitive threats. This is indeed the common thread among theories of harm revolving around the elimination of a potential rival, the suppression or taming of an alternative technology, or the increase in rivals' costs.

203. As a result, a pertinent question emerges: can a theory of harm be crafted to view a merger, through which an ecosystem enters a new market, not merely as a defensive maneuver but as an offensive strategy aimed at establishing dominance in that market without the expenses of independent product or service development? Recently, with the Meta–Within case, the FTC has attempted to move in this direction, albeit unsuccessfully.

V. The Acquisition of the Best Kid on the Block

204. As mentioned earlier, antitrust authorities typically do not intervene when ecosystems expand into new markets through internal growth processes. Protecting innovation is indeed a key aim of antitrust law, regardless of which firm is introducing new products and services. Therefore, even if a digital ecosystem gains a first-mover advantage by entering a new market, which goes hand-in-hand with all the benefits it already enjoys, such as economies of scale, scope, or winner-takes-all and tipping effects, antitrust law still prioritizes fostering innovation.

205. However, what if digital ecosystems enter a new market by acquiring a promising company – one that may not be dominant or possess significant market power but is expected to emerge as a major player? Considering that digital ecosystems are already corporate giants, should we not question whether granting them the ability to acquire promising companies, thereby securing a first-mover advantage in yet another market, is providing them with too much latitude? Especially in cases where they fail to establish themselves as first movers through their own initiatives? These questions seem to be even more pressing when considering the context of the genAI market and its peculiarities, which place it at the center of antitrust authorities' attention – a focus that, as mentioned, the Commission addresses together with the VR.

206. These concerns formed the basis of the FTC's action against the Meta–Within merger.[50] First, the FTC contended that Meta could have entered the VR fitness applications market independently, utilizing its existing applications like Beat Saber or potentially developing new ones, without needing to acquire Within's software (i.e., Supernatural). Second, the FTC

---

50 Complaint, Fed. Trade Comm'n v. Meta Platforms, Inc., No. 9411, Civil Action 20-3590 (JEB) (D.D.C. Apr. 26, 2023), https://www.ftc.gov/system/files/ftc_gov/pdf/D09411MetaWithinComplaintPublic.pdf; *see* Michael Moiseyev, Geneva Torsilieri Hardesty & Rachel Williams, *The US FTC abandons its challenge to the acquisition of a virtual reality fitness app by a social media giant following a District Court's denial of its request for a preliminary injunction (Meta / Within)*, e-Competitions Feb. 2023, art. No. 112748.

argued that Meta's acquisition of Within aimed to control a "killer app" (Supernatural) in a niche where Meta had previously failed to pioneer such a high-quality product. In essence, the FTC alleged that the merger posed a substantial risk of eliminating both current and future competition. In the absence of the merger, two scenarios could have unfolded: Meta could have independently entered the market for VR fitness apps, leveraging its financial resources, infrastructure, brand recognition, talent pool, and consumer base. Conversely, Supernatural could have continued to enhance its product and services, introducing new features, retaining employees, fostering innovation, and intensifying competition to outpace Meta's potential entry.[51] Thus, with Supernatural being a leading application in the VR Dedicated Fitness Apps market, its acquisition facilitated Meta's consolidation of its position within the Metaverse.

207. In light of the topic of this chapter, according to the FTC, the anticompetitive concern behind the Meta–Within merger was that Meta had the option to compete directly with Within based on merit. However, Meta chose to acquire Within instead. Specifically, the FTC asserted that Meta's motive for acquiring Within was not to eliminate a potential competitor, stifle alternative technology, or increase rivals' costs. Meta purchased Within with the intention of achieving complete ubiquity in "killer apps," thereby solidifying its early lead using the self-reinforcing mechanisms commonly found in digital markets.

208. Is not this always the case? In other words, is it not true that any company has the option to pursue either internal growth or external growth? If the FTC's criticism of the Meta–Within merger hinges on this point, one might infer that any company entering a new market through a merger is in violation of antitrust laws. By suggesting that Meta could have competed with Within directly, the FTC seems to be stating the obvious: internal growth processes are generally less detrimental to competition than external growth processes. However, is this observation alone sufficient to establish the anticompetitive nature of the Meta–Within merger? Should we not be looking for *quid pluris*, that is, any element indicating that the merger itself is anticompetitive, independently of the fact that any merger is inherently less procompetitive than internal growth processes?

209. The Court of the Northern District of California, which rejected the FTC's claim, chose not to address these questions. Instead, it stated that the Meta–Within merger would not significantly reduce competition in the relevant market for VR fitness apps, limiting its analysis to the fact that there was no reasonable likelihood of Meta entering the relevant market, with either a new VR fitness app or an improved version of Beat Saber. In essence, the counterfactual scenario presented by the FTC

---

51 In this sense, *see Meta Platforms, Inc.*, *supra* note 50, at ¶ 11: "Regardless of whether such a company actually intends to enter, the possibility that it may do so can spur other companies already in the market to proactively ramp up their own competitive efforts," *see* Moiseyev, Hardesty & Williams, *supra* note 50.

to demonstrate the anticompetitive effects of the Meta–Within merger lacked the necessary objective and subjective evidence.[52] The Meta–Within merger took place.

VI. Well Begun is Half Done

210. Why are tech giants under such intense scrutiny from antitrust authorities, especially as they venture into the genAI industry? Well, one might say that antitrust measures are crucial in preventing these giants from leveraging their market dominance to stifle competition and control innovation, particularly in sectors where new tech-driven players might challenge or even replace them in the future. After all, though not devoid of criticism, it is a historical pattern too; antitrust has always kept a close watch on those leading economic development, from big oil (like Standard Oil) to telecom (like AT&T), to hardware and software (think IBM and Microsoft). Today, one could maintain that the urgency is even more pronounced, as these giants operate vast ecosystems and can exploit the quirks of digital markets and consumers' cognitive biases to their advantage.

211. Therefore, we cannot help but welcome the investigations by the EC, CMA, and FTC. If for no other reason, than that they will *at least* shed light on how the genAI industry operates, thereby improving our understanding of its competitive dynamics.

212. The *at least* in the above paragraph is worth noting for two reasons. First, antitrust enforcers have specific tools and principles they must adhere to. Second, our stance does not advocate for the blanket prohibition of mergers involving digital ecosystems, acknowledging that public enforcement carries its own costs and limitations.[53] This is why this chapter, without aiming for comprehensive coverage, touches on potentially anticompetitive behaviors that these authorities might identify in their investigations. Specifically, the chapter has explored theories of harm that could be used to make the case that, within the genAI market, tech giants are curbing competition and shaping innovation to their benefit. The outcomes of such a preliminary analysis are as follows.

213. As long as tech giants invest in the genAI industry by acquiring minority shares in target companies without gaining control, authorities' options to prosecute are somewhat limited. Within the EU, this might result in charges of anticompetitive agreements under specific conditions, while in the US, these minority investments could also be seen as anticompetitive acquisitions. In the UK, they could be managed regardless of classification, thanks to carefully crafted commitments.

---

52 Fed. Trade Comm'n v. Meta Platforms, Inc., 654 F. Supp. 3d 892, No. 5:22-cv-04325-EJD (N.D. Cal. Jan. 31, 2023).

53 Geoffrey Manne & Joshua D. Wright, *Innovation and the Limits of Antitrust* (GEO. MASON LAW & ECON., Research Paper No. 09-54, 2009), where more extensive bibliographic references are provided.

214. However, if tech giants enter the genAI market through genuine mergers, acquiring control over the target companies, antitrust law faces a different challenge. We currently have numerous theories of harm that address scenarios where a company eliminates potential competitors, obtains an alternative technology, or defends its market position by raising rivals' costs. Nevertheless, we currently lack a valid theory of harm for a company failing to penetrate a market with its own products yet can do so by acquiring a promising company, which is not yet considered a potential competitor nor an alternative technology holder. Essentially, the antitrust community has not yet formulated a theory of harm that explains how and why competition and innovation are undermined when a dominant Big Tech firm, seeks to enter new markets, such as the markets for genAI and VR, by acquiring companies that have proven more adept than it.

215. Is this detrimental? As of today, answering this question is akin to a shot in the dark. On the one hand, concerns about the expansive reach of digital ecosystems and their potential to stifle innovation diversity are understandable. On the other hand, these ecosystems possess the resources and expertise to drive forward emerging technologies for the benefit of many, including current consumers. Adding to this dichotomy, one must also acknowledge an alternative path, diverging from the antitrust paradigm but often advocated, that suggests corrective measures aimed at reducing barriers to entry for new competitors, reinstating competition through regulation, or even supporting or financing players alternative to digital ecosystems via governmental resources or new tax law and IP regimes.

Artificial Intelligence, Uncertainty, and Merger Review

MARK J. NIEFER* AND AARON D. HOAG**
Antonin Scalia Law School, George Mason University
Antitrust Division, United States Department of Justice

Abstract

Merger analysis is inherently uncertain. When two firms merge, a competition authority typically must predict the state of competition with and without the merger, as well as predict the effect of any proposed remedy on competition. These predictions are uncertain because no one can perfectly forecast the evolution of competition, which may be determined by factors such as the number of firms in the industry, the technologies they employ, their business strategies, and the effects of entry and exit, among others, which may themselves evolve in uncertain ways. Uncertainty typically is heightened when a merger involves a rapidly innovating industry, which makes predicting the future state of competition particularly fraught. Rapid innovation can take firms, industries, and economies in unexpected directions, with unpredictable implications for competition. These concerns are particularly acute in the case of mergers involving artificial intelligence (AI), a technology for which substantial new developments in the underlying technology, its applications, and the business strategies employed by firms developing or using it occur almost daily. In this chapter, we draw on the existing literature on innovation, uncertainty, and mergers, as well as our own experience as competition law enforcers, to discuss the ways in which a competition authority might encounter and address uncertainty about the competitive effects of a merger involving AI.

---

* Adjunct Professor of Law, Scalia Law School – George Mason University; former Attorney, Antitrust Division, United States Department of Justice.

** Chief, Technology and Digital Platforms Section, Antitrust Division, United States Department of Justice. The views expressed in this chapter do not purport to represent those of the Department of Justice.

I. Introduction

216. Suddenly, it seems, artificial intelligence (AI) is everywhere. Every issue of the newspaper teems with articles about the latest developments in AI, from product launches for business and consumer use cases, to the underlying infrastructure needed to bring AI to the market at large. Trillions of dollars of stock market value are tied to the future of AI. At the same time, AI is currently one of the most active areas for venture capital investment.[1] Just about every public company has a strategy for harnessing AI to improve its business. This includes not just the technology companies one would expect to be at the forefront of such a revolution, but also firms in manufacturing, healthcare, and countless other industries.

217. The same phenomenon permeates competition law circles just as it does everywhere else in society. Conferences and panels on the competition issues raised by AI occur on a frequent basis. Enforcers around the globe have joined the conversation and noted a range of potential issues worthy of closer scrutiny.[2]

218. As attorneys with a couple decades of experience each enforcing antitrust laws, we are as fascinated as everyone else with how artificial intelligence will develop. What products and services using AI will break through and reshape markets? How deeply embedded will AI become in everyday business practices? How big of a first-mover advantage will accrue to the firms, such as NVIDIA and OpenAI, that are currently at the forefront of enabling and delivering AI? Will AI reinforce the market power of current dominant firms, or enable challengers to more readily top incumbents? Or will it create new instances of market power? We, alas, have no special skills that would enable us to deliver answers to these questions any more than any other pair of informed enforcers. But the questions themselves have intrigued us and led us to ponder how that very uncertainty may play out in antitrust.

219. Specifically, we want to focus here on the impact of that uncertainty in mergers involving artificial intelligence, as they raise particularly interesting questions given the predictive nature of merger review. To set the stage for a deeper exploration of this interaction, a brief discussion of the nature of merger review will be useful. In the United States,[3] mergers are typically reviewed under Section 7 of the Clayton Act, which proscribes

---

[1] *See, e.g.*, Berber Jin, *Investors Pour Cash Into AI Startups With Little Revenue*, WALL ST. J. (Apr. 30, 2024), https://www.wsj.com/tech/ai/investors-are-showering-ai-startups-with-cash-one-problem-they-dont-have-much-of-a-business-94534fc9.

[2] *See, e.g.*, JOINT STATEMENT ON COMPETITION IN GENERATIVE AI FOUNDATION MODELS AND AI PRODUCTS (July 2024) (issued by European Commission, U.K. Competition and Markets Authority (CMA), U.S. Dep't of Justice & Fed. Trade Comm'n); COMPETITION BUREAU CANADA, DISCUSSION PAPER: ARTIFICIAL INTELLIGENCE AND COMPETITION (Mar. 2024); COMPETITION & MARKETS AUTHORITY, AI FOUNDATION MODELS: INITIAL REPORT (Sept. 2023) (hereinafter, CMA FOUNDATION MODELS); AUTORIDADE DA CONCORRENCIA (PORTUGAL), ISSUES PAPER: COMPETITION AND GENERATIVE ARTIFICIAL INTELLIGENCE (Nov. 2023) (hereinafter PORTUGAL ISSUES PAPER); DIGITAL PLATFORM REGULATORS FORUM (AUSTRALIA), EXAMINATION OF TECHNOLOGY: LARGE LANGUAGE MODELS (Oct. 2023).

[3] While we discuss the statutory framework and case law from the United States that we are most familiar with, the same issues will arise in other jurisdictions as they are inherent to the nature of merger review.

mergers "where in any line of commerce or in any activity affecting commerce in any section of the country, the effect of such acquisition may be substantially to lessen competition, or to tend to create a monopoly."[4]

220. There are three interrelated features of merger review that are worth noting here because of how they interact with the assessment of mergers in fast-moving industries such as artificial intelligence. First, as the Supreme Court has emphasized, the Clayton Act was passed to halt trends towards concentration in any particular line of commerce in their "incipiency," before competitive conditions have already been eroded.[5] Second, as the text of the Clayton Act makes clear, the government is only required to show that a merger *may* substantially lessen competition or tend to create a monopoly, not that it will certainly do so. Merger analysis is, in short, an inherently *probabilistic* undertaking, where uncertainty abounds.[6] Third, by its very nature, merger review is a *predictive* exercise. As the Supreme Court has explained, merger review must look beyond the "immediate impact of the merger upon competition" and entail a "prediction of its impact upon competition conditions in the future."[7] Merger review therefore typically requires looking into the future to predict whether the merger poses a significant threat to competition.[8] Even when analyzing a merger that has already been consummated, there often is substantial uncertainty about what the world would have looked like absent the merger. The difficulty arises in isolating the effect of the merger in light of all the other factors that influence the evolution of markets and competition to predict what the effect of the merger was. Analysis of consummated mergers therefore requires the same type of prediction about what the world would have looked like absent the merger that we must undertake in predicting the future effect of planned mergers.

221. This analysis can be challenging in the simplest of circumstances. Even with the various presumptions and simplifications provided by the case law, it typically requires an in-depth assessment of market conditions to determine the lines of commerce that may be affected by the merger. There are no certain answers to the question of how markets will develop over time or how that development will be affected by the merger under review. While requiring too much certainty from the government

---

4   15 U.S.C. § 18.

5   Brown Shoe Co. v. United States, 370 U.S. 294, 317 (1962).

6   *See* United States v. Philadelphia Nat'l Bank, 374 U.S. 321, 362 (1963) (explaining that merger analysis under Section 7 of the Clayton Act "is not the kind of question which is susceptible of a ready and precise answer in most cases"); Hospital Corp. of America v. FTC, 807 F.2d 1381, 1389 (7th Cir. 1986) (Posner, J.) (noting that proof of competitive harm such as increased prices is not required, as "[a]ll that is necessary is that the merger create an appreciable danger of such consequences in the future").

7   *Philadelphia Nat'l Bank*, 374 U.S. at 362.

8   This discussion has been focused on the framework for analyzing mergers on Section 7 of the Clayton Act. Mergers involving firms with monopoly power can be considered not only under this statute, but also under Section 2 of the Sherman Act, which outlaws achieving or maintaining monopoly power through anticompetitive means. There is a strong argument that for a firm with monopoly power, any merger with an appreciable risk of entrenching and enhancing that monopoly power should be blocked. The "tend to create a monopoly" language in Section 7 itself may be another avenue for reaching the same principle.

challenging a merger could render the merger laws ineffectual, it is also true that courts are, quite properly, unwilling to block mergers based on the mere theoretical possibility of competitive harm. The fundamental question is, therefore, when does the risk of competitive harm go from merely theoretical to likely enough to warrant blocking a merger. There is substantial debate about the appropriate approach to dealing with the tension between the need to predict the future and the fundamental truth that the future cannot be predicted with certainty; rather, one can at most assess that certain outcomes are more or less probable. The challenges can be heightened when the merger takes place in a market that is, or that may be, undergoing rapid change. If a market is in flux, it renders the predictions of competition harm less certain. It is in this context that we turn to consider how this uncertainty may affect merger review in cases involving AI.

222. This chapter starts with a brief overview of artificial intelligence to orient the reader and provide a baseline. We then consider the various aspects of merger review that could be impacted by developments in AI and the uncertainty surrounding those developments. We conclude by considering various approaches to merger review that account for this fundamental uncertainty, and how those approaches may play out in the specific case of AI.

II. Artificial Intelligence: A Brief Overview

223. In this section, we loosely define artificial intelligence and characterize its development and deployment, which will serve as a foundation for our discussion of the role of uncertainty in merger review. There are many ways to characterize AI development and deployment; we chose an approach that we believe helps illuminate potential competition issues related to AI, uncertainty, and merger review.

1. AI and Foundation Models

224. There is no single widely-accepted definition of AI,[9] a term that encompasses an amorphous set of technologies and techniques that have a variety of applications across nearly all segments of society.[10] In its most basic

---

[9] *See, e.g.*, COMPETITION BUREAU CANADA, ARTIFICIAL INTELLIGENCE AND COMPETITION: DISCUSSION PAPER 5 (Mar. 2024) ("there is still no universal definition for AI.") (hereinafter CBC DISCUSSION PAPER); Harry Surden, *ChatGPT, AI Large Language Models, and Law*, FORDHAM L. REV. 1939, 1944 (2024) ("AI is notoriously difficult to define, and there is probably no single satisfactory definition that most researchers would agree to." (citations omitted)). The Organization for Economic Cooperation and Development (OECD) has been working on developing a consensus definition of AI; its most recent effort defines an "AI system" as "a machine-based system that, for explicit or implicit objectives, infers, from the input it receives, how to generate outputs such as predictions, content, recommendations, or decisions that can influence physical or virtual environments. Different AI systems vary in their levels of autonomy and adaptiveness after deployment." *See* Marko Grobelnik, Karine Perset & Stuart Russell, *What is AI? Can You Make a Clear Distinction Between AI and Non-AI Systems?* OECD.AI (Mar. 6, 2024), https://oecd.ai/en/wonk/definition.

[10] Ryan Calo, *Artificial Intelligence Policy: A Primer and Roadmap*, 51 U.C.D. L. REV. 399, 404 (2017) ("There is no straightforward, consensus definition of artificial intelligence. AI is best understood as a set of techniques aimed at approximating some aspect of human or animal cognition using machines.").

form, for example, AI might involve the use of a simple algorithm to sort data and perform simple calculations; in its more advanced form, it can involve complicated systems that approximate – or appear to approximate – human thought.[11] We focus on the more advanced forms of AI that allow "a machine to perform cognitive functions that we associate with human minds, such as perceiving, reasoning, learning, interacting with the environment, problem solving, and even exercising creativity."[12] In this chapter, we will focus on the development and deployment of foundation models, which have become a "new paradigm for developing AI systems,"[13] and which have attracted the attention of competition enforcers around the world.[14]

225. Foundation models (FMs) are "large-scale AI models that are pre-trained on vast amounts of general data and that can be adapted for downstream applications (e.g., by fine-tuning them through further training on application-specific data)."[15] Thus, FMs can form the basis for fine-tuned models that are suitable for specific applications. Large language models (LLMs) are a specific type of foundation model that is trained to generate text, code, images, video, and audio.[16] Examples of LLMs include OpenAI's GPT-4 and GPT-4o, Google's BERT, Claude (developed by Anthropic), Meta's Llama, and Microsoft's Orca. It is rare that a day passes without news regarding the development of a new foundation model, the refinement of an existing foundation model, or new applications for a foundation model.[17]

2. Foundation Model Development and Deployment

226. There are many ways to characterize foundation model development and deployment. We focus on the FM stack (which we also refer to as the FM ecosystem), which comprises various steps, activities, inputs, and applications that link the development of a FM to its application by end users. We adopt the taxonomy of the stack as described in a recent report by the UK's Competition and Markets Authority (CMA)[18] to

---

11 Thomas Hoppner & Luke Streatfield, *ChatGPT, Bard & Co.: An Introduction to AI for Competition and Regulatory Lawyers*, (Feb. 23, 2023), https://papers.ssrn.com/sol3/papers.cfm?abstract_id=4371681.

12 *Id.*

13 Johannes Schneider et al., *Foundation Models: A New Paradigm for Artificial Intelligence*, 66 BUS. INF. SYST. ENG. 221, 221 (2024) (citation omitted).

14 *See, e.g.*, CMA FOUNDATION MODELS, *supra* note 2; COMPETITION & MARKETS AUTHORITY, AI FOUNDATION MODELS: UPDATE PAPER (Apr. 11, 2024), https://assets.publishing.service.gov.uk/media/661941 a6c1d297c6ad1dfeed/Update_Paper__1_.pdf.

15 Schneider et al., *supra* note 13, at 224.

16 The models are underpinned by a technological leap known as the 'transformer.'" *See* Anouk van der Veer & Friso Bostoen, *Two Views on Regulating Competition in Generative AI*, NETWORK L. REV. (Feb. 20, 2024), https://www.networklawreview.org/veer-bostoen-generative-ai/.

17 A New York Times journalist recently described his experience using six different AI apps to make AI "friends." *See* Kevin Roose, *Meet My A.I. Friends*, N.Y. TIMES (May 9, 2024), https://www.nytimes.com/2024/05/09/technology/meet-my-ai-friends.html.

18 CMA FOUNDATION MODELS, *supra* note 2.

help identify and think about potential linkages and connections among various participants, which is a useful first step toward thinking about competition and merger review.[19]

Figure 1. Foundation Model Stack

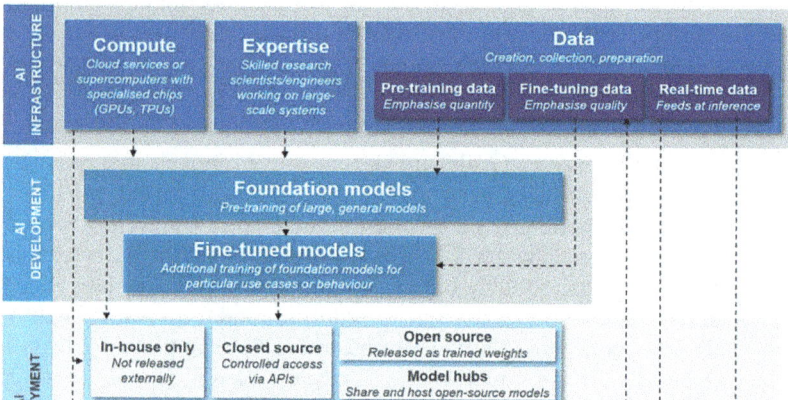

227. Figure 1, reproduced from the CMA's recent report,[20] shows three layers to the foundation model (FM) stack. Foundation models sit at the center of the stack as part of the AI development layer, which also includes fine-tuned models. The AI infrastructure layer encompasses the key inputs used to develop FMs, and the AI deployment layer encompasses the application and use of FMs by end users.

228. The key inputs into foundation and fine-tuned models can be grouped into three broad categories: (1) compute, (2) expertise, and (3) data.[21] Compute includes cloud services and specialized microchips.[22] Cloud services might be provided by firms such as Amazon, Microsoft, and Google; chips might be provided by firms such as Nvidia, Intel, and AMD. It appears as though relatively few firms are capable of

---

19 *See also* Friso Bostoen & Anouk van der Veer, *Regulating Competition in Generative AI: A Matter of Trajectory, Timing, and Tools* (2024), at 2 (identifying three generative AI layers: infrastructure, models, and applications), https://papers.ssrn.com/sol3/papers.cfm?abstract_id=4756641; *cf.* CBC DISCUSSION PAPER, *supra* note 9, at 8 (describing possible antitrust product markets in infrastructure, development, and deployment); Sebastian Lins et al., *Artificial Intelligence as a Service: Classification and Research Directions*, 63 BUS. INF. SYST. ENG. 441, 443 (2021) (describing the "AI as a service (AIaaS)" stack); *see also* Thibault Schrepel & Alex Sandy Pentland, *Competition between AI Foundation Models: Dynamics and Policy Recommendations* 4–6 (AMSTERDAM L. & TECH. INST., Working Paper No. 1-2023, 2023), https://papers.ssrn.com/sol3/papers.cfm?abstract_id=4493900 (proposing a general taxonomy for types of foundation models).

20 CMA FOUNDATION MODELS, *supra* note 2, at 10.

21 *See, e.g.*, Bostoen & van der Veer, *supra* note 19.

22 van der Veer & Bostoen, *supra* note 16.

providing cloud services and microchips,[23] which may have implications for certain vertical mergers, as we describe below. Data refers to the data necessary to train foundation and fine-tuned models.[24] Finally, expertise refers to the technical expertise possessed by individual AI researchers. Although these categories of inputs are not homogeneous (e.g., certain types of data may be suitable for certain types of foundation or fine-tuned models), they are useful for our general exploration of the role of uncertainty in merger analysis.

229. Foundation models can be in-house and closed or open source. In-house models are developed within a firm for use by the firm, which provides the firm with full control over the model.[25] Closed source models are developed within a firm but they may be shared with third parties on terms determined by the model developer.[26] Closed source model developers may, for example, license their models to third parties, permitting them to develop their own applications, but forbidding them from modifying the model.[27] Open source models are made public, which allows third parties to modify them in ways that may improve the model; however, developers of open source models often restrict permitted uses of the model.[28] The extent to which FM developers impose restrictions on model use may raise competition concerns.

3. Uncertainty About the Evolution of the FM Ecosystem

230. There are several sources of uncertainty about the future evolution of the foundation model ecosystem, including in-context learning by FMs, the type of business model adopted by FM developers, increasing returns, network effects, and likely industry participants. Each of these sources of uncertainty may contribute to uncertainty about how a merger affects competition.[29]

231. *In-context learning.* One key feature of foundation models is that they seem to be able to develop unanticipated capabilities through "in-context learning."[30] These unexpected capabilities contribute to uncertainty about how FMs will evolve and be utilized in the future, with implications for the

---

23 Christophe Carugati, *Competition in Generative Artificial Intelligence Foundation Models* 6–8 (Bruegel Working Paper No. 14/2023, June 2023).

24 van der Veer & Boeston, *supra* note 16; *see also* Schneider et al., *supra* note 13, at 225.

25 CMA FOUNDATION MODELS, *supra* note 2, at 55.

26 *Id.* at 14.

27 Carugati, *supra* note 23, at 2.

28 *Id.* For a more nuanced discussion of FM licensing and its competitive consequences, *see* Thibault Schrepel & Jason Potts, *Measuring the Openness of AI Foundation Models: Competition and Policy Implications* (SCI. PO DIGIT., Working Paper, 2024), https://papers.ssrn.com/sol3/papers.cfm?abstract_id=4827358.

29 For a comprehensive discussion of the uncertainties associated with foundation models, *see* CMA FOUNDATION MODELS, *supra* note 2, at 41–52.

30 Schneider et al., *supra* note 13, at 224 (citation omitted).

development and use of upstream inputs and downstream applications.[31] Different capabilities may also call for different business strategies regarding deployment and monetization, further contributing to uncertainty about how competition will play out among entities participating at the three levels of the FM stack.

232. *The rise of FM platforms.* Although the development and application of foundation models still seems to be in its early days, there are hints that we may see the rise of ecosystems centered on FMs that may be like the rise of the large digital platform ecosystems we have seen over the last two decades, which often have been centered on a core service. Some of the most prominent FM developers appear to be adopting a platform model wherein their model connects application developers with end users,[32] which may precipitate the rise of competition between platforms.[33] Indeed, many of the factors that led to the rise of digital platform ecosystems appear to be influencing the evolution of the FM stack. Scale economies, network effects, and the importance of data may influence the evolution of the FM stack and lead to the rise of ecosystems.[34] Uncertainty about whether FM developers will adopt a platform model or some other business model will contribute to uncertainty about the form competition will take among and between infrastructure providers, model developers, and app developers.

233. *Increasing returns in users.* In the case of foundation models, there may be "positive feedback loops (i.e., increasing returns)," such that the more users a FM can attract, the better the performance by a FM or a fine-tuned FM will be.[35] For example, suppose a FM developer licenses its FM to individual businesses, allowing each to adapt the model for its own particular purposes. The FM developer may realize increasing returns as the fixed costs of developing and maintaining the FM are spread across larger numbers of users. In addition, as the number of users increases, the FM developer may be able to learn from users to develop a higher quality model. Thus, a dominant FM developer may have an interest in acquiring users, and in preventing other developers from acquiring users (especially at the expense of the dominant developer). A merger that allows a dominant FM developer to acquire users, or that may give the developer the ability to prevent competing developers from acquiring

---

31 *Cf. id.* ("Emergence implies a substantial uncertainty over the capabilities, flaws, and limitations of foundation models, making it hard to understand, explain, and predict their behavior and potential failure modes.") (citation omitted).

32 Schneider et al., *supra* note 13, at 226 (2024) ("a shift from product to platform business is visible: OpenAI recently announced the launch of a platform, similar to the Apple App Store, to provide access to and the possibility to publish AI applications based on the OpenAI foundation models").

33 *Id.* at 224 (2024) ("Increasingly, AI applications will be the product of an emerging ecosystem comprised of foundation model providers, foundation model adapters and integrators, and end users... .").

34 Carugati, *supra* note 23, at 8 (noting that "[i]n data-driven markets, entry barriers faced by firms typically include data advantages, network effects and economies of scale and scope").

35 Schrepel & Pentland, *supra* note 19 at 11–12.

users, may have long-lasting effects that set the course of competition over a long period of time. At some point, however, one might expect that an increase in the number of users will yield diminishing returns. Just what that point is may be uncertain, and it may evolve over time as evolution in FM design may reduce the need for additional users and the data they provide.

234. *Network effects.* Another form of increasing returns relates to network effects. To the extent that foundation model developers adopt a platform model linking application developers and end users, there may also be direct and indirect network effects that may serve as a barrier to entry.[36] As has been seen elsewhere in the digital economy, application developers typically would prefer to develop apps for the most widely used FMs because larger numbers of users typically result in larger revenues for app developers. The barrier to entry that results from network effects may encourage FM developers to acquire as many users as possible, as quickly as possible, to freeze out competing FM developers.[37]

235. *The role of data.*[38] Although some of the most widely recognized foundation models were developed using large amounts of data, it appears as though FM design innovations may make it possible to train FMs of equal quality on substantially less data.[39] Recent developments suggest that FMs are beginning to "see diminishing returns to scale in the use of data, which means that large datasets are a necessary but not sufficient condition" to develop a high-quality foundation model.[40] Thus, innovations in FM design may reduce the data barrier to entry, making FM development less dependent on large volumes of data.[41] However, there is uncertainty about the relative importance of data, which may complicate the analysis of a merger of a data provider with other data providers or with firms participating in other layers of the FM stack. The uncertain role of data may also create some uncertainty about the ability of new FM developers to enter.

236. *Participants.* Many of the large digital platforms are intimately involved in all levels of the foundation model stack. Google, Amazon, Meta, and Microsoft are involved in the development of FMs, and they have begun to incorporate FM capabilities into their existing services. They also have

---

36   We adopt Joe Bain's approach to entry barriers rather than that of George Stigler. *See, e.g.*, Herbert J. Hovenkamp, *Whatever Did Happen to the Antitrust Movement?*, 94 NOTRE DAME L. REV. 583, 614–615 (2019) (explaining differences between Bain and Stigler on entry barriers).

37   *Cf.* Schrepel & Pentland, *supra* note 19, at 12.

38   It is, of course, the case that not all data is homogeneous or that all data is suitable for the development of all foundation models or fine-tuned models. Explicitly recognizing the complications that heterogeneous data might bring to the analysis is beyond the scope of this chapter.

39   *Id.* at 12.

40   *Id.* at 7.

41   *Id.* at 9 ("recent technical developments are increasing the importance of the efficiency of AI models, while proportionally decreasing the importance of ever-larger data sets.").

access to a large base of users of their existing services; as a result, they may be particularly attractive platforms for app developers. Thus, the large digital platforms may have a leg up in attracting app developers, which may serve as a barrier to entry for other FMs. In addition, dominant platforms such as Google and Microsoft already participate in the AI infrastructure level of the stack, providing computing services used by FM developers. Thus, an FM-related merger involving one of the existing dominant platforms may raise concerns about horizontal, vertical, or conglomerate (ecosystem) effects.

237. If we end up in a world with only a handful of foundation developers, all adopting a platform model, competition authorities may be forced to deal with many of the same competition and enforcement issues with which they have been forced to deal regarding large digital platforms. And they may be forced to deal with many of the very same players who currently dominate the digital world. However, there appears to be no small amount of uncertainty about whether there will be few or many foundation model developers in the future.[42] This uncertainty may have a profound effect on merger review and the development of merger policy related to FMs.

III. Merger Review Involving AI

238. Merger analysis involves developing theories of the ways in which a merger might harm competition and developing evidence to support those theories (or not). Competition authorities and others around the world have begun to articulate theories about the ways in which mergers involving AI might affect competition,[43] which often parallel the theories used to analyze mergers in more traditional, static industries; however, competition authorities have begun to articulate theories of harm better tailored to more dynamic industries. Indeed, some theories of harm regarding FMs have parallels in approaches to digital markets taken by many competition agencies around the world in recent years.[44]

1. Assessing Effects

239. In this section, we describe potential theories of harm involving mergers among various entities participating in the foundation model stack. It is, of course, impossible to identify in advance all the relevant markets that

---

[42] *See* Bostoen & van der Veer, *supra* note 19.

[43] PORTUGAL ISSUES PAPER, *supra* note 2; CMA FOUNDATION MODELS, *supra* note 2; OECD COMPETITION COMMITTEE, ARTIFICIAL INTELLIGENCE, DATA, AND COMPETITION: BACKGROUND NOTE 8 (May 2024); Benoit Coeure, President, French Competition Authority, Artificial Intelligence: Making Sure It's Not a Walled Garden, Address Before the Bank for International Settlements (Mar. 19, 2024).

[44] *See, e.g.*, Bostoen & van der Veer, *supra* note 19, at 2 (noting that foundation models show "platform characteristics, as independent developers can build their own products on top of this technological foundation." This suggests that theories of harm related to platforms may play an increasingly important role in assessing possible competitive effects from a merger involving foundation model developers. *See* U.S. Dep't of Justice & Fed. Trade Comm'n, *Merger Guidelines (2023)*, (Dec. 18, 2023), https://www.justice.gov/d9/2023-12/2023%20Merger%20Guidelines.pdf (hereinafter Merger Guidelines).

may arise in the context of a merger involving FMs. Market definition, although not strictly necessary to assess a merger's competitive effects, often is a cornerstone of merger analysis. Absent a detailed investigation into a merger, however, it is not possible to know or predict with much accuracy what relevant markets might come into play or how a merger might affect competition.[45] Rather than consider all the possible permutations of mergers involving participants involved in the FM stack, which would be impossible in this chapter, we simply describe in generic terms the types of theories of harm that might arise with respect to a small set of mergers involving FM model developers, with an eye toward illustrating the uncertainty inherent in merger analysis.

240. Mergers often are categorized as horizontal, vertical, or conglomerate. These categories can be somewhat vague and, if not properly understood, misleading – particularly if competition takes place within or between platforms or ecosystems. In the case of competition involving platforms or ecosystems, horizontal mergers may have competitive effects like that of a vertical merger, or vice versa; and a conglomerate merger may implicate competitive effects like those arising from horizontal or vertical mergers.[46] Despite the shortcomings of the horizontal–vertical–conglomerate taxonomy, we adopt it here to explain some of the more obvious competition concerns that may arise with respect to mergers related to AI. To introduce some of the uncertainties associated with merger review, we consider four simple hypotheticals: two involving a merger of unintegrated model developers, and two involving a merger of integrated model developers.

A. *Merger of Unintegrated Foundation Model Developers*

241. *Developers with established models.* Consider a simple horizontal merger between two standalone foundation model developers who do not participate in the infrastructure or deployment layers of the stack. If both developers have established models that compete against one another, the analysis of competition is relatively straightforward. In this case, a typical concern would be that the models are competing for the same users and the merger would reduce head-to-head competition, harming users by raising the cost of access to the model or by reducing the quality of the model relative to a world in which both developers remained independent; that is, the typical concern would be that the merger results in unilateral effects.[47] At least in principle, the tools and evidence needed to analyze the merger are reasonably well-developed and relatively uncontroversial.[48]

---

45 Some agencies have begun to try to identify potential markets related to AI. *See, e.g.,* CBC Discussion Paper, *supra* note 9, at 8.

46 For a brief discussion of horizontal, vertical, and conglomerate mergers regarding AI, *see id.* at 18–19.

47 There are, of course, many other possible effects of a merger. In this chapter, we focus on unilateral effects to illustrate the uncertainty associated with merger review involving artificial intelligence. There may be even greater uncertainty in the cases of other theories of harm.

48 *See* Merger Guidelines, *supra* note 44, § 2.2 (discussing unilateral effects analysis).

242. *Developers engaged in R&D to produce models.* Now consider a merger of standalone foundation model developers who have not yet brought a model to market, but who are engaged in research and development to produce a model for release to the public. In this case, the typical competition concern would be that each developer is engaged in research and development to produce models that customers would view as close substitutes. Once again, the typical concern would be that the merger would reduce the incentive for the merged firm to innovate, harming consumers through the development of a less desirable model relative to the situation in which both developers remain independent.[49] In this case, many of the types of tools and evidence relied upon in the case of a typical merger raising unilateral effects concerns still would apply. However, there likely will be greater uncertainty in this case than in the first one.

243. The difference between these two cases, of course, is that the second case involves developers engaged in ongoing innovative efforts. As a general matter, the effect of a merger on competition to develop foundation models (rather than on competition between established models) tends to be more uncertain. For example, there may be substantial uncertainty about whether, absent the merger, each model would be developed or ultimately be released; and if released, whether consumers would view them as close enough substitutes such that the merger might result in an inferior or more costly model being brought to market. In short, there will be relatively more uncertainty about the effects of a merger involving ongoing innovative efforts to produce a foundation model.

B. *Merger of Integrated Foundation Model Developers*

244. *Developers with established models.* Consider a merger between two vertically integrated foundation model developers that have established models. In addition to the usual horizontal unilateral effects arising because of lost competition between the models following the merger, there may be concerns that the merged firm would foreclose its rivals from access to critical inputs, such as compute (e.g., access to cloud computing resources) or data, or foreclose rival app developers from access to their model. Foreclosure concerns often are addressed using an ability and incentive analysis to determine whether the merged firm may cut off or otherwise make the terms of access to a critical input more onerous for competing foundation model developers, whether they have an established model or are in the process of developing a foundation model that will compete with the merging firms. The latter case – where the concern is with foreclosure of a nascent or potential competitor – involves some uncertainty

---

49 Our discussion of the effect of a merger on R&D and FM model development implicitly relies on the notion of an innovation market, which focuses on overlapping R&D activities of merging firms. *See, e.g.*, Richard J. Gilbert & Steven C. Sunshine, *Incorporating Dynamic Efficiency Concerns in Merger Analysis: The Use of Innovation Markets*, 63 ANTITRUST L.J. 569, 595–597 (1995). For criticism of the use of innovation markets in merger analysis, *see, e.g.*, Richard T. Rapp, *The Misapplication of Innovation Market Approach to Merger Analysis*, 64 ANTITRUST L.J. 19, 24–46 (1995).

about whether the innovative efforts will be successful, and whether the nascent or potential competitor will constrain the ability of the merged firm to exercise market power.

245. *Developers engaged in R&D to produce models.* Now consider the case of two vertically integrated developers who are engaged in research and development to bring a foundation model to market. Again, there may be concerns that the merger will result in the loss of head-to-head competition to innovate to bring a foundation model to market. As with a merger of standalone developers engaged in R&D to produce a FM, there will be uncertainty about whether the R&D produces a product that actually will be brought to market, and whether consumers will find the models of the two firms close enough substitutes that the merger will result in a loss of competition. And there may be traditional vertical concerns that the merger may foreclose competing developers from access to critical inputs.

246. In the case of a merger between firms in the process of developing a FM, however, uncertainty about the foreclosure effects of the merger may be heightened relative to that when competition is among existing, established FMs. When models are still in development, the infrastructure needed to produce a FM may depend on the type of model being developed: some models may require more or less data than others, and some models may require more or less computing capacity. Uncertainty about the form the foundation model will take and its infrastructure needs in turn may generate uncertainty about whether the merged firm has the ability and incentive to foreclose other firms, as well as the ultimate impact on consumers.

C. *Other Sources of Uncertainty in Assessing Effect*

247. There are, of course, many other possible combinations of entities operating at the three layers of the foundation model stack. It is possible that a foundation model developer might seek to acquire a fine-tuned model developer, or a firm providing infrastructure, or an application developer. Any one of these types of mergers (or any other merger among firms in the FM ecosystem) will involve some uncertainty. As the hypotheticals described above suggest, the further a foundation model is from reaching users (as when a foundation model still is in the development stage), the greater will be the uncertainty about the extent to which the merger produces anticompetitive effects. In addition to the uncertainty associated with FMs or fine-tuned models in the development stage, there are other sources of uncertainty that might affect a merger analysis. As noted above,[50] there is uncertainty about how the FM ecosystem will evolve such that agencies evaluating mergers in the near future will have to grapple with substantial uncertainty about the state of the world with and without a merger.

---

50 *See* discussion *supra*, Section II.3.

2. Defense Arguments

248. So far, we have been discussing how a merger involving AI may lead to anticompetitive effects. It is worth pausing for a moment to consider the ways in which developments in AI could be offered in defending a merger that otherwise may raise competitive concerns. We are thinking primarily of three situations: entry and repositioning, efficiencies, and remedies.

249. *Entry and repositioning.* Merging parties can rebut competitive concerns by demonstrating that other firms would enter the market or expand their existing presence in a manner that would counteract any risk of competitive harm. This analysis requires a prediction of how firms may react in the future to anticompetitive effects from a merger that are themselves uncertain. The Merger Guidelines in the United States, for example, require merging parties to show that such entry or repositioning meets three key tests. It must be *timely*, in the sense that the entry must be capable of being accomplished quickly enough to discipline firms in the market and prevent anticompetitive harm. Entry also must be *likely*. Merging parties cannot simply hypothesize that various companies could enter the market if they chose to; they must show that there is a reason to think this entry or repositioning will actually happen. Finally, entry or repositioning must be *sufficient* in scope to ensure that there will be no loss of competition due to the merger. This requires a projection not only as to what firms might enter, and how quickly, but also into how successful they are likely to be, and whether they can fully replace the competition eliminated by the merger.[51] As one can readily see, each element of this test involves uncertain predictions that become even more challenging the more dynamic the market.

250. An argument along these lines, involving developments in AI, was advanced in a recent case considered by the UK's Competition and Markets Authority (CMA). In September 2022, Adobe announced its planned acquisition of Figma for $20 billion.[52] Figma sold the leading product design software, Figma Design, which the CMA concluded had a "very strong position" in the market with a share far larger than that of any competitor.[53] Product design software is used to design interactive elements in mobile applications and websites.[54] Adobe sold the industry-leading Creative Cloud bundle for creative professionals. The bundle included two products that have long been dominant in their respective markets, Photoshop and Illustrator.[55] Photoshop is used for professional-level photo editing, while Illustrator is

---

[51] *See* Merger Guidelines, *supra* note 44, § 3.2.

[52] *See* COMPETITION & MARKETS AUTHORITY, ANTICIPATED ACQUISITION BY ADOBE INC. OF FIGMA, INC., PROVISIONAL FINDINGS REPORT ¶ 3.1 (Nov. 28, 2023) (hereinafter CMA PROVISIONAL FINDINGS). While one of the authors (Aaron Hoag) was involved in the US antitrust review of the *Adobe/Figma* merger, the discussion here is based solely on public information.

[53] *See id.* ¶ 8.62.

[54] *See id.* ¶¶ 5.10–5.11.

[55] *See id.* ¶ 9.117.

used to create non-photo-based images and designs, such as company logos that can scale to any size. Adobe's Creative Cloud bundle also included XD, a product design tool that competed with Figma Design.

251. The CMA found that the merger may have led to a substantial lessening in the product design, photo editing, and illustration markets.[56] In short, the CMA was concerned that Figma threatened Adobe's dominant positions in photo editing and illustration, while Adobe threatened Figma's leading position in product design. One of the arguments advanced by Adobe and Figma in defense of their merger was that while AI was not central to any of the markets at that time, future developments in AI could change the markets so radically that there would be no reason to be concerned about the merged firm's ability to maintain its leading market positions. With regard to the product design market, for example, the parties contended that tools using artificial intelligence could become capable of converting simple text prompts into complex digital assets in code form and that Figma's competitors were working to develop that functionality.[57] Similar arguments were advanced regarding photo-editing and illustration software; according to the parties, existing competitors in these markets were investing in AI to disrupt competition, as were a number of new entrants, such as Dall-E, that specialized in AI.[58] Perhaps unsurprisingly, especially given Adobe's longstanding dominant position, these arguments were rejected by the CMA with little discussion.[59] The parties ultimately abandoned their merger in December 2023, saying that they did not see a path to receiving regulatory approval from the CMA and the European Commission.[60] This case neatly highlights how uncertainty surrounding developments in AI can be difficult to harness as a defense by merging parties, even where the theories of harm themselves are very forward-looking in nature.[61]

252. *Efficiencies.* Merging parties often assert that a deal will enable them to achieve certain efficiencies that could not be readily captured through other means. While the Supreme Court has noted that "[p]ossible economies

---

56 *See id.* ¶ 1.

57 Response to Issues Statement of July 26, 2023 ¶ 2.24(c) (Aug. 9, 2023) (submitted by Adobe and Figma to CMA); *see also id.* ¶ 6.7(d) (arguing that AI was "both a competitive threat to Figma as well as an opportunity").

58 *See* CMA Provisional Findings, *supra* note 52, ¶¶ 11.26–11.28 (summarizing parties' contentions).

59 *See id.* ¶ 9.464 (noting that "the threat of entry and expansion" from AI and certain other tools "does not pose more than a weak competitive constraint on Adobe's product development in vector editing [i.e., illustration] for professional users over the short to medium term"); *id.* ¶ 9.557 (reaching same conclusion for photo editing tools).

60 Adobe Press Release, Adobe and Figma Mutually Agree to Terminate Merger Agreement (Dec. 18, 2023, 08:00 AM), https://news.adobe.com/news/news-details/2023/Adobe-and-Figma-Mutually-Agree-to-Terminate-Merger-Agreement/default.aspx.

61 Sullivan and Su offer a thorough discussion of the closely related issue of the disparate treatment of potential competition by agencies and courts when it is part of the theory of harm, as in the acquisition of a potential competitor, compared to when the prospect of potential entry is offered defensively by the merging parties. Sean P. Sullivan & Henry C. Su, *Antitrust Time Travel: Entry & Potential Competition*, 85 Antitrust L.J. 147 (2023).

cannot be used as a defense to illegality,"[62] the Merger Guidelines recognize the potential for efficiencies to rebut the initial showing of competitive harm if the parties demonstrate they are merger specific, verifiable, not resulting from an inherent harm to competition, and substantial enough in degree such that "no substantial lessening of competition is threatened by the merger in any relevant market."[63] Some commentators have claimed this places an unattainable burden of proof on merging parties generally. Whatever the merits of this complaint, the efficiencies defense, outlined in the Merger Guidelines, seems particularly unlikely to apply to cases involving rapidly developing technology such as AI. If there is substantial proof of anticompetitive harm, it is unlikely to be countervailed by the type of cost efficiencies that comprise a run-of-the-mill efficiencies claim. These markets are changing too fast to say with certainty that cost savings can be achieved only through a specific merger.

253. Moreover, the type of harms that we have discussed in this chapter relate more to innovation and dynamic competition, making it unlikely that cost savings alone would eliminate the risk to competition. It seems more plausible that arguments about pro-competitive efficiencies will be central to the underlying assessments of the risk of competitive harm from the merger. A compelling argument that a merger will enable the firms to develop or deploy new products together may in some cases provide an alternative narrative for the transaction to the theories of competitive harm considered by reviewing agencies. This could take the form of, for example, the combination of expertise within each firm's research and development teams. Alternately, the merging firms could have complementary technologies or other innovations that, when combined, would spark innovation. Or one firm may have the resources to exploit or deploy innovations from the other firm that it could struggle to market itself. These possibilities could paint the picture of a world in which the merger is not only competitively neutral, but also good for both the merging parties and for competition. Certainly, a story of merger-enabled innovation may have little persuasive value where the merger eliminates substantial direct competition between the merging parties and therefore the outcome of the merger seems relatively more certain. Mergers involving alternate theories of harm, such as foreclosure of rivals, may in some cases depend on more uncertain projections about future actions. In a fast-moving industry like AI, such theories may rest upon even more uncertain predictions about market outcomes and therefore be more subject to rebuttal by a narrative that the merger will enable more efficient innovation.

---

62   FTC v. Procter & Gamble Co., 386 U.S. 568, 580 (1967).

63   Merger Guidelines, *supra* note 44, § 3.3. One complication, which is beyond the scope of this chapter, is that existing US case law and agency enforcement policy generally does not credit out-of-market efficiencies. *Id.* (efficiencies must "prevent the risk of a substantial lessening of competition in the relevant market.") As a conceptual matter, an agency seeking to block a merger may have an incentive to define relevant markets narrowly enough that they exclude potentially large efficiencies that arise outside the market, increasing the likelihood that it will prevail in a court challenge. This may, of course, harm consumers who might be part of a broader relevant market. For a discussion of out-of-market efficiencies in antitrust and a proposed reform, *see* John M. Yun, *Reevaluating Out of Market Efficiencies in Antitrust*, 54 Ariz. St. L.J. 1262 (2022) (proposing the use of a "relevant" efficiencies approach).

254. *Remedies.* Identifying and assessing the effect of possible remedies is an inherently uncertain exercise. No agency will know with certainty the way in which a technology, market, or firm strategy may change in the future, which makes predicting the effect of a remedy uncertain. That uncertainty is heightened in the case of remedies involving dynamic or rapidly innovating areas, such as AI. Many of the uncertainties discussed above are relevant to an agency or court's analysis of potential remedies for mergers involving AI. In court, this issue can also arise when parties propose their own remedies – which could include both conduct commitments and divestitures – and try to "litigate the fix" at trial. In such cases, merging parties argue that their commitments are sufficiently robust and certain that they eliminate any potential competitive harm and therefore compel the court to rule against the government. Given recent successes with this strategy, it seems likely that parties would pursue it in cases involving AI, especially in concert with an argument that the market is too dynamic to give too much credence to the potential harms alleged by the government. While an in-depth discussion of the legal merits of allowing merging parties to design their own remedies for their anticompetitive transactions is beyond the scope of this chapter, we would observe that the uncertainty surrounding AI cuts both ways as it is equally challenging to determine if remedy proposals proffered by merging parties will truly be effective in forestalling any potential harm.[64]

IV. Accounting for Uncertainty in Mergers Involving AI

255. Although some competition agencies (and others) have begun to think about the ways in which a merger involving AI might affect competition,[65] little attention has been given to the way in which extreme uncertainty about the evolution of foundation models, their applications, and associated business strategies might affect competition, or the ways in which competition agencies ought to review mergers.[66] These uncertainties can

---

[64] Another complication, which is beyond the scope of this chapter, concerns the extent to which it might be desirable to allow for greater flexibility in the development and implementation of remedies. It may be desirable to allow the terms of a remedy to be altered as time passes and an agency revises or updates its beliefs about the efficacy of the remedy in light of new evidence; this may be particularly desirable when the agency faces great uncertainty about its analysis of a merger's competitive effects, as may be the case with a merger involving FMs. It also may be desirable to build into the remedy a mechanism that facilitates the collection of information about the evolution of competition in the relevant markets and the efficacy of the implemented remedy. In short, a dynamic industry may call for dynamic remedies that are flexible enough to be revised in light of changing market conditions and the resolution of uncertainty. *See infra*, Section IV.

[65] *See, e.g.,* CBC DISCUSSION PAPER, *supra* note 9, at 18–19.

[66] For example, in its recent discussion paper on AI and competition the Canadian competition authority did not raise the issue of uncertainty in merger analysis or in antitrust analyses involving AI. *See id.* Although the Portuguese Competition Authority noted that there are some uncertainties associated with analyzing competition issues related to AI in its recent issues paper, *supra* note 2, it did not address how to account for uncertainty in a competition analysis. The OECD in its note on AI and competition, *supra* note 43, identifies some uncertainties and the possible need to address uncertainty in a competition analysis but it does not propose an approach or method of dealing with uncertainty. The Federal Trade Commission (FTC) in 1996 issued a report that called for taking account of magnitudes and probabilities of merger-related efficiencies. However, the report did not state how those magnitudes and probabilities should be taken into account in merger analysis. *See* Michael L. Katz & Howard A. Shelanski, *Merger Analysis and the Treatment of Uncertainty: Should We Expect Better?*, 74 ANTITRUST L.J. 537, 542 (2007) (citing FED. TRADE COMM'N, ANTICIPATING THE 21ST CENTURY: COMPETITION POLICY IN THE NEW HIGH-TECH GLOBAL MARKETPLACE (1996)).

have a profound effect on the accuracy of predictions of a merger's effect on competition. We highlight this uncertainty not to suggest that it eliminates the need for antitrust enforcement in this space. Given the potential significance of AI across so many areas of the economy, the stakes are far too high to permit a wait and see approach that sits by, allowing all mergers to proceed while hoping for uncertainty to diminish below some arbitrary threshold. Such an approach risks allowing anticompetitive market conditions to become so fully entrenched that agencies will not be able to catch up.[67] Rather, we focus on these uncertainties because addressing them is necessary to ensure that agencies and courts can block mergers that do risk harm to competition in AI.

256. There has been substantial discussion in the academic literature about how to deal with uncertainty in merger analysis, even in the case of established, static industries.[68] That uncertainty is heightened in the case of dynamic, innovative industries, and there are concerns that the law and current agency approaches may be a less than perfect fit.[69] In particular, dynamic, innovative industries, which may involve substantial degrees of uncertainty, present the possibility that a very low probability event produces a large effect on competition. Perhaps no area of the economy currently is as dynamic and innovative as the development and deployment of foundation models.

257. One can imagine a situation in which uncertainty is so great with respect to a merger involving a new technology like foundation models that the plausible outcomes of a merger range from strong anticompetitive effects to strong procompetitive effects. Even if a competition authority knew the probabilities of all possible outcomes, it still is far from clear how the authority (or a reviewing court) should account for that uncertainty. What is an agency to do in the case of such extreme uncertainty? Especially in the case of a nascent technology like foundation models, or AI more generally, that may have substantial long-lasting effects on competition across a broad swathe of the economy, where getting it wrong (by blocking a procompetitive merger or failing to block an anticompetitive merger) may impose substantial harm on consumers?

---

[67] To be sure, many antitrust commentators – such as Barnett, elsewhere in this book – contend that there is a greater risk that premature or overly aggressive enforcement could interfere with innovation by reducing incentives for market participants to invest in ground-breaking technology. *See, e.g.*, Jonathan M. Barnett, *The Case Against Preemptive Antitrust in the Generative Artificial Intelligence Ecosystem*, in ARTIFICIAL INTELLIGENCE AND ANTITRUST POLICY (Alden Abbott & Thibault Schrepel eds., 2024).

[68] *See, e.g.*, Katz & Shelanksi, *supra* note 66, at 538 (noting that merger review "involves making predictions based on incomplete information about current or future facts, predictions that are almost inevitably subject to a high degree of uncertainty.").

[69] *Id.* at 539 ("The potential social and consumer costs of a failure to take explicit, systematic account of the magnitudes (as well as the probabilities) of a merger's predicted effects will only grow as federal competition agencies increasingly apply merger policy to technologically dynamic industries where effects on innovation, which are particularly subject to uncertainty, are central to the analysis.") (citing Michael L. Katz & Howard A. Shelanski, *Mergers and Innovation*, 74 ANTITRUST L.J. 1 (2007)). *See also* Matthew Jennejohn, *Innovation and the Institutional Design of Merger Policy*, 41 J. Corp. L. 167, 171 (2015) (noting that "the uncertainty problem in antitrust enforcement decision making is one of the most pressing issues in contemporary antitrust policy").

258. Academics and commentators have made various proposals for dealing with such uncertainty.[70] The most widely discussed academic proposals involve using some form of decision theory.[71] The approach can be as simple as accounting for the possibility that a low probability outcome can have a large effect on competition or that a high probability outcome can have a small effect on competition. More formal and complicated approaches call for an agency or court to identify the set of possible merger effects (e.g., by measuring the effect of the merger on consumer welfare or some other measure), assign a probability to each outcome, and then assess the net present value of the merger.[72]

259. The decision theoretic approach is straightforward in principle. We illustrate it here drawing on the work of Katz and Shelanski.[73] Suppose there is a set of possible effects of a merger, ranging from a reduction in competition to an increase in competition, represented by $\{X_1, X_2, \ldots, X_n\}$. One might measure X_i by, for example, the effect of the merger on prices to consumers, or by consumer welfare, or by some other measure of the effect of the merger on competition. Suppose that for each possible effect, there is an associated probability, ρ_i, that the effect will occur if the merger is approved. The expected value of the effect of the merger will be

$$E(X) = \Sigma \, \rho_i \, X_i.$$

260. Assuming that X_i is negative for bad outcomes (e.g., an increase in price) and positive for good outcomes (e.g., a decrease in price), the expected effect of the merger will be bad if $E(X) < 0$, e.g., if the expected value of the merger's effect on prices is negative, which might suggest a competition authority should challenge the merger.

261. The simple framework can be modified to incorporate other factors that may matter to an agency given its mission to protect the public interest. For example, an agency may believe that failing to challenge a merger will encourage anticompetitive mergers in the future, or an agency may be risk averse such that it seeks to avoid the effects of anticompetitive mergers relative to procompetitive mergers, or an agency may care about the distribution of harm across consumers. If an agency views a merger as undesirable if it has an adverse impact on low-income consumers, for example, then those outcomes might be weighted accordingly in assessing

---

70 *See, e.g.*, Steven C. Salop, *The Evolution and Vitality of Merger Presumptions: A Decision-Theoretic Approach*, 80 ANTITRUST L.J. 269 (2015). Competition agencies, on the other hand, largely seem to have remained silent about how they incorporate uncertainty into a merger analysis, other than to state that uncertainty does not bar the agency from finding a merger to be anticompetitive. *See, e.g.*, Competition & Markets Authority, *Merger Assessment Guidelines 2021*, (Mar. 18, 2021), https://assets.publishing.service.gov.uk/media/61f952dd8fa8f5388690df76/MAGs_for_publication_2021_--_.pdf (uncertainty "does not in itself preclude the CMA from finding competition concerns on the basis of all the available evidence where the CMA is satisfied that the relevant standard of proof is met.").

71 Katz & Shelanski, *supra* note 66, at 549.

72 *Id.*

73 We draw heavily on Katz & Shelanski, *supra* note 66, at 549–564.

the merger. If we assume that the value an agency places on a particular outcome can be represented by $u(X_i)$, then the expected "payoff" to the agency of allowing the merger will be

$$E(u(X)) = \Sigma\, u(X_i)\, \rho_i.$$

262. Assuming that $u(X_i)$ is positive for desirable outcomes from the agency's point of view, and negative for undesirable outcomes, the agency will view approving the merger as undesirable if $E(u(X)) < 0$, and desirable if $E(u(X)) > 0$.

263. We do not recommend that an agency adopt a formal decision-theoretic approach, which would call for the agency to specify its payoff function and assess the magnitude and probability of each possible outcome arising from a merger. That would be an impossible task in practice, and it would be especially difficult in the case of a merger involving a rapidly evolving technology like foundation models. Instead, we more modestly propose that an agency do its best to identify and account for the risk associated with possible future states of the world when deciding, for example, whether to allocate resources to investigating or challenging a merger involving foundation models or AI more generally. This more modest approach, which would draw inspiration from the more complicated models set out in the decision theory literature, would at least force agencies to confront and account for uncertainty when determining whether to investigate or challenge an AI-related merger, or when formulating merger policy regarding AI-related mergers.

264. For example, an agency might develop views on the relative likelihood that a merger of foundation model developers will harm competition. It might consider three possible states of the post-merger world: that the merger (1) will not affect competition, (2) will benefit competition, and (3) harm competition. Then the agency might structure and explain its merger analysis by assessing the probability of each scenario and the magnitude of the consequences of each; if there are other considerations (such as concerns about the distribution of effects or deterrence) then the agency might explain how these factors entered into its decision to investigate or challenge a merger. If the agency determined that the likelihood of no effect or a beneficial effect is high relative to a harmful effect, but that the magnitude of anticompetitive effects would greatly exceed that of any potential procompetitive effects, then it might have a reason to challenge the merger. As part of its analysis, the agency might describe how it assessed uncertainty associated with the various elements of its analysis, including entry and exit, and business models and strategy.[74]

---

74 There is a good argument – if untested to date – that the statutory framework in the United States is quite consistent with this framework. In particular, a court could properly take into account the *magnitude* of a potential risk of harm in addition to its *likelihood* in deciding whether there is a "reasonable probability that the merger will substantially lessen competition" and therefore violate Section 7, in accord with the Supreme

265. In an ideal world, the agency would articulate its reasons for assessing the probabilities and magnitudes of harm as it did, as well as identify any sources of uncertainty that went into the analysis. For example, if considering a merger at the FM infrastructure level, an agency might consider the extent to which demand for chips was driven by demand at the FM development level, and the way in which uncertainty about the evolution of the development of FMs would affect demand for chips. Thus, an agency might articulate its views about whether FMs will become more differentiated such that it might result in differentiation among chip developers, and the likelihood that chips will become differentiated in ways that affect its analysis of the merger.

266. Agency articulation of its views on uncertainty surrounding mergers involving AI would provide multiple benefits to the agency and to the public. First, it might provide greater guidance to merging parties and the public, reducing uncertainty about whether a contemplated or proposed merger is likely to attract antitrust scrutiny.[75] Second, it might prompt a dialogue with the public about whether the agency's approach is appropriate or the best possible or feasible approach. Third, it might prompt academic research on approaches to uncertainty and methods of assessing the probabilities and magnitudes of the effects of mergers involving AI. Greater dialogue and more research, in turn, might help an agency develop a better or more reliable approach to dealing with uncertainty, including remedies.

267. There are several ways in which an agency might better equip itself to deal with the uncertainty surrounding merger analysis involving a rapidly evolving technology like foundation models or artificial intelligence more generally. Continuous learning and monitoring of the evolution of the FM/AI ecosystem is critical. A typical merger investigation can last more than a year; given the fast-moving nature of FMs, it is important that an agency not waste any time unnecessarily. One way to do that is to engage in continuous learning, whether through market studies or other means.[76] As the FM ecosystem evolves over time, some uncertainties may resolve themselves, which might shed some light on whether prior judgments about the effects of a merger were warranted. This, in turn, may help an agency learn how to better predict the effects of a merger involving FMs. If the passage of time suggests that prior judgments were incorrect,

---

Court's decision in *Brown Shoe*, 370 U.S. at 325. In other words, whether a particular probability of harm is reasonable may depend on its magnitude. A twenty percent chance of an extremely small harm is categorically different than a twenty percent chance that a substantial market will be monopolized for decades, and there is no reason a court should have to ignore that difference in assessing what risks of harm are "reasonable."

75 *Cf.* Jennejohn, *supra* note 69, at 172 ("The uncertainty inherent in the modern antitrust analysis of mergers undermines the M&A market by hamstringing merging parties' ability to assess risk ex ante.").

76 It might be desirable for an agency to publish market studies or ex post evaluations of its analysis of mergers involving FMs or AI more generally. Although an agency might not have an incentive to publish such studies when they suggest the agency improperly evaluated a merger or the evolution of an industry, publication would advance the public's understanding of the strengths and weaknesses of the techniques and tools used by the agency, which may, in turn, promote the development of better approaches to evaluating mergers in the face of great uncertainty.

it might also call for remedial action to undo harm arising either from failing to challenge a harmful merger or challenging a beneficial merger. If the resolution of uncertainty with the passage of time suggests a remedy was unnecessary or unduly harsh (or perhaps not stringent enough), then it might be appropriate for the agency to modify its remedy. In short, the uncertainty associated with the rise of new technology like FMs or AI more generally puts a premium on continuous learning and the flexibility to respond to changing conditions.

V. Conclusion

268. Merger analysis is an inherently uncertain exercise, and that uncertainty often is heightened when a merger involves a new, dynamic, and innovative technology like AI. The introduction of a new technology like foundation models may upset established markets, create new markets, and generate new business models and strategies. Predicting the effect of a merger involving a new technology on competition can be a fraught exercise. We believe that competition authorities should address this uncertainty head on by expressly accounting for it and incorporating it into their decision-making processes. The analysis of uncertainty need not be especially sophisticated nor even quantitative in nature to be effective; it can be as simple as assessing the relative probability that a merger will be anticompetitive and the relative magnitude of its effects. Only by grappling with these probabilities explicitly and consciously can antitrust enforcers hope to contend effectively with mergers in an industry as dynamic and rapidly changing as artificial intelligence.

PART II
AI Challenges for Competition Law

What is the Relevant Product Market in AI?

LAZAR RADIC[*] AND KRISTIAN STOUT[**]

IE Law School, Madrid
International Center for Law & Economics, Portland

Abstract

AI has taken the world by storm, and competition law is no exception. Policymakers, academics, and commentators are struggling to make sense of how to apply competition law principles to burgeoning AI markets. The question is spurred by an impending sense that inaction is likely to lead to monopolistic outcomes that will later be impossible to revert. What is feared is that AI will become dominated by a few large technology companies and, more spuriously, that these will be the same companies that already control vast swathes of the so-called digital sphere. In other words: it will make big tech even bigger.

One difficulty with this narrative, however, is that, strictly speaking, there is no such thing as an "AI market" because AI is not a unitary, monolithic technology. Against this backdrop, this chapter argues that the first step to ensuring that antitrust law stays relevant in the age of AI is developing a principled approach to defining AI relevant markets; one that is legally, economically, and technologically sound. Relevant market definition is central to antitrust law because it is the starting point of most, if not all antitrust law cases. A relevant product market is typically comprised of all those products which consumers view as substitutable and which can therefore be said to compete against each other. By delineating the boundaries

---

[*] Senior Scholar for Competition Policy, International Center for Law & Economics and Assistant Professor of Law, IE Law School.

[**] Director of Innovation Policy, International Center for Law & Economics.

of competition between firms, relevant product market definition alerts of the presence of market power and, by extension, of the likelihood of anticompetitive effects. Despite its limitations and despite not being an end in itself, relevant product market definition is, and likely will remain, the main tool for thinking about the contours of competition between firms for the foreseeable future.

We suggest how a principled approach to relevant product market definition in AI can be developed by grasping the internal heterogeneity of AI and by understanding what makes AI similar to HI and HI-powered tasks, thus eschewing simplistic narratives about AI's supposed ubiquity and uniqueness that are bound to impede antitrust law from discharging its social role, which is to protect competition for the ultimate benefit of consumers.

I. Introduction

269. Artificial Intelligence (AI) has taken the world by storm, and competition law[1] is no exception.[2] Policymakers, academics, and commentators are struggling to make sense of how to apply competition law principles to burgeoning AI markets.[3] The question is spurred by an impending sense that inaction is likely to lead to monopolistic outcomes that will later be impossible to revert.[4] What is feared is that AI will become dominated by a few large technology companies and, more spuriously, that these will be the same companies that already control vast swathes of the so-called digital sphere.[5] In other words: it will make Big Tech even bigger.

270. One difficulty with this narrative, however, is that, strictly speaking, there is no such thing as an "AI market"[6] because AI is not a unitary,

---

1   The terms "competition law" and "antitrust law" will be used indistinctly throughout this article.

2   The "hype" surrounding AI has loomed over the field of competition law for some time now. See, for instance, Nicolas Petit, *Antitrust and Artificial Intelligence: A Research Agenda*, 8 J. COMPETITION L. & PRAC. 6, 361 (2017) (noting that the "hype" surrounding AI had reached the antitrust community in 2016).

3   The many contributions in this book are a prime example.

4   *See, e.g.*, CPI, *AI Boom to Fuel Anticompetitive Behavior in Big Tech, Warns German Antitrust Chief*, CPI (June 26, 2024), https://www.pymnts.com/cpi-posts/ai-boom-to-fuel-anticompetitive-behavior-in-big-tech-warns-german-antitrust-chief/; Margarethe Vestager, Speech, Competition in Virtual Worlds and Generative AI (Brussels, June 28, 2024), https://ec.europa.eu/commission/presscorner/detail/en/SPEECH_24_3550 (arguing that "now is the time to act"); Indeed, much of this is born out of antipathy to the growth of the internet ecosystem in general, and a fear that the "harm" that arose from the growth of early internet companies will recur. *See, e.g.*, Rana Foroohar, *The Great US-Europe Antitrust Divide*, FIN. TIMES (Feb. 5, 2024), https://www.ft.com/content/065a2f93-dc1e-410c-ba9d-73c930cedc14 (FTC Chair Lina Khan opining that "we are still reeling from the concentration that resulted from Web 2.0, and we don't want to repeat the missteps of the past with AI.")

5   The concerns of competition authorities over AI are well summarized in an op-ed by *The Economist*. "Broadly speaking, the authorities have two areas of concern. The first is whether the world's biggest companies are trying to tie businesses into their products in anticompetitive ways. The second is about control: are some of the largest generative-AI investments poorly disguised acquisitions intended to sidestep antitrust consideration?" Schumpeter, *Is Artificial Intelligence Making Big Tech too Big?*, THE ECONOMIST (June 23, 2024) https://www.economist.com/business/2024/06/23/is-artificial-intelligence-making-big-tech-too-big.

6   *Impact Assessment Accompanying the Proposal for a Regulation of the European Parliament and of the Council laying down harmonised Rules on Artificial Intelligence (Artificial Intelligence Act) and Amending Certain Union Legislative Acts*, COM (2021) 206 final, at 27, https://artificialintelligenceact.eu/wp-content/uploads/2022/06/AIA-COM-Impact-Assessment-1-21-April.pdf. ("This risk is significantly higher when the AI market is fragmented with individual Member States taking unilateral actions." (emphasis added))

monolithic technology.[7] On the contrary, the "AI Stack" is made up of several layers of technology with vastly different features that result in an almost infinite range of potential business and end-user functionalities. Granted, those calling for stringent preemptive antitrust intervention in AI are not oblivious to this fact. For example, Andreas Mundt, the head of Germany's Bundeskartellamt, worried that there is a "great danger that we'll see an even deeper concentration of digital markets and power increases *at various levels*, from chips to the front end where users interact with tech platforms."[8] Similarly, Margrethe Vestager, the EU's Competition Commissioner differentiated between foundational models and the rest of the value chain.[9] Indeed, in other regulatory contexts, lawmakers often differentiate between so-called "high risk" and "limited risk" AIs,[10] such as between AI techniques used to guide weapons systems or predictive policing on the one hand, and those used to power consumer products like search engines or cars, on the other. While this categorization is crude, it demonstrates that lawmakers are aware that "AI" as a regulatory concept needs to be decomposed into categories that are useful in the real world.

271. However, and despite these caveats, the full complexity of AI still eludes the antitrust community. Some – maybe even most – of these limitations stem from the habitual challenge of keeping theoretical and conceptual frameworks up to date with brisk technological developments. This is especially apposite in the field of AI, however, which has blossomed at break-neck speed in a variety of directions and across a plethora of fields. For example, as Jonathan Barnett points out in another contribution to this book, there is still no settled definition of "foundation model."[11] The European Commission's 2021 proposal for an AI Act did not even cover foundation models,[12] which shows just how quickly the market is evolving. Our understanding of AI is bounded in other fundamental ways, such as by the lack of clarity surrounding the relation between human cognition and the cognitive tasks that AI is increasingly capable

---

7   *See, e.g.*, Benedict Evans, *The Problems of AI Ethics*, BENEDICT EVANS (Mar. 23, 2024) https://www.ben-evans.com/benedictevans/2024/3/23/the-problem-of-ai-ethics-and-laws-about-ai (arguing that it is pointless to think about the ethics of AI or "regulating AI" because AI is not one thing, but a changing bundle of technology with many different uses which raise a plethora of different questions).

8   CPI, *supra* note 4 (emphasis added).

9   Vestager, *supra* note 4.

10 *See, e.g.*, Commission Regulation (EU) 2024/1689 of the European Parliament and of the Council of June 13, 2024 laying down harmonised rules on artificial intelligence and amending Regulations (EC) 300/2008, (EU) 167/2013, (EU) 168/2013, (EU) 2018/858, (EU) 2018/1139 and (EU) 2019/2144 and Directives 2014/90/EU, (EU) 2016/797 and (EU) 2020/1828 (Artificial Intelligence Act), at art. 6.

11 Jonathan M. Barnett, *The Case Against Preemptive Antitrust in the Generative Artificial Intelligence Ecosystem*, in 9 ARTIFICIAL INTELLIGENCE AND COMPETITION POLICY (Alden Abbott and Thibault Schrepel eds., Concurrences 2024).

12 Thibault Schrepel, *Decoding the AI Act: A Critical Guide for Competition Experts* 4–5 (AMSTERDAM L. & TECH. INST., Working Paper No. 3-2023, 2023), https://papers.ssrn.com/sol3/papers.cfm?abstract_id=4609947.

of performing.[13] Much of this ignorance stems from the fact that we do not know that much about how the human mind works, in the first place – let alone AI.[14]

272. It is naïve, some would even say hubristic, to expect antitrust law to solve most – or even *any* – of the major theoretical conundrums surrounding AI. As a mostly reactive discipline that functions through piecemeal doctrinal progression and the incorporation of mainstream insights from other fields – so far mostly from economics and industrial organization – the best antitrust law can hope to do is observe, learn, and adapt. The trillion-dollar question is how.

273. In this chapter, we argue that the first step to ensuring that antitrust law stays relevant in the age of AI is developing a principled approach to defining AI relevant markets; one that is legally, economically, and technologically sound. Relevant market definition is central to antitrust law because it is the starting point of most, if not all antitrust law cases.[15] A relevant product market typically comprises those products which consumers view as substitutable and which can therefore be said to compete against each other. By delineating the boundaries of competition between firms, relevant product market definition alerts to the presence of market power and, by extension, of the likelihood of anticompetitive effects.[16] Despite its limitations and despite not being an end in itself,[17] relevant product market definition is, and likely will remain, the main tool for

---

[13] *See, e.g.*, Oren Etzioni, *AI's Progress Isn't the Same as Creating Human Intelligence in Machines,* MIT TECH. REV. (June 28, 2022), https://www.technologyreview.com/2022/06/28/1054270/2022-innovators-ai-robots/. That is despite claims that AGI is on the horizon, *see* Leopold Aschenbrenner, *Situational Awareness: The Decade Ahead* (2024), https://situational-awareness.ai/, the mechanisms of what makes humans "generally intelligent" still elude us. Thus, the very nature of the task that current AIs perform similarly eludes a functional definition of how it is related to human-level intelligence. It is entirely possible that the cognitive tasks we are modeling are as to full human consciousness as having a thumb is to being an artist. The artist may employ his hands to make a beautiful work of art, or may direct automated tools, but the art is not in the thumb or the tool but in the direction of those tools. Similarly, we do not yet know whether the "thinking" the machines do is more like a tool or is somehow the first step toward consciousness, but it is entirely possible that these systems remain highly sophisticated tools with no internal consciousness.

[14] *Researchers Are Figuring Out How Large Language Models Work*, THE ECONOMIST (July 11, 2024), https://www.economist.com/science-and-technology/2024/07/11/researchers-are-figuring-out-how-large-language-models-work ("Because LLMs are not explicitly programmed, nobody is entirely sure why they have such extraordinary abilities ... LLMs really are black boxes."); *see also* ANIL ANANTHASWAMY, WHY MACHINES LEARN: THE ELEGANT MATH BEHIND MODERN AI 7–25 (2024) (Describing how, despite advances in the ability of machines to recognize patterns, the basic learning mechanisms of even simple biological creatures continue to elude AI researchers).

[15] In this chapter, we focus exclusively on relevant product markets. We do not discuss relevant geographic markets or relevant temporal markets.

[16] Jonathan B. Baker, *Market Definition: An Analytical Overview*, 74 ANTITRUST L.J. 1, 129 (2007) ("Throughout the history of U.S. antitrust litigation, the outcome of most cases has surely turned on market definition than on any other substantive issue. Market definition is often the most critical step in evaluating market power and determining whether business conduct has or likely will have anticompetitive effects."); *see, e.g.*, in the US, Eastman Kodak Co. v. Image Technical Servs. Inc., 504 U.S. 451, 469 n. 15 (1992) ("Because market power is often inferred from market share, market definition generally determines the outcome of the case").

[17] *See, e.g.*, Magali Eben, *The Antitrust Market Does Not Exist: Pursuit of Objectivity in a Purposive Process*, 17 J. COMPETITION L. & ECON. 3 586, 567 (2021). It is important to recognize that a dogmatic adherence to relevant market definition could, for instance, discount the pressure exercised by products defined as falling outside. *See, e.g.*, Case 1275-1276/1/12/17, Flynn Pharma Ltd v. Competition and Markets Authority [2018] CAT 11, ¶ 119. ("It is fallacious to regard as relevant to the competition analysis only those products defined as falling within the relevant market and to disregard entirely any competitive pressure from those products defined as falling outside it."), *see* Ian Giles & Clio Angeli, *The UK Competition Appeal Tribunal partly annuls*

thinking about the contours of competition between firms for the foreseeable future.[18]

274. At the time of writing, however, two interrelated issues stand in the way of a constructive understanding of relevant product market definition in AI. The first is the lack of sophistication in demarcating internal product market boundaries. Currently, the antitrust community seems to treat AI more like the enigmatic black monolith from Stanley Kubrick's 2001: A Space Odyssey (or Arthur C. Clarke's eponymous novel)[19] than the heterogenous, loosely connected "bundle" of technologies that it is. For example, in a recent joint statement, leading antitrust enforcement authorities identify shared competitive concerns across generative AI foundation models and "AI products."[20] But this is like making blanket statements about potential competitive problems in "food markets" or "technology." It is unclear what this level of generality adds to the conversation, or what can be gleaned from it except that "competition is good" – which can also be said about virtually any market that is not a natural monopoly. Similarly, in her speech, Margrethe Vestager mentioned that, if they were allowed to gain control of vital parts of the AI value chain, large tech companies could foreclose "AI competitors."[21] But competitors to whom, or to what? Does the provider of Large Language Models (LLMs) compete with a company developing computer vision solutions? In theory, both are active in AI. Take a set of products that are more similar: do autonomous drones compete with self-driving cars? Arguably, the core AI developed in those systems will be similar in many respects, yet the products and relevant consumer groups are highly distinct. Or what about AI developed to help radiologists sort through X-Rays and AI systems that help with protein folding problems in medical research? Both are using very similar core technologies, but for very different purposes and for different types of users.

275. The discussion is not purely academic, either. Other authors have pointed to how lumping different AI systems under the same regulatory framework can lead to questionable outcomes, such as unwittingly

---

the Competition Authority's decision that pharmaceutical companies abused their dominant position by setting excessive and unfair prices for an epilepsy drug (Pfizer / Flynn), e-COMPETITIONS June 2018, art. No. 90012; see also RICHARD WHISH & DAVID BAILEY, COMPETITION LAW 24 (10th ed. 2021). ("People must not be seduced by numbers [or relevant market shares] when determining whether a firm has market power").

18 For an opposite view, see, e.g., Daniel A. Crane, *Market Power Without Market Definition*, 90 NOTRE DAME L. REV. 1, 31 (2014–2015). Ten years after the publication of this paper, however, market definition remains the principal tool for inferring market power. For a defense of relevant market definition in antitrust law given managed expectations concerning its utility and function. See Eben, *supra* note 17.

19 Arthur C. Clarke, *2001: A Space Odyssey*, (Orbit & Abacus 1990).

20 U.S. Dep't of Justice & Fed. Trade Comm'n, *Joint Statement on Competition in Generative AI Foundation Models and AI Products*, (July 24, 2024) (issued by European Commission, UK Competition and Markets Authority (CMA), https://competition-policy.ec.europa.eu/about/news/joint-statement-competition-generative-ai-foundation-models-and-ai-products-2024-07-23_en.

21 Vestager, *supra* note 4.

favoring one technology over another and thereby stifling innovation.[22] In a similar vein, a flawed understanding of the limits of the several relevant product markets that make up AI is likely to lead to erroneous inferences about market power and to misconstrue competitive dynamics in the AI value chain – leading to sub-optimal antitrust enforcement.

276. The second hurdle to an effective antitrust approach to relevant market definition in AI is the misconception that AI does not compete with non-AI technology. This, too, is only partially true, and is most likely based on the more general perception that AI is idiosyncratic. However, AI was inspired by the human brain and designed with the specific goal of performing tasks that hitherto required the exertion of human intelligence.[23] While the intelligence of AI is not yet at the level of an average human adult,[24] it can already carry out a range of tasks that have traditionally been performed by humans, such as planning, personal assistance, classification, writing texts, composing songs, producing images, making videos, coding, research, strategizing, negotiating, designing ads – to name just a few. Often, the results are nigh indistinguishable from output generated by human intelligence (HI).[25] Part of this is arguably because AI acquires "knowledge" in a manner not too different from the way humans do; that is, through induction, deduction and, above all, the recognition of patterns. In fact, AI has even been known to reproduce human biases, such as uncritically citing papers in the field that are already highly cited or sputtering plausible-sounding gibberish. Just like humans sometimes do.[26]

277. And yet we often hear from enforcers and commentators that AI is unique. Some maintain that it is uniquely dangerous because it perpetuates

---

22 Schrepel, *supra* note 12, at 10–11 (arguing that the AI Act indirectly sanctions AI systems that are easier to control, thus effect tilting the market in favor of AI systems that behave predictably i.e., deterministic AI systems with low or no randomness, over more "creative" or unpredictable AI systems i.e., nondeterministic AI systems with high randomness. And concluding that "neutrality requires imposing different regulatory burdens on different designs").

23 THE ECONOMIST, *supra* note 14. LLMs, for instance, are built using "a technique called deep learning, in which a network of billions of neurons, simulated in software and modelled on the structure of the human brain"; Weijie Zhao, *Inspired but not Mimicking: A Conversation Between Artificial Intelligence and Human Intelligence* NAT. SCI. REV. 9, 3 (2022) ("[AI] is a tool inspired by the human brain and empowered by mathematical and computational methods that can realize multiple intelligent behaviors").

24 Yann LeCun of Meta and Francois Chollet of Google, two respected AI researchers, have said that current AI systems hardly merit being called "intelligence." *A New Lab and a New Paper Reignite an Old AI Debate*, THE ECONOMIST (June 27, 2024) https://www.economist.com/business/2024/06/27/a-new-lab-and-a-new-paper-reignite-an-old-ai-debate. However, according to a recent paper published by Leopold Aschenbrenner, a former OpenAI employee, AI will be as capable as humans at all intellectual tasks by 2027. Leopold Aschenbrenner, *Situational Awareness: The Decade Ahead*, 7 (June 2024) https://situational-awareness.ai/wp-content/uploads/2024/06/situationalawareness.pdf.

25 *See, e.g.*, Vinu Sankar Sadasivan et al., *Can AI-generated Text be Reliably Detected?*, COMP. SCI. (2024), https://arxiv.org/abs/2303.11156 (finding that detectors of LLM-generated text are largely ineffective).

26 An early adopter of this was the Corporate Gibberish Generator, http://www.andrewdavidson.com/gibberish/. What made the software fun and widely disseminated was that it generated phrases that sounded truthful, or at least plausible, despite not meaning anything. This, of course, was a parody of the way businesses sometimes communicate through their spokespeople.

falsehoods, fallacies, and stereotypes.[27] But so do humans. Others argue that AI will be able to perform tasks that humans cannot, either because these tasks are qualitatively distinct or because they operate at a scale impossible to replicate by HI. This vision of the distinctiveness of AI might ultimately be more persuasive. One thing is clear, however: treating AI as entirely distinct from tasks performed by humans severely downplays the extent to which the two are substitutable. Analogously to how the view that AI is monolithic obfuscates its heterogeneity, an excessive focus on what makes AI unique – as opposed to what makes it similar to HI and HI-powered tasks – is bound to result in erroneous conclusions about market power, competitive dynamics, monopoly, and, ultimately, thwart socially-optimal antitrust enforcement.

278. In this chapter, we argue that, in order to overcome these blind-spots, enforcers need to substitute hype and pre-existing biases – whether pro- or anti-enforcement – for careful, evidence-based analysis.[28] Ultimately, we believe that this is a more "future-proof" approach than committing to an enforcement agenda from the outset or regulating AI under blanket rules.[29] We suggest how this can be done by grasping the internal heterogeneity of AI and by understanding what makes AI similar to HI and HI-powered tasks, thus eschewing simplistic narratives about AI's supposed ubiquity and uniqueness that are bound to impede antitrust law from discharging its social role, which is to protect competition for the ultimate benefit of consumers.

279. The chapter is organized as follows. In Section II, we show that AI is internally heterogenous, which complicates any claims about an "AI market." This narrows relevant AI markets for the purpose of competition law. In Section III, we explain the similarities in how AI and HI acquire and reproduce knowledge. We use this as a basis to contend that some of the functionalities of AI are not as unique as they may initially appear and are substitutable for non-AI powered products, services, and inputs (including HI). This expands relevant AI markets for the purpose of competition law. In Section IV, we put forward a tentative set of principles to guide relevant market definition in AI, based on the insights from Sections II and III. We argue that the task of competition authorities will be to understand the different relevant product markets that comprise AI, on the one hand, and to separate the uses of AI reasonably substitutable for

---

27 Mekela Panditharatne & Noah Giansiracusa, *How AI Puts Elections at Risk – And the Needed Safeguards*, BRENNAN CTR. JUST. (July 13, 2023), https://www.brennancenter.org/our-work/analysis-opinion/how-ai-puts-elections-risk-and-needed-safeguards.

28 For example, we would classify blanket claims such as that AI is going to make "all competition problems worse" (CPI, *supra* note 4) as based on hype rather than evidence. There is no indication of how AI will affect competition, or whether the effect will be negative, neutral, or positive. "AI hype" is not exclusive to competition law, either. *See e.g.*, Anna Cooban, *AI Investment is Booming, How much is Hype?* CNN (July 23, 2023, 05:01 AM), https://edition.cnn.com/2023/07/23/business/ai-vc-investment-dot-com-bubble/index.html.

29 The former is what certain competition authorities seem to be doing. *See* Joint Statement, *supra* note 20; with the latter, we are referring to initiatives such as the AI Act. *See* Commission Regulation (EU) 2024/1689 (Artificial Intelligence Act).

tasks performed by HI from those that are not, on the other. To achieve this, enforcers should be guided by three questions:
- Who are the consumers?;
- What is the Product?; and
- Does AI Fundamentally transform a comparable product or service?

280. Section V concludes.

II. What Makes AI Markets Internally Heterogeneous

281. AI is not monolithic. On the contrary, the "AI Stack" is made up of several layers of technology with vastly different characteristics that result in an almost infinite range of potential business and end-user functionalities. As the National Security Commission on Artificial Intelligence has observed:

> AI is not a single technology breakthrough .... The race for AI supremacy is not like the space race to the moon. AI is not even comparable to a general-purpose technology like electricity. However, what Thomas Edison said of electricity encapsulates the AI future: "It is a field of fields ... it holds the secrets which will reorganize the life of the world." Edison's astounding assessment came from humility. All that he discovered was "very little in comparison with the possibilities that appear."[30]

282. It is overstated to claim that AI is not a general-purpose field, but directionally, the Security Commission is correct. AI is in fact a diverse collection of different techniques and technologies that are deployed to handle different tasks across many different industries. First, it is important to note that when we talk about AI we are not talking about "general artificial intelligence" (genAI), nor what we normally see in science fiction movies. Notwithstanding predictions that genAI is just around the corner,[31] we do not currently have anything operating on that level, but instead have a collection of technologies that depend upon statistical analysis to approximate human-like intelligence.[32]

283. More practically, understanding how the "AI Stack" works at a high level can help us begin to understand the difficulty with defining broad product markets around heterogeneous technologies. Arguably, the first layer to consider is hardware (semiconductors and raw computing hardware) and "XaaS" services. XaaS is an umbrella term for providers of virtualization

---

30 ERIC SCHMIDT ET AL, NATIONAL SECURITY COMMISSION ON ARTIFICIAL INTELLIGENCE, FINAL REPORT 7 (2021), https://www.dwt.com/-/media/files/blogs/artificial-intelligence-law-advisor/2021/03/nscai-final-report--2021.pdf.

31 *See, e.g.*, Aschenbrenner, *supra* note 13.

32 *See generally* THE STANDING COMMITTEE, ARTIFICIAL INTELLIGENCE AND LIFE IN 2030, ONE HUNDRED YEAR STUDY ON ARTIFICIAL INTELLIGENCE (2016), https://arxiv.org/pdf/2211.06318; *see also generally* ANANTHASWAMY, *supra* note 14.

284. Next, there is the data layer. The foundation of any AI system lies in the data it is trained on. Data can be categorized into structured data, which is highly organized and easily searchable in databases (e.g., spreadsheets with rows and columns), and unstructured data, which lacks a predefined format (e.g., text, images, videos).[34] The quality and quantity of this data crucially defines the efficacy for the performance of the ultimate AI systems. Data collection involves gathering relevant information from various sources, while data preparation includes cleaning (removing noise and inconsistencies), transforming (converting data into a usable format), and labeling (tagging data with appropriate labels for supervised learning). As the saying goes, garbage in, garbage out: This initial stage is fundamental because the accuracy and efficacy of an AI system are directly correlated with the quality of its training data. Thus, the techniques not just for gathering data, but curating (or even generating it in the case of synthetic data), is an enterprise unto itself.

or abstraction services for on-demand access to storage, processing, and a variety of types of software.[33]

285. Model training, on the other hand, is the process by which AI systems learn from data.[35] Several techniques are employed in this phase, each serving different purposes and applications. Supervised learning involves training the model on labeled data, where the input-output pairs are known, and are used for tasks like classification and regression.[36] By contrast, unsupervised learning involves training on unlabeled data, where the model must identify patterns and relationships within the data itself, commonly used for clustering and dimensionality reduction tasks.[37] Reinforcement learning stands out as a method where models learn through trial and error, making sequential decisions and receiving rewards or penalties based on their actions.[38] This is particularly effective for dynamic and complex decision-making environments.[39] Transfer learning, another critical approach, involves adapting a pre-trained model on one task to perform a related task, significantly reducing the resources

---

33   Romit Dey & George Korizis, *How Anything-As-A-Service (XaaS) Can Help Reinvent Business Models and Transform Outcomes Across Industries*, PWC https://www.pwc.com/us/en/services/consulting/business-transformation/library/use-xaas-to-reinvent-business-models.html (last visited July 31, 2024).

34   *See, e.g., Structured vs Unstructured Data*, IBM (June 29, 2021), https://www.ibm.com/think/topics/structured-vs-unstructured-data; Dongdong Zhang et al., *Combining Structured and Unstructured Data for Predictive Models: A Deep Learning Approach*, BMC MED. INFORMATICS DECISION MAKING 20, 280 (2020), https://link.springer.com/article/10.1186/s12911-020-01297-6 (describing generally the use of both structured and unstructured data in predictive models for health care).

35   ANANTHASWAMY, *supra* note 14, at 12.

36   *See Id.* at 12–13.

37   *See Id.* at 18–25.

38   *Id.*

39   *See Id.* at 24–25 (Describing how unsupervised learning models excel at finding solutions to multi-variate problems involving linear relations, such as learning how to predict housing prices based on a training set with many data points).

required compared to training a model from scratch.[40] Once again, training of models is an enterprise unto itself, involving a variety of hardware[41] and software components[42] and firms with different levels of specialization in each of these tasks.

286. Once trained, AI models must be deployed in an environment where they can operate and provide value. The deployment phase can occur in various environments, each with its own set of advantages and considerations. Cloud deployment involves hosting models on cloud platforms, providing scalability and easy access to computational resources, making it ideal for handling large-scale data and serving a global user base.[43] Edge deployment, on the other hand, involves placing models on local devices or edge servers closer to the data source, reducing latency and bandwidth usage, which is crucial for real-time applications like autonomous vehicles and IoT devices.[44] Finally, on-premises deployment entails hosting models on an organization's internal servers, offering greater control over data security and compliance, particularly important in industries with stringent data protection regulations such as healthcare and finance.[45] Each of these deployment methods is managed by different firms operating under different business models. For instance, ChatGPT deploys its service on cloud infrastructure but only allows consumers to access their trained models through a subscription front end. Hugging Face is an open source project that hosts a wide variety of models, and allows individuals to deploy them in any environment, including either locally[46] or online in a cloud.[47]

287. Beyond the stack, there is heterogeneity in how AI is deployed and employed. While large LLMs have garnered significant attention for their capabilities in natural language processing, it is crucial to recognize that the field of AI encompasses a multitude of other technologies, each excelling in different domains. Computer vision models, for example, utilize Convolutional Neural Networks to interpret and analyze visual data.[48]

---

40 *What is Transfer Learning?*, AWS, https://aws.amazon.com/what-is/transfer-learning/ (last visited July 31, 2024).

41 *See, e.g.*, *TPUs vs. GPUs: What's the Difference?*, PureStorage (May 2, 2024), https://blog.purestorage.com/purely-educational/tpus-vs-gpus-whats-the-difference/; Josh Schneider & Ian Smalley, *What is a Field Programmable Gate Array (FPGA)?*, IBM (May 8, 2024), https://www.ibm.com/think/topics/field-programmable-gate-arrays.

42 *See, e.g.*, *Why TensorFlow*, TensorFlow, https://www.tensorflow.org/about (last visited July 31, 2024); *Learn the Basics*, PyTorch, https://pytorch.org/tutorials/beginner/basics/intro.html (last visited July 31, 2024); *About Keras 3*, Keras, https://keras.io/about/ (last visited July 31, 2024). These software packages are often developed as a complex mix of open-source and proprietary efforts that cross-inform each other.

43 *See, e.g.*, *Amazon SageMaker*, AWS, https://aws.amazon.com/sagemaker/ (last visited July 31, 2024).

44 *What is Edge AI?*, IBM, https://www.ibm.com/topics/edge-ai (last visited July 24, 2024).

45 *See, e.g.*, *NVIDIA ChatRTX*, NVIDIA, https://www.nvidia.com/en-us/ai-on-rtx/chatrtx/ (last visited July 31, 2024).

46 *See, Stable Diffusion Web UI*, GitHub, https://github.com/AUTOMATIC1111/stable-diffusion-webui (last visited July 31, 2024).

47 You can, for example, run a number of models from Hugging Face in Google Collab notebooks. *See, Transformers Notebooks*, Hugging Face, https://huggingface.co/docs/transformers/en/notebooks (last visited July 31, 2024).

48 *Convolutional Neural Networks (CNNs), Deep Learning, and Computer Vision*, Intel, https://www.intel.com/content/www/us/en/internet-of-things/computer-vision/convolutional-neural-networks.html (last visited July 31, 2024)

These models are indispensable in applications like medical imaging, where they assist radiologists in detecting abnormalities in X-rays and MRIs, and in autonomous vehicles, where they enable the car's system to recognize and respond to road conditions and obstacles.[49]

288. As noted above, reinforcement learning is another pivotal AI technology, distinct in its approach to training models through a system of rewards and penalties. This technique is particularly effective in environments requiring sequential decision-making and adaptability. For instance, reinforcement learning has been instrumental in developing advanced robotics, where machines learn to perform complex tasks by optimizing their actions based on continuous feedback. A prominent example is AlphaGo, which mastered the game of Go, a strategy board game, by learning from millions of games and refining its strategy through self-play.[50] Similarly, reinforcement learning is being applied in dynamic resource management and real-time strategy games, showcasing its versatility and effectiveness in scenarios where adaptability and strategic planning are paramount.[51]

289. Moreover, Graph Neural Networks (GNNs) represent a burgeoning area of AI that excels in handling data structured as graphs.[52] These models are particularly adept at capturing relationships and interactions within complex networks, making them invaluable in fields such as social network analysis, molecular biology, and recommendation systems.[53] For example, GNNs can analyze social media connections to identify influential users or detect communities, and in molecular biology, they can predict the properties of molecules based on their structural relationships, aiding in drug discovery.[54]

III. What Makes AI Similar to Non-AI Products and Services

290. In early history, the humanoid giant Talos of Greek mythology and the Golem of Jewish folklore exemplified one of humanity's longest-standing fantasies: a helper capable of performing tasks with the same proficiency as a human.

---

49 *See,* Georgios Kourounis et al., *Computer Image Analysis with Artificial Intelligence: A Practical Introduction to Convolutional Neural Networks for Medical Professionals,* 99 Postgraduate Med. J. 1178 (2023).

50 *See, Alpha Go,* Google DeepMind, https://deepmind.google/technologies/alphago/ (last visited July 31, 2024)

51 *See,* Ying Chen et al., *Deep Reinforcement Learning-Based Dynamic Resource Management for Mobile Edge Computing in Industrial Internet of Things,* 17 IEEE Transactions Ind. Informatics 4925 (2021), https://ieeexplore.ieee.org/document/9214878; *see also,* Harshit Sethy, Amit Patel & Vineet Padmanabhan, *Real Time Strategy Games: A Reinforcement Learning Approach,* 54 Procedia Comput. Sci. 257 (2015), https://www.sciencedirect.com/science/article/pii/S1877050915013354X.

52 Rick Merritt, *What Are Graph Neural Networks?,* Nvidia (Oct. 24, 2022), https://blogs.nvidia.com/blog/what-are-graph-neural-networks/.

53 *Id.*

54 *Id.*

291. Today, AI can carry out tasks which have typically required the exertion of HI,[55] to the extent that it is often very difficult to tell if output that has been generated by AI or by "the wonderful computers in our head."[56] According to one paper, detectors of LLM-generated text are woefully incapable of detecting AI-generated content.[57] In turn, universities and other institutions are putting in place onerous disclosure rules about the use of AI; tacitly admitting that detection is impossible – or prohibitively complicated – unless disclosed voluntarily.[58] One possible explanation is that AI and HI-generated output are so similar because the two systems acquire and utilize knowledge in comparable ways. As John Searle famously said, "if you can exactly duplicate the causes, you could duplicate the effects."[59] However, the question of what humans know or can know is a difficult one, a dispute that is further complicated by the introduction of AI. Epistemology has been one of the four pillars of philosophy since the time of the ancient Greeks,[60] and it is unlikely that we will contribute to the debate in any significant way here. Suffice it to say that there are epistemological theories which were (obviously) developed with HI in mind but which nevertheless also appear to fit AI, at least to some extent. According to Kissinger, Schmidt and Huttenlocher, for example, AI is more Wittgenstein than Plato.[61] By this they mean that the "knowledge" of AI results from observing the particular qualities of things.[62]

---

[55] One definition of AI is "the ability of software to perform tasks that traditionally require human intelligence." See Michael Chui et al., *The Economic Potential of Generative AI: The Next Productivity Frontier*, McKinsey Co. 3 (2023), http://dln.jaipuria.ac.in:8080/jspui/bitstream/123456789/14313/1/The-economic-potential-of-generative-ai-the-next-productivity-frontier.pdf.

[56] *At Least 10 Percent of Research May Already be Co-Authored by AI*, The Economist (June 26, 2024), https://www.economist.com/science-and-technology/2024/06/26/at-least-10-of-research-may-already-be-co-authored-by-ai (academic policies on LLM use are in flux. Some journals ban it outright. Others have changed their minds. Up until Nov. 2023, Science labelled all LLM text as plagiarism, saying: "Ultimately the product must come from – and be expressed by – the wonderful computers in our heads").

[57] Sadasivan et al., *supra* note 25.

[58] *See, e.g.*, European Commission, *Living Guidelines on the Responsible Use of Generative AI in Research*, Era Forum Stakeholders Document 6 (2024), (recommending that researchers detail the generative ai tools used in the research process, and how these tools have been used); IE University, *Guidelines for Faculty Use of AI Tools for Academic Work*, (Feb. 2024), https://sites.google.com/view/teachingwithai/ie-statement-policies#h.vl5yg05kvnmq (suggesting that students disclose a list of prompts used and how they were used in producing the relevant output).

[59] John Searle, *Minds, Brains, and Programs*, 3 Behav. Brain Sci. 417, 422 (1980).

[60] The other three are logic, ethics, and metaphysics.

[61] Henry A. Kissinger, Eric Schmidt III & Daniel Huttenlocher, The Age of AI 59 (2021).

[62] As indicated earlier, this is a complex debate. But, oversimplifying, according to Platonic epistemology, humans perceive imperfect reflections of perfect things that exist in the realm of ideas. knowledge is formed by grouping observed phenomena under pre-existing "ideal" or "perfect" categories. For instance, no one has ever seen a perfect circle or a perfect line, yet everyone knows what a perfect circle and a perfect line are. See Plato, Cratylus, ¶ 389. The early enlightenment refined these ideas further, with a continued focus on classifying observable phenomena according to mechanistic, rational rules. Kissinger, Schmidt & Huttenlocher, *supra* note 61, at 61. Kant argued in *The Critique of Pure Reason* that the mind structures incoming sensory "data" according to certain impositions that exist a priori (before experience) and are thus metaphysical. Earlier, Descartes had rejected the notion that a posteriori claims based on observation could be the basis for knowledge because we did not know whether our perceptions were accurate. From the maxim of *cogito ergo sum* – the only principle that could be ascertained as absolutely true – he concluded that material things were knowable not based on sensorial experience, but on their substance. Contrast this with Hume, who broke away from Plato's "substance." See David Hume, A Treatise on Human Nature, Book I, Part I, Sect. VI (1739). In contrast

292. But why does it matter how AI and HI "think" for the purpose of relevant product market definition? Is relevant product market definition not about product, rather than process substitutability – so that the only thing that matters is that two products are interchangeable, regardless of how they are made? Mostly, yes.

293. From the perspective of the consumer – which generally also includes the intermediary or business customer under competition law[63] – two systems that think or operate similarly could reasonably be expected to perform the same, or similar tasks, and thus be substitutable. When a potential buyer is deciding whether to hire a human creative or invest in an AI powered generator of text, for instance, he or she will want to know what each one can do. Since it is impossible to test for the full universe of output which that buyer (or employer) will require in the future, nor is it likely that the buyer even knows what those needs will be, a useful heuristic is to know what the AI or HI is capable of doing. Part of this is gauging how they think. Thus, an employer evaluates a candidate in an interview based on their credentials (the knowledge that has been "put" into him or her), portfolio (previous examples of output), and an interview – which is essentially an intelligence or competence test where the employer tries to estimate how successful a candidate is likely to be at resolving current or future work-related problems. The same applies to AI. The potential buyer might want to know what input the LLM has been "trained" on, would want to see examples of previous output (perhaps some texts, lyrics, or articles – depending on the "job"), and would want to test the AI's ad hoc response to a series of prompts.

294. Does this mean that an understanding of AI and HI epistemology is enough to inform relevant product market definition in AI? No, it is only one tool for understanding substitutability. Furthermore, not all AIs are trained in the same way,[64] which can have implications for what they can

---

to Plato's rationalism (i.e., the idea that knowledge results from absolute, immutable principles that are not learned through experience but implicit in reasoning), Hume argued that knowledge resulted from the observance of similarities. For instance, gold, Hume argued, was just the collection of certain ideas of color, weight, malleableness, fusibility, etc. *Id.* Even the concept of self, Hume contended, is a "heap or collection of different perceptions united together by certain relations and suppos'd, tho' falsely, to be endow'd with a perfect simplicity or identity." *Id.* Part IV, Sect. II. Hume's epistemology laid the basis for successive philosophers to argue, like Wittgenstein did, that knowledge resulted from generalizations about similarities across phenomena, which he termed "family resemblances." KISSINGER, SCHMIDT & HUTTENLOCHER, *supra* note 61, at 49; see also, generally, LUDWIG WITTGENSTEIN, PHILOSOPHICAL INVESTIGATIONS (2009); *See also,* KISSINGER, SCHMIDT & HUTTENLOCHER, *supra* note 61, at 48–49: "In the late twentieth century and early twenty-first, this thinking informed theories of AI and machine learning. Such theories posited that AI's potential lay partly in its ability to scan large data sets to learn types and patterns – e.g., groupings of words often found together, or features most often present in an image when the image was of a cat – and then to make sense of reality by identifying networks of similarities ... even if AI would never know something in the way a human mind could, an accumulation of matches with the patterns of reality could approximate and sometimes exceed the performance of human perception and reason."

63 Case C-377/20, Servizio Elettrico Nazionale SpA v. Autorita Garante della Concorrenza e del Mercato, ECLI:EU:C:2022:379, ¶ 46; In the EU and UK context, see John Vickers, *Competition Policy and the Consumer Welfare Standard*, 1 J. ANTITRUST ENF'T 3–4 (2024) ("In line with the jurisprudence I will take it that [business customers are 'consumers']"), see Anne Wachsmann, Nicolas Zacharie & Daniel Green, *Notion of abuse: The European Court of Justice specifies the criteria to be used to qualify the abuse of a dominant position characterised by exclusionary practices (Servizio Elettrico Nazionale)*, CONCURRENCES No. 3-2022, art. No. 108177.

64 Stefan Feuerriegel et al., *Generative AI*, 66 BUS. INFO. SYS. ENG'G. 111 (2024).

do and how they can do it. The point here is to underscore that there are overlaps between AI and HI which can also help explain why AI and HI can be used to perform the same, or similar tasks, and why the final output is sometimes so difficult to distinguish. Epistemological considerations can also serve to predict and understand the extent of the substitutability between the two. Ultimately, however, the question of whether there is substitutability will depend on the objective characteristics of the product and, above all, whether consumers *view* them as interchangeable.

295. And, sure enough, AI can today perform many tasks which would appear to be substitutable for their HI-powered counterparts in terms of style, performance, and quality.[65] As Feuerriegel et al put it:

> For a long time in history, it has been the prevailing assumption that artistic, creative tasks such as writing poems, creating software, designing fashion, and composing songs could only be performed by humans. This assumption has changed drastically with recent advances in [AI] that can generate new content in ways that cannot be distinguished anymore from human craftsmanship.[66]

296. "Creativity" is no longer a human-only endeavor.[67] There are hundreds of genAI applications that are able to produce text (ChatGPT), images (Dall-E), speech (Speechify), music (Suno), code (GitHub Copilot), and video (Runway). Some of these are generalist (horizontal), while others are specialized (vertical). For example, ChatGPT is a generalist multimodal model ideal for text and image generation, while Reword specializes in writing blogs and SocialBee specializes in social media posts. Other specific applications include marketing, innovation management, scholarly research, and education.[68] Generative AI can "increase efficiency and productivity by automating many tasks that were previously performed by humans, such as content creation, customer service, code generation, etc."[69] What a genAI application is "good" at will depend on the foundation model and input it has been trained with. Thus, some text generators will be better at producing one type of output than another – even within a specialized category – depending on their training and subsequent "fine-tuning."[70] The possibilities borne of different combinations of training,

---

65 *Id.* at 166 (stating that generative AI is a "human-task technology.").

66 Feuerriegel et al., *supra* note 64, at 111.

67 For a discussion on whether there even is such a thing as creativity, *see* Marvin L. Minsky, *Why People Think Machines Can't*, 3 AI MAGAZINE 4, 5. (Arguing that creativity is just a shorthand for better-knitted, ordinary virtues. The upshot is that machines can at least be creative in the same way as humans. In other words, if machines cannot be "creative," neither can humans); *see also*, Feuerriegel et al., *supra* note 64, at 116. (Arguing that while AI was in the past mostly understood to be analytic, whereas now AI has gained the capability to perform generative tasks suitable for content creation by combining elements in novel ways.)

68 Feuerriegel et al., *supra* note 64, at 111, and the literature cited therein.

69 *Id.* at 120.

70 *Id.* at 117. (Arguing that the correctness of generative AI models is highly dependent on the quality of training data and the according leaning process.)

fine-tuning, and the underlying technology used are almost endless. In addition to creating new content based on learned patterns, AI can also be used to assist humans such as, for instance, by checking style, punctuation, grammar, clarity, engagement, and spelling (i.e., Grammarly). If it were human, Grammarly would be an editor. A recent paper showed that a personalized GPT-4 powered model was 82 percent more persuasive than a human in debates, suggesting that AI could also play the part of the debater (or sophist?).[71] AI can also serve to improve not just human output, but input (skills) by exposing HI to new ways of thinking.[72]

297. For instance, all of the tasks outlined above can and still are performed by humans: by a research assistant, a clerical assistant,[73] a musician, a translator, a singer-songwriter, software developer, editor, coach, a blogger, etc. The list of tasks a foundation model can perform, for example, is vast.[74] In time, generative AI applications will likely transition from dispatching mundane tasks to more sophisticated ones, including passing moral judgment[75] (assuming this is not already the case).[76] As a result, the line between HI and AI output is likely to become even more blurred. Some studies even estimate that AI could replace 300 million jobs of knowledge workers, with 900 occupations exposed to some degree of automation by AI (most jobs, however, are likely to be complemented, rather than substituted by AI, the study points out).[77]

298. But let us not understate the counterfactual, either. We have so far talked about instances in which AI and HI might be similar. But there are also important differences between AI and HI.[78] On the one hand, AI can

---

[71] Francesco Salvi et al., *On the Conversational Persuasiveness of Large Language Models: A Randomized Controlled Trial* (Working Paper, Mar. 21, 2024), https://arxiv.org/pdf/2403.14380.

[72] Minkyu Shin et al., *Superhuman Artificial Intelligence Can Improve Human Decision-Making by Increasing Novelty*, 120 PROC. NAT'L ACAD. SCI. 12 (2023).

[73] One study in fact finds that it is this group that will be the most exposed to automation from generative AI. Paweł Gmyrek, Janine Berg & David Bescond, *Generative AI and Jobs: A Global Analysis of Potential Effects on Job Quantity and Quality* (ILO Working Paper No. 96, 2023).

[74] *See*, Thibault Schrepel & Alex Sandy Pentland, *Competition between AI Foundation Models: Dynamics and Policy Recommendations* (AMSTERDAM L. & TECH. INST., Working Paper No. 1-2023, 2023), n. 2, https://papers.ssrn.com/sol3/papers.cfm?abstract_id=4493900. Also, at n. 1, arguing that "recent advances in deep learning have given rise to foundation models that underpin an infinite number of generative AI applications."

[75] Feuerriegel et al., *supra* note 64, at 116.

[76] Eyal Aharoni et al., *Attributions Toward Artificial Intelligence Agents in Modified Turing Test*, 14 SCI. REP. 8458 (2024) (Finding that people rated AI's moral reasoning as superior in quality to humans' along almost all dimensions, including virtuousness, intelligence, and trustworthiness).

[77] Joseph Briggs & Devesh Kodnani, *The Potentially Large Effects of Artificial Intelligence on Economic Growth*, GOLDMAN SACHS (Mar. 26, 2023, 09:05 PM EDT), https://www.gspublishing.com/content/research/en/reports/2023/03/27/d64e052b-0f6e-45d7-967b-d7be35fabd16.html. *See also, id.* at 7, stating that "our estimates intuitively suggest that […] 18 percent of work globally could be automated by AI on an employment-weighted basis."; *see also*, Xiang Hui, Oren Reshef & Loufeng Zhou, *The Short-term Effects of Generative Artificial Intelligence on Employment: Evidence from the Online Labor Market* (CESifo Working Papers, 2023) (Finding that "freelancers in highly affected occupations suffer from the introduction of generative AI, experiencing reductions in both employment and earnings," and that "in the short term generative AI reduces overall demand for knowledge workers of all types").

[78] *See*, in general, Johan Egbert Korteling et al., *Human- versus Artificial Intelligence*, FRONTIERS A.I. (2021); Zhao, *supra* note 23, at 4.

"produce phenomena that are truly new, not simply more powerful or efficient versions of things past."[79] This could come about, for instance, through a qualitatively different functionality that cannot be replicated by human capabilities at all, such as a way of processing data that is fundamentally alien to humans. Or it could come about as a result of compressing the time frame in which results are achieved to an extent that renders tasks performed by humans obsolete, or relegates them to a different category (e.g., AI's fast data processing might be able to automate output personalization across various product categories with a speed that is impossible to replicate by HI, thus creating a new type of product altogether).[80] Indeed, signals from AI systems propagate much quicker than human nerves, which operate at the speed of at most 120/ms.[81] Similarly, the amount of cognitive information we can retain and our ability to "multi-task" is severely limited.[82] Ultimately a sufficiently large quantitative improvement can become indistinguishable from a qualitative leap. Furthermore, an AI-enabled reduction in production costs could lead to new capabilities, not just increased output.

299. For instance, according to one study, robotic process automatization will not only improve handcrafted processing rules but "enable entirely new types of automatization by retrofitting and thus intelligentizing legacy software."[83] The authors anticipate the development of a new generation of process guidance systems in business process management:

> While traditional system designs are based on static and manually crafted knowledge bases, more dynamic and adaptive systems are feasible on the basis of large enterprise-wide trained language models (internal citations omitted).[84]

300. By the same token, it could be that AI is incapable (or incomparably worse) at performing tasks that can be easily carried out by HI, thus limiting substitutability in exactly the opposite direction. In 1988, Moravec wrote:

> It is comparatively easy to make computer models exhibit adult level performance on intelligence tests or playing checkers, and difficult or impossible to give them the skills of a one-year-old when it comes to perception and mobility.[85]

---

79 Chui et al., *supra* note 55, at 50.

80 Think, for example, of advertising, summarizing web pages for mobile devices, creating songs or playlists, generating relevant instant social media feeds, or even adapting political narratives instantaneously depending on the audience. These are all things that HI is capable of, it just generally requires more time. In addition, humans can retain less information that AI due to memory constraints.

81 Korteling et al., *supra* note 78, at 4–5. But note, for instance, that human brains are millions of times more efficient in energy consumption than computers. *Id.*

82 *Id.* at 4.

83 Feuerriegel et al., *supra* note 64, at 118, and the papers cited therein.

84 *Id.*

85 Hans Moravec, MIND CHILDREN (1988).

301. As Korteling et al have pointed out, "Moravec's Paradox implies that HI and AI are intelligent in different ways."[86] Generally, people are better at carrying out a broader spectrum of cognitive and social tasks under uncertainty.[87] Recent developments in image recognition however challenge this view. Foundation models especially have been trained to recognize images with deep learning technology, "which is based on some principles of biological neural networks."[88] As a result, one of the major "weaknesses" (or differences) of AI compared to HI, i.e., the difficulty of interpreting human language, symbolism and context – which typically requires an extensive frame of reference[89] – is potentially being bridged.

302. There is no hard and fast rule, and substitutability will need to be studied on a case-by-case basis for every AI product or service. But, at the most general level, the question that competition law will have to answer is which type of AI are we dealing with? The type that is reasonably substitutable for human capabilities, or the type that is not. In the end, this is what is going to inform the boundaries of relevant product market definition.

IV. Tentative Principles for Market Definition in AI

303. Jonathan Barnett has observed that in the genAI market, a preemptive approach to antitrust is dangerous to consumer welfare:

> At the early stages of a market's development, uncertainty concerning the competitive effects of certain business practices is likely to be especially high, which supports concerns that preemptive intervention would result in significant false-positive error costs by potentially suppressing practices that are either innocuous or yield procompetitive efficiencies. This suppressive effect arises both by constraining existing practices and limiting the future range of transactional innovation. Hence, without grounds to anticipate a future anticompetitive outcome at a sufficiently high level of confidence, there would seem to be a strong presumption against preemptive intervention in the genAI market at its current nascent stage of development. It remains to consider whether there are sufficiently compelling factual or other grounds to overcome that presumption.[90]

304. Barnett's observation is part of a larger exhortation to enforcers to refrain from jumping to antitrust remedies for feared anticompetitive harms that have yet to materialize. Barnett is correct in general, and specifically if

---

86 Korteling et al., *supra* note 78, at 6. *See also* Anathaswamy's description of the intelligence of a duckling, and how even that basic level of automatic learning continues to elude computer scientists. ANANTHASWAMY, *supra* note 14, at 7–8.

87 *Id.* at 7.

88 *Id.*

89 *Id.* at 7; Zhao, *supra* note 23, at 4.

90 Barnett, *supra* note 11, at 8.

we apply his observations to market definition for AI; as not only are the relevant AI markets nascent, as Barnett notes, but what will even count as a relevant product is as-yet undefined.[91] As we have repeated above: there is no single "AI" technology, even within areas we currently think of as unified. In Barnett's paper, for example, he notes numerous possible divisions within antitrust-relevant markets just for generative AI.[92]

305. What is needed at the moment, therefore, is not an authoritative definition of what counts as a relevant AI market for antitrust purposes, but a good set of questions that can be asked to help understand when a relevant market emerges.

306. In competition law, a relevant market comprises the relevant product market and the geographic market. In this chapter, we have focused on the former. An authoritative competition law textbook defines a relevant product market thus:

> The definition of the market is essentially a matter of interchangeability. Where goods or services can be regarded as interchangeable, they are within the same product market.[93]

307. Accordingly, the definition of the relevant product market hinges on product substitutability. The question that is asked is "would consumers view these two products as substitutable?" If the answer is "yes", then both products form part of the same relevant product market. If the answer is "no", then they do not. The sort of evidence that may be used in defining relevant product markets includes evidence of substitution in the recent past, quantitative tests (e.g., own-price elasticities and cross-price elasticities), views of customers and competitors, market studies and consumer surveys, barriers (including regulatory barriers), and switching costs.[94]

308. The problem, of course, is that none of this data may be readily available today.[95] The process of market definition in AI is further complicated by the fact that even AI-based companies do not always seem know what the technology is and what it is for.[96] The default fallback marketing

---

91 *Id.* at 5.

92 *Id.* at 3–6.

93 WHISH & BAILEY, *supra* note 17, at 26; *see also*, Case 6/72, Europemballage Corp v. Commission of the European Communities, ECLI:EU:C:1973:22, ¶ 32; Case T-321/05, AstraZeneca AB v. European Commission, ECLI:EU:T:2010:266, ¶¶ 30–31, see Thomas Graf, *The EU General Court fines a company for abuse of a dominant position in the pharmaceutical sector addressing the issues of market definition and dominance analysis (AstraZeneca)*, e-COMPETITIONS July 2010, art. No. 35645.

94 *Id.* at 32.

95 This is an overarching problem in relevant product market definition, not just in the field of AI. As Richard Whish and David Bailey note, in practice, the measurement of interchangeability can be difficult for several reasons: there may not be data on the issue, or the data that is available might be unreliable, incomplete or deficient in some other way. *Id.* at 26.

96 Benedict Evans, *The AI Summer*, BENEDICT EVANS (July 9, 2024), https://www.ben-evans.com/benedictevans/2024/7/9/the-ai-summer.

trope that AI "is for everything"[97] will, if taken literally (or almost literally), likely lead to erroneous antitrust outcomes. A more principled and informed approach is needed.

1. Who Are the Consumers? What is the Product or Service?

309. The first, and most basic questions are: who are the consumers and, relatedly, what is the product? In the rapidly evolving field of AI these questions remain elusive. For example, the market for semiconductors involves advanced processors and chipsets, but the market for those chipsets can have multiple consumers. High-end chips are useful not just for AI, but also for gaming, advanced (non-AI) mathematical modeling, crypto mining, etc. Moreover, even if we try to restrict the market consideration just to AI-relevant consumers, as Barnett notes, there are endemic make-or-buy considerations throughout the AI value chain.[98] A firm can stand up a completely vertically integrated solution, purchasing chipsets, designing hardware, setting up data centers, etc., or it can outsource a number of its operations to external firms, like cloud providers. The market for different types of hardware components consists of a large number of heterogeneous parties with different demand elasticities (indeed, some of the parties view their potential competitors for purchasing hardware as potential partners for providing hosted access to hardware).

310. Moving further up the stack, there are a number of firms that provide a wide variety of services. For example, firms specialize in providing curated data,[99] testing/red-teaming services,[100] and other types of services that could otherwise be developed in-house as part of a full-stack solution (e.g., providing low-level open-source libraries for use in larger machine learning systems[101]). Generally speaking, the consumers of these products will be other firms developing AI tools, but spread out across a wide array of industries and applications. But, again as Barnett notes, the presence of the make-or-buy decision is relatively complicated in these markets because many of the tools that are currently in use can be deployed internally within a firm using publicly available scientific papers and tools, or they can be purchased where efficiencies are greater.

311. At the same time, firms can provide the development of foundation models, fine-tuning of models, or both. The consumers here will be even more heterogeneous. Some will be firms using models as modules within their

---

97   *Id.*

98   Barnett, *supra* note 11, at 5–6.

99   *See, e.g.*, DATABRICKS, https://www.databricks.com/ (last visited July 31, 2024); SCALE, https://scale.com/ (last visited July 31, 2024).

100 *See, e.g.*, *AI Red Teaming, Protect Your AI Systems*, REPLY, https://www.reply.com/en/cybersecurity/ai-red-teaming (last visited July 31, 2024).

101 *See, e.g.*, PYTORCH and TENSORFLOW, *supra* note 42.

own consumer-facing products or to improve their own processes. Those same models can be deployed to consumers through web interfaces (e.g., ChatGPT, Claude). But the same basic model can power both the firm-facing and the consumer-facing experience or can be the same model but tailored to the needs of a particular customer on demand. Which leads to the question: is it the model that is forming the relevant market? Or is it the particularly customized model prepared by a specific party that forms the relevant market? The further you get toward restricting the relevant markets based on user-customization, the more you converge on extremely restricted product markets.

312. By the same token, as you pull back from the customizations of individual consumers, the more generalized the model, and the larger both the consumer pool as well as the pool of potential substitutes becomes. That is to say, if, for argument's sake, we assume a relevant market is, e.g., a customized version of ChatGPT trained on a firm's internal data, the relevant market in that case might be exactly that instance of ChatGPT with exactly one relevant consumer. Hardly an analytically useful market.

313. Pulled back slightly you could construe the market as the market for model customization services, and pull in possible alternative providers, like Anthropic and Mistral,[102] as well as open-source alternatives. Such a move would bring in a larger number of potential providers and would be more analytically interesting but then you begin to introduce further complications. What exactly is the service being provided? Is it the training services or the end-consumer product of having some usefully indexed and accessible compendium of their relevant data packaged together with other retrieval services? Which leads to the next question: what exactly is unique about the AI contribution relative to its unique costs.

314. Pull-back even further to the domain of so-called "foundation models." From the perspective of the customer, these foundation models may not be fully interchangeable, or interchangeable at all. As Thibault Schrepel and Alex Pentland have pointed out:

> Foundation models are commonly observed by policymakers and social scientists at the species level (i.e., "foundation model" as a class), but these lenses fail to see the inherent diversity within the species.[103]

315. There is no accepted taxonomy of foundation models. The tentative classifications that have been attempted, however, point to how foundation models can have different characteristics with potentially different uses ultimately serving customers. For example, depending on their training data, foundation models could be general purpose or domain

---

[102] Mistral is an interesting example because they provide their models open weight (meaning that anyone can fine-tune them using their own infrastructure or through a cloud provider), but they also provide a fine-tuning service (like OpenAI does for their models).

[103] Schrepel & Pentland, *supra* note 74, at 3.

specific.[104] The former are trained on a large variety of data with the aim of performing tasks in all possible domains (e.g., ChatGPT and Google Bard). The latter are designed for a specific task or topic, and are thus trained on more granular, specialized data. If someone is looking to produce a generalist generative AI application, they may not consider domain specific foundation models to be substitutable for general purpose foundation models. Furthermore, someone looking to make a specialized generative AI application on one topic – e.g., finance or cooking – might not find a foundation model trained on specialized data in another field to be interchangeable. The authors make further subdivisions, such as personal foundation models (typically pre-trained on large data sets and fine-tuned on individual's private data) and ecosystem foundation models (trained on data that is not publicly available such as data from different companies in the same industry).[105] These models all have different returns and limits,[106] and they experience various competitive dynamics.[107]

316. It seems reasonable to assume that these foundation models underpin different end-uses and are thus not fully interchangeable at the intermediary stage, either, indeed:

> Access to *unique* data sets is critical. There are two reasons for this. First, access to unique datasets may be necessary to provide the specific answer that users of foundation models are looking for. [...] Second, these datasets may play a critical role in the overall training of foundation models.[108]

317. Moreover, inputs and outputs of foundations models vary. Some foundation models are unimodal while others are multimodal. "Unimodal models take instructions from the same input type as their output (e.g., text). On the other hand, multimodal models can take their input from different sources and generate output in various forms."[109] For example, GPT-4o, which underpins OpenAI's ChatGPT, accepts both image and text output to generate text. MusicLM is text-to-music; AlphaCode is text-to-code, and so on.[110]

318. Clearly, there are limits to the substitutability between foundation models that produce different output at the modal level, such as text, image, code, or audio. Customers who want images may not want text, audio or code. But a customer may also not view different inputs as substitutable

---

104 *Id.*
105 *Id.* at 4–6
106 *Id.* at 12.
107 *Id.* at 23.
108 *Id.* at 9.
109 Feuerriegel et al., *supra* note 64, at 113; *see also* Barnett, *supra* note 11, at 3–4.
110 Feuerriegel et al., *supra* note 64, at 113.

319. based, for instance, on the availability of either input. The upshot is that foundation models could, in principle, also be broken down at the modal level depending on their respective input and output data modalities.[111]

319. But if we take another step, the systems level of generative AI is also quite diverse. Generative AI systems comprise not only generative AI models but also the "underlying infrastructure, user-facing components, and their modality as well as the corresponding data processing (e.g., prompts)."[112] Generative AI systems make the underlying mathematical model "usable" across real-world cases by enabling user interaction through a practical interface.[113] For example, generative AI that produces text often use conversational agents and search engines, image generating AI may use bots, etc. How easy generative AI is to use may be a crucial contributing factor to its success,[114] and thus a relevant competitive parameter vis-à-vis other systems and products.

320. Other important questions arise across the genAI technology stack. For example, are large datasets substitutable for small datasets? Some suggest that they are.[115] Are open-source and closed-source foundation models substitutable? The answers to these, and similar questions will depend on who the consumers are and what they want the AI product or service for i.e., on the product's function and utility. Evidently, this is not something that can be resolved in the abstract; instead, it will require analysis on a case-by-case basis.

2. Does AI Fundamentally Transform the Product or Service?

321. The third question is going to be the hardest for enforcers to grapple with. The overarching theme is that, if AI furnishes a product or service that is essentially not reasonably replicable by HI, it constitutes a separate product market. Today, AI is shiny and new (at least in the popular imagination) but, with time, we will inevitably become intimately familiar with the limitations of AI-powered products. Indeed, already we are beginning to come to grasp the awesome power needs of generative AI.[116]

---

111 See also Barnett, *supra* note 11, at 11. (Arguing that "there is a strong likelihood that the models layer of the genAI ecosystem will devolve into multiple differentiated [foundation model] segments tailored for particular industries or uses .... Foundation model markets may disaggregate based on output or input modalities or specific uses.") (internal parenthesis omitted for clarity).

112 Feuerriegel et al., *supra* note 64, at 113.

113 *Id.*

114 For example, ChatGPT's ease of use, especially for non-experts, was a core contributing factor to its worldwide adoption. Feuerriegel et al., *supra* note 64, at 114–115.

115 Thibault Schrepel, *Alternatives to Data Sharing*, THE REGUL. REV. (Feb. 21, 2022), https://www.theregreview.org/2022/02/21/schrepel-alternatives-data-sharing/; Igor Susmelj, *Optimizing Generative AI: The Role of Data Curation*, LIGHTLY, https://www.lightly.ai/post/optimizing-generative-ai-the-role-of-data-curation (last visited June 15, 2024) (discussing the importance of data curation to ensure that datasets are devoid of noise, irrelevant instances, and duplications, thus maximizing the efficiency of every training iteration).

116 Beth Kindig, *AI Power Consumption: Rapidly Becoming Mission-Critical*, FORBES (June 20, 2024, 04:13 PM EDT), https://www.forbes.com/sites/bethkindig/2024/06/20/ai-power-consumption-rapidly-becoming-mission-critical/.

The costs for chips are also high and growing as the need for more data centers increases.[117] Cooling costs are also very large for the data centers powering consumer AI applications.[118]

322. At the moment, many of these costs may be dwarfed by the enthusiasm of investors, but if the scale of the potential power demand is at all in line with reality, very quickly many of these services will need to be more adequately priced into the cost of accessing AI services. This does not mean that AI will go away; to the contrary, we foresee it being a major component in the future economy. What it does mean is that the tradeoffs inherent in using AI for a particular application will become more apparent and will force consumers of all types to evaluate AI not on its own, but in comparison to other possible products and services that, while inferior on some dimensions, will be effective substitutes.

323. Returning to the example of the custom GPT that a firm wishes to train on its own internal documents: That could be a service operating in a market for "GPT customization services" if we look at it from the provider's perspective. But from the consuming firm's perspective it could be comparing alternative document archival systems, some of which will be based on "good enough" traditional relational databases, some of which on semi-intelligent unstructured retrieval systems, and some on AI-powered LLMs. Arguably, a MySQL-powered traditional application based on well-trodden search algorithms is not as powerful as an LLM-powered system that allows you to interactively "chat" with your document history. But the substitutability is based not just on what the more powerful application is but, given the tradeoffs necessary, which products are good enough alternatives. The cost and time involved in standing up an LLM, including coding data to be the most effective for training, and the direct or indirect power and cooling costs of having constantly available LLMs may not make sense when well-understood (but maybe boring) traditional alternatives exist. It is not clear that such traditional applications *should* exist in the same product market as an LLM, but our goal here is to merely point out that is it not clear yet that they do not do so.

324. On the other hand, there might genuinely be AI functionalities that bring entirely novel products to the market. For example, it is possible to imagine that in litigation, AI processing of documents, and interactive discussion with an LLM about the corpus of documents produced in discovery will simply be so much more effective than HI-driven traditional solutions that it does not make sense to treat them as substitutes. Large litigations can be extremely expensive, and, indeed, AI may drive down the marginal cost of document review in a way that makes it possible for more parties

---

117 Angus Loten, *Rising Data Center Costs Linked to AI Demands*, WALL ST. J. (July 13, 2023, 07:00 AM ET), https://www.wsj.com/articles/rising-data-center-costs-linked-to-ai-demands-fc6adc0e.

118 David Berreby, *As Use of A.I. Soars, So Does the Energy and Water It Requires*, YALE ENV'T 360 (Feb. 6, 2024), https://e360.yale.edu/features/artificial-intelligence-climate-energy-emissions.

to engage in litigation. We are not passing judgment on whether this is good or bad, but merely observing that such an augmentation offered by AI could be a relevant product market in this case with which traditional HI-driver alternatives simply would not be able to compete.

325. Furthermore, a sufficiently notable quantitative leap can produce a qualitative different product that is not interchangeable for the "old" version. For instance, although HI can also analyze data patterns, AI can analyze vastly bigger swathes of data much more quickly and possibly uncover correlations that would escape the human mind. AI can also work 24/7, without the need for rest, which applies across all tasks. Similarly, AI can perform repetitive tasks without fatigue or degraded performance. Granted, this is also true for traditional software, but in the context of AI–HI substitutability, it might tilt the decision in one direction or the other depending on the extent to which the ability to work non-stop constitutes an important competitive dimension of the product at hand (e.g., it might be decisive in the context of HI vs. AI powered security systems, but less important when it comes to editing blog posts).

326. Thus, AI can not only create qualitatively new uses but improve existing ones so much that they effectively become a new relevant product market. Competition law recognizes that products belonging to different quality segments can constitute a variety of relevant product markets, similar to how, for example, "premium smartphones" are in a separate category from other smartphones.[119]

V. Conclusion

327. Abstraction usually follows the observation of similar phenomena. As such, it is the basis for all deductive reasoning and a fundamental pillar of human epistemology. However, abstractions can sometimes lead to erroneous inferences by ascribing general properties to phenomena that are, in reality, distinct. In the social sciences and in popular discourse, we refer to erroneous abstractions as "generalizations." Obviously, the opposite is also possible, such as when we fetishize small differences and confound novelty – or minor idiosyncrasies – with uniqueness.[120] The same applies to AI. From the outside, AI is an imposing block – not unlike the black monolith from Stanley Kubrick's 2001: A Space Odyssey. Upon closer examination, however, it turns out that AI is neither homogenous nor as unique as would initially seem.

328. The lack of a proper understanding of the outward and inward boundaries of AI markets has practical implications for antitrust policy and regulation because it may lead to inaccurate assessments of market concentration and

---

119 U.S. Dept. Justice, *Justice Department Sues Apple for Monopolizing Smartphone Markets*, (Mar. 21, 2024), https://www.justice.gov/opa/pr/justice-department-sues-apple-monopolizing-smartphone-markets.

120 One manifestation of this is, arguably, Freud's "narcissism of the small differences." *See generally*, SIGMUND FREUD, CIVILIZATION AND ITS DISCONTENTS (1929).

market power, resulting in both under and over-enforcement of competition law compared to the social optimum. For example, it is likely – or at the very least plausible – that as soon as one accounts for the substitutability of AI and non-AI product, the concentration in some of those markets that hitherto appeared to be monopolistic withers away. What changes is not the observed phenomena, but the level of abstraction: we understand that the market is more competitive than initially envisioned because competitive pressures are exerted both from within and without AI-specific products.

329. Conversely, a failure to account for the internal heterogeneity of AI could lead to an underestimation of market concentration and market power, by artificially expanding the universe of products that comprise the same relevant market. This currently does not seem to be what is happening, however, as the predominant narrative is, somewhat paradoxically, that AI markets are both extremely broad – encompassing a range of different products and technologies – and extremely concentrated. Ultimately, this may be due more to a combination of the lack of sophistication in demarcating AI markets, technological anxiety, pro-enforcement bias, and prejudice against large technological firms, than the articulation of a principled creed. And yet it must be noted that the upshot of this distorted perception, if it ever translated into enforcement, will be a distorted antitrust policy.

330. But the problem is not just that the over and underestimation of the relevant product market mystifies the appraisal of market power. Confounding the boundaries of product markets obfuscates the real competitive dynamics in those markets, rendering the enforcer myopic to incentives, the direction of competitive threats, the potential for procompetitive benefits, and the possibility and likelihood of entry and expansion. Such an unprincipled approach would force authorities to rely on tasseography and pre-existing biases to address potential competition problems in a rapidly changing environment that is inherently hostile to such unwavering assumptions. In the world of AI, that is called "hallucination."

Defining AI Markets: Who is Afraid of Digital Ghosts?

STEPHEN DNES[*]

Royal Holloway, University of London

Abstract

"Digital ghosts" from perceived missteps with Web 2.0 technologies pervade antitrust analysis of artificial intelligence. Chief amongst these is a concern that orthodox requirements to define markets led to undue precaution in relation to network effects. This chapter analyzes the relationship between market definition and competitive effects to reveal the continuing relevance of at least one of the two core concepts. It considers the likely evidence requirements relating to demand and supply side factors. The analysis identifies several areas of likely contention in future AI-related antitrust analysis.[1]

---

[*] Lecturer in Law, Royal Holloway, University of London. Partner, Dnes & Felver PLLC. Disclaimer: The author has advised on merger clearance issues in AI-related merger clearance matters before the UK CMA and was engaged in the reform process relating to the new UK competition law which includes aspects of merger clearance relevant to the chapter. All views are the author's own.

[1] I am grateful for the comments of Thibault Schrepel and Matthew Sinclair. However, all errors and omissions are my own.

I. Introduction

331. Increasingly, competition agencies are called on to review activities in artificial intelligence (AI)-related markets.[2] It is sometimes said that digital ghosts from Web 2.0 haunt competition agencies.[3] This is usually taken to mean that there is concern that agencies were insufficiently interventionist in relation to the growth of the largest technology companies, prompting an increase in precautionary analysis in relation to AI-related innovation and other emergent technologies.

332. The underarticulated question is: which ghosts? This chapter will consider several precedents and aspects of agency guidance relevant to the boundary with Web 2.0 analysis. It draws on a range of precedents from around the world to consider what lessons might be drawn in relation to market definition. This is acutely relevant to AI analysis because the approach taken to supply side entry and expansion will strongly determine the granularity of analysis relating to entry and expansion dynamics.

333. There is already active debate about the correct role of market definition – if any – in fast-paced technology markets.[4] The well-known debate around whether market definition is logically necessary, or even helpful, is newly relevant, especially following legislative reforms relaxing the requirement for relevant markets to be proven.[5] There is a close relationship between market definition requirements and the requirement to prove effects, and the exact nature of this relationship under the new law remains ambiguous.

334. The question is also highly relevant. What exactly takes the place of conventional market definition analysis will alter the outcomes in relation to vertical supply chains. There is also the important question of how far the existing analysis will be carried over to new regimes adjacent to merger review, including business conduct cases and market investigation powers.

---

2   *See* especially Thibault Schrepel, *A Database of Antitrust Initiatives Targeting Generative AI*, NETWORK L. REV. (Jan. 23. 2024), https://www.networklawreview.org/antitrust-generative-ai/.

3   *See, e.g.*, Alec Stapp, *The Ghosts of Antitrust Past: What the IBM-AT&T-Microsoft Trilogy Can Teach Us About Calls to Break Up Google, Amazon, Facebook and Apple*, (ICLE ISSUE BRIEF FOR ANTITRUST & CONSUMER PROT. RSCH. PROGRAM, Apr. 28, 2020); *see* most recently comments of Christophe Caraguti, Concurrences Innovation Law and Economics Conference (Apr.23, 2024), https://www.concurrences.com/en/evenement/innovation-law-economics-conference; *See also* in this volume, CHRISTIAN BERGQVIST & CAMILA RINGELING, *Finding the Ghost in the Shell: EU and US Antitrust Enforcement of AI Collusion*, *in* ARTIFICIAL INTELLIGENCE & COMPETITION POLICY (Alden Abbott & Thibault Schrepel eds., 2024).

4   JACQUES CRÉMER, YVES-ALEXANDER DE MONTJOYE & HEIKE SCHWEITZER, COMPETITION POLICY FOR THE DIGITAL ERA, REPORT TO THE EUROPEAN COMMISSION 110 (2019); FIONA MORTON ET AL., COMMITTEE FOR THE STUDY OF DIGITAL PLATFORMS: MARKET STRUCTURE AND ANTITRUST SUBCOMMITTEE REPORT (Sept. 16, 2019). *See also*, US HOUSE JUDICIARY COMMITTEE, SUBCOMMITTEE ON ANTITRUST, COMMERCIAL AND ADMINISTRATIVE LAW, INVESTIGATION OF COMPETITION IN DIGITAL MARKETS (Oct. 2020); cf. William E. Kovacic & D. Daniel Sokol, Understanding the House Judiciary Committee Majority Staff Antitrust Report, CPI ANTITRUST CHRON. (Jan 2021), https://www.competitionpolicyinternational.com/wp-content/uploads/2021/01/UNDERSTANTING-THE-HOUSE-JUDICIARY-COMMITTEE-1.pdf (noting issues with divergent focus and lack of clear recommendations in report: "The case studies are elaborate and rich in detail; the discussion of the doctrinal reforms many with great significance for the entire US antitrust system, is slim by comparison." *2).

5   Louis Kaplow, *Why (Ever) Define Markets?*, 124 Harv. L. Rev. 437 (2010); compare Gregory J. Werden, *Why (Ever) Define Markets? An Answer to Professor Kaplow*, 78 ANTITRUST L.J. 729 (2013).

335. After surveying the history with the definition of markets, this chapter analyses three prominent instances of agency intervention to assess the role of market definition analysis in AI-related settings:
   – The US Federal Trade Commission's challenge to Meta's acquisition of Within ("Meta/Within");
   – The UK Competition and Markets Authority's investigation of Microsoft's partnership with OpenAI and its putative challenges to AI-related investments by Amazon and Microsoft, respectively in Anthropic and Mistral ("Amazon/Anthropic" and "Microsoft/Mistral"); and
   – The information gathering by both agencies into the artificial intelligence sector.

336. It considers how existing laws approach market definition questions, before noting the potential impact of new laws. It concludes by considering important future questions related to the "digital ghosts" perceived to haunt agency intervention from Web 2.0 antitrust analysis, and how AI analysis might yet differ from that which went before.

II. (How) Should Market Definition Approach Innovation Markets?

1. Are Relevant Markets Relevant to AI?

337. Since at least the late 2000s, there has been significant debate about the role of market definition in the context of innovation. In a prominent exchange, Louis Kaplow and Gregory Werden debated the roles of market definition on the pages of the Harvard Law Review.[6] Under the banner of the provocative question, "Why (ever) define markets?", Kaplow challenged the orthodoxy by which antitrust analysis starts with market definition. The essence of the critique was that affirmative evidence of market power effects could displace relevant market analysis. Advanced economic modelling, especially merger simulation techniques such as upward pricing pressure analysis, do not logically depend on specification of a market.[7] This implies that there could be a helpful shift from emphasis on relevant market definition to analysis of competitive effects, provided that the analysis is also robust and quantitative.

338. Significant advantages could accrue if departing from the potential blind spots of rigid market definition, notably attention to non-average consumers whose preferences might depart from the average. A classic example exists in the market for cars: some might pay extra for red over blue cars, or a particular model or brand. These economic effects exist even

---

6 Id.

7 For further discussion of modern merger simulation methods, see Oliver Budzinski & Victoriia Noskova, *Prospects and Limits of Merger Simulations as a Computational Antitrust Tool*, 58(Stan. Computational Antitrust 56–77 (2022).

though market definition would exist at a higher level of abstraction (e.g., a market for sports cars or sedans). Considering direct evidence of pricing impacts, where available, would help to identify and address pricing pressure, whatever its origin.

339. The genesis of this line of argument in modern econometric methods meant that it came as some surprise that a degree of opposition to omitting market definition came from one of the most prominent voices for sophisticated, modern economic techniques. Werden argued that, for all the considerable utility of the new methods – a number of which he had helped to introduce into modern agency analysis of mergers[8] – there was also administrative utility in the market definition concept. Werden pointed to advantages in evidence gathering and analysis which suggested that analysis of effects, while a very helpful end point, might not be the right starting point for the analysis of a merger.[9]

340. In a closely related but earlier vein of writing, the rigidities of antitrust categorization were already challenged as early as the 1980s. In a seminal contribution, Michael Porter introduced what is now called the ecosystems critique,[10] noting complexities in management strategies in multi-business firms. The critique cuts both ways: if there is competition *for* a future market, then analysis within an obsolescent market is less relevant. Equally, barriers to innovation in a current market might prevent significant future growth despite not themselves showing up as immediate competitive effects on an orthodox approach.[11]

2. Market Definition in Innovation Markets

341. More recent comments have drawn attention to the need for flexibility in market definition analysis where innovation is involved. The UK CMA's Chief Economist, Mike Walker, went so far as to state recently that "if market definition is the focus of a case, then something is wrong."[12] If taken to mean that effects rather than market definition should be the

---

8 *See, e.g.*, the DOJ's comments surrounding the award of the US Department of Justice's Mary C. Lawton Lifetime Service Award to Werden, noting his pivotal role in expanding the role of quantitative methods in antitrust analysis.

9 *See* most recently, Gregory J. Werden, *Contribution of Gregory J. Werden on Commission Notice on the Definition of the Relevant Market for the Purposes of Union Competition Law*, (Nov. 26, 2022), https://papers.ssrn.com/sol3/papers.cfm?abstract_id=4287552.

10 *See* NICOLAS PETIT, BIG TECH AND THE DIGITAL ECONOMY 74–75 (Oxford Univ. Press 2020) (noting a range of seminal contributions including: Harvey Leibenstein, "Bandwagon, Snob, and Veblen Effects in the Theory of Consumers' Demand 64 Q. J. Econ. 183 (1950); HAL VARIAN, JOSEPH FARRELL & CARL SHAPIRO, ECONOMICS OF INFORMATION TECHNOLOGY (Cambridge Univ. Press 2004); CARL SHAPIRO & HAL VARIAN, INFORMATION RULES (Harv. Bus. Rev. Press 1998); Nicholas Economides, *The Economics of Networks*, 14 INT'L. J. IND. ORG. 678 (1996); Michael L. Katz & Carl Shapiro, *Systems Competition and Network Effects*, 8 J. ECON. PERSPS. 106 (1994). As early as 1981, Michael Porter noted the complex relationships between multi-business firms: Michael Porter, *The Contributions of Industrial Organization to Strategic Management*, 6 ACADEMY MGMT. REV. 617 (1981).

11 Michael G. Jacobides & Ioannis Lianos, *Ecosystems and Competition Law in Theory and Practice*, (UCL CTR. L., ECON. & SOC'Y, Research Paper Series No. 1/2021, 2021).

12 Comments of Mike Walker, Concurrences Innovation Law and Economics Conference (Apr. 23, 2024).

end point, rather than the starting point of analysis, then the comment is neither as acerbic as it first seems, nor out of line with the debate between Kaplow and Werden outlined above. All that is said is that the market definition should not become the focal point of analysis, but rather, a step within it.

342. The point is particularly significant coming from Walker as his book, co-authored with Simon Bishop, is the canonical source for supply side substitution in a UK competition law context.[13] The authors use the example of specialist paint production to show supply side substitution: provided that mainstream paint producers can switch to creating the specialist paint at low cost, they are likely to constrain the specialist paint producer even where the current market share for the specialist producer is 100 percent.

343. In summary, the authors note discussion of whether supply side substitution should be considered at the market definition stage, or whether it can safely be left to the competitive effects stage. Although they note that there is an argument that it does not matter when the effects are considered, provided that they are considered at some point, they ultimately conclude that there is scope for misleading market shares if a significant supply side aspect is not considered when defining the relevant market.

344. This seemingly cerebral debate has significant implications for the analysis of AI services in merger cases. If supply side switching is credited, it will tend to dilute findings of unilateral effects.

A. *From Paperclips to Hal 9000: What AI Analysis Can Learn from a 1997 Case about Legal Pads*

345. Since at least the 1990s, a trend emerged towards narrow market definitions driven by unilateral effects analysis. The classic example is the 1997 challenge *FTC v. Staples*.[14] There, an innovative retail format of dedicated office supplies superstores had effectively created its own demand through innovation. The FTC had identified competition between superstores using scanner data and could show risks of unilateral effects from upward pricing pressure, even though in principle many different types of store could sell office supplies. Although the superstore-specific market definition was ultimately credited, this was done with significant reservations based on concerns that the parties were effectively punished for innovation.

346. Memorably, Judge Hogan likened the exercise to the scene in *Hamlet* in which Hamlet mistakenly kills Polonius by stabbing him through the curtain under the impression that he was in fact Claudius. Said Judge Hogan:

> In light of the undeniable benefits that Staples and Office Depot have brought to consumers, it is with regret that the Court reaches the decision that it must in this case. This decision will most likely

---

13 Simon Bishop a& Mike Walker, The Economics of EC Competition Law ¶ 4-012–4-015 (2010).

14 FTC v. Staples, Inc., 970 F. Supp. 1066 (D.D.C. 1997).

kill the merger. The Court feels, to some extent, that the defendants are being punished for their own successes and for the benefits that they have brought to consumers. In effect, they have been hoisted with their own petards. *See* William Shakespeare, *Hamlet,* act 3, sc 4. In addition, the Court is concerned with the broader ramifications of this case.[15]

347. The reference is not at all flattering to the use of narrow market definitions in innovation contexts. As noted above, the controversy arose from the difference between the intuition that office supplies can be purchased in many different retail stores. Thus, a market defined in terms of a particular superstore presentation risks overlooking possible constraints from alternative, non-superstore presentations. Nonetheless, pricing data evidence supported intervention at the narrower level.

348. The memorable *Hamlet* reference reveals an important fulcrum for debate: for all the advances in economics, it will still be necessary to define a market to satisfy judges that legal tests are applied, and that evidence is organized in a recognized format. In this sense, there is almost a due process aspect to the requirement. Thus, an agency playing fast and loose with market definition risks losses in court.

349. Therefore, market definition arguments become an important focus for some of the thorniest antitrust questions of the AI era. For example, is competition *for* the market adequate even if there are limitations to competition within it? How should a court approach the existence of scale barriers to entry for a time, but that may not endure? Implicitly, the *Staples* court contemplated space for competition for the market rather than within it. Judge Hogan's comment has aged well and become more relevant over time: the challenge will always remain that a narrow market definition based on innovation has the possible unintended consequence of undermining the very competition it was supposed to promote.

B. *Innovation Analysis since Staples (1997)*

350. In the years since *Staples*, three main scenarios have emerged in which innovation might alter market definition analysis. In recent comments, Walker noted three overlapping instances in recent merger review decisions:

1. *Innovation as a necessary metric of competition.* Drawing on *Dow/Dupont,* Walker commented that some consumer-facing products require companies to compete on innovation. In some contexts, that would be a generic statement, but Walker pointed to specific instances where the consumer demand *necessarily* implies competition over innovation. For example, the chemical compounds at issue in *Dow/Dupont* must be changed every few years as pesticide resistance

---

15 *Staples* *1093.

increases. The ability to outpace obsolescence is a metric of competition and implied by the market definition – since it is not possible to meet demand without a future innovation plan. So, evidence would exist of these plans if the company is to be present at all. Thus, analysis of future competition can take place, at least on a qualitative basis, reflecting those plans. This *necessarily innovative* aspect of certain markets distinguishes such cases from the more generic sense in which a degree of innovation is present in many competitive markets.

2. *Pipeline products markets.* Innovation may show up clearly in pipeline products markets. The distinction between innovation as a metric (as in *Dow/Dupont*) and pipeline product markets is that pipeline products are already better defined. For instance, a pipeline pharmaceutical product is already under regulatory review. It is possible for an agency to review the confidential plans of the merging parties and to compare this with the pricing dynamics of products already in the market. These are the most concrete future innovation markets as the pipeline already aims to meet a clear consumer demand; it is just that this exists in the future (e.g., after all regulatory approvals are in place).

3. *Capabilities analysis.* The broadest conception of a future market is based on a looser concept of capabilities. This is a striking theory because it could amount to the elevation of an economic input to the level of a relevant market. If the analysis is of the purchase of a capability, then the analysis effectively reaches into the vertical integration decisions of suppliers and equates these to a consumer demand – here, for the capability (e.g., engineers; chips). This is a significant change from the orthodox starting point which is to consider consumer demand. If that consumer-facing demand does not yet exist, then the purchase of a capability is a possible proxy for the future product. If the view is taken that merger control should apply to prospective competition, then the attraction in the capabilities framework is clear.[16] However, so are the risks: the familiar balance between false positives and false negatives shifts towards the former.

III. Application to Emerging Markets Including AI

351. Capabilities-based analysis is the most relevant of the three frameworks for analysis above. Unfortunately, it is also the vaguest. Whereas direct evidence of competition over innovation, or of a product pipeline, is testable, all that can be said in relation to capabilities is that they exist and *might* later lead to competition. The link to consumer demand is at best weak, because the demand is in the future. It is possible to argue that the

---

16 *See* further: David J. Teece. *The Dynamic Competition Paradigm: Insights and Implications*, 1 COLUM. BUS. L. REV. 374 (2023).

capabilities are consumed by their buyer, e.g., of chips or of labor, but strictly speaking that would be a different level of the supply chain and thus not the relevant market for the issue at hand. Therefore, if capabilities analysis replaces analysis of consumer demand for end products, a major relaxation in relevant market doctrines results.

352. For instance, and as explored further below, in the recent *Microsoft / Mistral* merger, chips and labor would be inputs to the overlapping services (if such they were) rather than the overlapping services themselves. The more obvious overlap would be the foundation models. A capabilities analysis may be helpful to identify competitive dynamics such as entry and expansion, but if looking at capabilities alone, there are risks of missing substitution patterns and constraints from them if the capabilities analysis predominates over the analysis of consumer demand.

353. It very much remains to be seen whether courts will accept any such relaxation, especially as elements of relevant market analysis are still present in the new generation of laws. For example, following amendments to the UK's Digital Markets, Competition and Consumers Act at its Bill stage, it is necessary to prove material market power changes in relation to boundaries between activities before certain conduct rules can bite.[17] Thus, unrelated entry is protected from regulation, as can also be seen in relation to new Pro-Competitive Intervention powers.[18] It follows that even the more permissive, new generation legislation will require a statement of some sort as to the relationship between products and services and how competition dynamics emerge between them. If that sounds familiar, it is because it is not a new question at all: the balance between proving definitions and proving effects stretches back at least to the 1970s and perhaps before.

1. Which Ghosts Did You Have in Mind?

354. As noted in the introduction, agencies are said to be beset by "digital ghosts". Interestingly, there is a strong argument that market definition doctrines have long identified the issues with the trade-off between market definition specificity and direct evidence of effects in innovation markets. That the issue is far from new would suggest that market definition – or at least a surrogate for it – will remain relevant in the new generation of competition laws. This reflects the fact that the underlying issue with defining relevant markets arises from issues of principle and not merely the requirement by courts to see market definition proven.

355. So, it might be said that the so-called digital ghosts take different characters and that the haunting is richer and more diverse than a singular focus on perceived excesses in precaution in relation to Web 2.0 products.

---

17 Digital Markets, Competition and Consumers Act 2024, § 19(3)(c) (material market power changes must be proven *in relation to* a digital activity).

18 Digital Markets, Competition and Consumers Act 2024, § 46(5) (perceived adverse effect on competition must arise in connection with a relevant designated activity before intervention can occur).

To examine this point, the chapter will now turn to some classic points of analysis as to supply side changes and how market definition doctrines have approached them.

2. Guidelines position

356. The underlying question as to segmentation of consumer demand, and how far it must be proven before independent analysis of competitive effects can take place, dates back well before even the 1997 *Staples* case. Early analysis in a 1979 Virginia Law Review article noted several US Courts of Appeals cases in which supply side substitution had broadened the market.[19] The issue is particularly acute in several contexts:

- *Telecoms mergers:* The long-standing position in the EU that there is a relevant market for mobile telecommunications encompassing both private and business customers is driven in part by supply side substitution between the different customer groups.[20] Although the analysis is quick to point out that competitive constraints would still need to be analyzed, and do not necessarily follow from supply-side substitution in market definition, the focus of the competitive analysis will then be framed around a wider market with wider entry and expansion characteristics.[21]

- *Agricultural mergers:* A significant seam of UK merger review case law considers whether particular types of agricultural products exist in their own market segment. For example, markets for grain are not segmented by type,[22] and supply side analysis takes place to determine the boundary between organic and non-organic supply using the same farmland.[23]

- *Transportation services:* There is precedent for supply side substitution altering market definition in relation to transportation markets, in Stagecoach/Eastbourne Buses took account of supply-side factors in determining market definition, at paragraphs 5.26–29. As buses are mobile and relatively flexible, the possibility of moving them between

---

19 Bruce A. Karsh *The Role of Supply Substitutability in Defining the Relevant Product Market*, 65 VA. L. REV. 129–151 (1979). The note refers to Telex Corp. v. Int'l Business Mach. Corp., 510 F.2d 894 (10th Cir. 1975) (Supply side substitution considered relevant where it allowed conversion of peripheral computing products to work with IBM equipment); Yoder Bros. v. California-Florida Plant Corp., 429 U.S. 1094 (1977) (A relatively wide market for "ornamental cuttings" was found in part on the basis of supply substitutability between different types of plants).

20 Case M.6497 (*Hutchison 3/Orange Austria*), 2012, O.J. (C 202) 4, at ¶ 34–35.

21 *Id.*, n.16 (noting that separate analysis of competitive constraints would then occur). The same approach can be seen in Case M.7018 (*Telefonica Deutschland/EPlus*), 2013, O.J. (C 351) 37, at ¶ 30–31. *See also* Simon Vande Walle & Julia Wamback, *The Commission's Review of Mobile Telecoms Mergers*, (COMPETITION MERGER BRIEF No. 1/2015, 2015), 10–11 ("The main reason underlying this wide definition is the relative ease of supply-side substitution, meaning mobile operators can easily switch from offering one service to offering another.")

22 UK CMA, Case No. ME/3850/08 *Grainfarmers/Centaur* (Oct. 20, 2008), para 13 to 14, https://www.gov.uk/cma-cases/grainfarmers-centaur-grain (noting supply side flexibility between provision of different types of grain).

23 UK CMA, Case ME/6737/18 *Arla Foods Ltd / Yeo Valley Dairies Merger Inquiry,* (May 17, 2018), https://www.gov.uk/cma-cases/arla-foods-limited-yeo-valley-dairies-merger-inquiry.

depots was taken into account in determining relevant markets, subject to relevant constraints such as the location of depots. This is a close analogy to new product market definition issues, because customer demand may be induced by the movement of the transportation line to a new location. So, a market definition based entirely on current patterns of customer demand will miss possibilities that consumer demand will expand *including into new products and services* based on rearrangement of existing supply.

357. So, if there are ghosts of the past, they do not only relate to missed interventions. There are in fact several species of ghost. They at least encompass three major markets in which supply side substitution has altered market definition in recent years.

358. The point is particularly clear in relation to earlier generations of telecoms analysis. Not only are there the merger cases noted above, but there are also guidelines providing clear demarcation of supply and demand side characteristics in relation to market definition. For instance, the Hong Kong Telecommunications Authority's Guidelines state:

> 3.43 Confusion sometimes arises between what is considered to be a supply-side substitution (a factor relevant to market definition) and entry into a market (a factor usually taken into account in competition analysis once the market has been defined).
>
> 3.44 Supply-side substitution concerns the ability of firms to switch their production lines at relatively short notice in response to a price increase and supply a close substitute to the product in question. Firms with this ability may not actually be in the market. However, if they are considered likely to enter rapidly in response to a price increase, they are considered to be market participants because of the constraints that their rapid entry places on a hypothetical monopolist.
>
> 3.45 On the other hand, market entry may involve significant sunk costs of entry and exit. Sunk costs are capital costs that can only be used in the production of the product in question and which, once incurred, cannot easily be recouped. An example in telecommunications is the cost of network facilities, which cannot easily be recouped if the investing firm decides to exit the market.[24]

A. *Relationship with Consumer Analysis*

359. If the position of the sector-specific guidance above is clear, it still remains to be seen in relation to general antitrust law. The same underlying question pertains, which is the relationship between supply side capacities and

---

[24] Hong Kong *Telecommunications Authority Guidelines*, CB(1)2416/10-11(02), https://www.legco.gov.hk/yr09-10/english/bc/bc12/papers/bc120607cb1-2416-1-e.pdf.

consumer demand. Slightly shifting emphasis on the consumer can be seen reviewing iterations of the US Horizontal Merger Guidelines:

- 1997: Market definition focuses solely on demand substitution factors – i.e., possible consumer responses. Supply substitution factors – i.e., possible production responses – are considered elsewhere in the Guidelines in the identification of firms that participate in the relevant market and the analysis of entry.[25]

- 2010: Market definition focuses solely on demand substitution factors, i.e., on customers' ability and willingness to substitute away from one product to another in response to a price increase or a corresponding non-price change such as a reduction in product quality or service. The responsive actions of suppliers are also important in competitive analysis. They are considered in these Guidelines in the sections addressing the identification of market participants, the measurement of market shares, the analysis of competitive effects, and entry.[26]

360. It is significant that there is less emphasis on consumers in the new 2023 Guidelines. The strong statement of a consumer focus from the 1997 and 2010 Guidelines has gone, but there is still a strong customer demand focus in the detailed sections on competitive analysis:

- 2023: Customer Substitution. Customers' willingness to switch between different firms' products is an important part of the competitive process. Firms are closer competitors the more that customers are willing to switch between their products, for example because they are more similar in quality, price, or other characteristics.[27]

361. The same focus can be seen in the updated UK Merger Assessment Guidelines. Even if these are now more accommodating to intervention within supply chains following a 2021 update, the analysis is still couched in terms of customer impacts:

> Unilateral effects giving rise to an SLC can occur in relation to customers at any level of a supply chain, for example at a wholesale level or retail level (or both), and is not limited to end consumers.[28]

---

[25] Fed. Trade Comm'n. and U.S. Dep't of Justice, *Horizontal Merger Guidelines (1997)*, 1.0 Overview (1997), https://www.justice.gov/archives/atr/1997-merger-guidelines.

[26] Fed. Trade Comm'n. and U.S. Dep't of Justice, *Horizontal Merger Guidelines (2010)*, § 4 (2010). The same approach is visible in the Commentary to the earlier 1992/7 Guidelines: "Product market definition depends critically upon demand-side substitution – i.e., consumers' willingness to switch from one product to another in reaction to price changes. The Guidelines' approach to market definition reflects the separation of demand substitutability from supply substitutability – i.e., the ability and willingness, given existing capacity, of firms to substitute from making one product to producing another in reaction to a price change. *Under this approach, demand substitutability is the concern of market delineation, while supply substitutability and entry are concerned with current and future market participants.*" (emphasis added)

[27] Fed. Trade Comm'n. and U.S. Dep't of Justice, *Merger Guidelines (2023)*, § 4.2 (Dec. 18, 2023), https://www.ftc.gov/system/files/ftc_gov/pdf/P234000-NEW-MERGER-GUIDELINES.pdf

[28] CMA, *UK Merger Assessment Guidelines (2021)*, ¶ 4.1 (Mar. 18, 2021), https://assets.publishing.service.gov.uk/media/61f952dd8fa8f5388690df76/MAGs_for_publication_2021_--_.pdf

362. The point is even more emphatic in the case of the EU. Even following the 2024 update to the Notice on Market Definition, the canonical statement of product market definition from *Hoffman La Roche* is still applied verbatim:

> The relevant product market comprises all those products that customers regard as interchangeable or substitutable to the product(s) of the undertaking(s) involved, based on the products' characteristics, their prices and their intended use, taking into consideration the conditions of competition and the structure of supply and demand on the market.[29]

363. Thus, despite the rewriting of crucial aspects of each of the US, UK and EU merger review frameworks, there is still a strong consumer focus as a starting point for analysis in all cases.

IV. Instances of Agency Intervention

364. As explored above, the Guidelines contained a statement of orthodoxy: analysis should proceed based on consumer demand in the first instance. Supply side changes would be considered later as an entry or expansion dynamic. Although the emphasis differed across geographies and timeframes, all of the guidelines above still retain a strong emphasis on consumer-centric analysis. As is particularly clear from the 1997 version of the US Guidelines, there is a close relationship between potential competition and market definition in innovation contexts.

365. This section will assess how market definition has developed in recent cases against this historic backdrop. It will start with the analysis of asserted future competition in the UK CMA's review of *Facebook/Giphy* before considering the market definition in the US FTC's challenge to *Meta/Within*. It will then consider the latest proposed challenges to AI-related investments. It will conclude by considering the AI-related information gathering underway at the US FTC and UK CMA.

A. *Facebook/Giphy*

366. In *Facebook/Giphy*, the UK CMA required Meta (as it had since become) to divest the GIF provider Giphy. The essential underlying theory of harm was that Giphy was thought a potential competitor in the sale of advertising. Even though Giphy did not yet have UK revenue, concerns arose that marketing sales could arise which would compete with Meta. The claim is driven by the strong desire in some quarters to see competition emerge despite the scale barriers to expansion perceived in "tipped" technology markets.[30]

---

[29] Commission Notice on the definition of relevant market for the purposes of Community competition law, 1997, O.J. (C 372) 5, at § 1.3.

[30] Case C-226/11, Expedia Inc., v. Autorité de law concurrence, ECLI:EU:C:2012:795, applied a seemingly per-se approach to exclusionary conduct in the context of strong network effects, *see* Marie Koehler de Montblanc, *De minimis Communication: The Court of Justice recalls that the de minimis Communication is not intended to bind the competition authority nor the national jurisdictions (Expedia)*, Concurrences No. 2-2012,

367. The case was primarily analyzed in jurisdictional terms, especially relating to the question whether vertical supply can confer jurisdiction within the meaning of the UK's Enterprise Act 2002, which historically had been thought not to allow a challenge relating to a vertical input.[31] The critical point for future AI analysis is a different one. It is that candidate market definition can include future effects. Indeed, on the crucial policy point of what must be proven for there to be a plausible theory of harm, the Competition Appeal Tribunal strongly deferred to the agency.

368. There is accordingly little analysis of the relationship between tipped and untipped markets, let alone analysis of the relationship between demand and supply side characteristics in them.[32] So, it will fall to future cases to define what exactly is meant by perceived issues with future competition and the relationship between supply and demand side characteristics there. This will be a priority area for analysis under the merger reforms of the UK's new Digital Markets, Competition and Consumers Act 2024 which allows significantly more scope to call in transactions, but applies the same underlying Enterprise Act test – so, there will be plenty of scope for review, but also a need to apply the same framework as in Giphy to the new transactions.

B. *Meta/Within*

369. In 2023, the US FTC lost its attempt to challenge Meta's acquisition of virtual reality app developer Within Unlimited. The case provides a mixed precedent. On the one hand, the FTC prevailed on its candidate market definition: "VR dedicated fitness apps, meaning VR apps 'designed so users can exercise through a structured physical workout in a virtual setting.'" Meta had contended that the market should instead encompass: "fitness apps on gaming consoles and other VR platforms, and non-VR connected fitness products and services." So, on the critical unilateral effects question alluded to above, the FTC succeeded – as expected given the precedents – on the crucial question of carrying a narrow market definition at least at the preliminary injunction stage.

---

art. No. 50969. Petit notes that per se unbundling may be apt where strong direct network effects are thereby diminished in an already-tipped market, using *Expedia* as an example: Petit, *supra* note 10, at 217.

[31] Sabre Corporation v. Competition and Markets Authority [2021] CAT 11, ¶¶ 302–307, *see* Megan Yeates, Theodore Souris, Thomas McGrath & Martin McElwee, *The UK Competition Appeal Tribunal, in a merger case involving two companies providing technology solutions to the travel industry, confirms the Competition Authority's broad discretion to review deals with limited UK nexus (Sabre / Farelogix)*, e-Competitions May 2021, art. No. 101018.

[32] Meta Platforms v. CMA [2022] CAT 26, ¶ 110 (noting stages of analysis but not defining evidence in relation to them: (i) potential competition; (ii) the timeframes in which the perceived impairment will take place; (iii) an empty statement to "keep well in mind the particular positions of each merging party" in both static and potential terms; (iv) a requirement to consider the dynamic element of competition from the introduction of new products and new processes; (v) a requirement to consider that some investments will be "duds", such that dynamic competition may not actually manifest itself; and (vi) where there is a conclusion that a merger may substantially lessen dynamic competition, a cross-check of considering the competitive dis-benefits of preventing or unwinding the merger. Absent from the list is any quantitative or qualitative assessment of the evidence in relation to the new product, which instead was remitted to the agency under principles of judicial review).

370. The court also credited the scope for potential competition theories to sound within the meaning of lost competition referred to in Section 7 of the Clayton Act. Critically, however, the court noted that such a future competition theory is only likely in concentrated markets. The issue was instead a factual one: there was no evidence that Meta was, actually, entering the market. The point is striking in relation to *Facebook/Giphy*, where Facebook was also purchasing inputs which it was not yet creating. Thus, it seems that potential competition theories are still subject to a plausibility hurdle from market structure in the US, whereas the requirement appears to have been relaxed in the UK.

371. For AI deals, this may explain why the UK CMA has proven the more active party: if there is, as it would appear, a degree of enforcer consensus as between the US FTC and UK CMA on challenges to incipient consolidation, then the UK provides the more permissive venue for a challenge.

C. *Microsoft/OpenAI; Microsoft/Mistral*

372. It was therefore relatively unsurprising to find that the focus of merger control review of AI deals shifted to the UK CMA in 2024. The FTC faced the limitations above, and the EU arguably lacks jurisdiction because of difficulties in assessing future competition harms under jurisdictional tests even after reforms.

373. The CMA invited comment on two Microsoft AI-related transactions: *Microsoft/OpenAI*, and *Microsoft/Mistral*. The OpenAI case follows the prominent board dynamics of OpenAI in late 2023. The Mistral case followed a small investment by Microsoft in Mistral.

374. As of this writing, there are no details on substantive merger analysis. The OpenAI case page remains open, apparently indefinitely, following a January 2024 invitation to comment. This seems to imply, but not affirmatively state, that the CMA will not challenge the case: as there is no formal case open, one can infer no notification, as a notification would have triggered a formal investigatory clock (at least if accepted as complete). Thus, the CMA's lookback power applies, but this expires after four months. It would be helpful to clarify this aspect of the case file.

375. There is more clarity relating to *Microsoft/Mistral*. The authority declined to act there after seeking views. The procedural posture of this case is curious. The parties received notice that a formal investigation had been opened on May 17, 2024, with a notional deadline of July 4, 2024 based on the four-month lookback deadline. Just days later, on May 21, 2024, the parties received notice that the transaction did not qualify for merger review because Microsoft had not obtained material influence.[33]

376. The obvious question is why the case was formally opened if only to close it days later. Normally a reasonable likelihood of a finding of influence

---

[33] CMA, *Microsoft Corporation's partnership with Mistral AI: Decision on Relevant Merger Situation*, (May 21, 2024), https://assets.publishing.service.gov.uk/media/664c6cfd993111924d9d389f/Full_text_decision.pdf.

would be checked as a precursor to formal investigation. The open-then-closed posture seems to be, perhaps, a warning shot that any larger future transaction would be investigated. As with *Facebook/Giphy*, this gives little indication as to the future substantive analysis since the case was decided on procedural grounds.

D. *Amazon/Anthropic*

377. The same point recurs with *Amazon/Anthropic*.[34] The outcome is, as of this writing, the same as for the OpenAI case: there was an invitation to comment, and there is no formal investigation as of this writing approximately six weeks after the comment period closed. Therefore, there is as yet no indication of substantive analysis.

E. *UK CMA Foundation Models Paper*

378. For a clearer guide to the pertinent substantive analysis, the UK CMA's "initial review" of the AI Foundation Models segment provides some commentary. In May 2023, the UK CMA opened an "initial review" which seems to be an exercise short of a formal Market Study; thus, is not subject to the same procedural rules under the Enterprise Act. The matter was last updated in April 2024.

379. This major update noted 90+ partnerships and a variety of vertically integrated, quasi-integrated and non-integrated approaches to AI investment.[35] A prominent spider's web diagram identified ten rivals in "foundation models" (FM) development.[36] The report also seems to understate the role of non-big tech partnerships, which are not materially assessed.[37] It may prove significant that the paper refers to *both* competition and consumer protection analysis given the dual charge of the CMA as a consumer and competition regulator.[38] That theme is already seen in the prominent Google Privacy Sandbox case, in which consumer-facing analysis of data protection, e.g., as to browser interfaces, is taking place alongside competition analysis.[39]

380. The essential competition argument in the paper is conventional: there is a concern about lost contestable share from lost multi-homing.[40] So, although the paper does not provide any specific comment on market definition, it seems to contemplate the application of doctrines within

---

34 Disclaimer: the author was consulted on the *Amazon/Anthropic* partnership. Therefore, comments on the case are kept relatively limited to avoid potential conflicts of interest in comment.

35 CMA, AI FOUNDATION MODELS: UPDATE PAPER (Apr. 11, 2024), https://assets.publishing.service.gov.uk/media/661941a6c1d297c6ad1dfeed/Update_Paper__1_.pdf.

36 *Id.* Figure 7.

37 I am grateful to Thibault Schrepel for this insight.

38 *Id.* at 1.3.

39 CMA, REVIEW OF GOOGLE PRIVACY SANDBOX: UPDATE PAPER (Apr. 26, 2024), https://assets.publishing.service.gov.uk/media/65ba2a504ec51d000dc9f1f5/A._CMA_Q1_2024_update_report_on_Google_Privacy_Sandbox_commitments_24.4.24.pdf.

40 *Id.* at 5.11.

which market definition arises. Implicitly, the climbdown on substantive merger control in the UK Digital Markets, Competition and Consumers Act process, such that the same competition analysis remains for mergers albeit with enhanced jurisdictional scope, means that market definition for the contestable share argument would still remain.

381. It will be interesting to see how the new Digital Markets Unit chooses to use the merger power in relation to new powers to set codebook rules (Section 20 DMCC) and to order pro-competitive interventions to address failures in competition (Section 46 DMCC). If choosing to challenge using the mergers powers, then a market will have to be defined in relation to foundation models. This may encourage the CMA towards regulation rather than merger review, especially considering the new powers to regulate data use and business activity in adjacent markets in the Digital Markets, Competition and Consumers Act (DMCC) rulebook provisions.[41]

F. US FTC AI-related Information Request

382. Finally, we can note the questions sent by the US FTC to Alphabet, Inc., Amazon.com, Inc., Anthropic PBC, Microsoft Corp., and OpenAI, Inc. on January 25, 2024.[42] The five questions asked are all geared to the relationships between companies in the partnerships also seen in the UK CMA matter. Thus, as with the UK CMA matter, there is no direct information on relevant market definition. However, there is a very clear warning shot and as with the UK CMA position, there will yet be a need to define markets, not least following *Meta/Within*. If there is a merger challenge to an AI transaction following the information request, it will be interesting to see how far the FTC pushes for a narrow market definition considering its win on market definition but loss on competitive effects in *Within*.

V. Conclusion

383. It is well known that merger control cannot and is not required to define the future. As crisply noted in the 2010 iteration of the US Horizontal Merger Guidelines: "certainty about anticompetitive effect is seldom possible and not required for a merger to be illegal."

384. Even if there is no crystal ball, there is now an indication of where the analysis will focus in any future challenge to AI mergers. In particular, the CMA Update Paper suggests a strong focus on contestable share analysis. Likewise, the US Federal Court's position in *Meta/Within* shows that analysis of market shares and market dynamics will both continue to be relevant.

---

41 Digital Markets, Competition and Consumers Act 2024, § 20(3)(c) and (f).

42 FTC Press Release, FTC launches FTC Launches Inquiry into Generative AI Investments and Partnerships Agency Issues 6(b) (Jan. 25, 2024), https://www.ftc.gov/news-events/news/press-releases/2024/01/ftc-launches-inquiry-generative-ai-investments-partnerships.

385. That is so despite the refresh to the underlying competition law in the case of the UK. In that process, the former CMA Chairman Lord Tyrie noted that there was no need for a crystal ball to understand risks from market tipping: "we have the history."[43] The comment cuts both ways: not only should the perceived shortfalls of the Web 2.0 antitrust analysis be avoided, but there is also continuing relevance in some doctrines, notably market definition.

386. It is not yet clear however, how contestable share will map to the possibility of large open-source platforms without a power to exclude rivals. So, it will be especially interesting to see how competition agencies approach the web of interconnection in AI-related markets. If these represent a creative commons, then contracts to harness the commons are particularly valuable.[44] Competition may even take the form of open contracting to prevent others from creating lock-in, and if so, one large platform with open governance may be the optimal solution. If that is the case, then effects analysis can logically support even a 100 percent market share in the short term, as there is no prospect of lock in – while attaining efficiencies from large scale platform operations.[45]

387. Part of the enterprise, then, is simply to apply the existing frameworks in order to identify market power risks on an evidenced basis. What exactly that means will be evolutionary, and not revolutionary, there having been no revolution in merger clearance law, at least so far, whether in the US, EU or UK. So, however much some may have criticized the logic, the orthodoxy requiring a statement of market definition remains even in AI merger cases. What exactly that definition will be is a fulcrum of debate in the coming cases – that is, if the continuing need to answer the question in merger cases does not tilt cases towards new powers to regulate conduct instead.

---

[43] Lord Tyrie, HL Deb (Jan. 22, 2024) (835) col. 355GC.

[44] Thibault Schrepel & Jason Potts, *Measuring the Openness of AI Foundation Models: Competition and Policy Implications* (Sci. Po Digit., Working Paper, 2024), https://papers.ssrn.com/sol3/papers.cfm?abstract_id=4827358.

[45] *Id.* at 29.

Finding the *Ghost in the Shell*: EU and US Antitrust Enforcement of AI Collusion

"There's nothing sadder than a puppet without a ghost, especially the kind with red blood running through them."
  —Batou

CHRISTIAN BERGQVIST[*] AND CAMILA RINGELING[**]
University of Copenhagen
The George Washington University

Abstract

In recent years, considerable focus has been directed toward AI collusion and whether the proliferation of AI-driven decision-making could enable entities to coordinate in anti-competitive ways beyond the scope of enforcement. While the perceived risk may seem exaggerated – antitrust law is well equipped to deal with these questions, and AI may not lead to unavoidable collusive outcomes – the question deserves careful consideration. Notably, US enforcers face challenges in contrast to their EU colleagues, as US case law has developed a narrow notion of "agreement" or "understanding" under

---

[*] Dr. Christian Bergqvist is an Associate Professor University of Copenhagen and Senior Fellow at GW Competition and Innovation Lab at The George Washington University.

[**] Camilla Ringeling is an Associate at Hausfeld and a Senior Fellow at GW Competition and Innovation Lab at The George Washington University. Hausfeld represents clients in litigation concerning AI collusion. The views expressed in this chapter do not represent Hausfeld or its clients.
The authors have no conflict of interest to declare and can be reached at cbe@drbergqvist.dk and milaringeling@gmail.com. The authors thank Christopher Leslie, Eleanor Fox, Steve Salop, Hazel Berkoh, Philip Hanspach, Carina Dall, Alden Abbott, Giorgio Monti, Joseph E. Harrington, Jonathan Rubin, and Cristina Volpin for their comments, suggestions, and contributions.

Finding the *Ghost in the Shell*: EU and US Antitrust Enforcement of AI Collusion

high and inconsistent evidentiary burdens. Despite the odds, AI-assisted collusion is currently being challenged more aggressively through ex-post action in the US and discussed in the EU. The outcomes and effects of this new wave of litigation remain to be seen. This chapter explores antitrust enforcement of AI collusion to identify potential enforcement gaps and available remedies on both sides of the Atlantic.

Since the launch of Ezrachi's and Stucke's book[1] on virtual competition in 2016, enforcers and practitioners have pondered[2] if auto-generated price adjustments and AI-supported price systems could undermine effective antitrust enforcement. No consensus has emerged,[3] and in addition to facilitating the formation and maintenance of traditional cartels, the fear is that AI can lead to tacit coordination that escapes enforcement under antitrust laws. This could happen when AI scans the internet for competitors' prices and makes autonomous decisions, including competitors' possible reactions.[4] Under certain circumstances, this could reduce competition and create parallelism outside enforcement's reach.

Enforcers on both sides of the Atlantic have sought to address this issue by identifying the *ghost behind the shell*. This is, the human will and responsibility in designing and implementing the AI, as well as the *meeting of the minds* to use it in a way that leads to collusive outcomes. Ultimately identifying the spirit of AI collusion, where skilled programmers and developers, like puppeteers, harness the power of algorithms and big data to influence market outcomes.

This chapter proceeds as follows. Section I presents the AI-supported decision-making issue and concludes that antitrust in the EU and the US has successfully addressed facilitating and implementing conduct. Section II explores the more difficult application of the notions of concerted practice or action, highlighting potential gaps in the legal frameworks and concluding the US requires a more principled approach to the evidentiary burdens for proving collusion through indirect evidence and alternatively if cases fall under the rule of reason, a quick look to condem. Section III briefly references proposals for *ex-ante* regulatory interventions, recommending careful consideration of empirical evidence and identifying market failures. Section IV sums up the key differences between the EU and US and offers conclusions.

---

1   Ariel Ezrachi & Maurice E. Stucke, Virtual Competition: The Promise and Perils of the Algorithm-driven Economy (Harv. Univ. Press 2016).

2   *See, e.g.,* Competition and Markets Authority, *Pricing Algorithms: Economic Working Paper on the Use of Algorithms to Facilitate Collusion and Personalized Pricing*, (Oct. 2018), https://assets.publishing.service.gov.uk/government/uploads/system/uploads/attachment_data/file/746353/Algorithms_econ_report.pdf; OECD, *Algorithms and Collusion Competition Policy in the Digital Age*, (May 17, 2017), https://www.oecd-ilibrary.org/finance-and-investment/algorithms-and-collusion-competition-policy-in-the-digital-age_258dcb14-en; OECD, *Algorithmic Competition – OECD Competition Policy Roundtable Background Note,* (May 10, 2023), https://www.oecd-ilibrary.org/finance-and-investment/algorithmic-competition_cb3b2075-en; Bundeskartellamt Press Release, The French Autorité de la concurrence and the German Bundeskartellamt Present Their Joint Study on Algorithms and Competition, (2019), https://www.bundeskartellamt.de/SharedDocs/Meldung/EN/Pressemitteilungen/2019/06_11_2019_Algorithms_and_Competition.html.

3   *See, e.g.,* Cento Velkanovski, *What Do We Know about "Machine Collusion"*, 13 J. Eur. Competition L. & Prac. 47–50 (2022) (challenging the alarmist calls from other academics).

4   *See generally*, Ai Deng, *What Do We Know About Algorithmic Tacit Collusion?,* 33 Antitrust 88 (2018).

I. Does AI-Supported Decision-Making Raise Concerns?

388. There is no consensus on whether AI-supported pricing decisions are anti-competitive. The obvious reason is that AI as a tool, like any other, can be used legally or illegally and may lead to efficient or inefficient outcomes. AI-supported decision-making has evident advantages. By facilitating a more profound understanding of the market and customer preferences, not only through price discrimination[5] but, more broadly, through quality and demand fluctuations,[6] AI can reduce costs and allow for more informed decisions on the supplier side.[7] AI can also reduce transactional costs by matching supply and demand across nominal independent providers (a classic example is Uber) and lowering entry barriers.[8] These efficiencies may be passed on to consumers as benefits of a more competitive service in the form of higher quality and/or lower prices. AI can also empower consumers[9] by reducing shifting costs, highlighting alternatives, and allowing them to form buyer groups. Despite these advantages, pricing algorithms may also increase entry barriers when relying on proprietary data.[10] Moreover, AI-supported decision-making can reduce competition by facilitating, implementing, or creating a cartel.[11] For example, if the algorithm predicts how competitors will respond to price reductions or increases, it could lead to an understanding of a mutually beneficial course of action and facilitate coordination.[12] In particular, if the AI was programmed with self-learning abilities, "there is growing experimental

---

5   For more, *see, e.g.,* Terrell McSweeny & Brian O'Dea, *The Implications of Algorithmic Pricing for Coordinated Effects Analysis and Price Discrimination Markets in Antitrust Enforcement*, 32 Antitrust 75 (2017).

6   For a short introduction to AI and algorithmic competition, *see, e.g.,* Avigdor Gal, It's a Feature, Not a Bug: On Learning Algorithms and What They Teach Us, OECD Roundtable on Algorithms and Collusion (June 21–23, 2017); OECD, *Algorithmic Competition, supra* note 2, at 9–12; *see also* Karsten T. Hansen, Kanishka Misra & Mallesh M. Pai, *Collusive Outcomes via Pricing Algorithms* 12 J. Eur. Competition L. & Prac. 334–337 (2021).

7   This probably explains why 53 percent of the responding retailers confirmed relying on software to track competitors' prices in European Commission, Staff Working Document accompanying the document report from the Commission to the Council and the European Parliament final report on the E-commerce Sector Inquiry, SWD (2017) 154 final, at 149; *see also* Commission Decision of July 24, 2018 relating to a proceeding under Article 101 of the Treaty on the Functioning of the European Union and Article 53 of the EEA Agreement Case AT.40182 (*Pioneer (vertical restraints)*), 2018, O.J. (C 338), at ¶ 136.

8   This partly explains why the Brazilian Competition Authority decided not to proceed with a case against Uber *cf.* OECD, *Algorithmic Competition, supra* note 2, at 8.

9   *See, e.g.,* Michael S. Gal & Niva Elkin-Koren, *Algorithmic Consumers*, 30 Harv. J. L. & Tech. (2017).

10 *See, e.g.,* Daniel L. Rubinfeld & Michal S. Gal, *Access Barriers to Big Data*, 59 Ariz. L. Rev. 339, 373 (2016).

11 *See, e.g.,* Salil K. Mehra, *Price Discrimination-Driven Algorithmic Collusion: Platforms for Durable Cartels*, 26 Stan. J. L. & Bus. Fin. 171, 177 (2021); Salil K. Mehra, *Antitrust and the Robo-Seller: Competition in the Time of Algorithms*, 100 Minn. L. Rev. 1323, 1328 (2016) ("The Sherman Act contains a gap in its coverage under which oligopolists that can achieve price coordination interdependently, without communication or facilitating practices generally escape antitrust enforcement, even when their actions yield supracompetitive pricing that harms consumers." (footnote omitted)).

12 Actual examples of this are limited, but one regarding the German retail gasoline sector might be cited in Stephanie Assad, Robert Clark, Daniel Ershov & Lei Xu, *Algorithmic Pricing and Competition: Empirical Evidence from the German Retail Gasoline Market* 4–5 (CESifo, Working Paper No. 8521, 2020). Another involves selling second-hand copies of a book (The Making of A Fly) on Amazon, *cf.* John D. Sutter, *Amazon Seller Lists Book at $23,698,655.93 – Plus Shipping*, CNN (Apr. 25, 2011), http://edition.cnn.com/2011/TECH/web/04/25/amazon.price.algorithm/index.html. Finally, OECD, *Algorithmic Competition, supra* note 2, at 14–15, refers to a French investigation into auto spare parts.

evidence that an algorithm can be designed to collude tacitly."[13] Before analyzing the applicable legal theories, we discuss the anti-competitive potential of AI decision-making.

1. What Does AI-Supported Decision-Making Do?

389. The advantages of AI-supported decision-making[14] are apparent but cannot be isolated from their potential negative impact on competition if used for anti-competitive coordination.[15] This could happen in at least the following scenarios.[16]

   1. *Monitoring an existing cartel.* The AI is used to monitor cartels, removing their normal instability and enforcing compliance,[17] e.g., a central enforcer and parallel exchange of data showing members' compliance with agreed quotas, prices, allocated markets, or customers. Delegating these monitoring tasks to an AI can secure the cartel's longevity.

   2. *Facilitate entering into a new cartel.* The AI is used to facilitate entering into a collusive agreement. Where the parties indicate interest in entering an anti-competitive partnership but struggle to reach a consensus on the cartel's operative parts, including specific details such as prices or customer allocation AI can assist the parties in understanding and accepting the benefits of cooperating and even eliminate internal mistrust, as the AI may eliminate cheating.

   3. *Facilitate entering into a cartel without direct contact.* The AI facilitates the exchange of commercially sensitive information that allows the parties to collude without direct contact. Either as a variation of 1.) or 2.), where the parties have already indicated an interest in forming a cartel or as part of a meeting of the minds regarding this outcome. Moreover, this might involve a third party acting as an intermediary or facilitator, e.g., by providing the algorithm that induces collusion.

   4. *Facilitate tacit collusion.* The AI facilitates tacit collusion, where the parties or their AI system reach an unspoken understanding of prices

---

13 Deng, *supra* note 4. For an empirical study, rebutting the risk as material, *see* Philip Hanspach, Geza Sapi & Marcel Wieting, *Algorithms in the Marketplace: An Empirical Analysis of Automated Pricing in E-Commerce* (NET INST., Working Paper No. 21-06, Mar. 11, 2024), https://papers.ssrn.com/sol3/papers.cfm?abstract_id=3945137

14 According to the OECD, a distinction can be made between *machine learning* and *deep learning*. The former covers supervised learning, unsupervised learning, and reinforcement learning, where the algorithm learns based on the provided data and thus involves human decisions. In contrast, deep learning is more autonomous, and the human involved is minimized. *See* OECD, *Algorithmic Competition*, *supra* note 2, at 9–10.

15 AI could also be abused to provide vertical and horizontal leverage. Here, the owner of an AI could engage in self-favoring directed at monopolizing adjacent markets and services or perfect price discrimination, maximizing producer welfare at the expense of consumer welfare. None of these theories of harm will be explored in this chapter.

16 *See, e.g.*, OECD *Algorithms and Collusion*, *supra* note 2, at 33–40; *see also* EZRACHI & STUCKE, *supra* note 1, at 39–81.

17 For a more detailed explanation, *see, e.g.*, Joseph E. Harrington Jr., *How Do Cartels Operate* in FOUND. & TRENDS MICROECON. 43–72 (2006).

or market/customer allocation without an explicit or tacit agreement.[18] Tacit collusion typically becomes a concern only in highly concentrated markets,[19] where a degree of interdependence may develop, enabling tacit coordination. AI's capacity to analyze large amounts of data and increase transparency may replicate this situation in non-oligopolistic and fragmented markets making, them more prone to collusive outcomes.

390. The anti-competitive risk of algorithms ultimately depends on how advanced AI systems are and the available data they draw from. It also arguably depends on what safeguards are put in place in their design and implementation. Commentators highlighted this challenge in 2017, finding that rather than more advanced pricing AI, it is the proliferation of big data that has increased AI capabilities.[20] The use of computers and algorithms to optimize prices has been utilized by airlines since the 1980s but did not gain attention from enforcers until the 1990s.[21] In particular, hypothesis 4.), where AI induces tacit collusion through autonomous decision-making, might require very advanced AI.

2. Enforcers Have Addressed Coordination through AI

391. Although AI auto-generated price adjustments and recommendations could theoretically increase or facilitate anti-competitive conduct, this behavior may not present novel legal challenges. In other words, the legal challenges seem circumscribed to the US policy choice of adopting a narrow concept of "agreement" or "understanding" as opposed to its broad conception in EU competition law. This, of course, is not new.

392. Additionally, as we explain below, enforcers in both the EU and the US have investigated and successfully prosecuted AI pricing conduct facilitating or implementing collusion, which suggests their enforcement toolkit is generally well-equipped to address these challenges.[22] Moreover, enforcers on both sides of the Atlantic have taken a "compliance by design"[23] approach under which defendants may not hide behind the "shell" of the automation but are responsible for designing and overseeing that their algorithms do not fix prices.

---

18 This scenario could also potentially consider a situation where the parties unilaterally utilize the same AI and end up pursuing parallel behavior if there was no agreement between them.

19 For more on this *see e.g.* MARC IVALDI, BRUNO JULLIEN, PATRICK REY, PAUL SEABRIGHT & JEAN TIROLE, THE ECONOMICS OF TACIT COLLUSION IDEI, FINAL REPORT FOR DG-COMP. (Eur. Comm'n, Mar. 2003). For an attempt to apply this to AI, *see e.g.,* OECD, *Algorithms and Collusion, supra* note 2, at 19–32.

20 *See, e.g.,* Ulrich Schwalbe, *Algorithms, Machine Learning, and Collusion,* 14 J. COMP. L. & ECON. 568–607.

21 *See* United States v. Airline Tariff Publ'g Co., 836 F. Supp. 9 (D.D.C. 1993).

22 *See* EZRACHI & STUCKE, *supra* note 1, at 39–42.

23 *See* Margarethe Vestager, Speech, Bundeskartellamt 18th Conference on Competition (Mar. 16, 2017) ("What businesses can – and must – do is to ensure antitrust compliance by design."); *See* Hannah Garden-Monheit & Ken Merber, *Price Fixing by Algorithm is Still Price Fixing,* FTC (Mar. 1, 2024), https://www.ftc.gov/business-guidance/blog/2024/03/price-fixing-algorithm-still-price-fixing ("your algorithm can't do anything that would be illegal if done by a real person.").

Finding the *Ghost in the Shell*:
EU and US Antitrust Enforcement of AI Collusion

A. *The EU Has Successfully Prosecuted AI Monitoring and Facilitating Conduct*

393. The Directorate General for Competition of the European Commission (DG Comp) has investigated infringements of Article 101 of the Treaty on the Functioning of the European Union (TFEU) (Article 101) involving AI in *Asus*,[24] *Philips*,[25] *Pioneer*,[26] and *Denon & Marantz*.[27] The four cases, decided on the same day in 2018, involved using AI to monitor compliance with an illegal retail price maintenance policy.[28] Because the parties had already reached an (illegal) understanding[29] and relied on AI only to secure compliance, there was no controversy in finding an agreement, as expressed by DG COMP:

> ... firms involved in illegal pricing practices cannot avoid liability on the ground that their prices were determined by algorithms. Just like an employee or an outside consultant working under a firm's "direction or control,", an algorithm remains under the firm's control, and therefore the firm is liable even if its actions were informed by algorithms.[30]

394. This is an essential qualification, as cartels tend to collapse unless enriched with compliance and monitoring tools. AI can facilitate this, but this question alone presents no legal novelty, even when decision powers are delegated to the AI, because parties have entered into a traditional agreement, relying only on AI to secure compliance. The same should apply when parties intend to coordinate but struggle to reach a consensus on the operative parts of the agreement, using an AI to implement the cartel. In *VM Remonts*,[31] the European Court of Justice found that an undertaking[32] could be held liable for an external service provider when the service provider acted under its direction. This situation is easily applicable to scenarios in which decision-making powers are delegated to an AI whether directly operated by the firms or a

---

24 Commission Decision of July 24, 2018 relating to a proceeding under Article 101 of the Treaty on the Functioning of the European Union Case AT.40465 (*Asus (vertical restraints)*), 2018, O.J. (C 338), at ¶ 27.

25 Commission Decision of July 24, 2018 relating to a proceeding under Article 101 of the Treaty on the functioning of the European Union Case AT.40181 (*Philips (vertical restraints)*), 2018, O.J. (C 340), at ¶ 64.

26 Case AT.40182, *Pioneer (vertical restraints)*, *supra* note 7, at 155.

27 Commission Decision of July 24, 2018 relating to a proceeding under Article 101 of the Treaty on the Functioning of the European Union Case AT.40469 (*Denon & Marantz (vertical restraints)*), 2018, O.J. (C 335), at ¶ 95.

28 As the cases involved illegal vertical restraints, they neither qualify as cartels nor pertain to all of the four scenarios outlined initially. Regardless, they do speak to DG COMP's awareness of the risk presented by AI.

29 In the EU, agreements on price coordination, including very indirect arrangements such as the exchange of sensitive business information, would almost per se be an infringement of Article 101 and thus illegal.

30 *See, e.g., Guidelines on the Applicability of Article 101 of the Treaty on the Functioning of the European Union to Horizontal Co-Operation Agreements*, at ¶ 379, COM (2023) 3445 final (June 1, 2023).

31 Case C-542/14, SIA VM Remonts v. Konkurences padome, ECLI:EU:C:2016:578, ¶¶ 27–33, *see* Michel Debroux, *Purely internal situation: The Court of Justice of the European Union clarifies the limited conditions under which a company can be held liable for the anticompetitive behavior of a service provider (SIA "VM Remonts")*, Concurrences No. 4-2016, art. No. 82037.

32 "[t]he concept of an undertaking encompasses every entity engaged in an economic activity, regardless of the legal status of the entity and the way in which it is financed ..." Case C-41/90, Klaus Höfner and Fritz Elser v. Macratron GmbH, ECLI:EU:C:1991:161, ¶ 21.

B. *Enforcement of AI Monitoring and Facilitating Conduct Has Also Been Successful in the US*

395. The US Federal Trade Commission (FTC) and the US Department of Justice (DOJ), (Agencies)[33] have joined their EU counterparts in their focus on AI. Lina Khan, FTC Chair, stated in 2023 that: "Although [AI] is novel, [it is] not exempt from existing rules, and the FTC will vigorously enforce the laws we are charged with administering."[34] "… the A.I. tools that firms use to set prices for everything from laundry detergent to bowling lane reservations can facilitate collusive behavior that unfairly inflates prices (…)."[35]

396. The FTC has provided more insight into its concerns regarding AI and how it could be anti-competitive, explaining that businesses may not: "(i) use an algorithm to evade the law banning price-fixing agreements" and that (ii) "an agreement to use shared pricing recommendations, lists, calculations, or algorithms can still be unlawful even where co-conspirators retain some pricing discretion or cheat on the agreement."[36]

397. In other words, while AI may be, in effect, automated conduct, companies remain liable given their decision to design, deploy, and/or utilize AI tools.

398. While these statements are recent, and there is renewed and heightened interest in the issue, the Agencies have investigated and prosecuted similar conduct since the 1990s.[37] In 1993, the DOJ brought a seminal case concerning collusion through a centralized pricing system in *Airlines*.[38] The case, which ended with a settlement by all defendants, concerned allegations that six airlines had utilized a collectively owned computerized online booking system known as the Airline Tariff Publishing Company

---

[33] The DOJ enforces Section 1 of the Sherman Act, a criminal and civil statute. The FTC may prosecute all types of behavior that violate the Sherman Act pursuant to Section 5 of the FTC Act, a civil statute (prohibiting "unfair methods of competition"), which courts have suggested may reach beyond the ambits of the Sherman Act. The FTC has entered into Section 5 settlements whereby firms agreed not to issue "invitations to collude," which would not qualify as agreements under Section 1 of the Sherman Act. Hardcore cartel price-fixing cases are normally only brought by the DOJ because it has criminal prosecution powers not possessed by the FTC. Indeed, the FTC typically turns a matter over to DOJ when it sees evidence of hard-core collusion.

[34] *See* Fed. Trade Comm'n, *Policy Statement of the Federal Trade Commission on Biometric Information and Section 5 of the FTC Act,* (May 18, 2023), https://www.ftc.gov/legal-library/browse/policy-statement-federal-trade-commission-biometric-information-section-5-federal-trade-commission.

[35] *See* Lina Khan: *We Must Regulate A.I. Here's How*, N. Y. TIMES (May 3, 2023), https://www.nytimes.com/2023/05/03/opinion/ai-lina-khan-ftc-technology.html.

[36] Fed. Trade Comm'n, *supra* note 23.

[37] *See* Maureen K. Ohlhausen, *Should We Fear the Things That Go Beep in the Night? Some Initial Thoughts on the Intersection of Antitrust law and Algorithmic Pricing*, FTC (May 23, 2017), https://www.ftc.gov/news-events/news/speeches/should-we-fear-things-go-beep-night-some-initial-thoughts-intersection-antitrust-law-algorithmic, ("[j]ust as the antitrust laws do not allow competitors to exchange competitively sensitive information directly in an effort to stabilize or control industry pricing, they also prohibit using an intermediary to facilitate the exchange of confidential business information.")

[38] United States v. Airline Tariff Publ'g Co., 836 F. Supp. 9 (D.D.C. 1993).

(ATP) to collude regarding airline fares. While ATP served as a platform for disseminating fare information to the public, it also allowed airlines to conduct discreet discussions on fare strategies.

399. A more recent example was *Topkins,* where the DOJ charged a relatively small e-commerce poster retailer and two executives who sold and distributed their products through Amazon with price fixing.[39] The reason for the DOJ's attention to this rather small matter was the use of a pricing algorithm. To align their prices, retailers configured their algorithm to identify the lowest price a non-conspiring competitor offered for a specific poster. Through this system, their prices appeared higher on searches and eliminated the competition between them, which would have lowered prices. Once the system was put in place, the conspiracy was largely self-executing. The case ended with a plea agreement.[40]

400. These precedents suggest that, like DG Comp, the DOJ has successfully investigated and enforced antitrust laws in cases involving centralized systems or AI to monitor or facilitate cartels, aka scenarios 1.) and 2.) above.

3. Potential Concerns Are Limited to the Boundaries of Concerted Practices

401. Prosecuting AI conduct that eliminates the challenges of running a cartel as a facilitation or implementation tool presents no material legal obstacles.[41] In the EU, this prosecutorial scope is broader. It extends, even to undertakings partly informed of the original cartel, who later adopt the AI and may be presumed full cartel members unless they can successfully establish ignorance of the link between the cartel and the AI.[42] In the US, as will be discussed,[43] the burden is inverted, and it's the plaintiff who must prove that a member later adopting the AI did it to facilitate or implement a collusive agreement. The same would apply to scenarios where the undertakings delegate decision-making powers to the AI or use the AI to facilitate indirect contact, as the parties have accepted membership in a partnership to reduce competition between them. Interestingly, although EU precedent offers a more pro-enforcement framework, the prosecution of AI-assisted cartels has been more prevalent in the US. The outcomes and effects of this new wave of US litigation remain to be seen.

402. Against this background, concerns regarding potential underenforcement should be focused on collusive outcomes achieved through "tacit agreements"

---

39 United States v. Topkins, No. CR 15-00201, (N.D. Cal. Apr. 30, 2015); DOJ Press Release, Online Retailer Pleads Guilty for Fixing Prices of Wall Posters (Aug. 11, 2016).

40 *See id.* Plea Agreement (Doc. No. 7).

41 Naturally, detecting the infringement will always present challenges, but this is more of a factual inquiry than a legal question. Some thought on this is offered in A. Ezrachi & M. E. Stucke, Algorithmic Collusion: Problems and Counter-Measures, Note by OECD Roundtable on Algorithms and Collusion (June 22–25, 2017). *See also* OECD, *Algorithmic Competition, supra* note 2, 28–35.

42 The matter is discussed in detail in Section II.1.

43 The matter is discussed in detail in Section II.2.

and analyzing interactions falling short of the notions of an *understanding* or agreement. There is a grey area between illegal collusion (explicit or tacit agreement) and legal conscious parallelism, where the latter represents an intelligent adaptation to the prevailing market conditions. Courts in the EU and the US have been unwilling to condemn this parallel behavior when decision-making is unilateral. Some degree of AI scanning of the market is accepted as nothing more than a sophisticated way of market monitoring and adaptation, no longer confined to anonymous calls or "mystery shopping." The true fear is, therefore, not only that AI coordination may fall outside the reach of enforcers but that AI will expand the existing grey area.[44]

II. The Notion of an Understanding in EU and US Antitrust

403. While enforcement of Article 101 TFEU and Section 1 of the Sherman Act differ in important respects, neither condemns *conscious parallelism*, as their infringement requires a *concurrence of wills*[45] or a *meeting of the minds*, expressing the parties' shared interest. When independent undertakings unilaterally adopt parallel behavior, their conduct does not fall afoul of either Article 101 or Section 1. Enforcers' ability to react against any anti-competitive aspect of AI-assisted decision-making mainly depends on the reach of the notions of a meeting of the minds or agreement and the evidentiary burden required to prove them. In this regard, of the four scenarios presented initially,[46] only scenarios 3.) *Facilitate entering into a cartel without direct contact* and 4.) *Facilitate tacit collusion*, where the AI induces collusion without an explicit agreement, warrant consideration, since we can assume that 1.) *Monitoring an existing cartel*, and 2.) *Facilitate entering into a new cartel* would be adequately covered by Article 101 and Section 1. In these scenarios, the parties have directly or indirectly expressed interest in entering an (illegal) anti-competitive agreement. In contrast, there is no direct contact, or potentially no contact at all, between the colluding parties in scenarios 3.) *Facilitate entering into a cartel without direct contact*, and 4.) *Facilitate tacit collusion*, as these either rely on third parties to coordinate using the same AI or their parallel but unilateral use of AI to bring about the desired coordination.

---

44   *Cf.* OECD, *Algorithmic Competition*, *supra* note 2, at 25.

45   *See, e.g.,* Case C-211/22, Super Bock Bebidas SA v. Autoridade da Concorrencia, ECLI:EU:C:2023:529, ¶ 49; *see also* Victor Levy & Pierre Chellet, *The EU Court of Justice delivers a judgment to clarify the status of resale price maintenance under EU Competition Law (Super Bock)*, E-COMPETITIONS June 2023, art. No. 113173, and Case T-41/96, Bayer AG v. Comm'n of the Eur. Cmtys., ECLI:EU:T:2000:242, ¶¶ 69; *see also* Annette Kliemann, *The EU Commission files an appeal before the EU Court of Justice against the annulment by the Court of First Instance of its decision to fine a German pharmaceuticals company for prohibiting exports (Bayer)*, E-COMPETITIONS Oct. 2000, art. No. 39124 ; *See* Brooke Grp. Ltd. v. Brown & Williamson Tobacco Corp., 509 U.S. 209, 227 (1993) (holding that "conscious parallelism" is "not in itself unlawful"); Theatre Enters., Inc. v. Paramount Film Distrib. Corp., 346 U.S. 537, 541 (1954) ("Circumstantial evidence of consciously parallel behavior may have made [h]eavy inroads into the traditional judicial attitude toward conspiracy; but 'conscious parallelism' has not yet read conspiracy out of the Sherman Act entirely.").

46   *See supra*, Section I.1.

Finding the *Ghost in the Shell*:
EU and US Antitrust Enforcement of AI Collusion

1. Conspiracy under EU Competition Law

404. Under EU Competition law,[47] companies are prohibited from engaging in anti-competitive coordination through agreements, decisions, or concerted practices. Agreements and decisions are well-established legal concepts encompassing situations in which parties convey their intent and commitment to be legally bound.[48] In contrast, concerted practices represent an abnormality that significantly expands the reach of Article 101, to encompass interactions that don't normally create (legal) obligations. The lines between the concepts are blurred and overlapping but include all forms of coordination and collusion that faithfully express the parties' intentions.[49] In practice, agreements and concerted practices are only distinguishable by their intensity and forms.[50] A decision might also result from a former agreement and be implemented by a concerted practice. As a practical matter, larger and more complex infringements might consist of a mix of agreements and concerted practices,[51] perhaps blending in decisions that bind the members together, as only the original perpetrators are privy to the formal agreement establishing the infringements. While desirable, enforcers are not legally obligated to attribute the (potentially) anti-competitive actions unambiguously to one specific agreement,[52] making it sufficient that the actions, as a minimum, qualify as a concerted practice.

405. To answer whether AI collusion can be effectively enforced under Article 101, only the notion of a concerted practice needs to be explored, as this represents the lower threshold for finding a "meeting of the minds."

A. *Concerted Practice*

406. It follows from the wording of Article 101 that an understanding can also come in the form of a *concerted practice*. While this term is not defined in

---

[47] The EU applies a uniform competition code to cartels, making Article 101 directly applicable in all member states and prohibiting the adoption of more restrictive national measures for actions covered by Article 101. The latter precludes national laws overlapping with Article 101 unless they conform with it.

[48] "Decisions" were included to prevent undertakings from evading on account of the form they structure their anti-competitive coordination *cf.* Case T-111/08, MasterCard Inc. v. Eur. Comm'n, ECLI:EU:T:2012:260, ¶ 243, but particularly found usefulness against trade organizations, *see* Michel Debroux, *Multilateral interchange fees: The General Court confirms a Commission's decision against an international payment organisation (Mastercard)*, CONCURRENCES No. 3-2012, art. No. 48294.

[49] *Cf.* Joined Cases C-2/01 P and C-3/01 P, Bayer AG v. Comm'n of the Eur. Cmtys., ECLI:EU:C:2004:2, ¶ 97.

[50] Case C-49/92P, Comm'n of the Eur. Cmtys. v. Anic Partecipazioni SpA, ECLI:EU:C:1999:356, ¶ 131–133.

[51] *Cf.* Case T-305/94, Limburgse Vinyl Maatschappij NV v. Comm'n of the Eur. Cmtys., ECLI:EU:T:1999:80, ¶ 698.

[52] Case C-8/08, T-Mobile Netherlands BV v. Raad van Bestuur, ECLI:EU:C:2009:343, ¶ 23–24; *see* Dominique Ferré, *Anticompetitive object : The ECJ decides on the criterions to assess the anticompetitive object of an exchange of informations (T-Mobile Netherlands)*, CONCURRENCES No. 3-2009, art. No. 29662 ; *Anic Partecipazioni*, *supra* note 50 and Case C-238/05, ASNEF-EQUIFAX v. AUSBANC, ECLI:EU:C:2006:734, ¶ 31–32; *see* Michel Debroux, *Exchange of information: The ECJ contributes to the antitrust analysis of exchange of information between competitors, and clears the inter-bank systems of exchange of information on clients' solvency (Asnef-Equifax c/ Ausbanc)*, CONCURRENCES No. 1-2007, art. N° 12789.

the EU Treaty, the European Court of Justice[53] has explained that: "... such a practice is a form of coordination between undertakings by which, without it having been taken to the stage where an agreement properly so-called has been concluded, practical cooperation between them is knowingly substituted for the risks of competition." The European Court of Justice[54] has further explained that concerted practices: "... are intended to catch forms of collusion having the same nature and are only distinguishable from each other by their intensity and the forms in which they manifest themselves."

407. Concerted practices are best understood negatively, as they do not constitute a formal agreement or decision but still allow the parties to act in concert. Moreover, an infringement can encompass several parallel concerted practices involving different members with gaps or periods of inaction or a mix of agreements, decisions, and concerted practices.[55] In contrast, unilateral parallel behavior, even if it involves conscious replies to the prevailing market condition and competitors, remains legal, even in markets prone to collusion, uniform prices, etc.

408. A meeting of minds must be established to help delineate the delicate boundary between lawful conscious parallelism and illicit coordination.[56] A meeting of the minds requires three elements: (1) contact between undertakings, (2) directed at coordinating, and (3) parallel commercial behavior. The following section elaborates on the three requirements and then examines whether a distinct approach is necessary for vertical concerted practices.

(1) Contact between undertakings

409. Contacts between competitors must be established to find illegal coordination distinct from legal parallel behavior. The European Court of Justice[57] has set a low barrier for this, stating that: "[this] condition is met where one competitor discloses its future intentions or conduct on the market to another when the latter requests it or, at the very least, accepts it."

410. Accordingly, courts have accepted a single meeting[58] as sufficient, provided that commercially sensitive matters are discussed, thereby reducing

---

53 *T-Mobile Netherlands, supra* note 52; *see* Ferré, *supra* note 52. See also Case C-48/69, Imperial Chemical Industries (ICI) Ltd. v. Comm'n of the Eur. Cmtys., ECLI:EU:C:1972:70, ¶ 64.

54 *Anic Partecipazioni, supra* note 50, ¶¶ 131–133. *See also T-Mobile Netherlands, supra* note 52, ¶ 23; *see also* Ferré, *supra* note 52.

55 *See e.g., Limburgse Vinyl Maatschappij, supra* note 51; and Case T-7/89, Hercules Chemicals NV v. Comm'n of the Eur. Cmtys., ECLI:EU:T:1991:75, ¶ 264.

56 It's possible to read Case T-587/08, Fresh Del Monte Produce Inc v. Eur. Comm'n, ECLI:EU:T:2013:129, ¶ 300 as rebutting meeting of mind as a requirement, but this does not comport well with what has been applied in practice, *see* Nathalie Jalabert-Doury, *Cooperation: The General Court confirms the qualification of concerted practice by object and of single infringement in the banana cartel case (Fresh Del Monte Produce)*, Concurrences No. 2-2013, art. No. 52299.

57 Case T-25/95, Cimenteries CBR SA v. Comm'n of the Eur. Cmtys., ECLI:EU:T:2000:77, ¶ 1849. Confirmed on appeal as Case C-204/00P, Aalborg Portland AS v. Comm'n of the Eur. Cmtys., ECLI:EU:C:2004:6.

58 *T-Mobile Netherlands, supra* note 52, ¶¶ 42–43; *see* Ferré, *supra* note 52.

uncertainty regarding competitors' future conduct.[59] A formal reply or acquiescence is unnecessary, rendering the question of whether an undertaking remained passive during the meeting irrelevant.[60] Even indirect communications through public means, e.g., the press,[61] are considered contacts between competitors if proved to have been used as an invitation to collude to, for example, raise prices. However, stripped of any such contact, the legal position differs as undertakings have a right to adapt to prevailing market conditions, including competitors' decisions,[62] supporting the notion of (legal) conscious parallelism. In concentrated markets prone to (genuine) parallel behavior, enforcers must be mindful of this and refrain from intervening unless parties' intent to collude is established.

411. While case law has been very lenient regarding the quantitative nature of the interaction, finding, for instance, that a single meeting is sufficient to find an agreement, there is an implicit qualitative requirement, as the contact must be directed at securing an anticompetitive objective. This would typically be inferred from unveiling internal, confidential, and business-sensitive information. In the context of larger and more complex cartels, a single meeting might not even be sufficient to ensure compliance and, thus, the longevity of the infringement, adding an unspoken quality requirement to the nature of the agreement.[63] The inability to satisfy the "multiple meetings" standard might even compromise the value of the evidence and, ultimately, the case, as the alleged infringements must be proved, and unsubstantiated assumptions do not suffice.

a) Indirect contacts using an intermediary

412. Contact can also take an indirect form through an intermediary,[64] for example, an IT system or an auto-generated e-mail provided it can reasonably

---

59 *See, e.g.*, Case T-1/89, Rhône-Poulenc SA v. Comm'n of the Eur. Cmtys., ECLI:EU:T:1991:56, ¶¶ 122–123; and Case C-286/13P, Dole Food Co. Inc. v. Eur. Comm'n, ECLI:EU:C:2015:184, ¶ 128–138; *see* Kyriakos Fountoukakos & Kristien Geeurickx, *The EU Court of Justice considers that the bilateral exchange of pre-pricing informations to a concerted practice with the object of restricting competition is an anticompetitive practice (Dole)*, E-COMPETITIONS Mar. 2015, art. N° 79540.

60 Case T-202/98, Tale & Lyle Plc v. Comm'n of the Eur. Cmtys., ECLI:EU:T:2001:185, ¶ 54. Confirmed on appeal as Case C-359/01P, British Sugar Plc. v. Comm'n of the Eur. Cmtys., ECLI:EU:C:2004:255. See also Case T-303/02, Westfalen Gasen Nederland BV v. Comm'n of the Eur. Cmtys., EU:T:2006:374, ¶ 484.

61 Commission Decision of Apr. 23, 1986 relating to a proceeding under Article 85 of the EEC Treaty Case IV/31.149 (*Polypropylene*), 1986, O.J. 1986(L 230), at ¶ 67

62 *Cf.* Joined Cases 40-48/73, C-50/73, C-54-56/73, C-111/73, C-113/73 and C-114/73, Cooperatieve Vereniging Suiker Unie UA v. Comm'n of the Eur. Cmtys., ECLI:EU:C:1975:174, ¶ 174. See also *Tale & Lyle Plc*, *supra* note 60, ¶ 56; and *T-Mobile Netherlands*, *supra* note 52, ¶ 33; *see also* Ferré, *supra* note 52.

63 *See, e.g.*, Case T-240/17, Campine NV v. Eur. Comm'n, ECLI:EU:T:2019:778, ¶ 308, focusing on content and not the number of meetings, *see also* Etienne Thomas, *Fine: The General Court of the European Union partially annuls the European Commission decision regarding an anticompetitive agreement in the purchasing market of car battery recycling and lowers significantly the fine (Campine)*, CONCURRENCES No. 1-2020, art. No. 93291. *See also T-Mobile Netherlands*, *supra* note 52, ¶¶ 60–61, outlining how more complex infringements might require several meetings, *see also* Ferré, *supra* note 52.

64 *Guidelines on the Applicability of Article 101 of the Treaty on the Functioning of the European Union to Horizontal Co-Operation Agreements*, at ¶¶ 401–404, COM (2023) 3445 final (June 1, 2023). See also Case C-74/14, Eturas UAB v. Lietuvos Respublikos Konkurencijos Taryba, ECLI:EU:C:2016:42, ¶¶ 42–44, *see also* Alexandre Lacresse, *Presumption of innocence: The Court of Justice of the European Union considers*

be assumed that the recipient has read the e-mail and decided to follow suit. DG COMP has also identified[65] how the use of category management systems, where most actors in an industry allow a single supplier to manage their inventory, can lead to collusion. The same[66] would apply, according to DG COMP, if several parties rely on a shared optimization algorithm, that recommends a course of action based on commercially sensitive data received from competitors. No actual cases have been pursued, demoting the value of the statements, but it remains an example of how information shared indirectly through an intermediary can constitute a concerted practice, even when stripped of any prior dialogue.[67]

413. The statement also comports neatly with, e.g., *VM Remonets*,[68] where an undertaking was held liable for the acts of an external service provider when either: a) the latter acted under the direction of the former, b) was aware of the anti-competitive objectives pursued by its competitors and willing to contribute, or c) could reasonably have foreseen the anti-competitive acts of the competitors and the services provider. Under the notion of a single and continuous infringement,[69] companies can also be involved in an infringement without knowing all the details as long as they are aware of the overall plan and have accepted its main components. While DG COMP's stance has merit, it may reflect more of an enforcement priority aimed at steering case law in a particular direction rather than adhering to well-established legal principles.

b) Indirect contact through hub-and-spoke agreements

414. In addition to direct contact during a meeting or phone call, consideration has been given to whether various forms of indirect contact and vertical information exchange, referred to as *hub-and-spoke cartels*, fall under Article 101.[70] Under the notion of hub-and-spoke cartels, the perpetrators will (ab)use nominal benign vertical restraints to secure horizontal collusion.[71] More specifically, this can involve either suppliers exchanging

---

that the principle of the presumption of innocence prevents to deduct the undertaking's participation in the agreement from the sole sending of an email (Eturas), CONCURRENCES No. 2-2016, art. No. 79522 , and *SIA VM Remonets, supra* note 31, ¶¶, *see* Debroux, *supra* note 31.

65 *See Guidelines on vertical restraints*, at ¶ 387, COM (2022) O.J. (C 248) 1.

66 *Guidelines on the Applicability of Article 101 of the Treaty on the Functioning of the European Union to Horizontal Co-Operation Agreements*, at ¶ 402, COM (2023) 3445 final (June 1, 2023).

67 As outlined below, the case law does not offer full support for DG COMP's suggestions. It's doubtful that shared use of the same inventory management system would be considered an infringement of Article 101 unless masking an underlying illegal arrangement or involving sharing of commercially sensitive information outside the legitimate object of the system.

68 *SIA VM Remonets, supra* note 31, ¶¶ 27–33, *see* Debroux, *supra* note 31.

69 *See generally*, Christian Bergqvist, *Single and continuous infringement*, 5 EUR. COMPETITION & REGUL. L. REV. 380–393 (2021).

70 This appears to be the position of DG COMP *cf. Guidelines on the Applicability of Article 101 of the Treaty on the Functioning of the European Union to Horizontal Co-Operation Agreements*, at ¶ 402, COM (2023) 3445 final (June 1, 2023).

71 Hub-and-spoke cartels are not confined to information sharing but can involve RPM or MFN clauses. As the former is usually considered close to a *per-se* anti-competitive in the EU, the vertical restraints might not be benign.

information through their distributors or distributors exchanging information through their suppliers. In both cases, such information exchanges seek collusive outcomes but do not have a direct agreement or explicit understanding. The information exchange is vertical, but the effect is horizontal, making it resemble a bicycle wheel, hence the name.

415. Traditionally, vertical information exchanges have not been a priority in the EU,[72] and case law has consistently rejected their falling under Article 101, even in instances involving suspicious parallel public price announcements and coordinated behavior in highly concentrated and transparent.[73] Only when an underlying horizontal understanding is the only plausible explanation for the parallelism has case law applied Article 101, de facto establishing a form of reverse burden of proof that allows the undertakings to rebut the existence of a meeting of mind. This makes sense, as any other position would make it difficult to operate in such a market and would extend Article 101 to cover conscious parallelism. However, recently, DG COMP has returned to the matter of dual distribution.[74] Here, the manufacturer also operates in the downstream retail market, thus competing with its distributor, adding a horizontal twist to what otherwise would be a vertical information exchange. Moreover, DG COMP has identified the risk associated with hub-and-spoke arrangements and the abuse of vertical information exchange to facilitate horizontal collusion,[75] but it remains unclear how a meeting of mind can be demonstrated in the context of indirect information exchange unless the arrangements are part of a larger plan.

(2) Directed at coordinating

416. Article 101 requires that the (allegedly) problematic contact between undertakings be directed at influencing future behavior concerning central competition parameters. However, this does not have to follow from a (master) plan,[76] nor be in the mutual interest.[77] In contrast, whether a level of reciprocity in intent is required remains unsettled, though this interpretation seems to be the more compelling. The (fine) line between legal conscious parallelism and illegal coordination is marked by mutual *interests* in pursuing coordination. The ability to identify some form of contact predominantly serves evidentiary

---

72 *See, e.g.,* Commission Decision of Dec. 2, 1981 relating to a proceeding under Article 85 of the EEC Treaty Case IV/25.757 (*Hasselblad*), 1982, O.J. (L 161/18), ¶ 49, rebutting an infringement of Article 101. The case was partly overturned in Case 86/82, Hasselblad (GB) Ltd. v. Comm'n of the Eur. Cmtys., ECLI:EU:C:1984:65, but not regarding information exchange.

73 *See, e.g.,* Joined Cases C-89, 104, 114, 116-117, 125-129/85, A. Ahlström Osakeyhtio v. Comm'n of the Eur. Cmtys., ECLI:EU:C:1993:120, ¶¶ 59–65. *See also* Case T-442/08, International Confederation of Societies of Authors and Composers (CISAC) v. Eur. Comm'n, ECLI:EU:T:2013:188, ¶¶ 134–13,182, *see also* Peter L'Ecluse, *The EU General Court partially annuls a Commission decision on anti-competitive conduct among copyright collecting societies (CISAC)*, e-Competitions Apr. 2013, art. No. 57206.

74 *See, e.g., Guidelines on vertical restraints*, at ¶ 97–103, COM (2022) O.J. (C 248) 1.

75 *Guidelines on the Applicability of Article 101 of the Treaty on the Functioning of the European Union to Horizontal Co-Operation Agreements*, at ¶ 402, COM (2023) 3445 final (June 1, 2023).

76 *See Tale & Lyle Plc, supra* note 60, ¶ 55. Confirmed on appeal as *British Sugar Plc., supra* note 60.

77 *See Bayer AG, supra* note 49, at ¶¶ 97, 98.

purposes, but as case law has formulated an unspoken quality requirement, it also serves to make the coordination risk and shared interests plausible. During physical meetings,[78] the absence of an adverse reaction to, e.g., the disclosure of commercially sensitive information, often indicates acceptance of the implicit invitation to collude. In contrast, when information is sent by mail or email, how the recipient will react[79] remains unclear to the sender until later. Because of the mutual consent requirement, certain actions, especially unilateral statements through public media, may not be considered a cartel agreement unless they are found to be veiled attempts to communicate with competitors.

a) Unilateral declarations through public media

417. Undertakings intentionally communicating indirectly through public statements can fall within Article 101 if anti-competitive. Such arrangements could either involve an agreement, if pre-arranged, or a concerted practice, if more ad-hoc.[80] In contrast, genuine unilateral statements typically fall short of a violation in the absence of an element of reciprocity, even if made through an industry publication, and only read by direct competitors.[81] DG COMP recently stated[82] a different position, suggesting that unilateral expressions (e.g. in meetings or on websites) fall under Article 101 if (clearly) directed at a competitor. This was more clearly expressed in *Container Shipping*,[83] a case brought against fourteen shipping companies' practice of announcing suggested price increases with three to five weeks' notice through press releases. Due to the unusually long notice and non-binding nature of the announcements, DG COMP did not accept these as directed at the customers but more akin to testing the competitor's appetite for a price increase. No formal decision was issued, as the parties opted to modify the policy, leaving unresolved the question of whether the actions constituted a concerted

---

78   *See, e.g., Cimenteries CBR SA*, supra note 57, at 1889. Confirmed on appeal as *Aalborg Portland SA*, supra note 57.

79   It follows from cases such as *SIA VM Remonets*, supra note 31, ¶ 33, *see* Debroux, supra note 31, and *Eturas UAB*, supra note 64, ¶¶ 26–50, that it can normally be presumed that the e-mail has been read. However, the matter must be considered and possible to rebut, *see* Lacresse, supra note 64.

80   *See, e.g.,* Case IV/31.149 (*Polypropylene*), supra note 61, ¶ 67. Upheld on appeal in, i.e., Case T-10/89, Hoechst AG v. Comm'n of the Eur. Cmtys., ECLI:EU:T:1992:32, at 78–81, and Case C-227/92 P, Hoechst v. Eur. Comm'n, ECLI:EU:C:1999:360.

81   It might be deduced from *Suiker Unie*, supra note 62, and *Tale & Lyle Plc*, supra note 60, ¶ 56 that attempt to communicate own information can be seen as a concerted practice. On the other hand, did the Court, in, e.g., Case T-249/17, Casino, Guichard-Perrachon & AMC v. Eur. Comm'n , ECLI:EU:T:2020:458, ¶¶ 263–267, appear very adamant in rebutting unilateral public announcement as problematic in isolation, *see* Enzo Marasà & Irene Picciano, *The EU General Court annuls partially the Commission's decision ordering dawn raids on the premises of French supermarkets and their joint purchasing alliance (Casino, Guichard-Perrachon / Achats Marchandises Casino) (Intermarché Casino Achats) (Les Mousquetaires / ITM Entreprises)*, E-COMPETITIONS Oct. 2020, art. No. 99870.

82   *Guidelines on the Applicability of Article 101 of the Treaty on the Functioning of the European Union to Horizontal Co-Operation Agreements*, at ¶¶ 396–400, COM (2023) 3445 final (June 1, 2023).

83   Commission Communication published pursuant to Article 27(4) of Council Regulation (EC) 1/2003 in Case AT.39850 (*Container Shipping*), 2016, O.J. (C 60), at ¶¶ 40–43, *see* Christophe Lemaire, *Commitments: The European Commission accepts commitments offered by 14 container shipping companies to remedy competition concerns regarding alleged concerted practices (Container Shipping)*, CONCURRENCES No. 4-2016, art. N° 82138.

practice. DG COMP's position seems to be that such announcements fall under Article 101, however, case law does not provide clear support for this proposition.[84]

(3) Parallel conduct

418. In addition to contact between competitors, a common interest in coordinating must be established.[85] This requires specifying how uniform market behavior follows from contact between competitors. In practice, this is established through a system of presumptions[86] which include, for example, accepting a causal link if the undertakings remain active on the market and finding it immaterial that the contact did not involve senior manager or sales representatives[87] involved in the actual price adjustments. Undertakings then have the burden of rebutting these presumptions,[88] including showing that the information did not influence their conduct.[89] Conversely, it is not necessary for the behavior to be implemented. What matters is the reduction of strategic uncertainty in the market[90] and the fact that the company has become aware of the information.[91]

B. *Vertical Concerted Practice*

419. As outlined initially, concerted practices cover situations where undertakings substitute competition for cooperation, providing broad application to all conscious coordination between *competitors*. This means that the concept does not apply to vertical interactions,[92] which was confirmed by *Bayer*.[93] In that case, the European Court of Justice rejected claims of collusion based solely on the supplier's (a pharmaceutical manufacturer) attempt to enter into an agreement when the distributor had rejected this in all but name. *Bayer* interpreted the concept of an agreement, but its principle should also apply to concerted practices, suggesting that the notion of vertical concerted practice might have a more limited scope or that it might not apply to vertical cases.

---

84 *See, e.g., Casino, Guichard-Perrachon & AMC, supra* note 81, ¶¶ 263–267, *see* Marasà & Picciano, *supra* note 81.

85 *See e.g. Suiker Unie, supra* note 62.

86 *See, e.g., Anic Partecipazioni, supra* note 50, ¶¶ 118–119, where the Court did not find it substantiated.

87 Commission Decision of Oct. 15, 2008 relating to a proceeding under Article 81 of the EC Treaty Case COMP/39188 (*Bananas*), (2009) O.J. (C 189), at ¶¶ 159, 309. Confirmed on appeal as *Dole Food Co. Inc.*, *supra* note 59, see Fountoukakos & Geeurickx, *supra* note 59.

88 *Cf. Cimenteries CBR SA*, *supra* note 57, ¶ 1865. Confirmed on appeal as *Aalborg Portland SA*, *supra* note 57.

89 Case C-199/92P, Hüls AG v. Comm'n of the Eur. Cmtys., ECLI:EU:C:1999:358, ¶ 167.

90 *Cf. Cimenteries CBR SA*, *supra* note 57, ¶ 1852. Confirmed on appeal as *Aalborg Portland SA*, *supra* note 57.

91 This must be deduced from *Eturas UAB, supra* note 64, ¶¶, *see* Lacresse, *supra* note 64.

92 For further, *see* JONATHAN FAULL & ALI NIKPAY, THE EC LAW OF COMPETITION 223–224 (3d ed. 2014).

93 *See Bayer AG, supra* note 49, at ¶¶ 101–102. *See also Super Bock Bebidas SA, supra* note 45, at ¶¶ 48–49; *see also* Levy & Chellet, *supra* note 45.

2. Can AI Collusion be Effectively Prosecuted under Article 101?

420. As presented initially, EU courts and enforcers have interpreted Article 101 as encompassing an expanded notion of understanding, including anything with a flavor of a meeting of the minds between undertakings. This also includes interactions falling short of traditional concepts of an agreement, and it seems DG COMP wants to expand this to include communication through intermediaries or public sources. EU courts are not unsympathetic to DG COMP's suggestions, but some of DG COMP's statements appear to be more intent to move or develop case law than a faithful expression of judicial precedent. For example, the European Court of Justice has not endorsed DG COMP's position regarding indirect information exchanges through an intermediary. This creates legal uncertainties and undertakings should be mindful of the different risks associated with acts that run counter a policy statement and could trigger an investigation versus conduct that has been established as illegal by EU courts.

421. Translated to AI, of the four scenarios presented initially, Article 101 would cover scenarios: 1.) *Monitoring an existent cartel*, 2.) *Facilitate entering into an existent cartel*, and 3.) *Entering into a cartel without direct contact*, and some of 4.) *Facilitate tacit coordination*, but a grey area remains regarding the use of AI to remain informed about market developments and quickly react to changes. If done unilaterally, this would not be found to fall afoul of Article 101 by EU courts, indicating a possible enforcement gap. DG COMP is trying to target this conduct by suggesting that unilateral expressions, if (clearly) directed at a competitor, would be sufficient evidence of collusion.[94] DG COMP's close follow-up of the issue is positive and indicates that proactive enforcement actions can be expected should the need emerge.

A. *Possible Options to Close the Gaps*

422. In some cases, AI-assisted pricing inevitably falls outside the scope of Article 101 enforcement due to the absence of a finding of a *meeting of the minds* among competitors. There are, however, other alternatives to enforce competition law against this conduct. For example, in *Langnese-Iglo*,[95] DG COMP took issue with a series of parallel vertical agreements, locking up a large portion of the market. The anti-competitive effect, infringing Article 101did not follow from a single (horizontal) agreement but the compounded effect of many (vertical). Translated to AI, would this allow DG COMP to build the case around the compounded effect of multiple vertical AI licensing software agreements, indirectly facilitating

---

[94] *Guidelines on the Applicability of Article 101 of the Treaty on the Functioning of the European Union to Horizontal Co-Operation Agreements*, at ¶¶ 396–400, COM (2023) 3445 final (June 1, 2023).

[95] Case T-7/93, Langnese Iglo GmbH v. Comm'n of the Eur. Cmtys., ECLI:EU:T:1995:98, ¶¶ 99–105, *see* Panayotis Adamopoulos, *The EU Court of First Instance partially confirms the Commission's decision refusing to grant individual exemptions for exclusive purchasing agreements in the ice cream market (Langnese-Iglo / Schöller Lebensmittel)*, e-Competitions June 1995, art. No. 39596.

horizontal collusion between undertakings.[96] EU competition law also has a well-developed theory of liability for cartel facilitators[97] that allows for the inclusion of undertakings not directly involved in the market if instrumental in bringing the infringement case.

423. However, relying on *Langnese-Iglo* is not a free pass for easy intervention. Under this doctrine, DG COMP must establish how the AI has an anti-competitive effect, and undertake a substantial market analysis. Moreover, as the restrictions follow from the compounded effect of parallel agreements, and not a single one, these must dominate the market, making it daunting to intervene outside a very limited window of circumstances. Finally, under the system of block exemptions[98] Article 101 is normally declared non-applicable provided none of the direct parties' market shares exceed certain thresholds, typically 20–30 percent.[99] Intervention relying on *Langnese-Iglo* would, therefore, also require repealing the block exemption.

424. Naturally, some residual conduct may fall outside this framework, such as instances where the AI is developed internally or by a limited number of undertakings, rendering reliance on *Langnese-Iglo* unfeasible. In the same vein, relying on the concept of cartel facilitation would require expanding this notion, as case law so far has been limited to situations where third parties facilitate a traditional cartel and then were considered a member, and not the other way around. Regardless, alternative enforcement options are available in the EU to address these lacunas.[100]

3. Conspiracy under US Antitrust

425. To establish a conspiracy under Section 1,[101] plaintiffs must show: (1) "a combination or some form of concerted action between at least two legally distinct economic entities,"[102] (2) affecting interstate commerce,

---

96 Some forms of tacit collusion might also fall under Article 102 EU, but no further considerations are offered on this. For discussions, see, e.g., NICOLAS PETIT, *The Oligopoly Problem in EU Competition Law*, in RESEARCH HANDBOOK IN EUROPEAN COMPETITION LAW, (I. Liannos and D. Geradin eds., Edward Elgar 2013.

97 Case C-194/14 P, AC-Treuhand AG v. Eur. Comm'n, ECLI:EU:C:2015:717, ¶ 36. Here liability was imposed upon an undertaking facilitating the collusive behavior but not active in the market.

98 For further information on the system of block exemptions, see RICHARD WHISH & DAVID BAILEY, COMPETITION LAW 176–179 (10th ed., 2021).

99 Depending on the arrangement, this would either be Regulation 2022/720 or one of the specialized IP exemptions, accounting for the 20–30 percent spreads.

100 For further options, *see, e.g.*, Francisco Beneke & Mark-Oliver Mackenrodt, *Remedies for algorithmic tacit collusion*, 9 J. Antitrust Enf't 152–176 (2021). For regulatory amendments, *see, e.g.*, Vasileios Tsoukalas, *Should the New Competition Tool be Put Back on the Table to Remedy Algorithmic Tacit Collusion? A Comparative Analysis of the Possibilities under the Current Framework and under the NCT, Drawing on the UK Experience*, 13 J EUR. COMPETITION L. & PRAC. 234–248 (2022).

101 15 U.S.C. § 1.

102 United States v. Apple Inc., 952 F. Supp. 2d 638, 687 (S.D.N.Y. 2013), aff'd, United States v. Apple Inc., 791 F.3d 290 (2d Cir. 2015), *see* Jean-Christophe Roda, *United States: The US Court of Appeals for the Second Circuit considers an elaborate agreement in the e-books sector as anticompetitive (US / Apple)*, CONCURRENCES No. 4-2015, art. No. 76589 (citing Primetime 24 Joint Venture v. Nat'l Broad. Co., 219 F.3d 92, 103 (2d Cir. 2000).

that (3) "constituted an unreasonable restraint of trade either *per se* or under the rule of reason."[103] Overall, "[c]ircumstances must reveal a unity of purpose or a common design and understanding, or a meeting of minds in an unlawful arrangement."[104]

426. The existence of an agreement must always be considered separately from the question of legality.[105] A plaintiff may prove the existence of a contract combination or conspiracy through: (1) direct evidence that "explicitly refer[s] to an "understanding" between the alleged conspirators,[106] or (2) circumstantial evidence that "tends to exclude the possibility [of independent action]."[107] Evidence of conduct that is "as consistent with permissible competition as with illegal conspiracy [cannot support] an inference of ... conspiracy."[108] However, a plaintiff is not required to disprove all non-conspiratorial explanations for the defendant's conduct; rather, one must provide sufficient evidence to show that the conspiratorial explanation is more likely than not.[109] To prove a price-fixing agreement through circumstantial evidence, plaintiffs must show that: (1) defendants engaged in similar conduct referred to as "conscious parallelism," and (2) plus factors suggesting the conduct is the result of collusion and not independent decision-making.

427. The following sections explore the concept of concerted action and its requirements, with a focus on how federal courts address plus factors, followed by an analysis of vertical and hub-and-spoke agreements

A. *Concerted Action*

428. The Supreme Court defined the concept of "concerted action," and thus the most extensive scope of an agreement under Section 1, in a series of cases between the 1930s and 1950s.[110] These are *Interstate Circuit,*

---

103 *Primetime 24 Joint Venture*, 219 F.3d at 103.

104 Monsanto Co. v. Spray Rite Serv., 465 U.S. 752, 764 (1984); Apex Oil Co. v. DiMauro, 822 F.2d 246, 252 (2d Cir. 1987).

105 PHILLIP E. AREEDA & HERBERT HOVENKAMP, ANTITRUST LAW: AN ANALYSIS OF ANTITRUST PRINCIPLES AND THEIR APPLICATION 1400b. (Wolters Kluwer eds., 4th ed. 2020).

106 Golden Bridge Tech., Inc. v. Motorola Inc., 547 F.3d 266, 271 (5th Cir.2008) (internal citations omitted); see also In re Baby Food Antitrust Litig., 166 F.3d 112, 118 (3d Cir. 1999) ("Direct evidence in a Section 1 conspiracy must be evidence that is explicit and requires no inferences to establish the proposition or conclusion being asserted.").

107 Matsushita Elec. Indus. Co. v. Zenith Radio Corp., 475 U.S. 574, 588 (1986); *Monsanto*, 465 U.S. at 768; *see also* Dickson v. Microsoft Corp., 309 F.3d 193, 202 (4th Cir. 2002); see also In re Text Messaging Antitrust Litig., 630 F.3d 622, 629 (7th Cir. 2010), ("Direct evidence of conspiracy is not a sine qua non (...) Circumstantial evidence can establish an antitrust conspiracy."), *see also* Jeffrey May, *The US Court of Appeals for the 7th Circuit upholds the plausibility of a claim for alleged conspiracy in the telecommunications sector under the Twombly standard (Text messaging antitrust litigation)*, E-COMPETITIONS Dec. 2010, art. No. 35933.

108 *Matsushita*, 475 U.S. at 588.; *Brooke Grp. Ltd.*, 509 U.S. at 227 (holding that "conscious parallelism" is "not in itself unlawful"); *Theatre Enters., Inc.*, 346 U.S. at 541 ("Circumstantial evidence of consciously parallel behavior may have made [h]eavy inroads into the traditional judicial attitude toward conspiracy; but 'conscious parallelism' has not yet read conspiracy out of the Sherman Act entirely.").

109 *Apple Inc.*, 952 F. Supp. 2d at 696–97 (citing In re High Fructose Corn Syrup Antitrust Litig., 295 F.3d 651, 655–56 (7th Cir. 2002)).

110 *See* William E. Kovacic et al., *Plus Factors and Agreement in Antitrust Law*, 110 MICH. L. REV. 393, 401 (2011).

Inc. v. United States,[111] *American Tobacco Co. v. United States*,[112] *United States v. Paramount Pictures, Inc.*,[113] and *Theatre Enterprises*.[114] The following principles flow from these precedents: (i) courts may infer an agreement in the absence of a direct exchange of assurances, (ii) this is proved through circumstantial evidence showing that the conduct *more likely than not* resulted from concerted action, and finally, (iii) courts may not infer an agreement merely from defendants recognizing their interdependence and following each other.[115]

429. Subsequent Supreme Court cases have attempted to follow these principles in new formulas. In *Monsanto Co. v. Spray-Rite Service Corp.*,[116] a case concerning a resale price maintenance conspiracy, the court restated the standard, stating that "[t]he correct standard is that there must be evidence that tends to exclude the possibility [of independent action by the parties]."[117] Neither *Monsanto* nor the prior precedents establish what is needed for a finding of concerted action[118] beyond showing that the concept of agreement encompasses more than a direct exchange of assurances distinct from parallel conduct.

430. In *Matsushita*,[119] the US Supreme Court extended *Monsanto*'s demanding conspiracy standard and applied it to horizontal agreements.[120] Quoting *Monsanto*, the court specified plaintiffs' burden of proof to defeat summary judgment when only circumstantial evidence is introduced to involve a showing "that tends to exclude the possibility" that the alleged conspirators acted "independently."[121] The US Supreme Court later extended *Matsushita*'s plausibility screen to the pleading stage of antitrust litigation

---

111 Interstate Circuit, Inc. v. United States, 306 U.S. 208, 227 (1939) ("[a]cceptance by competitors, without previous agreement, of an invitation to participate in a plan, the necessary consequence of which, if carried out, is restraint of interstate commerce, is sufficient to establish an unlawful conspiracy under the Sherman Act.").

112 Am. Tobacco Co. v. United States, 328 U.S. 781, 810 (1946) (the finding of conspiracy is justified "[w]here the circumstances are such as to warrant a jury in finding that the conspirators had a unity of purpose or a common design and understanding, or a meeting of minds in an unlawful arrangement").

113 United States v. Paramount Pictures, Inc., 334 U.S. 131, 142 (1948) ("[i]t is not necessary to find an express agreement in order to find a conspiracy. It is enough that a concert of action is contemplated and that the defendants conformed to the arrangement.").

114 *Theatre Enters.*, 346 U.S. at 541 ("[c]ircumstantial evidence of consciously parallel behavior may have ... conspiracy; but 'conscious parallelism' has not yet read conspiracy out of the Sherman Act entirely.").

115 *See Brooke Grp. Ltd.*, 509 U.S. at 227 (describing conscious parallelism as "not in itself unlawful"); see also Reserve Supply Corp. v. Owens Corning Fiberglas Corp., 971 F.2d 37, 50 (7th Cir. 1992) (discussing why interdependent pricing is not unlawful); In re Text Messaging Antitrust Litig., 782 F.3d 867, 874 (7th Cir. 2015) (holding that tacit collusion is not a violation of the Sherman Act and "probably shouldn't be."), *see also* May, *supra* note 107.

116 465 U.S. 752.

117 *Id.* at 768.

118 William H. Page, *Communication and Concerted Action*, 38 Loy. U. Chi. L.J. 405, 417 (2007); *see also* Kovacic et al., *supra* note 110, at 400.

119 *Matsushita*, 475 U.S. at 574.

120 *Id.* at 588.

121 *Id.* (citation omitted) (quoting *Monsanto*, 465 U.S. at 764).

in *Bell Atlantic Corp. v. Twombly*.[122] The court stated that at the pleading stage, "an allegation of parallel conduct and a bare assertion of conspiracy will not suffice,"[123] the plaintiffs must present "enough facts to state a claim to relief that is plausible on its face."[124]

431. The heightened standard for pleading concerted action was motivated by perceived excesses caused by plaintiffs' private rights of action and mandatory trebling of damages.[125] As outlined, finding concerted action traditionally requires parallel conduct and plus factors. The following sections elaborate on these tests and examine various types of agreements, including hub-and-spoke and vertical arrangements.

(1) Parallel conduct and plus factors

432. Case law has not provided a clear operative test for establishing concerted action inferred from parallel conduct and "plus factors."[126] Plus factors considered by courts include: (i) economic market characteristics and factual evidence including the opportunity to conspire,[127] (ii) inter-competitor communications,[128] (iii) invitations to collude,[129] (iv) exchanges of commercially sensitive information,[130] and (v) and actions against interest,[131] among others.[132] "... [T]he character and effect of a conspiracy are not to be judged by dismembering it and viewing its separate parts, but only

---

122 550 U.S. 544 (2007), *see* Frédérique Daudret-John & François Souty, *Class action: The US Supreme Court imposes stricter standard of proof for antitrust class action (Bell Atlantic/William Twombly)*, CONCURRENCES No. 4-2007, art. No. 14326.

123 *Twombly*, 550 U.S. at 556, *see* Daudret-John & Souty, *supra* note 122.

124 *Id.* at 570.

125 *Id.* at 559–60.

126 *See* In re Chocolate Confectionary Antitrust Litig., 801 F.3d 383, 398 (3d Cir. 2015), *see* Daniel J. Boland & Michael J. Hartman, *The US Court of Appeals for the Third Circuit holds that the courts must carefully consider the nature of the industry and whether the actions of defendants can be equally attributed to independent conduct as to a conspiracy in antitrust cases involving concentrated markets (Chocolate Confectionary)*, E-COMPETITIONS Sept. 2015, art. No. 118187.

127 Petruzzi;s IGA Supermarkets, Inc. v. Darling-Delaware Co., 998 F.2d 1224, n.17 (3d Cir. 1993) ("[T]he testimony relating to the defendants' opportunity to conspire and the solicitation of others to partake in common action is also relevant.").

128 *See, e.g.*, SD3, LLC v. Black & Decker (U.S.) Inc., 801 F.3d 412, 432 (4th Cir. 2015) ("Allegations of communications and meetings among conspirators can support an inference of agreement because they provide the means and opportunity to conspire."), *see also* Deirdre McEvoy-Cappock & Taylor J. Kirklin, *The US Court of Appeals for the Fourth Circuit turns on the interpretation of the Twombly plausibility standard and the application of the Supreme Court's precedent on pleading standards to antitrust actions at early stages of litigation (Sawstop / Black & Decker)*, E-COMPETITIONS Sept. 2015, art. No. 76729.

129 *See* In re Delta/Airtran Baggage Fee Antitrust Litig., 245 F. Supp. 3d 1343, 1372 (N.D. Ga. 2017), aff'd sub nom. Siegel v. Delta Air Lines, Inc., 714 F. App'x 986 (11th Cir. 2018) (per curiam) ("Numerous cases have recognized that an invitation to collude can serve as evidence of a conspiracy.") (internal citations omitted).

130 *See e.g.,* Ash v. Hack Branch Distrib. Co., 54 S.W.3d 401, 419 (Tex. Ct. App. 2001) ("[Plus] factors traditionally include price parallelism, product uniformity, exchange of price information, and the opportunity for the alleged conspirators to meet to formulate illegal policies.").

131 *Chocolate.*, 801 F.3d at 398 ("evidence of actions against self-interest means there is evidence of behavior inconsistent with a competitive market."), *see* Boland & Hartman, *supra* note 126.

132 *See* Christopher R. Leslie, *The Probative Synergy of Plus Factors in Price-Fixing Litigation*, 115 Nw. U. L. REV. 1581 (2021) (providing a comprehensive typology of plus factors).

by looking at it as a whole."[133] Relying on plus factors provides a flexible framework in which no minimum number of plus factors is required[134] as the plus-factor inquiry is not intended to be rigid or formulaic and there is no one plus factor that is "strictly necessary."[135]

433. Despite the long-standing rules that circumstantial evidence alone can be sufficient to prove a *per se* violation of Section 1 and that courts should not compartmentalize a plaintiff's evidence of conspiracy, courts often inappropriately isolate individual plus factors.[136] In practice, many courts have diminished plus factors' probative value to the point that plaintiffs are required to present direct evidence of price fixing.[137]

434. A high standard to prove plus factors undoubtedly makes it more difficult to prosecute cases where AI facilitates or even eliminates the need for direct interaction. The following sections examine examples of these challenges and their potential impact on AI-driven collusion.

a) Invitations to collude

435. Invitations to collude are typically regarded as unilateral acts that fall outside the scope of Section 1 unless they are found to have been accepted. These invitations can be either explicit[138] or implicit.[139] Federal courts may also consider them a plus factor to prove a conspiracy.[140]

436. The challenge of prosecuting an alleged agreement formed via acceptance of invitations to collude under Section 1 lies in the difficulty of persuading the court that the parallel behavior observed in the market was a consequence of such invitations and not merely the result of conscious parallelism. The same is true for cases concerning the use of pricing AIs. *Delta/AirTran Baggage Fee Antitrust Litigation,* exemplifies this challenge. In this case, Plaintiffs brought a Section 1 claim against Delta and AirTran, alleging that the airlines conspired to fix prices. Among other things, plaintiffs alleged that AirTran invited Delta, on quarterly earnings calls and at industry conferences, to decrease capacity and impose baggage fees, which Delta allegedly accepted, creating an agreement.[141] The district court found that the plaintiffs' alleged Section

---

133 Continental Ore Co. v. Union Carbide & Carbon Corp., 370 U.S. 690, 699 (1962).

134 *In re High Fructose Corn Syrup Antitrust Litig.*, 295 F.3d at 655.

135 In re Flat Glass Antitrust Litig., 385 F.3d 350, 361 n.12 (3d Cir. 2004).

136 *See generally*, Leslie, *supra* note 132.

137 *See* Christopher R. Leslie, *The Decline and Fall of Circumstantial Evidence in Antitrust Law*, 69 Am. U. L. Rev. 1713, 1715 (2020).

138 United States v. Am. Airlines, 743 F.2d 1114 (5th Cir. 1984) (where a CEO called the CEO of its competitor and suggested the two companies agree on prices).

139 In the Matter of Precision Moulding Co., Inc., 122 F.T.C. 104, 105–107 (F.T.C. Sept. 3, 1996) (complaint).

140 *See, e.g., SD3, LLC*, 801 F.3d at 432 ("Allegations of communications and meetings among conspirators can support an inference of agreement because they provide the means and opportunity to conspire."), *see also* McEvoy-Cappock & Kirklin, *supra* note 128 (citing Evergreen Partnering Grp., Inc. v. Pactiv Corp., 720 F.3d 33, 49 (1st Cir. 2013); *see also*, Leslie, *supra* note 137, at 1734).

141 733 F. Supp. 2d 1348, 1352–56 (N.D. Ga. 2010).

1 violation was sufficiently substantial to survive a motion to dismiss, i.e., this was a plausible claim.[142] However, following discovery and under the more stringent summary judgment standard, the court found that the defendants had only engaged in conscious parallelism.[143]

437. There are alternative avenues for prosecuting invitations to collude. The FTC can challenge such invitations under Section 5 of the FTC Act.[144] Private plaintiffs can also challenge invitations to collude under state unfair competition laws, often referred to as 'Little FTC Acts,' since unfair or distortive competition rules are not limited to the federal level. This conduct may also fall under Section 2 of the Sherman Act as monopolization or attempted monopolization.[145] Unlike the Sherman Act, the FTC Act can reach collusive conduct without the showing of an agreement.[146] The FTC could also argue that weighing efficiencies under the rule of reason is not required under Section 5, especially when the conduct under scrutiny lacks plausible efficiency justifications. In support of these actions, the FTC may cite its multiple settlements regarding invitations to collude.[147] This approach is, however, not without challenges, given that the extent to which Section 5 reaches beyond the Sherman Act has not been litigated,[148] and the available remedies under Section 5 are limited to cease-and-desist orders.

b) Information exchanges

438. Information exchanges between competitors are typically analyzed under the rule of reason,[149] except in markets that are more prone to

---

142 *In re Delta/AirTran*, 733 F. Supp. 2d at 1362–63.

143 *In re Delta/AirTran Baggage Fee Antitrust Litig.*, 245 F. Supp. 3d 1343.

144 15 U.S.C. § 45.

145 15 U.S.C. § 2.

146 *See e.g.*, Aneesa Mazumdar, *Algorithmic Collusion: Reviving Section 5 of the FTC Act*, 122 COLUM. L. REV. 449 (2022).

147 *See e.g.*, In re UHaul Int'l Inc., F.T.C. No. 0810157 (2010) (concerning a specific intent to collude due to public communications detailing pricing strategy, urging competitors to increase price); In re Valassis Commc'ns Inc., F.T.C. No. 0510008 (2006) (concerning earnings call statements describing future pricing plans and customer strategy,"); In re Stone Container Corp., 125 F.T.C. 853 (1998) (concerning press releases and published interviews communicated a firm's intentions to lower output and draw down industry inventory levels and invite competitor industry-wide price increases).

148 Moreover, two recent Supreme Court decisions suggest federal courts may not be inclined to extend the reach of § 5. *See* AMG Cap.l Mgmt., LLC v. FTC, 141 S. Ct. 1341 (2021) (holding that the FTC cannot obtain equitable monetary relief, such as disgorgement or restitution, when it pursues district court litigation directly under § 13(b) of the FTC Act to obtain such relief, FTC must first follow its administrative adjudication procedures under § 5 of the Act), *see also* Steven A. Reed, Scott A. Stempel & Daniel S. Savrin, *The US Supreme Court holds that the FTC lacks the authority to seek equitable monetary relief in cases brought in federal court under FTC Act Section 13(b) (AMG Capital Management)*, E-COMPETITIONS Apr. 2021, art. No. 100669 ; and Axon Enter. , Inc. v. FTC 143 S. Ct. 890 (2023) (holding that district courts have jurisdiction to resolve constitutional challenges to the structure of the FTC and the Securities and Exchange Comm'n without any prior agency hearing or determination), *see also* Ben Gris, David Higbee, Ryan Shores & Jacob Coate, *The US Supreme Court gives a unanimous ruling which endorses early challenge to the FTC proceedings in Federal Courts (Axon I FTC)*, E-COMPETITIONS Apr. 2023, art. No. 112180.

149 *See* United States v. U.S. Gypsum Co., 438 U.S. 422 (1978) (considering the structure of the industry involved and the nature of the information exchanged).

coordination.[150] This standard poses additional challenges for prosecuting AI-assisted collusion. In particular, the need to prove a case concerning the exchange of current and future pricing information between competitors under the rule of reason is troubling. There are compelling economic and legal grounds to condemn horizontal price information exchanges under a *per se* standard.[151] Horizontal price exchanges are the hallmarks of *per se illegality*. First, they have a "pernicious effect on competition," and put upward pressure on prices even in the absence of an underlying agreement on prices.[152] Additionally, exchanges of price information between competitors, "lack ... any redeeming virtue,"[153] and do not serve procompetitive ends.[154] Horizontal price exchanges do not improve market efficiency or increase price competition. Indeed, they have the opposite effect.[155]

439. Information exchanges between competitors are considered by federal courts as plus factors to prove an overarching conspiracy.[156] However, federal courts have interpreted or applied this plus factor in a manner that deprives it of its probative value and, in practice, requires an actual agreement.[157] For example, the district court in *Chocolate*[158] quoted Third Circuit precedent for the proposition that: "[c]ommunications between competitors do not permit an inference of an agreement to fix prices unless those communications rise to the level of an agreement, tacit or otherwise."[159] In practice, this means plaintiffs and prosecutors are required to provide evidence of an overarching agreement to prove this plus factor, defeating its purpose. Similarly, in *In re Citric Acid*, plaintiffs detailed the time and participants of the commercially sensitive communication, but the court required even more "specific details regarding illegal discussions," essentially asking for direct evidence of an illegal agreement.[160]

---

150 *See* United States v. Container Corp., 393 U.S. 333 (1969) (prohibiting price verification practices in a concentrated industry).

151 *See generally* Joseph E. Harrington Jr. & Christopher R. Leslie, *Horizontal Price Exchanges*, 44 CARDOZO L. Rev. (Dec. 8, 2022).

152 *See id.*

153 *Id.*

154 *Id.*

155 *Id.*

156 *See* Am. Column & Lumber Co. v. United States, 257 U.S. 377, 411–12 (1921) (finding that the plan adopted by 365 companies to share sensitive information about their hardwood lumber production and sales was " simply an expansion of the gentleman's agreement of former days, skillfully devised [...] to evade the law"); *In re Flat Glass Antitrust Litig.*, 385 F.3d. 350 (a jury could infer that the exchange of pricing information was a concerted action designed to fix prices).

157 *See* Leslie, *supra* note 137, at 1734.

158 999 F. Supp. 2d 777 (M.D. Pa. 2014), aff'd, 801 F.3d 383 (3d Cir. 2015).

159 *See* Leslie, *supra* note 137, at 1734-35(quoting *Baby Food*, 166 F.3d at 126).

160 191 F.3d 1090, 1103 (9th Cir. 1999).

440. This framework has led to AI collusion cases being prosecuted separately for their information exchange component under the rule of reason, making it increasingly difficult to rely on this plus factor to infer a conspiracy. The DOJ's 2023 complaint against Agri Stats Inc. is a recent example of prosecuting information exchanges as stand-alone conduct.[161] The complaint alleges that Agri Stats infringed Section 1 by collecting, integrating, and distributing competitively sensitive price, cost, and output information among competing meat processors. The case is pending in a US district court in Minnesota.

441. Finally, it's important to note that the DOJ's repeal of several information exchange guidelines in the health sector also illustrates the heightened focus on information exchanges driven by advances in AI technology.[162] This suggests there will be challenges for AI-assisted information exchanges in these markets.

c) Indirect contact through hub-and-spoke agreements

442. A common way to analyze conspiracies involving the use of AI is under the classic hub-and-spoke framework. These conspiracies have four elements: (1) a "hub," which is the facilitating firm; (2) "spokes," which are upstream or downstream firms; (3) vertical restraints that connect the hub and the spokes; and (4) the "rim" that connects the spokes.[163] Hub-and-spoke conspiracies have a narrow meaning that refers to the practices that facilitate the collusion among the spokes at the rim, not to the arrangement of vertical relationships between the hub and the spokes.[164] The most challenging aspect in these cases is inferring the horizontal agreement from circumstantial evidence because courts have not expressly articulated how to use vertical agreements as plus factors.[165]

443. The Supreme Court opinions in *Interstate Circuit v. United States*,[166] *United States v. Masonite, Corp.*,[167] *Klor's, Inc. v. Broadway-Hale Stores, Inc.*,[168] *United States v. Parke, Davis & Co.*,[169] and *United States v. General Motors*

---

161 United States v. Agri Stats, Inc., 0:23CV03009 ECF No. 1

162 *See* DOJ Press Release, Principal Deputy Assistant Attorney General Doha Mekki of the Antitrust Division Delivers Remarks at GCR Live: Law Leaders Global 2023 (Feb 2. 2023).

163 *See* Barak Orbach, *Hub-and-Spoke Conspiracies* 1 (ARIZ. LEGAL STUDIES, Discussion Paper No. 16-11 Apr. 15, 2016); ABA SECTION OF ANTITRUST LAW, ANTITRUST LAW DEVELOPMENTS 20 (7th ed. 2012).

164 *Id.* at 14.

165 *Id.*

166 306 U.S. at 217 (finding that knowledge by each distributor about the same request by an exhibitor to all its competitors prohibiting "double features" constituted accepted under Section 1 of the Sherman Act).

167 316 U.S. 265 (1942) (finding that each licensee's awareness at different points of licensing agreements between a patent holder and wallboard manufacturers setting the price for products constituted an overarching agreement).

168 359 U.S. 207 (1959) (finding a *per se* illegal collusive agreement to refuse to deal when a retailer required several suppliers to refrain from selling or selling on highly unfavorable terms to a store that operated next to one of its branches).

169 362 U.S. 29 (1960) (finding that the defendant gained adherence to its policies and created unanimity among competitors).

Corp., laid the foundation for modern hub-and-spoke agreements.[170] These decisions allow an inference of a *per se* unlawful horizontal agreement from an arrangement involving vertical relationships. The decisions, except *Klor's*, emphasize that it is not the vertical relationships themselves but the totality of the communication and circumstances that may provide circumstantial evidence (namely, the plus factors) proving a conspiracy.[171]

444. Today, courts interpret *Interstate Circuit* to mean that a conspiracy may be inferred where: (1) two or more competitors enter into vertical agreements with a single upstream or downstream firm, (2) the vertical agreements could benefit each competitor only if its rivals enter into similar agreements, and (3) the firm that facilitates all the vertical agreements persuades each competitor that its competitors will take similar action.[172]

445. Modern cases dealing with hub-and-spoke conspiracies are *Toys "R" Us v. FTC (TRU)*,[173] *Dickson v. Microsoft*,[174] *PepsiCo v. Coca-Cola Co.*,[175] *United States v. Apple (eBook)*,[176] *In re Musical Instruments and Equipment Antitrust Litigation (Guitar Center)*[177] *Meyer v. Uber (Meyer)*.[178]

446. In *Dickson*, a software manufacturer argued that the distribution agreements between Microsoft and three original equipment manufacturers (OEMs) established a conspiracy among the OEMs.[179] The district court granted the defendants' motion to dismiss, holding that a rimless wheel antitrust conspiracy is not actionable.[180] The Fourth Circuit upheld, firmly rejecting the proposition that rimless wheel theories may establish antitrust conspiracies.[181]

447. Similarly, in *PepsiCo.*,[182] Coca-Cola's vertical agreements with distributors included loyalty clauses that forced distributors to choose between the company and PepsiCo. PepsiCo challenged the legality of these clauses by

---

170 384 U.S. 127 (1966) (holding that where vertical relationships facilitate a conspiracy, their pro-competitive effects are irrelevant to the legality of the horizontal agreement).

171 *Id.*

172 *See, e.g.*, Toys "R" Us, Inc. v. FTC, 221 F.3d 928, 935–36 (7th Cir. 2000); *Apple*, 791 F.3d at 319–20.

173 221 F.3d 928.

174 127 F. Supp. 2d 728 (D. Md. 2001), *aff'd, Dickson*, 309 F.3d 193. 309 F.3d 193 (4th Cir. 2002).

175 114 F. Supp. 2d 243 (S.D.N.Y. 2000), *aff'd, PepsiCo*, 315 F.3d 101 (2nd Cir. 2002).

176 *Apple Inc.*, 952 F. Supp. 2d 638, *aff'd, Apple Inc.*, 791 F.3d 290.

177 WL 3637291, at *3 (S.D. Cal. Aug. 20, 2012), *aff'd sub nom. Guitar Center*, 798 F.3d 1186 (9th Cir. 2015) ..., *see also* Jean-Christophe Roda, *United States: The US Court of Appeals for the Ninth Circuit rejects allegations of conspiracy concerning the electric guitars market and clarifies analysis of hub-and-spoke agreements (Musical Instruments and Equipments)*, Concurrences No. 4-2015, art. No. 76584.

178 Meyer v. Kalanick, 200 F. Supp. 3d 408, 410 (S.D.N.Y. 2016), *vacated sub nom.* Meyer v. Uber Techs., Inc., 868 F.3d 66 (2d Cir. 2017).

179 *See Dickson*, 309 F.3d at 203–04.

180 *Id.*

181 *Id.* at 203–05.

182 *PepsiCo*, 315 F.3d at 110–11.

arguing that they established a hub-and-spoke conspiracy, among other things. The district court granted summary judgment for Coca-Cola, and the Second Circuit upheld that decision, stressing the rim requirement.[183]

448. In *eBook*, the defendants centered their defense on *Monsanto*, arguing that the evidence did not tend to exclude unilateral lawful conduct.[184] In finding a conspiracy agreement, the court considered documents showing that Apple's and the publishers' interests in raising prices and defeating Amazon's pricing policy were aligned.[185] The court also considered identical emails in which a publisher expressed that "[a]fter talking to all the other publishers and seeing the overall book environment, here is what I think is the best approach for e-books" and then explained Apple's agency model.[186]

449. In *Guitar Center*, plaintiffs alleged that a hub-and-spoke conspiracy orchestrated by Guitar Center had pressured manufacturers into setting a minimum advertised price.[187] Plaintiffs argued that: (i) agreements were adopted around the same time, (ii) these actions were against their self-interest, and (iii) there were frequent meetings and public signaling. The district court found the complaint did not answer basic questions of who did what, to whom (or with whom), where, and when.[188] While acknowledging how "the line between horizontal and vertical restraints can blur" in hub-and-spoke conspiracies, the Ninth Circuit declined to find an agreement through the compliance of each vertical agreement.[189] Arguably, this reasoning required direct evidence of horizontal coordination at the pleading stage, creating a significant impediment to prosecuting hub-and-spoke conspiracies.[190]

450. In *Meyer*, the hub-and-spoke theories were tested for AI pricing schemes.[191] The district court for the Southern District of New York found that a class of plaintiffs successfully pled a hub-and-spoke conspiracy against Uber and its co-founder and CEO, Travis Kalanick.[192] The complaint's factual basis was that Uber drivers – who were independent contractors according to Uber – had conspired to raise their prices simultaneously through the Uber algorithm's "surge pricing" model.[193] Notably, the court relied

---

183 *Id.* at 110–11.

184 *Monsanto*, 465 U.S. at 762–63.

185 952 F. Supp. 2d at 648, *aff'd*, 791 F.3d 290.

186 *Id.* at 661.

187 WL 3637291, at *3 (S.D. Cal. Aug. 20, 2012), *aff'd sub nom Guitar Center*, 798 F.3d 1186 (9th Cir. 2015), *see also* Roda, *supra* note 177.

188 *Id.*

189 *Guitar Center*, 798 F.3d at 1192, *see also* Roda, *supra* note 177.

190 *See* Orbach, *supra* note 163, at 11

191 *See* Meyer v. Uber, No. 1:2015-cv-09796-JSR, Doc. 37 (S.D.N.Y. Mar. 31, 2016).

192 *See id.* at 22–23.

193 *See id.* at 820–22.

on *Interstate Circuit* to hold that plaintiffs plausibly alleged a conspiracy in which drivers "sign[ed] up for Uber precisely 'on the understanding that the other [drivers] were agreeing to the same' pricing algorithm."[194]

451. *Meyer* is an important precedent for finding an agreement through a pricing AI at the pleading stage, although its merits were never decided. After an appeal to the Second Circuit on arbitration issues and a remand to the Southern District of New York,[195] the case was sent to arbitration, effectively barring the plaintiffs from appealing the antitrust theory.

452. As we discuss below, these are precisely the issues that recent cases regarding AI collusion have grappled with. While plaintiffs allege an overall agreement to use the same AI to raise prices, courts often find individual vertical agreements with the AI platform but no horizontal agreement.

B. *Vertical Conspiracies*

453. While the *Monsanto* standard, requiring evidence of a conscious commitment to a common scheme, theoretically applies to horizontal, vertical, and mixed relationships, its application to non-horizontal agreements is dramatically different.[196] Federal courts[197] have acknowledged that: "[o]ne conspiracy can involve both direct competitors and actors up and down the supply chain, and hence consist of both horizontal and vertical agreements."

454. In such cases, the court must determine the orientation of the alleged conspiracy.[198] To evaluate hybrid conspiracies, courts may break down the conspiracy "*into its constituent parts*"[199] parsing allegations of vertical and horizontal agreements. The court can then analyze "the respective vertical and horizontal agreements ... under the rule of reason or as violations per se."[200]

455. As discussed in the next section, in the recent AI collusion cases, federal courts have broken down claims of a horizontal agreement into several vertical parts, finding insufficient plus factors to establish an overarching agreement falling under the *per se* standard.

---

194 *Id.* at 824.

195 *See* Meyer v. Uber Techs., Inc., 868 F.3d 66 (2d Cir. 2017).

196 *See* Ohio v. Am. Express Co., 138 S. Ct. 2274, 2284 (2018*);* Leegin Creative Leather Prods. v. PSKS, Inc., 551 U.S. 877, 893 (2007), *see* Peter J. Carney & Kristen J. McAhren, *The US Supreme Court overturns its long-standing prohibition against vertical agreements between manufacturers and their dealers setting minimum resale prices (Leegin Creative)*, E-COMPETITIONS June 2007, art. No. 38050 ; *see also*, ABA SECTION OF ANTITRUST LAW, PROOF OF CONSPIRACY UNDER FEDERAL ANTITRUST LAWS 69–92 (2010) (collecting authorities).

197 *Guitar Center*, 798 F.3d at 1192, *see also* Roda, *supra* note 177; *see also Apple, Inc.*, 791 F.3d at 314 ("Although th[e] distinction [between horizontal and vertical restraints] is sharp in theory, determining the orientation of an agreement can be difficult as a matter of fact and turns on more than simply identifying whether the participants are at the same level of the market structure[.]"), *cert. denied*, 136 S. Ct. 1376 (2016); *Meyer*, 174 F. Supp. 3d 817 (S.D.N.Y. 2016) (holding that Uber drivers' agreement to Uber terms was sufficient to allege a horizontal agreement among the drivers).

198 *See, e.g., Apple, Inc.*, 791 F.3d at 314.

199 *Guitar Center*, 798 F.3d at 1192, see also Roda, *supra* note 177.

200 *Id.* at 1192–93 (internal citations omitted).

4. Can AI Collusion Be Effectively Prosecuted in the US under Section 1 US under the *Per Se* Standard?

456. Following the review of US cases involving proof of conspiracy through circumstantial evidence, it is perhaps unsurprising to find a stark difference in the evidence needed to prove a *per se* cartel through AI collusion in the US versus the EU. This issue is well illustrated by the recent cases filed by private plaintiffs and strongly supported by the Agencies.

457. In *Realpage,* multifamily and student plaintiffs alleged that RealPage and its clients formed a price-fixing cartel among lessors of student housing properties to artificially inflate student housing prices across the United States, including near college campuses.[201] Similarly, in *Duffy v. Yardi Systems, Inc.*,[202] a proposed class action by tenants alleged that landlords colluded to use the pricing algorithms provided by Yardi Systems to inflate multifamily rental prices artificially. The Agencies filed Statements of Interest[203] in both cases.[204] Relying on *United States v. Socony-Vacuum Oil Co.*, the Agencies urged the *RealPage* and *Yardi* courts to apply the *per se rule* because "[a]greeing to use a common pricing formula is *per se* unlawful."[205]

458. In *Realpage,* citing *Interstate Circuit,* the DOJ argued that it was "enough" that the landlords, "knowing that concerted action was contemplated and invited," adhered "to the scheme and participated in it."[206] The DOJ explained that the alleged conspiracy was horizontal because the facilitator, RealPage, was not a player in a vertical distribution chain such as a vertically oriented distributor, wholesaler, or retailer. Even if some elements of RealPage's business could be described as vertical, where a vertical market player "conceptualized" or "orchestrated" the conspiracy among horizontal entities, then the restraint at issue remains horizontal in nature, and therefore, defendants "cannot escape the *per se* rule."[207]

---

201 In re RealPage, Inc., Rental Software Antitrust Litig. (No. II), No. 3:23-md_03071, 2023 WL 9004808, at *2 (M.D. Tenn. Dec. 28, 2023). The complaint explains that in a competitive market, these companies would compete with one another to attract student renters and maximize occupancy of their properties. However, following RealPage's introduction of its "Revenue Management" software defendants effectively stopped making independent pricing and supply decisions.

202 Duffy v. Yardi Sys., Inc., No. 2:23-cv-01391-RSL (W.D. Wash. Mar. 1, 2024).

203 Under 28 U.S.C. § 517, the Attorney General may submit a Statement of Interest (SOI) in any lawsuit, even when the United States is not a party. This is one of the many advocacy tools available to the DOJ and are used to explain to a court the interests of the United States in litigation.

204 *See* U.S. Dep't of Justice, Memorandum of Law in Support of the Statement of Interest of the United States, *In re RealPage*, at 17–18 (Nov. 15, 2023); U.S. Dep't of Justice & Fed. Trade Comm'n, Memorandum of Law in Support of the Statement of Interest of the United States, *Yardi*, at 3 (Mar. 1, 2024).

205 *See* U.S. Dep't of Justice, Memorandum of Law in Support of the Statement of Interest of the United States, *RealPage*, at 13 (citing Socony-Vacuum, 310 U.S. 150, 222 (1940)).

206 U.S. Dep't of Justice, Memorandum of Law in Support of the Statement of Interest of the United States, *RealPage,* at 13 (citing Interstate Circuit, 306 U.S. at 226–27).

207 U.S. Dep't of Justice, Memorandum of Law in Support of the Statement of Interest of the United States, In re RealPage, Inc., Rental Software Antitrust Litigation (No. II), No. 3:23-MD-3071, 20–21 (M.D. Tenn. Nov. 15, 202.3) (internal citations omitted).

The statement of interest added that at least two characteristics highlighted in the complaint show that the alleged scheme falls within the class of *per se* unlawful price-fixing restraints: (1) first, agreeing to use a common pricing formula is unlawful,[208] and (2) second, the complaints allege that defendants acted in concert by knowingly sharing "competitively sensitive" and "non-public" pricing information with RealPage, which can play an important role in facilitating price fixing.[209]

459. In December 2023, the Tennessee district court ruled on the defendants' motions to dismiss in *Realpage* and found that neither of the two plaintiffs' classes, the multifamily and student, had sufficiently alleged at the pleading stage a conspiracy justifying the application of the *per se* standard. For multifamily, the court found that the plaintiffs "clearly alleged vertical agreements" as well as "some level of horizontal conspiracy among those RMS Client Defendants to each contribute their commercially sensitive pricing and supply data for use by RealPage to calculate their horizontal competitors' pricing recommendations." In turn, RealPage "used horizontal competitors' commercially sensitive pricing and supply data to calculate their own pricing recommendations."

460. However, the plaintiffs' allegations fell short of the *per se* standard because they had not: (1) included the dates on which the different competitors had joined the agreement, (2) alleged any direct communications (absent through trade associations – which the court deemed indirect), (3) alleged sufficient lack of deviation (the court found that even a 10–20 percent deviation meant there was no absolute delegation for their price-setting to RealPage), (4) described punishment mechanisms for deviators. The court reserved judgment on whether, after discovery, proof of these or other additional facts beyond those alleged in the complaint would justify the application of the *per se* rule to the conduct challenged by the plaintiffs.

461. The court used similar reasoning to discard the sufficiency of the students' claims to satisfy the applicability of the *per se* standard, which it also dismissed under the rule of reason standard. Therefore, the case is continuing for the multifamily plaintiffs, and both the *per se* and the rule of reason claims will be analyzed again following discovery at summary judgment.

462. The Agencies referenced their submission in *Realpage* in their Statement of Interest for *Yardi*, adding that "competitors may not agree to fix the starting point of pricing (e.g., agree to fix advertised list prices) even if the actual charged prices vary from the starting point."[210] The ruling on the motion to dismiss was still pending at the time of this chapter.

---

208 *Id.* at 19 (citing *Socony-Vacuum*, 310 U.S. at 222).

209 *See id.* (internal citations omitted); *In re Flat Glass Antitrust Litig.*, 385 F.3d at 369.

210 U.S. Dep't of Justice & Fed. Trade Comm'n, Memorandum of Law in Support of the Statement of Interest of the United States, in *Yardi* at 20.

463. More recently, while the DOJ had opened a criminal investigation into RealPage's conduct,[211] the agency finally accused RealPage of civil Sherman Act violations for unlawful information sharing, vertical agreements to align prices, and monopolization of commercial revenue management software.[212]

464. Similar cases have been brought in the market for booking hotel rooms. In *Gibson v. MGM Resorts International*,[213] the plaintiffs alleged that hotel operations on the Las Vegas Strip entered into a *per se* price-fixing agreement in violation of Section 1 by agreeing to rely on the same pricing software. The Nevada District Court rejected the plaintiffs' allegations of an illegal *per se* agreement and a hub-and-spoke agreement but granted leave to amend.[214] The court rejected the applicability of the *per se* standard on a finding that the plaintiffs had not sufficiently pled parallel conduct by failing to prove that the defendant hotels used the same pricing algorithm simultaneously.[215] Following the plaintiff's amended complaint, the court granted the defendants' motion to dismiss with prejudice because pricing recommendations were not always accepted. Unfortunately, the court focused on the wrong "agreement." Instead of addressing the plaintiffs' allegation regarding a horizontal agreement, the court focused on each independent agreement to use the same pricing system.

465. In *Cornish-Adebiyi v. Caesars*, plaintiffs allege the same conduct as in *Gibson*, concerning price fixing through the use of the "Rainmaker" algorithm of defendant Cendyn Group of casino-hotel guest rooms in Atlantic City, New Jersey.[216] The Agencies once again filed a Statement of Interest, which referenced the one filed in *Realpage* and highlighted that: (1) plaintiffs need not allege direct communications between defendants, nor (2) that pricing recommendations were adhered to plausibly state a claim for concerted practices under Section 1.[217] Defendants' motions to dismiss are still pending (2024) at the New Jersey District Court.

466. A similar case, *Dai et al v. SAS Institute Inc. et al.*, was filed on April 26, 2024, against the operators of a revenue management system (SAS Institute Inc. and IDeaS Inc.) and the largest hotel operators in the US (Choice

---

211 *See* Josh Sisco, *DOJ escalates price-fixing probe on housing market*, Politico, Mar. 20, 2024.

212 *See United States et al v. Realpage, inc. et al.*, No. 1:24-cv-00710 (M.D.N.C. 2024), ECF No. 1.

213 *See* First Amended Class Action Complaint, Gibson et al. v. MGM Resorts Int'l et al., No. 2:23-cv-00140-MMD-DJA (D. Nev. Nov. 27, 2023).

214 *See id.*

215 *Id.* at 4.

216 *See* First Amended Class Action Complaint, Cornish-Adebiyi, et al. v. Caesars Ent., Inc., et al., No. 1:23-cv-02536 (D.N.J. 2024) ECF No. 80 at 1, *see* also Creighton Macy, Jeffrey Martino, Ashley Eickhof, Andrea Rivers & Mark G. Weiss, *The US FTC and DoJ submit a joint Statement of Interest to confirm that algorithmic price fixing is a per se antitrust violation (Cornish-Adebiyi / Caesars Entertainment)*, e-Competitions Mar. 2024, art. No. 118017.

217 *See* U.S. Dep't of Justice & Fed. Trade Comm'n, Memorandum of Law in Support of the Statement of Interest of the United States, in *Cornish-Adebiyi et al.*, ECF No. 96, *see* also Macy et al., *supra* note 215.

Hotels International Inc., Wyndham Hotels & Resorts, Inc., Hilton Worldwide Holdings, Inc., Four Seasons Hotels and Resorts US Inc., Omni Hotels & Resorts, and Hyatt Hotel Corporation).[218] Notably, the complaint references how IDeaS Inc. promotes its business model on its website:

> We are responsible for the performance of our solutions and, ultimately, the success of our clients. That's why we don't cut corners, and *we certainly don't leave anything to chance or human intuition – no offense, humans*. Because of these guiding principles, an IDeaS RMS is future-proof, fully automated, and truly science-backed. And even if all of that still isn't convincing enough for you, the real proof is in the ROI.[219]

467. *Gibson* was dismissed, and the plaintiffs in *Realpage* may still meet the *per se* standard at the summary judgment stage; *Yardi*, *Cornish*, and *Dai* are all pending at the pleading stage. Given the ubiquity of online pricing mechanisms and the Agencies' heightened attention to these cases, it is likely that many similar cases will be litigated. The outcome of these cases could establish important precedents and present and present a key opportunity for courts to clarify the boundaries of concerted practices, potentially opening the door to more effective prosecution of AI collusion.

A. What does US Case Law Tell us about AI Collusion?

468. *Airlines* and *Topkins* exemplify classic cases in which a centralized payment system or AI is used as a facilitating or communication tool to execute a classic cartel. Unfortunately, these cases do not shed much light on how courts evaluate more complex forms of concerted practices. Recent decisions in *Realpage* and *Gibson* depart from the 2016 *Meyer* decision, suggesting that significant hurdles must be overcome to persuade courts of a *per se* infringement under Section 1 when companies delegate their pricing decisions on the same platform or algorithm. This is largely explained by the development of US federal case law concerning concerted practices. It may also, arguably, be explained by courts' caution in deeming, per se, illegal commercial practices in the absence of certainty that there are no procompetitive justifications for them.

469. Despite this challenge, the rule of reason standard should not present an insurmountable burden under the fact pattern alleged in these cases, especially given the evidence of harm presented by plaintiffs, in the words of Judge Posner:

> [A] plaintiff who proves that the defendants got together and agreed to raise the price (whether directly or by restricting output, which would have the same effect) that he pays them for their products – which is

---

218 *See* Dai et al v. SAS Inst., Inc. et al., No. 3:24-cv-02537 (N.D. Cal. 2024), ECF No. 1.

219 *Id.* at 82 citing IDEAS, https://ideas.com/science-behind-g3-rms/.

what the plaintiffs in this case would have had to prove under the *per se* rule to establish liability and obtain damages – has made a prima facie case that the defendants' behavior was unreasonable. He need not prove market power; even though by definition without it a firm or group of firms can't harm competition, it is not a part of the prima facie case of illegal *per se* price fixing.[220]

470. The burden would then shift to defendants to show that the behavior is not anticompetitive.[221]

471. Regardless of the standard ultimately applied by courts, the outcome of the pending AI collusion cases will be crucial for effectively addressing and deterring AI-driven collusion among competitors. In doing so, it is key that courts pay attention to the key question of whether the agreement centralizes decision-making in a way inconsistent with individual profit maximization, in line with *American Needle*.[222]

472. Translated to the four scenarios presented initially, a plaintiff may prove a Section 1 case in scenario 1.) *Monitoring an existent cartel* and 2.) *Facilitate entering into an existent cartel*, and depending on how the above-described cases are decided, 3.) *Entering into a cartel without direct contact*, albeit under higher evidentiary burdens, than a similar case brought in the EU. In contrast, proving a case in scenario 4.) *Facilitate tacit coordination* would be more challenging.

B. *Possible Options to Close the Gaps*

473. While conduct falling short of a *per se* agreement may still be tried under the rule of reason, and individual components of a conspiracy may be federally prosecuted under Sections 1 and 2 of the Sherman Act, and Section 5 of the FTC Act, rule of reason theories require proof of the alleged effects of these agreements, a standard that does not seem adequate for concerted practices.

474. Additionally, from the perspective of the analysis of circumstantial evidence, as recommended by former FTC Chair Kovacic and his co-authors in 2011, a more principled approach to considering plus factors is key. This includes identifying "super plus factors"[223] and considering the heightened

---

220 In re Sulfuric Acid Antitrust Litig., 703 F.3d 1004, 1007 (7th Cir. 2012) (internal citations omitted), *see* Michelle Fischer, Thomas Demitrack, Paula W. Render & Brian Grube, *A US Federal Appeals Court reaffirms flexible legal standard for restraints in competitor collaborations (Sulfuric acid antitrust litigation)*, e-Competitions Dec. 2012, art. No. 50575.

221 *See id.* at 1007–08.

222 See Am. Needle, Inc. v. Nat'l Football League, 560 U.S. 183, 197 (2010), see also Anna Chehtova, *The US Supreme Court finds that a football league and its members should not be treated as a single entity* (American Needle / National Football League), e-Competitions May 2010, art. No. 33292.

223 *See* Kovacic et al., *supra* note 110, at 435 (listing as a super plus factor "3. A reliable predictive econometric model that accounts for all material non-collusive effects on price, estimated using benchmark data where conduct was presumed non collusive, produces predictions of prices that do not explain the path of actual prices in the period or region of potential collusion, at a specified high confidence level.")

probative effect of multiple plus factors (e.g., under potential theories of "constellation"[224] or "synergy"[225]). Similarly, it is key that the standard adopted does not consider plus factors as elements, i.e., as a prerequisite to finding collusion, but as evidence that should be considered holistically and that they do not translate, in practice, in a requirement of direct evidence.[226]

475. Nevertheless, under the rule of reason, a plaintiff's showing of effects in the market should be sufficient to invert the burden of proof onto defendants to show their alleged conduct is pro-competitive.[227]

476. This view is consistent with a quick-look approach to where defendants would bear the burden to overcome the presumption that their sharing of real-time or future price information is not anti-competitive[228] and would be more appropriate for cases dealing with horizontal exchanges of price information.

477. Finally, the FTC's investigation of these conducts may have an important deterrent effect, incentivizing businesses to operate and utilize AI pricing in a way that avoids collusive outcomes.

III. The *Ex-Ante* Approach

478. When it comes to rulemaking, as with any *ex-ante* market intervention, careful consideration is essential before addressing the issue through regulation. Regulatory interventions should be considered following empirical evidence of anticompetitive conduct that escapes *ex-post* enforcement and identification of a market failure. Some proposals discuss utilizing existing regulations, potential legal amendments, and introducing new legislation.[229]

---

224 *See id.* at 426.

225 *See generally*, Leslie, *supra* note 137.

226 Christopher R. Leslie, *The Factor/Element Distinction in Antitrust Litigation*, 64 Wm. & Mary L.Rev. 585 (2023).

227 *See id.*

228 See United States v. Apple, Inc., 791 F.3d 290, 330 (2d Cir. 2015) ("This 'quick look' effectively relieves the plaintiff of its burden of providing a robust market analysis by shifting the inquiry directly to a consideration of the defendant's procompetitive justifications." (citation omitted), *see* Roda, *supra* note 102; Metro. Intercollegiate Basketball Ass'n v. Nat'l Collegiate Athletic Ass'n, 337 F. Supp. 2d 563, 572 (S.D.N.Y. 2004) ("Under a 'quick look' analysis, a plaintiff is relieved of its initial burden of showing that the challenged restraints have an adverse effect on competition because the anticompetitive effects of the restraint are obvious." (citing Cal. Dental Ass'n v. F.T.C., 526 U.S. 756, 770 (1999))).

229 *See, e.g.,* Preventing Algorithmic Collusion Act of 2024 (S. 3686) a bill introduced by Senator Klobuchar (D-MN). The proposed legislation would create a presumption that a defendant entered into an agreement, contract, or conspiracy in restraint of trade in violation of the antitrust laws if the defendant: (i) distributed a pricing algorithm to two or more persons with the intent that the pricing algorithm be used to set or recommend a price, or (ii) used a pricing algorithm to set or recommend a price or commercial term of a product or service and the pricing algorithm was used by another person to set or recommend a price. The Act also would require companies using algorithms to set prices to provide transparency and would prohibit the use of "nonpublic competitor data" to train any pricing algorithm. *See also* HB24-1057 to Prohibit Algorithmic Devices Used for Rent Setting, rejected by the Colorado Senate. Under the proposed text, it would be considered an unfair or deceptive trade practice under the Colorado Consumer Protection Act for a landlord to set the amount of rent to be charged to a tenant for the occupancy of residential premises, employing or relying on an algorithmic device that uses, incorporates, or was trained with nonpublic competitor data.

Finally, the FTC may decide to regulate AI price fixing. However, the FTC's authority to regulate remains highly controversial.[230] Other *ex-ante* options include creating the right incentives for online platforms to adopt measures preventing AI collusion. These authors propose that the focus should be placed on reducing evidentiary burdens for *ex-post* prosecution.

IV. Finding the Ghost in the Shell

479. One of the key concerns regarding AI pricing algorithms is that they have the ability – and have often been found – to replicate and automate the characteristics of highly concentrated markets where each firm realizes that the effect of its actions depends on the actions of its rivals. This interdependence may lead to parallel market conduct. Whether this parallel conduct triggers liability under antitrust depends on other factors that would persuade a court of the existence of an agreement, "meeting of the minds," or "concerted action."

480. Stakeholders should be acutely aware that enforcers on both sides of the Atlantic have taken a "compliance by design"[231] approach under which companies or undertakings may not hide behind the "shell" of automation but are responsible for designing and overseeing the use of their algorithms to ensure that they are neither designed to fix prices nor learn to do this.

481. The key divide between EU and US enforcement is rooted in their different policy choices regarding the degree of *ex-post* intervention for conscious parallelism. In practice, this is reflected in how courts evaluate and weigh "extra" or "plus factors" to prove a conspiracy absent direct evidence. US courts require a significantly higher evidentiary standard. On the other hand, EU courts have reversed the burden of proof requiring defendants to prove an absence of an agreement when parallel conduct and some plus factors are present. Against this background, US enforcers and plaintiffs will need to overcome higher hurdles when prosecuting AI collusion, especially when conduct goes beyond facilitating or implementing a cartel through AI. Despite its relative advantages, the EU has also stated the need to explore options beyond its current toolbox to prosecute this conduct.

482. While the differences between the EU and US approaches to concerted practices are in part due to key differences in the enforcement systems (administrative versus judicial) and available remedies (civil versus criminal), the US would benefit from a more principled approach. The following paragraphs summarize our main findings.

483. Under EU competition law, a single meeting and parallel behavior is sufficient to infer a collusive agreement under Article 101 TFEU. This provision has been purposely interpreted to include an extended notion of "understanding," including interaction falling short of an agreement.

---

230 The current discussion and litigation surrounding the FTC's rule banning non-compete clauses nationwide, issued on Apr. 23, 2024, may be informative regarding this possibility.

231 *See supra* note 23.

In contrast, under US antitrust law, finding concerted action through indirect evidence under Section 1 has traditionally depended on a finding of plus factors, in addition to parallel behavior, to prove an agreement through circumstantial evidence. US federal courts have broad discretion in determining and weighing these factors and in determining how much evidence is required for each one to be considered, leading to significant legal uncertainty.[232] The different ways in which plus factors are considered and analyzed by courts[233] lead to the unsatisfactory result that the likelihood of the court's finding of an agreement in each specific case will depend on "the courts' unarticulated intuition about the likely causes of observed parallel behavior."[234] Against this background, it seems advisable for US courts to adopt a more principled approach to plus factors.

484. Despite these uncertainties, the problem presented by AI assisted coordination is not new.[235] Both the DOJ and private litigants have successfully brought Section 1 cases under similar fact patterns when they have been able to persuade courts of the existence of the "extra elements."

485. Traditional frameworks for analyzing hub-and-spoke cartels and invitations to collude provide useful guidance, but it remains somewhat unclear whether self-learning algorithms leading to collusive outcomes could be effectively prosecuted in the EU and the US despite enforcers' broad statements of "enforcement by design." The likelihood of this conduct escaping enforcement hinges on how quickly pricing AI develops and what safeguards are put in place to prevent these outcomes.

486. Regarding the four scenarios presented initially, 1.) *Monitoring an existing cartel*, and 2.) *Facilitate entering an existent cartel* involves the use of AI to secure compliance with an already established understanding; both the EU and US frameworks seem adequately equipped to address this type of AI-assisted collusion. On the other hand, regarding scenarios 3.) *Facilitate entering into a cartel without direct contact*, and 4.) *Facilitate tacit collusion*; involving indirect or no contact, depending on how the pending cases are decided, US plaintiffs may face higher burdens proving a case under scenario 3.) and both EU and US prosecution of scenario 4.) seems challenging. Therefore, enforcers on both sides of the Atlantic should take proactive steps to explore solutions, especially directed at situations in which there is collusion without direct or indirect contact.

487. Finally, the authors of this chapter believe that any *ex-ante* measure to regulate this conduct should be based on careful empirical observations of market failures.

---

232 *See* Kovacic et al., *supra* note 110, at 936.

233 *See, e.g.*, Louis Kaplow, *On the Meaning of Horizontal Agreements in Competition Law*, 99 Calif. L. Rev. 683, 749–50 (2011) ("[T]here is no readily accepted principle that determines what counts as a sufficient plus factor and what does not (or what combinations might be jointly sufficient).").

234 *See* Kovacic et al., *supra* note 110, at 407.

235 Olhausen, *supra* note 37.

What About Bob? Revisiting the Intersection of Antitrust Law and Algorithmic Pricing in 2024

MAUREEN OHLHAUSEN, TAYLOR OWINGS AND CORA ALLEN[*]
Wilson Sonsini Goodrich & Rosati LLP

Abstract

Courts have begun to flesh out the relevant questions for when competitors can use the same algorithmic pricing tools as each other. The US DOJ and FTC are intervening to try to push the law in a more plaintiff-friendly direction, showing skepticism when companies use pricing algorithms to capture more of consumers' willingness to pay. But the touchpoint for answering the tricky legal questions seems to be guided by analogies to human decision-making, at least for now. Courts are asking "What About Bob?" – that is, wouldn't this be fine if a human did it? – and that question should therefore guide companies when they consider three risk factors in making or using algorithmic pricing tools: (1) what sort of independent discretion do users of the tool exercise, (2) is the algorithm trained on public or non-public data, and, (3) does the algorithm have the power to manipulate market conditions to "learn" inferences?

---

[*] The authors are attorneys in the Antitrust and Competition practice at Wilson Sonsini Goodrich & Rosati LLP. This chapter is for informational purposes only. It expresses the view of the authors, not the firm or any of its clients.

What About Bob? Revisiting the Intersection of Antitrust Law and Algorithmic Pricing in 2024

I. Introduction

488. A test offered by Maureen Ohlhausen (one of the authors here), while Acting Chairman of the US Federal Trade Commission (FTC), in a 2017 speech has become a widely cited standard for the legality of using algorithms under the antitrust laws. What has become known as the "Bob" Test is as follows:

> Everywhere the word "algorithm" appears, please just insert the words "a guy named Bob". Is it ok for a guy named Bob to collect confidential price strategy information from all the participants in a market, and then tell everybody how they should price? If it isn't ok for a guy named Bob to do it, then it probably isn't ok for an algorithm to do it either.[1]

489. Seven years – and many technology cycles – later we ask, "What About Bob?" – how have companies been using pricing algorithms, and is the "Bob" Test still a reliable way of judging a business's antitrust risk from using or creating an algorithm? We think the answer is that the test is still apt, and the courts are following it despite plaintiffs' and enforcers' attempts to distinguish it. The US Department of Justice (DOJ), FTC, and private plaintiffs have recently argued that machines are fundamentally distinct from employees – in their view, technological tools that may help businesses more effectively maximize profits should receive special antitrust scrutiny. Some of the typical elements of Section 1 of the Sherman Act case should be relaxed in this context, according to this view.

490. With this heightened level of scrutiny on pricing algorithms, we think the time is right to a provide a roadmap of key factors for evaluating the antitrust risk from algorithmic pricing tools that: (1) have users who compete against each other in their market, (2) ingest large amounts of data (perhaps eclipsing the abilities of "Bob"), or (3) both.

491. The most important factors discussed in case law to date are taken in turn – what sort of independent discretion do users of the tool exercise, is the algorithm trained on public or non-public data, and does the algorithm have the power to manipulate market conditions to "learn" inferences?

II. First Factor: Does the User Exercise Independent Discretion Over the Output of the Algorithm?

492. One factor in assessing antitrust risk when an algorithm is involved in price-setting looks at how the output of that algorithm is used. Are the results of the algorithm mere non-mandatory pricing recommendations? Or have the parties agreed – independently or in coordination with one another – to delegate decision-making authority to the algorithm? Even where competitors have not clearly communicated or agreed, do they face pressure to adopt the pricing recommendations, and how often do they do so in practice?

---

1   Maureen K. Ohlhausen, *Should We Fear the Things That Go Beep in the Night? Some Initial Thoughts on the Intersection of Antitrust law and Algorithmic Pricing*, FTC (May 23, 2017), https://www.ftc.gov/news-events/news/speeches/should-we-fear-things-go-beep-night-some-initial-thoughts-intersection-antitrust-law-algorithmic.

493. The answers to these questions determine the binding or non-binding nature of reliance on the algorithm and constitute arguably the greatest risk factor in assessing how the use of an algorithm to set pricing will be scrutinized under an antitrust analysis.

494. At one end of the spectrum of risk, a business might decide to employ technology that cedes independent control of pricing to a third-party algorithm. Investigations and litigations in this area show that binding agreements to price according to an algorithm, whether by contractual agreement or by a less formal agreement, comes with risks.

495. *United States v. Topkins* is an early and relatively straightforward example of how agreeing to be bound by the results of an algorithm can violate Section 1 of the Sherman Act.[2] *Topkins* is commonly cited as the first US Department of Justice (DOJ) case to allege that a pricing algorithm was a tool to further an antitrust violation.[3]

496. In *Topkins*, defendant David Topkins sold posters through Amazon Marketplace, Amazon's Website for third-party sellers.[4] Topkins pleaded guilty to participating in a conspiracy with other poster sellers to fix the prices of certain posters sold on Amazon Marketplace.[5] The DOJ had evidence that Topkins and other poster sellers engaged in pricing discussions with each other during which they agreed to fix the prices of certain posters.[6] To implement these agreements, Topkins and his co-conspirators "agreed to adopt specific pricing algorithms for the sale of the agreed-upon posters with the goal of coordinating changes to their respective prices."[7] According to the DOJ, Topkins actually wrote the computer code that instructed his company's algorithm-based software to set prices of the agreed-upon posters in conformity with the conspirators' agreement.[8] Ultimately, Topkins pleaded guilty to violating Section 1 of the Sherman Act and agreed to pay a $20,000 criminal fine.[9]

497. As *Topkins* shows, an agreement between competitors to be bound by the prices generated by an algorithm is treated no differently from any other hardcore price-fixing agreement that violates Section 1. This is the

---

2   U.S. Dep't of Justice, Plea Agreement, United States v. Topkins, 3:15-cr-00201-WHO (N.D. Cal. 2015), https://www.justice.gov/atr/case-document/file/628891/dl; 15 U.S.C. § 1.

3   See, e.g., *AI and Antitrust – When Does an Algorithm Become an Agreement?*, JDSUPRA (May 19, 2023), https://www.jdsupra.com/legalnews/ai-and-antitrust-when-does-an-algorithm-6819337/.

4   *Id.*

5   U.S. Dep't of Justice, Plea Agreement, *Topkins*, 3:15-cr-00201-WHO, at ¶ 4(b) (N.D. Cal. Apr. 30, 2015).

6   *Id.*

7   *Id.*

8   *Id.* information at ¶ 8(d). *See also* Robert E. Connolly, *The US DoJ Secures the Guilty Plea of a Former E-Commerce Executive Following the Prosecution of the First E-Commerce Price Fixing Case (Topkins)*, E-COMPETITIONS April 2015, art. No. 73395.

9   *Former E-Commerce Executive Charged with Price Fixing in the Antitrust Division's First Online Marketplace Prosecution*, U.S. DEP'T OF JUSTICE (Apr. 6, 2015), https://www.justice.gov/opa/pr/former-e-commerce-executive-charged-price-fixing-antitrust-divisions-first-online-marketplace.

What About Bob? Revisiting the Intersection of Antitrust Law and Algorithmic Pricing in 2024

original premise of the "Bob" Test after all: if it isn't ok for Topkins to reach pricing agreements with his competitors without an algorithm, it is not ok to do so with an algorithm.

498. This basic concept was a touchpoint of the "Bob" Test and continues to be reinforced in agency positions and actions. For example, a recent FTC blog post reiterated: "[Y]our algorithm can't do anything that would be illegal if done by a real person... When you replace once-independent pricing decisions with a shared algorithm, expect trouble. Competitors using a shared human agent to fix prices? Illegal. Doing the same thing but with an agreed upon, shared algorithm? Still illegal."[10]

499. Moving on to a closer set of cases, agreements to assist, rather than end, independent pricing judgment are being scrutinized by the current Administration as well as private plaintiffs.

500. Recent statements from the FTC and FTC leadership have expressed the position that the use of price-setting algorithms can violate the US antitrust laws even if the competitors using the algorithm never "directly communicated and retained some pricing discretion."[11] So long as competitors "each agree to use [an algorithm] knowing the others are doing the same in concert[,]" the conduct may be scrutinized.[12] These statements show that there is some antitrust risk in using the same algorithm as your competitors, even if there is no separate agreement to be bound by the results of an algorithm, as was present in *Topkins*. We can learn this Administration's position from its recent Statements of Interest submitted in private litigation.[13] They argue that delegating "key aspects" of pricing to an algorithm could be per se illegal, "even if [parties] retain some authority to deviate from the algorithm's recommendations."[14]

1. In re: RealPage, Inc. Rental Software Antitrust Litigation (No. II)[15]

501. In November 2023, the DOJ filed a statement of interest in *In re: RealPage, Inc. Rental Software Antitrust Litigation (No. II)*, a federal multidistrict litigation out of Tennessee, arguing that an algorithm that

---

10 Hannah Garden-Monheit & Ken Merber, *Price Fixing by Algorithm is Still Price Fixing*, FTC BUSINESS BLOG (Mar. 1, 2024), https://www.ftc.gov/business-guidance/blog/2024/03/price-fixing-algorithm-still-price-fixing.

11 Lina Khan, TWITTER (Mar. 29, 2024, 10:45 AM), https://x.com/linakhanFTC/status/1773768439720509738.

12 Garden-Monheit & Merber, *supra* note 10.

13 As this chapter was going to print, the U.S. Department of Justice filed a civil lawsuit directly challenging the practices it criticized in the Statements of Interest we discuss here. See United States v. RealPage, Inc., No. 1:24-cv-00710, Complaint (M.D.N.C. Aug. 23, 2024) ECF No. 1. This development demonstrates that the principles articulated in this chapter are an enforcement priority for the U.S. antitrust agencies, and confirms that this is an important area to watch for businesses using modern pricing technologies.

14 U.S. Dep't of Justice, Statement of Interest of the United States of America, Duffy v. Yardi Systems, Inc., 2:23-cv-01391-RSL, at 3 (W.D. Wash. Mar. 1, 2024), ECF No. 149.

15 The authors' firm, Wilson Sonsini Goodrich & Rosati, represents one of the defendants in the RealPage litigation.

sets non-mandatory prices could still be per se illegal under the US antitrust laws.[16]

502. The *RealPage* litigation began in October 2022 when renters of multi-family residential real estate brought a class action case against RealPage and certain property owners and managers who use RealPage's offerings.[17] RealPage offers a revenue management software that collects property owners' and managers' sensitive pricing and supply data, applies an algorithm across this data, and then generates price recommendations for each rental unit.[18] Plaintiffs allege that by co-mingling their sensitive pricing and supply data within RealPage's revenue management software, property owners and managers – who are horizontal competitors – have been facilitated by RealPage to conspire to fix prices in student and multi-family rental housing markets throughout the United States.[19]

503. In a November 2023 Statement of Interest the DOJ argued that "the alleged scheme [met] the legal criteria for per se unlawful price fixing"[20] because "the common delegation of decision-making to a common entity allows its decisions to affect actual or potential competition – even without any additional subsequent agreement or coordination among the parties."[21] According to the filing, this kind of delegation, even if some deviation or non-conformity with the algorithm's pricing recommendations remained, represented "the joining together of separate actors with separate economic interests characteristic of concerted action that Section 1 of the Sherman Act reaches."[22]

504. Analogizing to the Supreme Court's decision in *State Bar. Goldfarb v. Virginia State Bar* as well as cases holding that fixing advertised list prices is per se unlawful, even if firms are free to ultimately charge lower prices to customers, the DOJ reasoned that fixing non-mandatory prices by way of an algorithm could still be per se illegal.[23]

505. The correct focus – according to the DOJ – is on how the challenged price-fixing scheme "disrupt[s] the competitive process," rather than the ultimate success of the scheme or, in this case, complete adherence to the algorithm's pricing.[24] According to the DOJ, it seems, when a seller uses

---

16 U.S. Dep't of Justice, Statement of Interest of the United States of America, In re: RealPage, Inc. Rental Software Antitrust Litigation (No. II), 3:23-md-03071 (M.D. Tenn. Nov. 15, 2023), ECF No. 628.

17 Edward Rogers, Elizabeth Weissert & Haesun Burris-Lee, *Algorithmic Pricing Programs Caught in Antitrust Crosshairs*, Law360 (Feb. 2, 2024, 2:28 PM), https://www.law360.com/articles/1791730/algorithmic-pricing-programs-caught-in-antitrust-crosshairs?copied=1.

18 Memorandum Opinion, *In re: RealPage*, at 2 (Dec. 28, 2023).

19 *Id.*

20 U.S. Dep't of Justice, Statement of Interest of the United States of America, *In re: RealPage*, at 2-3.

21 *Id.* at 5.

22 *Id.* at 5.

23 *Id.* at 22.

24 *Id.* at 20–21.

a pricing algorithm knowing its competitors will also do so, this could subject the party to per se liability even if the seller does not adhere to the prices generated by the algorithm on every occasion, and has made no agreement (with the algorithm provider or with its competitors) to do so.

2. Duffy v. Yardi Systems, Inc.

506. In March 2024, the DOJ – joined this time by the FTC – took this position again even more pointedly in a Statement of Interest filed in a separate class action involving a rent-setting algorithm, *Duffy v. Yardi Systems, Inc.*[25] In *Duffy v. Yardi*, a putative class of renters alleged that Yardi and a group of property management companies that used Yardi's RENTmaximizer revenue management software to "maximize rental income" engaged in price coordination in violation of Section 1 of the Sherman Act.[26]

507. The DOJ and FTC submitted a Statement of Interest specifically "to address an incorrect legal position in defendants' motion to dismiss: that the landlords' retention of some pricing discretion dooms a price-fixing claim."[27] The agencies explained that the case law is "clear" that competitors "may not agree to fix the *starting point* of pricing (e.g., agree to fix advertised list prices) even if the actual charged prices vary from the starting point."[28] Pointing to per se treatment for schemes that "fix advertised *list* prices or sticker prices,"[29] the agencies noted that "[t]he same principle holds in cases involving joint delegation of pricing recommendations to a common algorithm. By altering the starting point of prices, such agreements among competitors are analogous to agreements to fix list prices – distorting the competitive pricing process that the per se rule protects."[30]

508. According to the agencies, adherence to an agreed price is not a condition of per se illegality: "Price deviations don't immunize conspirators… Just because a software recommends rather than determines a price doesn't mean it's legal. Setting initial starting prices or recommending initial starting prices can be illegal, even if conspirators deviate from recommended prices. And even if some of the conspirators cheat by starting with lower prices than those the algorithm recommended, that doesn't necessarily change things."[31]

509. As of the publication of this chapter, no order on defendants' motion to dismiss has been issued in *Duffy v. Yardi*. Yet between the two statements of interest in these ongoing rent-setting algorithm litigations, US antitrust

---

25 U.S. Dep't of Justice, Statement of Interest of the United States of America, *Yardi Systems, Inc.*

26 *See* Complaint, *Yardi Systems, Inc.*, 2:23-cv-01391-RSL at 2 (W.D. Wash. Sept. 8, 2023), ECF No. 1.

27 U.S. Dep't of Justice, Statement of Interest of the United States of America, *Yardi Systems, Inc.*, at 2 (internal citation omitted).

28 *Id.* at 2 (internal citation omitted).

29 *Id.* at 4–5.

30 *Id.* at 6.

31 Garden-Monheit & Merber, *supra* note 10.

enforcement agencies have clearly put a stake in the ground for the proposition that "[j]ust because a software recommends rather than determines a price doesn't mean it's legal."[32]

510. While US enforcement agencies have taken the position that using the same algorithm as your competitors to generate non-binding price recommendations may subject a party to per se liability, the few court decisions that analyze this factor have tended to take a softer view – requiring evidence of something more for a per se Section 1 claim to withstand a motion to dismiss. A small (but perhaps growing) body of case law suggests that choosing to use the same third-party algorithm that competitors have used for mere non-binding pricing recommendations does not constitute a per se violation of Section 1 unless paired with evidence: (i) of a direct agreement or communications between competitors, or (ii) that the recommendations were somehow "binding or enforceable."[33]

511. For example, in the January 2024 motion to dismiss opinion in *RealPage*, the court found that multi-family plaintiffs had "not alleged a straightforward conspiracy justifying application of the per se standard."[34] A key defect noted by the court was that, "while Plaintiffs allege that RMS Client Defendants 'delegate[d]' their pricing decisions to RealPage, they also allege that as much as 10–20% of the time, RealPage's clients deviate or override those pricing recommendations."[35] Under these facts, the court could not find the plaintiffs alleged "an absolute delegation of their price-setting to RealPage."[36] While the plaintiffs had alleged "an aggressive scheme created by RealPage to monitor acceptance of its pricing recommendations,"[37] the court found such allegations insufficient without evidence that RealPage could "enforce acceptance of price recommendations" by, for example, "removing an uncooperative member from the conspiracy or applying some other form of punishment."[38] Absent this, the plaintiffs had not sufficiently alleged a per se price-fixing conspiracy.[39]

512. The court noted similar defects in the student plaintiffs' complaint: "the Student Complaint does not allege that the RealPage pricing recommendations were in any way binding or enforceable on Lessors… RealPage's pricing recommendations rel[y] on RealPage's monitoring and Lessors for

---

32 *Id.*

33 U.S. Dep't of Justice, Memorandum Opinion, *In re: RealPage*, at 48; *see also* Gibson v. MGM Resorts International, No. 2:23-CV-00140-MMDDJA, 2023 WL 7025996 at *2 (D. Nev. Oct. 24, 2023) ("[T]he Court cannot plausibly infer from the allegations in the Complaint that Hotel Operators are required to accept the recommendations provided by a particular software pricing algorithm. This is a fatal deficiency in the Complaint.")

34 U.S. Dep't of Justice, Memorandum Opinion, *In re: RealPage*, at 45.

35 *Id.* at 46 (cleaned up).

36 *Id.* (cleaned up).

37 *Id.* (cleaned up).

38 *Id.* (cleaned up).

39 *Id.* (cleaned up).

513. In late 2023, the court in another ongoing algorithmic pricing litigation came to a similar conclusion. In granting the defendants' motion to dismiss, the court in *Gibson v. MGM Resorts International*, took a similar position on the importance of alleging mandatory pricing recommendations in stating a per se Section 1 claim. The *Gibson* plaintiffs challenged "an unlawful agreement among Defendants to artificially inflate the prices of hotel rooms on the Las Vegas Strip above competitive levels,"[41] alleging that hotel operators "agreed to use a shared set of pricing algorithms" offered by Rainmaker "that recommend supra-competitive prices to the hotel operators."[42]

514. The court reasoned that it could not "plausibly infer from the allegations in the Complaint that Hotel Operators [were] required to accept the recommendations provided by a particular software pricing algorithm."[43] According to the court, this was a "fatal deficiency" to plaintiffs' allegations "as without an agreement to accept the elevated prices recommended by the pricing algorithm, there is no agreement that could either support Plaintiffs' theory or otherwise make out a Sherman Act violation[.]"[44]

515. The reasoning used in both *RealPage* and *Gibson* comports with the well-established principle in cases involving Manufacturer's Suggested Retail Prices that a manufacturer may suggest resale prices to dealers, and no agreement will result if a dealer "independently decides to observe specified resale prices."[45] Manufacturers have generally been permitted to provide suggested price lists to dealers, to advertise suggested resale prices to dealers' customers, and to print suggested resale prices on the product or a price tag without being found to have entered into resale price agreements with dealers.[46] Though these are cases involving vertical restraints, they address directly the question of whether a suggestion can satisfy the element of agreement under Section 1 of the Sherman Act. Where the decision to make a suggestion is a unilateral one, it seems that this crucial element of the Sherman Act may be missing.

516. When faced with the issue of how aggressively the manufacturer can press its dealers to abide by its pricing suggestions, courts have generally found that anything short of coercion – including "exposition, persuasion and

---

40 *Id.* at 48.

41 Gibson v. MGM Resorts Int'l, No. 2:23-CV-00140-MMDDJA, 2023 WL 7025996 at *1 (D. Nev. Oct. 24, 2023) (quoting complaint). Accord *Id.* (D. Nev. May 8, 2024) (reaching the same conclusion on motion to dismiss amended complaint).

42 *Id.*

43 *Id.* at *3.

44 *Id.*

45 United States v. Parke, Davis & Co., 362 US 29, 44 (1960); *see also* Isaksen v. Vermont Castings, Inc., 825 F.2d 1158, 1164 (7th Cir. 1987) (fact of adherence does not establish agreement to adhere).

46 *See generally*, AMERICAN BAR ASSOCIATION, 1-1 ANTITRUST LAW DEVELOPMENTS 1D-1-a-(3)(a) (2021).

argument" to encourage dealers to charge the suggested prices – does not constitute the sort of breakdown in independent decision-making that runs afoul of Section 1 of the Sherman Act.[47]

517. While so far the cases involving algorithmic pricing have not relied on a "coercion" standard, they suggest evidence of more than a mere unenforceable recommendation is required.

III. Factor Two: Does the Algorithm Use Non-Public Data to Determine a Price?

518. In addition to considering whether the output of the algorithm is binding on those using it or a mere recommendation not subject to an agreement or enforcement mechanism, investigations and litigations involving pricing algorithms have also focused on whether the algorithm generates pricing recommendations based solely on public information, or whether the algorithm uses non-public, competitively sensitive information provided by groups of competitors.

519. Notably, this public/non-public factor encompasses both what information is *fed into* the algorithm as well as what information *comes out* of the algorithm in each pricing recommendation. For example, even if an algorithm collects non-public pricing data from multiple competitors in a market, it likely makes a difference under a Section 1 analysis whether the pricing recommendation generated for one competitor takes into consideration only that competitor's non-public data, or also considers the pricing data collected from other competitors.

520. Parsing one level further, consider a scenario in which a pricing algorithm collects non-public data from multiple competitors and uses this to observe a demand-side trend – such as softening demand. The algorithm then uses that demand-side trend observation to make an individualized pricing recommendation to one competitor but *does not* use data related to how the other competitors will price in making the recommendation.

521. Despite the fuzzy boundaries and many complications of the public/non-public factor, one principle appears consistent: the "exchange" of non-public information between competitors through an algorithm creates substantial Section 1 risk.[48] This principle again harkens back to our old friend Bob. If it isn't ok for a guy named Bob to exchange non-public information with competitors, then it probably isn't ok for an algorithm to do it either.

522. What is less clear, however, is what exactly it means to "exchange" non-public information by means of an algorithm. Is it enough that multiple competitors feed their non-public information *into* an algorithm, or does Section 1 require that data to be fed into an algorithmic "melting pot" and then *produced out* of the algorithm by way of its pricing recommendations?

---

47   Gray v. Shell Oil Co., 469 F.2d 742, 748 (9th Cir. 1972).
48   *See e.g., In re: RealPage,* Memorandum Opinion, at 34 (Dec. 28, 2023).

For a Section 1 "exchange" to have occurred, must the pricing recommendation given by an algorithm to Competitor A be based on the pricing information fed into the system by Competitor B? What if non-public information is combined in training an algorithm to predict the shape of the demand curve in the market, but it does not factor in information about the competitors' plans for supply?

523. The picture emerging from the various cases to consider this factor is that the highest risk involves an algorithm that collects non-public sensitive data from a set of competitors, mixes that data together to forecast supply conditions, and uses it to make individualized recommendations to each competitor. The courts in both *RealPage* and *Gibson* appear to agree that this may constitute an "exchange" of non-public information by means of an algorithm in violation of Section 1.

524. In *Gibson*, for example, the court dismissed the plaintiffs' original complaint on the grounds that their allegations failed to support a hub-and-spoke Section 1 theory. The court explained: "Plaintiffs never quite allege (though they suggest by implication) that Hotel Operators *get* non-public information from other Hotel Operators by virtue of using insufficiently specified algorithmic pricing software."[49] While the plaintiffs alleged

> that confidential information is fed in, but less clearly out, of the algorithms… [plaintiffs do] not explicitly say that one Hotel Operator ever receives confidential information belonging to another Hotel Operator. Moreover, it is unclear whether the pricing recommendations 'generated' to Hotel Operators include that confidential information fed in; perhaps they only get their own confidential information back, mixed with public information from other sources… This does not quite say that the Rainmaker algorithm itself exchanges non-public information.[50]

525. Accordingly, the court found: "Plaintiffs attempt[ed] to create an inference of the exchange of nonpublic information in their Complaint without actually alleging such an exchange."[51]

526. Several months later, the *RealPage* court grabbed on to this distinction, noting this was the "critical difference" between the two cases, and the reason certain claims in *RealPage* survived while all claims were dismissed in *Gibson*. The court noted that unlike in *Gibson*, "[h]ere, the Multifamily Complaint unequivocally alleges that RealPage's revenue management software inputs a melting pot of confidential competitor information through its algorithm and spits out price recommendations based on that private competitor data."[52] The court held this was sufficient

---

49 2023 WL 7025996 at *4.

50 *Id.* at *5.

51 *Id.* at *4.

52 See *In re: RealPage*, Memorandum Opinion, at 34 (Dec. 28, 2023).

to allege a plausible "exchange" in violation of Section 1. This exchange of non-public, commercially sensitive information – along with the common motive to conspire – "taken together… support[ed] a 'reasonable expectation that discovery will reveal evidence of [an] illegal agreement.'"[53]

527. Even when the plaintiffs in *Gibson* amended their complaint to include additional allegations that non-public information had improved the algorithm's pricing predictions over time through machine learning, the court rejected that this constituted the sort of coordinated use of a competitor information prohibited by the antitrust laws. The court specifically appealed to a version of the Bob Test in deciding whether a hub having access to confidential information constitutes the illegal exchange of information between the spokes:

> Defense counsel persuasively analogized the pricing algorithms to an attorney's practice at the Hearing. He argued you can think of Plaintiffs' machine learning theory as to GuestRev and GroupRev as no different than an attorney improving her skills over time with the benefit of experience and access to confidential client information she gains with each client engagement. The attorney does not share one client's confidential information with another, but over time, she (ideally) gets smarter because of what she has learned from each client engagement she has successfully completed. And in time, clients seek her out because she has, for example, developed expertise in antitrust law. But that does not plausibly suggest that each new client who seeks out the attorney is entering into an agreement with every client she has ever worked with. How could it? And the same goes for Plaintiffs' machine learning theory. Thus, mere use of algorithmic pricing based on artificial intelligence by a commercial entity, without any allegations about any agreement between competitors – whether explicit or implicit – to accept the prices that the algorithm recommends does not plausibly allege an illegal agreement, or 'raise a reasonable expectation that discovery will reveal evidence of illegal agreement' sufficient to survive the Motion.[54]

528. In short, the courts in both *RealPage* and *Gibson* have not stopped the inquiry with what is fed *into* an algorithm, finding instead that an "exchange" of non-public information by competitors only occurs when there is proof that "the pricing recommendations '*generated*'… include th[e] confidential information fed in."[55] While "melting pot" algorithms present the highest risk, even the act of feeding non-public data to a third-party algorithm operator – without the additional step of the algorithm

---

53 *Id.* at 33.

54 Gibson v. MGM Resorts Int'l, No. 2:23-CV-00140-MMDDJA, Slip Op. at 12 (D. Nev. May 8, 2024) (quoting Kendall v. Visa U.S.A., Inc., 518 F.3d 1042, 1047 (9th Cir. 2008)).

55 2023 WL 7025996 at *5 (emphasis added).

mixing that information with competitors' information and using it to inform individualized pricing recommendations – appears to carry some antitrust risk.

529. In fact, for the DOJ, it appears the collection of non-public data alone may be sufficient. This was the focus in the recent *RealPage* Statement of Interest, in which the DOJ argued that when "competitors knowingly *combine* their sensitive, non-public pricing and supply information in an algorithm... with the knowledge and expectation that other competitors will do the same" they have violated Section 1.[56] For the DOJ, that competitors were alleged to have "knowingly *shar[ed]* 'competitively sensitive' and 'non-public' pricing information with RealPage" was sufficient to suggest a per se violation had occurred, without further inquiry into what RealPage ultimately did with the data and how it specifically informed its pricing recommendations.[57]

530. While this position does not yet appear to have been adopted by a court, it is an issue to monitor as the investigations and litigation over this use of data continue to play out in the courts.

531. Offering or using a pricing algorithm that collects and/or uses sensitive, non-public data from multiple competitors carries some amount of antitrust risk. However, the question remains whether an algorithm that collects only public data can also come under scrutiny.

532. The complaint brought against Amazon by the FTC in late 2023 appears to suggest so. The FTC alleges that Amazon created an algorithmic tool codenamed "Project Nessie" that allowed Amazon to track and observe price changes of other online retailers at such high frequency that it could "predict[] the likelihood that the online store or stores offering the lowest price for a given product would follow an Amazon price increase."[58] According to the FTC, Amazon could comfortably raise its own prices and, in doing so, induce other online retailers to raise their prices as well.[59]

533. The FTC complaint does not allege a Section 1 violation,[60] but rather that Project Nessie constitutes an unfair method of competition in violation of the FTC Act.[61] This is an area to watch and one in which the basic premise of the "Bob" Test may reach its limit. Though Bob has always been allowed under the antitrust laws to react to competitors' public price changes, even if the reaction was a "punishment" for deviating from a preferred price by steep (but not predatory) discounting, there remains a question whether the use of a technology tool changes the fundamental

---

56 Statement of Interest of the United States of America, *In re: RealPage*, at 15.

57 *Id.* at 20.

58 FTC v. Amazon.com Inc., No. 2:23-cv-01495, Complaint (Public Redacted Version) at 120 (W.D. Wa. Nov. 2, 2023) ECF No. 114.

59 *Id.* at 120.

60 *Id.* at 1.

61 *Id.* at 126–28.

nature of the practice – morphing it into "unfair." On the one hand, Bob cannot do what the technology allegedly does – it is the high frequency nature of the observations and reactions that lessen competitive pressure, according to the FTC. On the other hand, antitrust law has never stood in the way of technology to improve upon human processes before, and the human process of price comparison, and reacting to competitors' publicly available information, is the *sine qua non* of competition.

534. Indeed, all the technologies at issue in the cases we discuss here make their users better at pricing, which means: (1) capturing more of the value that customers are willing to pay – a prize that is fundamental to the profit-motive underpinning a market-based economy, and (2) levelling the playing field among the companies who use the technology – there is no longer a competitive advantage from employing the best price-comparison team to research competitor prices at the most relevant times. Technological shifts always have the risk of levelling playing fields and shifting the vector along which businesses need to compete. Pricing algorithms, when they do not involve agreements among competitors to refrain from striving to out-do one another, are no different.

IV. Factor Three: Does the Algorithm Manipulate Market Conditions On Its Own?

535. Just as an algorithm that only uses public data may raise certain concerns under an antitrust analysis that humans exchanging public data do not raise, some have suggested that certain forms of learning algorithms may violate the antitrust laws in new and unexpected ways and therefore deserve particular scrutiny. While this factor has not yet played out significantly in the case law, it is a concern discussed in academic literature and has been noted as an issue to look for as algorithms become more sophisticated.

536. The technology behind the types of pricing algorithms we see in antitrust cases and investigations has come a long way in the past decade. Some commentators have observed that we may not be far off the point from which an algorithm may *itself* decide to collude without human instruction. For example, the 2023 OECD Competition Policy Roundtable Background Note on Algorithmic Competition stresses the unique risks of self-learning autonomous algorithms, including that such algorithms can "decide to collude (or at least avoid reaching a competitive outcome) without information sharing or explicit coordination."[62]

537. While not yet the subject of case law, this is an area to watch. There have been early accounts of the potential for algorithmic autonomous tacit collusion by legal scholars, and even more recently economists have started to work on this topic.[63] Despite considerable research on algorithmic

---

62 OECD, *Algorithmic Competition* (2023) 13–14, https://www.oecd.org/en/topics/competition.html.

63 *Id.*

collusion, the OECD Background Note reports that "its feasibility and scale in practice are still relatively unclear. While the adoption of pricing algorithms has grown considerably, they are not yet universal, never mind the use of self-learning pricing algorithms. Even if firms use self-learning pricing algorithms, there is not conclusive evidence that algorithmic autonomous tacit collusion is a significant issue. Nonetheless, competition authorities should remain vigilant."[64]

V. Conclusion

538. When large technology shifts cause businesses to implement new ways of competing and maximizing profit, there are always questions about how antitrust law will apply. The use of algorithms to price using large amounts of data is no different. The antitrust agencies and private plaintiffs have seized on this moment of uncertainty to push new theories for how this use of technology is suspect. Courts are just beginning to grapple with the task of analogizing to analog practices; in short, they are asking, "What about Bob?" During this period of uncertainty, we recommend that businesses assess whether their pricing tools might come under antitrust scrutiny because they perform a function that a human would not be allowed to perform under the antitrust laws, or because they perform a function that no human would be able to accomplish, no matter how much time and diligence they exercised. In particular, businesses should analyze the three factors most important to an antitrust risk profile in this area: does the user retain independent discretion to price, does the algorithm use non-public data, and is the algorithm able to manipulate market conditions?

---

64 *Id.*

Korean Competition Rules on Algorithmic Discrimination

Yo Sop Choi[*]

Graduate School of International and Area Studies,
Hankuk University of Foreign Studies

Abstract

Discrimination matters. The principle of non-discrimination, pursuant to fair competition, is often important in the implementation of competition law in Korea. The Korean competition rules on unilateral conduct, which cover a wide variety of types of business practice, prohibit discriminatory conduct related to pricing and terms and conditions. As a result of increasing concerns about algorithmic discrimination, Korean policymakers seem to be trying to develop legislation that can prevent "Big Tech" companies from discriminating against business users and end users. The discussion of a new digital law in Korea stems from the theory of the algorithmic harm of self-preferencing. The Korean competition authority has applied the competition provisions on abuses of market dominance and on unfair trade practices to self-preferencing. Moreover, personalized pricing is becoming one of the most important competition issues in Korea. Therefore, it is timely to discuss the current approaches to self-preferencing and

---

[*] BA, MA in Economics, LLM, LLM, PhD in Law. Professor of Law, Graduate School of International and Area Studies, Hankuk University of Foreign Studies, Seoul. Professor Choi specialized in Competition Law and EU Law. His research interests mainly focus on comparative studies of competition law and digital policies related to data protection, artificial intelligence, and consumer protection. He is a member of the Academic Society of Competition Law (ASCOLA) and the Korean Competition Law Association (KCLA) and also an affiliated scholar of the Dynamic Competition Initiative (DCI). His articles have been published in peer review journals, including *Asia Pacific Law Review, Computer Law & Security Review, IIC-International Review of Intellectual Property and Competition Law, European Competition Law Review, European Business Organization Law Review, Journal of African Law, Queen Mary Journal of Intellectual Property,* and *World Competition: Law and Economics Review.*

algorithmic discriminatory pricing, and to argue against a new digital law in Korea on the basis that the existing competition rules are sufficient to tackle conduct involving algorithmic discrimination.

I. Algorithms and Competition Law in Korea

539. A lack of transparency and the possibility of discrimination have come to be among the major concerns of competition policy on artificial intelligence (AI). Over the last decade, the fear of misuse through the dominance of gatekeepers has triggered discussions about certain types of algorithmic conduct. In particular, the anxiety about AI's "social manipulation superpower"[1] has led to several legal and policy measures against tech giants whose focus is on AI developments. Responding to arguments for new theories of harm in the digital sector, the European Union (EU) has adopted several regulations, such as the General Data Protection Regulation (GDPR),[2] the Digital Services Act (DSA),[3] and the Digital Markets Act (DMA).[4] In addition, the European legislature has just approved a proposal for an AI Act.[5] A series of digital laws in the EU reflects the concerns about harm from the misuse of digital technologies, especially when those technologies are incorporated into the business models of Big Tech companies. There are numerous works setting out different views on algorithms – the benefits of dynamic efficiency and rigorous competition for the market, and the shortcomings of exclusionary and exploitative abuse, including discrimination related to different prices or terms and conditions.[6] Algorithmic discrimination by Big Tech companies has often been suggested when Big Tech holds a market-dominant position and has a dual role at the downstream level.[7]

---

[1] NICK BOSTROM, SUPERINTELLIGENCE 117 (2017). *See also* Joanna J. Bryson, *The Artificial Intelligence of the Ethics of Artificial Intelligence, in* OXFORD HANDBOOK OF ETHICS OF AI 3 (Markus D. Dubber et al. eds., 2021); Mariana Mazzucato et al., *Reshaping Platform-Driven Digital Markets, in* REGULATING BIG TECH 18, 25–26 (Martin Moore & Damian Tambini eds., 2022); FRANK PASQUALE, THE BLACK BOX SOCIETY 61 (2015); STUART RUSSELL, HUMAN COMPATIBLE 103–131 (2020).

[2] Commission Regulation (EU) 2016/679 of the European Parliament and of the Council on the protection of natural persons with regard to the processing of personal data and on the free movement of such data, and repealing Directive 95/46/EC (General Data Protection Regulation), 2016, O.J. (L 119) 1.

[3] Commission Regulation (EU) 2022/2065 of the European Parliament and of the Council on a Single Market for Digital Services and amending 2000/31/EC (Digital Services Act), 2022, O.J. (L 277) 1.

[4] Commission Regulation (EU) 2022/1925 of the European Parliament and of the Council on contestable and fair markets in the digital sector and amending Directives (EU) 2019/1937 and (EU) 2020/1828 (Digital Markets Act), 2022, O.J. (L 265) 1.

[5] Regulation (EU) 2024/1689 of the European Parliament and of the Council of 13 June 2024 laying down harmonised rules on artificial intelligence and amending Regulations (EC) 300/2008, (EU) 167/2013, (EU) 168/2013, (EU) 2018/858, (EU) 2018/1139 and (EU) 2019/2144 and Directives 2014/90/EU, (EU) 2016/797 and (EU) 2020/1828 (Artificial Intelligence Act), 2024, O.J. (L Series). *See generally* Thibault Schrepel, *Decoding the AI Act: A Critical Guide for Competition Experts* 4–5 (AMSTERDAM L. & TECH. INST., Working Paper No. 3-2023, 2023), https://papers.ssrn.com/sol3/papers.cfm?abstract_id=4609947.

[6] *See* OECD, *Algorithmic Competition*, 6 (2023), https://web-archive.oecd.org/temp/2024-01-22/652359-algorithmic-competition.htm.

[7] This chapter focuses mainly on the competition law approaches in a civil law system as in Korea. Price discrimination is legally ambiguous in the common law system as of the United States. For further discussion, *see also* Daniel L. Weisman & Robert B. Kulick, *Price Discrimination, Two-Sided Markets, and Net Neutrality Regulation* 13 TUL. J. TECH. & INTELL. PROP. 81 (2010).

540. The first landmark case involving algorithmic discrimination was the European *Google Shopping* case.[8] In *Google Shopping*, the European Commission decided that Google had violated Article 102 of the Treaty on the Functioning of the European Union (TFEU) because it had abused its market dominance. The Commission asserted that Google had leveraged its dominant position in the search market in the comparison-shopping service market, and the appellate court of the EU, the General Court, largely affirmed the Commission's decision.[9] This case has affected the development of "algorithmic theories of harm"[10] in relation to so-called *self-preferencing*, which has enabled competition authorities in other countries to investigate types of unilateral algorithmic discrimination conduct. Algorithmic discrimination can include self-preferencing through the manipulation of the rankings of search results and also price discrimination.[11] In particular, the recently developed theories of self-preferencing blur the boundaries between exclusionary and discriminatory abuse in the EU.[12] The EU legislature has adopted the DMA, whose substantive provisions prevent designated gatekeepers from self-preferencing,[13] and the Commission initiated its first investigation into compliance with the rule on self-preferencing in March 2024.[14]

541. The European Commission's active enforcement of the regulation against self-preferencing seems to be affecting the development of case law and the adoption of new digital laws in other jurisdictions, including Korea. Like EU competition law, Korean competition law prohibits abuses of market dominance, rather than monopolization, and prohibits both exploitative and exclusionary abuse. In addition, it contains a unique legal measure that deals with unfair trade practices (UTPs), which makes it different from competition law in the EU and the United States.[15] Being equipped

---

8   Commission Decision of June 27, 2017 relating to a proceeding under Article 102 of the Treaty on the Functioning of the European Union and Article 54 of the EEA Agreement Case AT.39740 (*Google Search (Shopping)*) 2017 O.J. (C 9), *see* Nicholas Banasevic, Beatriz Marques & Aurelien Portuese, *The Google Shopping decision*, Concurrences No. 2-2018, art. No. 86714.

9   Case T-612/17, Google and Alphabet v. Eur. Comm'n ECLI:EU:T:2021:763, *see* Frédéric Pradelles & Mary Hecht, *The EU General Court confirms the Commission's decision to fine a Big Tech company for abusing its dominant position in online search by discriminating against comparison shopping services to favour its own offering (Google Shopping)*, e-Competitions Nov. 2021, art. No. 106676.

10 OECD, *supra* note 6, at 6.

11 Some argue that personalized discriminatory pricing may cause harmful effects when it is carried out by a dominant platform and consumers are not aware of the targeted pricing. *See* Suzanne Rab, *Artificial Intelligence, Algorithms and Antitrust*, 18 Comp. L.J. 141, 146 (2019).

12 Alison Jones et al., EU Competition Law 1224 (8th ed. 2023).

13 Article 6(5) DMA prevents the gatekeeper from treating "more favourably, in ranking and related indexing and crawling, services and products offered by the gatekeeper than similar services or products of a third party". *See e.g.*, Philipp Hacker et al., *Regulating Gatekeeper Artificial Intelligence and Data*, 15 Eur. J. Risk Reg. 49, 61 (2024).

14 European Commission Press Release IP/24/1689, Commission Opens Non-compliance Investigations against Alphabet, Apple and Meta under the Digital Markets Act (Mar. 25, 2024), https://ec.europa.eu/commission/presscorner/detail/en/ip_24_1689.

15 *See* Yo Sop Choi, *The Evolution of Fair and Free Competition Law in the Republic of Korea*, *in* Research Handbook on Asian Competition Law 65, 77–78 (Steven Van Uytsel et al. eds., 2020); Mark Furse, Antitrust Law in China, Korea and Vietnam 259 (2009).

with robust provisions, the Korea Fair Trade Commission (KFTC) has shown enthusiasm in its applications to self-preferencing of both the rule on abuse of market dominance and the rule on UTPs.[16] In summary, the self-preferencing cases in Korea and under the EU's DMA have provoked discussion on a digital regulation that can easily catch algorithmic conduct. In December 2023, the KFTC announced that it was initiating a new digital law.[17] After the general elections of April 2024, the KFTC appeared to resume the process of adopting a DMA-type regulation which prohibits self-preferencing, tying, restrictions on multihoming, and most-favored-nation (MFN) clauses. Considering the recent debates on case law and the legislative proposal in Korea, this chapter aims to examine the possible development of the rules on algorithmic discrimination under Korean competition law and to critique the possible digital law for the Korean digital markets.

II. Substantive Provisions of Korean Competition Law

542. The competition regime of Korea differs from those of other countries. Because of the increasing convergence of competition policies across the globe, it is certainly true that the substantive provisions and enforcement of Korean competition law have become similar to those of other competition regimes.[18] However, when considering the number of decisions made by the KFTC over the last forty years,[19] it is not difficult to find that it has vigorously enforced competition law. The background of *economic democratization* under Article 119(2) of the Korean Constitution and the mature competition culture have influenced the rapid development of the Korean competition regime.[20] Supported by the people, the KFTC has developed various competition theories and legal techniques relating to the digital sector, including AI. In particular, the KFTC has sanctioned two Korean Big Tech companies – *Naver Shopping*[21] and *Kakao Mobility*[22] – for self-preferencing. These cases seemed to trigger the consideration of a new digital regulation. The KFTC announced a bill for the Platform Competition Promotion Act (PCPA) in December 2023 and has initiated the process of reviewing

---

16 See Yo Sop Choi, *"Decision to Leave": From Traditional Competition Law to New Digital Regulation and Potential Problems* 13 KLRI J. L. & LEG. 7, 38–39 (2023).

17 KFTC Press Release, The Initiation of Policies on Dominant Platform (Dec. 19, 2023) (in Korean), https://www.ftc.go.kr/www/selectReportUserView.do?key=10&rpttype=1&report_data_no=10360.

18 See e.g., Yo Sop Choi, *The Choice of Competition Law and the Development of Enforcement in Asia*, 22 ASIA PAC. L. REV. 131 (2014).

19 The number of decisions made by the KFTC between 1981 and 2022 is over 15,000. See KFTC, *Statistical Yearbook of 2022*, (Apr. 27, 2023) (in Korean), at 5, https://www.ftc.go.kr/www/cop/bbs/selectBoardList.do?key=190&bbsId=BBSMSTR_000000002317&bbsTyCode=BBST01#.

20 Yo Sop Choi, *The Rule of Law in a Market Economy*, 15 EUR. BUS. ORG. L. REV. 419, 429 (2014).

21 KFTC Decision No. 2021-027 (Jan. 27, 2021); Seoul High Court Judgment 2021 *Nu*36129 (Dec. 14, 2022).

22 KFTC Decision No. 2023-093 (June 13, 2023).

the contents of the draft by gathering opinions from stakeholders and commentators.[23] Importantly, the PCPA bill contains a provision on self-preferencing. As the draft of the bill has not been disclosed to the public, it is uncertain whether the bill will embrace a self-preferencing rule. However, it is highly likely that the PCPA, if it is adopted, will include rules to regulate algorithmic discrimination as a discriminatory treatment that can be described as anti-competitive or unfair under the Monopoly Regulation and Fair Trade Act (MRFTA).[24]

543. Article 1 MRFTA emphasizes that the objective of the regime is to promote *fair and free competition*. The priority given to fair and free competition means that the MRFTA contains two substantive provisions on unilateral conduct that extensively cover not only exclusionary but also exploitative abuse.[25] Unlike competition laws in the EU and the United States, the MRFTA includes a clear presumption of market dominance based on market share and entry barriers. For example, Article 6 MRFTA stipulates that, when an undertaking's market share is 50 percent or more or the total market share of the three largest undertakings reaches 75 percent (except if the market share of the particular undertaking is less than 10 percent or its yearly turnover or value of purchases is less than four billion KRW), an undertaking is presumed to hold market dominance.[26] Article 5(1) MRFTA, then, provides an exhaustive list of types of prohibited abusive conduct by a market-dominant undertaking. Article 5 prohibits: (i) unreasonably determining, maintaining or changing prices, (ii) unreasonably controlling sales of goods or services, (iii) unreasonably interfering with the business activities of other undertakings, (iv) unreasonably impeding the market entry of a new competitor, and (v) unreasonably excluding a competitor or substantially undermining consumer interests. The provisions on the presumption of dominance and the prohibitions indicate the regime's preference for legal certainty.[27] Importantly, the prohibitions of (i) *pricing abuse* and (iii) *interfering abuse* can be interpreted as the prohibition of exploitative and discriminatory conduct.[28]

544. In addition to Article 5, the MRFTA includes the rule on UTPs. Article 45 MRFTA broadly covers various types of unfair conduct: (i) unfairly refusing to deal, (ii) unfairly discriminating against other traders, (iii) unfairly

---

23 *See e.g.*, Charles McDonnell, *KFTC Commits to Proper Consultation before Passing DMA-style Bill*, GCR (Feb. 1, 2024), https://globalcompetitionreview.com/article/kftc-commits-proper-consultation-passing-dma-style-bill.

24 Korean Law No. 19990 (amended and effective July 10, 2024).

25 *See generally* Meong-Cho Yang, *Competition Law and Policy of the Republic of Korea*, 54 ANTITRUST BULL. (2009).

26 Under Article 2(3) MRFTA, the KFTC needs to assess market share and entry barriers to decide if there is a market-dominant position.

27 Yo Sop Choi, *The Enforcement and Development of Korean Competition Law*, 33 W. COMP. 301, 312 (2010).

28 This provision is mainly applicable to excessive pricing rather than predatory pricing. The provision on interfering conduct can be applied to discriminatory pricing or terms and conditions. *See* OHSEUNG KWON & MYEONGSU HONG, ECONOMIC LAW 163, 170–171 (15th ed. 2024) (in Korean).

excluding a competitor, (iv) unfairly inducing a competitor's customer to trade, (v) unfairly forcing a competitor's customer to trade, (vi) unfairly abusing a bargaining position, (vii) unfairly restricting business activities, and (viii) unfairly interfering with business activities, and other activities.[29] The list of UTP focuses on the unfairness of the transactions, and the provisions against the conduct of *discriminating against trading parties* (e.g., (ii) above), *excluding competitors (predatory pricing)* ((iii) above),[30] and *abusing a bargaining position (imposing disadvantages)* ((vi) above) can be applicable to algorithmic discrimination. We have seen very few KFTC decisions on price discrimination,[31] and the Supreme Court of Korea has established a particularly high standard of proof for concluding that there has been price discrimination.[32] Despite the difficulties in regulating pricing abuse, it is not impossible to apply the rules above to price discrimination, and the KFTC often relies on Article 45 because its application does not require market dominance.[33]

545. According to a paper by the OECD, algorithmic unilateral conduct can be categorized as: (i) exclusionary abuse, which means conduct such as self-preferencing, predatory pricing, rebates, and tying,[34] or (ii) exploitative abuse, which includes excessive pricing, unfair trading conditions, and price discrimination.[35] In effect, the two substantive provisions within the MRFTA can effectively cover algorithmic exclusionary and exploitative conduct, as described in the OECD article, if the undertaking uses a large volume of data on consumer behavior or characteristics.[36] In particular, the Korean agency and the Korean courts have dealt with quite a number of cases of exclusionary and exploitative abuse, although there has been no algorithmic pricing case but only self-preferencing cases.[37] As will be discussed in Section IV, the Korean self-preferencing cases involved the application of the rule on abuse of market dominance and the rule on UTPs, which is different from the EU *Google Shopping* case.[38]

---

29 Article 45 also prohibits other types of UTP like unfairly assisting certain "related parties" or undertakings and other types of conduct that can impede fair trade.

30 KFTC Decision No. 94-205 (July 28, 1994); KFTC Decision No. 98-39 (Feb. 24, 1998); KFTC Decision No. 2001-31 (Feb. 14, 2001).

31 KFTC Decision No. 93-241 (Oct. 28, 1993); KFTC Decision No. 2004-091 (Mar. 12, 2004).

32 Supreme Court of Korea Judgment 2004 *Du*4703 (Dec. 7, 2006); Supreme Court of Korea Judgment 2004 *Du*4697 (Dec. 8, 2006); Supreme Court of Korea Judgment 2004 *Du*9338 (Dec. 7, 2006). *See also* Dong Kweon Shin, Antitrust Law 178–180 (4th ed. 2023) (in Korean).

33 Christian Bergqvist & Yo Sop Choi, *Controlling Market Power in the Digital Economy*, 50 Comp. L. & Sec. Rev. 1, 8 (2023).

34 Some argue that algorithmic targeting can exclude competitors. *See* Thomas K. Cheng & Julian Nowag, *Algorithmic Predation and Exclusion*, 25 U. Penn. J. Bus. L. 41 (2023).

35 *See* OECD, *supra* note 6, at 7.

36 *See e.g.*, Oren Bar-Gill et al., *Algorithmic Harm in Consumer Market*, 15 J. Leg. Anal. 1, 12 (2023). Some also argue that data protection laws can control and limit personalized pricing through the usage of cookies. *See* Peter Seele et al., *Mapping the Ethicality of Algorithmic Pricing*, 170 J. Bus. Ethics 697, 706 (2021).

37 *See* OECD, *supra* note 6, at 17.

38 Commission Decision and Banasevic, Marques & Portuese, *supra* note 8.

III. Korean Competition Policy on Self-Preferencing and Algorithmic Pricing: The Guidelines

546. The KFTC has recently given decisions against several Big Tech companies by finding infringements of both Article 5 and Article 45 MRFTA.[39] These cases include *Naver Shopping*[40] and *Kakao Mobility*[41] (both self-preferencing), *Google Android* (anti-fragmentation agreement),[42] and *Google Play Store* (app distribution).[43] Facing difficulties in digital cases, the KFTC has also issued its Guidelines for Assessing Abuses of Market Dominance by Online Platform Undertakings (hereinafter, the Guidelines).[44] The Guidelines stipulate how the agency assesses conduct by platforms that may fall within Article 5 MRFTA. There are some similarities, and some minor differences, between the Korean Guidelines and the European DMA. On the one hand, the Korean Guidelines provide a list of *online platform services*,[45] which have much in common with the *core platform services* in the DMA. On the other hand, the Guidelines emphasize the need to define the relevant market in digital cases,[46] which is not part of the DMA. The criteria for presuming digital dominance include entry barriers from network effects, the impacts of a gatekeeper, the advantages of data, the potential competition, and the market share. The Guidelines then set out a catalog of prohibited conduct, such as restrictions on multi-homing, the imposition of MFN clauses, self-preferencing, and tying.[47] This statement clarifies the possible application of Article 5 MRFTA to self-preferencing.

547. The Guidelines define self-preferencing and provide an assessment benchmark.[48] They explain that self-preferencing takes place when an undertaking treats its own (or its affiliate's) products or services more favorably than those of its competitors in an online ranking. Therefore, the conduct relates to algorithmic discrimination in favor of an undertaking's own services to the detriment of its rivals' services.[49] The Guidelines also

---

39 *See* Dong Kyu Lee, 36 Key Antitrust Issues 350–371 (2023).

40 KFTC Decision No. 2021-027 (Jan. 27, 2021), *see* Sangyun Lee, *The Seoul High Court upholds the South Korean FTC's decision to sanction "self-preferencing" as an abuse of dominance (NAVER Shopping)*, e-Competitions Dec. 2022, art. No. 112163; Seoul High Court Judgment 2021 *Nu*36129 (Dec. 14, 2023).

41 KFTC Decision No. 2023-093 (June 13, 2023).

42 KFTC Decision No. 2021-329 (Dec. 30, 2021).

43 KFTC Decision No. 2023-103 (July 20, 2023).

44 KFTC Guidelines No. 418 (Jan. 12, 2023). *See* Bergqvist & Choi, *supra* note 33, at 9.

45 According to the KFTC Guidelines pt I, art. 3, online platform services include online platform intermediation services, online search engines, social networking services, digital content or video streaming services, operating systems, online advertising services, and other types of platform services involving transactions, among others.

46 KFTC Guidelines pt II, art. 3.

47 KFTC Guidelines pt III.

48 KFTC Guidelines pt III, art. 2.*Da*.(3).

49 *See* Massimo Motta, *Self-preferencing and Foreclosure in Digital Markets*, 90 Int'l J. Ind. Org. 1, 2 (2023).

discuss the anti-competitive effects under the condition of a dual role at the downstream level and a leverage of market power. According to the Guidelines, the KFTC may also consider a trade-off between foreclosure and efficiency in self-preferencing when deciding if there has been an infringement of Article 5 MRFTA.[50] Furthermore, the Guidelines provide a list of factors to be considered when examining whether there has been anti-competitive self-preferencing, as follows: (i) intention or purpose, (ii) period of time of conduct and characteristics of products and services; (iii) means and specific measures for implementing self-preferencing, (iv) effects and degrees of access to the products or services through self-preferencing, (v) transparency and predictability of online ranking criteria, (vi) the undertaking's position and the competitive conditions in the online platform market, and (vii) those in the leveraged market, (viii) entry barriers, (ix) effects on diversity and innovation, and (x) efficiency and consumer welfare. The list of assessment factors above indicates its notable difference from the DMA because unlike the Korean Guidelines, the DMA does not provide any clear objective justification clause. Understanding the Guidelines will be important, as the KFTC may duplicate the contents of the Guidelines when designing the structure of the PCPA bill.

548. Lastly, while the Korean Guidelines provide details about how self-preferencing is assessed, the provision does not discuss algorithmic discrimination related to personalized pricing. However, Article 5 MRFTA can cover price discrimination.[51] In the *Qualcomm I* case, the Korean courts confirmed that discriminatory fees can constitute discriminatory conduct that falls within Article 5.[52] Besides, Article 45 MRFTA can be directly applicable to the unfair conduct of discriminatory pricing affecting either undertakings or consumers.[53] Therefore, the KFTC can apply both provisions to algorithmic discrimination including pricing or terms and conditions. According to the settled case law on price discrimination, the Korean courts will examine the overall effects, such as the degree of price discrimination, the impact on other undertakings' activities and the anti-competitive effects, and the objective justification.[54] To conclude, the Korean competition regime has various antitrust weapons in its arsenal to allow it to ban any anti-competitive or unfair harm from algorithmic discrimination.[55] The antitrust toolkit in Korea can be used when an algorithm poses antitrust threats.

---

50 OECD, *supra* note 6, at 18.

51 KFTC Guidelines No. 2021-18 (Dec. 30, 2021), Guidelines on Assessing Abuses of Market Dominance, Section IV.3.Ra.(3). *See also* SHIN, *supra* note 32, at 178–80.

52 Seoul High Court Judgment 2010 *Nu*3932 (June 19, 2013); Supreme Court of Korea Judgment 2013 *Du*14726 (Jan. 31, 2019). *See also* BONG-EUI LEE, FAIR TRADE ACT 268–269 (2d ed. 2023) (in Korean); Choi, *supra* note 16, at 36.

53 OHSEUNG KWON & JEONG SEO, ANTITRUST LAW 439 (6th ed. 2023) (in Korean).

54 Supreme Court of Korea Judgment 2004 *Du*4703 (Dec. 7, 2006).

55 Antitrust can be an essential weapon in a government's arsenal. *See* LAWRENCE A. SULLIVAN ET AL., THE LAW OF ANTITRUST 824 (4th ed. 2023).

IV. Korean Competition Law on Algorithmic Discrimination: Fairness in Online Rankings

549. There has been no case about algorithmic pricing in Korea, but algorithmic price discrimination by an online retailer was discussed a few years ago.[56] Algorithmic price discrimination can take place through pricing algorithms in order to carry out personalizing or to target consumers. Tailoring prices for consumers may be based on data on the consumers' purchasing behavior.[57] However, it appears to be difficult to collect evidence of personalized pricing in practice. In effect, consumers are not willing to accept personalized pricing because they often perceive that different prices indicate *economic discrimination* when there is a lack of algorithmic *transparency*. Therefore, to attract consumers in a competitive market, a platform often refrains from setting personalized pricing that can be used to discriminate between consumers because the platform's reputation is crucial.[58] Moreover, as personalized pricing can also bring about the redistribution of wealth, it cannot always be concluded that there is consumer harm.[59] Therefore, it would not be easy for agencies to bring a case against personalized pricing.

550. Antitrust cases against personalized pricing seem to be uncommon not only in Korea but also in other countries,[60] but platforms can set their prices or their terms and conditions by monitoring the prices in the market and identifying consumers' behavior.[61] Despite the complexity in the appraisal, it is possible that the so-called catch-all rule on UTP can regulate personalized pricing in Korea if the gap between the different prices is noteworthy. It is usually important to conduct economic analyses, such as a price-cost test and an as-efficient-competitor test, for pricing abuse cases.[62] Nonetheless, the Korean UTP rule does not require a complicated assessment, and the KFTC may not face a high burden of proof for regulating *unfair* personalized pricing, especially when a platform manipulates its algorithms. In particular, the possibility

---

56 Sung-Hwan Kim, *Consumer Data and Personalized Pricing*, 29 KOR. J. IND. ORG. 49, 53–54 (2021) (in Korean).

57 OECD, *supra* note 6, at 20.

58 *Id.* at 11–12.

59 *Id.* at 21.

60 *See generally* Marco Botta & Klaus Wiedemann, *To Discriminate or Not to Discriminate? Personalised Pricing in Online Markets as Exploitative Abuse of Dominance*, 50 EUR. J. L. & ECON. 381, 400 (2020); Axel Gautier et al., *AI Algorithms, Price Discrimination and Collusion*, 50 EUR. J. L. & ECON. 405 (2020); Salil K. Mehra, *Algorithmic Competition, Collusion, and Price Discrimination, in* CAMBRIDGE HANDBOOK OF THE LAW OF ALGORITHMS 199, 207 (Woodrow Barfield ed., 2021).

61 Critics argue that the number of companies using pricing algorithms is increasing, as shown in the use of algorithms for travel ticketing, hotel reservations, insurance, entertainment and e-commerce. A platform can use its algorithms to monitor competitors' prices and adjust its prices to respond to competitors. *See* Bar-Gill et al., *supra* note 36, at 5; Inge Graef, *Algorithms and Fairness*, 24 COLUM. J. EUR. L. 541 (2018). There has been a case against personalized pricing in Europe. *See also* Friso Bostoen, *Artificial Intelligence and Competition Law* (Dec. 27, 2023), at 18, https://papers.ssrn.com/sol3/papers.cfm?abstract_id=4655894.

62 *See e.g.*, Christopher R. Leslie, *Predatory Pricing Algorithms*, 98 N.Y.U. L. REV. 49, 56 (2023).

of the violation of competition rules related to algorithmic conduct has recently been discussed in Korea.[63] It appears that the KFTC is initiating an investigation in the AI sector at the time of writing.[64] The KFTC's experience from its probes into self-preferencing cases, such as *Naver Shopping*[65] and *Kakao Mobility*,[66] has become the foundation for opening antitrust investigations in the area of AI. Therefore, it is necessary to discuss the self-preferencing cases in Korea in order to comprehend the KFTC's approaches to AI.

551. Unlike the situation in the EU and the United States, several Korean platforms hold market power in the Korean digital markets.[67] One of the large platforms in Korea is Naver, which offers numerous services, such as a search engine, mobile maps, online shopping, and others. As in the EU *Google Android* case,[68] the KFTC concluded that Naver manipulated its algorithms to favor its trading partners over its competitors, doing this to leverage its market power in the adjacent market. According to the KFTC, Naver's biased rankings did not stem from competition on the merits.[69] The KFTC asserted that Naver's market position in the online comparison-shopping service should be regarded as a *gateway*,[70] and that its market share of over 70 percent revealed its market-dominant position.[71] The KFTC also examined other factors of entry barriers including network externalities and economies of scale.[72] The KFTC concluded that Naver had infringed Article 5 MRFTA as it had favored the sellers in its own online store over those in its rivals' marketplaces by giving higher rankings in the search results for its trading partners.[73] It also decided that Naver's conduct violated Article 45 MRFTA, which prohibits treatment that discriminates in favor of trading partners. The KFTC argued

---

[63] Most Korean research on the anti-competitive concerns arising from algorithmic pricing are about algorithmic collusion. *See e.g.*, Nansulhun Choi, *Collusion Based on Artificial Intelligence (AI) and Competition Law's Response*, 38 J. Kor. Competition L. 83 (2018) (in Korean).

[64] Min-jeong Kim & Mi-geon Kim, *S. Korean Government to Investigate Monopolistic Practices by Foreign AI Firms*, Chosun Daily (Apr. 8, 2024, 14:08 PM), https://www.chosun.com/english/national-en/2024/04/08/L5EPMOINVZBFZOKWGIRC6EFNUQ/.

[65] KFTC Decision No. 2021-027 (Jan. 27, 2021), *see* Lee, *supra* note 40.

[66] KFTC Decision No. 2023-093 (June 13, 2023).

[67] *See e.g.*, Sangyun Lee, *A Cursory Overview of Self-Preferencing in Korea: NAVER Shopping*, Kluwer L. Blog (Feb. 15, 2023), https://competitionlawblog.kluwercompetitionlaw.com/2023/02/15/a-cursory-overview-of-self-preferencing-in-korea-naver-shopping/.

[68] Commission Decision of July 18, 2018 relating to a proceeding under Article 102 of the Treaty on the Functioning of the European Union and Article 54 of the EEA Agreement Case AT.40099 (*Google Android*), 2018, O.J. (C 402), *see* Frédéric Marty, *Exclusivity payments: The European Commission publishes its decision on the practices implemented by the dominant operator on the market for open licensable smart mobile operating systems (Google Android)*, Concurrences No. 1-2020, art. No. 93304.

[69] OECD, *supra* note 6, at 18.

[70] KFTC Decision No. 2021-027, ¶¶ 143, 146, *see* Lee, *supra* note 40.

[71] *Id.* ¶¶ 251–52.

[72] *Id.* ¶¶ 257–69.

[73] *Id.* ¶ 294.

that Naver's manipulation of the algorithm had hindered the distribution of resources by steering consumers' choice,[74] which involved asymmetric information.[75] The KFTC held that Naver had violated both Article 5 and Article 45 MRFTA by manipulating the algorithms, and imposed a cease-and-desist order and a surcharge of 26.7 billion KRW (approximately 23 million USD).[76] The Seoul High Court upheld the agency's decision,[77] and as of July 2024 the case is pending before the Supreme Court of Korea.

552. The KFTC also made an analogous decision against Kakao, which is another of the large platforms in Korea. Kakao provides several services, mainly messenger and banking services. In particular, Kakao Mobility is a popular platform that provides a mobile app for ride-hailing services that match passengers with taxis. In this case, the KFTC concluded that Kakao had manipulated its taxi distribution algorithms to favor its affiliated (or franchised) taxi drivers over non-affiliated (or non-member) taxis, which leveraged its market dominance.[78] It argued that Kakao's manipulation fell within Articles 5 and 45 MRFTA and imposed a surcharge of 25.7 billion KRW (approximately 20 million USD).[79] In the *Kakao Mobility* case, the KFTC defined the relevant market and assessed market shares and entry barriers, such as network externalities and lock-in effects.[80] It asserted that Kakao manipulated its algorithms to allocate passengers to its affiliated taxis before its non-affiliated ones and to assign less profitable calls, such as those for short journeys, to non-affiliated taxi drivers, which incentivized the non-affiliated drivers to join the Kakao franchise.[81] According to the KFTC, Kakao engaged in discriminatory treatment and imposed disadvantages, which constituted an abuse of its superior bargaining position pursuant to Article 45 MRFTA.[82] This case is important as it is the first case to find an abuse of a superior bargaining position in a self-preferencing algorithmic manipulation. As of July 2024, this case is pending before the Seoul High Court.

---

74 *Id.* ¶ 424.

75 *Id.* ¶¶ 446–51.

76 There has been notable criticism about the absence of considerations of objective justifications in the *Naver* case. *See e.g.*, Bong-Eui Lee, *Competition Law Issues on Self-Preferencing by a Dominant Digital Platform*, 30 YONSEI L. REV. 365 (2020) (in Korean); Sinsung Yun, *A Thought on Regulation on Self-preferencing by Online Platforms under Korean Competition Law*, 27 EWHA L.J. 295 (2023) (in Korean), *see also* Lee, *supra* note 40.

77 Bergqvist & Choi, *supra* note 33, at 7; Lee, *supra* note 67. Only a summary of the Seoul High Court's judgment was published. *See also Seoul High Court Judgment 2021Nu36129 Cancellation of Corrective Order and Fine Payment Order*, L. TIMES (Jan. 30, 2023, 07:13 AM) (in Korean), https://www.lawtimes.co.kr/news/184894.

78 *See* OECD, *supra* note 6, at 19.

79 *KFTC Sanctions Kakao Mobility for Giving More Calls to Its Affiliated Taxis*, KFTC (Feb. 14, 2023), https://www.ftc.go.kr/solution/skin/doc.html?fn=c5345c36473713d9875ccde12ebaf5596bf8cfb554bc639ef11d01700a955991&rs=/fileupload/data/result/BBSMSTR_000000002402/.

80 KFTC Decision No. 2023-093, ¶¶ 117–18.

81 OECD, *supra* note 6, at 19.

82 *See* FURSE, *supra* note 15.

553. In conclusion, self-preferencing is the best-known of the algorithmic theories of harm. It is certain that the EU *Google Shopping* case triggered self-preferencing investigations in Korea, but the Korean cases differ from the EU case as the Korean agency and court examined both the anti-competitive and the unfair effects of self-preferencing. The Korean UTP rule, which can be applied without proving dominance or anti-competitive outcomes, may effectively deal with self-preferencing. At the time of writing in July 2024, it is uncertain whether the Supreme Court of Korea will issue clear guidance on the factors like indispensability, leveraging, and preferential treatment that should be assessed in order to conclude that self-preferencing constitutes a violation of the MRFTA.[83] The new theories of *unfair harm* should also be examined at the Korean Court, and this will eventually influence the design of the Korean digital regulation. In particular, the problem of imposing fairness duties in self-preferencing, even when market power or competitive harm are not present, will be discussed in Korea.[84] Moreover, the Court needs to consider its potential impacts on AI ecosystems as excessive applications of competition rules like the UTP provision may disregards the significant benefits from innovation in AI ecosystems.[85]

V. A Suggestion for Competition Policy on Algorithmic Discrimination

554. The Korean Government recently issued a "Digital Bill of Right" which pursues "the values and principles for a digital shared prosperity society"[86] and contains various goals, including the promotion of *fair competition*. Article 12 of the Bill articulates the importance of proper measures to prohibit harm by monopolies in information and technology and unfairness in algorithms in order to create a *fair and competitive* environment. Of course, this Bill is not legally binding, but this statement underpins the government's focus on the role of competition law and on a digital law that can actually prevent the Big Tech companies from using AI to abuse their algorithmic powers. The overall development of Korean government policy on AI and the digital economy, especially as it relates to competition, indicates an attempt to establish legal measures to solve digital market failures.

555. Most of all, the current approaches of the KFTC have highlighted its enthusiastic application of the UTP rule in the context of algorithmic

---

83 In the EU, the element of indispensability seems to be important for a conclusion that there has been a violation of Article 102 TFEU, but this is not so in the Korean cases. *See also* PABLO IBÁÑEZ COLOMO, THE NEW EU COMPETITION LAW 269 (2023).

84 For further discussion about self-preferencing and the relevant legislation, *see* Herbert Hovenkamp, *Antitrust and Self-Preferencing*, 38 ANTITRUST 5 (2023).

85 For discussions on the issue related to innovation ecosystems, *see e.g.*, Alden F. Abbott & Daniel F. Spulber, *Antitrust Merger Policy and Innovation Competition*, 19 J. BUS. & TECH. L. 265, 284 (2023).

86 Ministry of Science and ICT, *Digital Bill of Right*, MSIT (Feb. 13, 2024), https://www.msit.go.kr/eng/bbs/view.do?sCode=eng&mId=10&mPid=9&bbsSeqNo=46&nttSeqNo=19.

discrimination. A platform can discriminate against consumers by personalizing prices only if it holds notable market power. However, even if a platform uses algorithms to set various prices for different consumers without holding market dominance, the UTP provision can be applicable. The KFTC has a notable urge to regulate algorithmic discrimination, as shown in the self-preferencing cases. Because the self-preferencing cases are pending in front of the courts, it is rather too early to say that self-preferencing can fall within the UTP rule. Nonetheless, it is possible that algorithmic discrimination may still infringe the UTP provision, although it does not constitute an abuse of market dominance. In summary, unfairness in rankings or online discrimination matters in Korea,[87] as shown in the self-preferencing cases.

556. To conclude, this chapter argues against a digital regulation to address algorithmic conduct. First, anti-competitive or even unfair algorithmic conduct can be sufficiently captured by the existing competition law in Korea. For example, the UTP provision can allow the KFTC to initiate a probe into personalized pricing if this pricing significantly harms consumer welfare through imposing discriminatory pricing or terms and conditions. Although the welfare outcome of personalized pricing is ambiguous,[88] when the algorithmic discrimination constitutes *unfair* treatment, it can be caught by the UTP rule. Second, the Korean digital markets are competitive at the moment. The Korean platforms, in their development of AI, are competing rigorously with multinational platforms. Given the unpredictable innovation, algorithmic competition and uncertainty about the welfare results of algorithms,[89] it is not wise to establish an *ex ante* regulation for the Korean digital markets. There is a proliferation of fears around the world about a "digitized hand"[90] which causes social and economic discrimination, and this has led to multiple antitrust investigations against the manipulation of algorithms and to discussions about digital laws. However, it is time to adopt appropriate measures to balance the potential harms of algorithms with dynamic innovation in Korea's own market economy.[91]

---

[87] Competition rules on the abuse of market dominance are the well-known legal measures on "fairness of rankings." *See* Hacker et al., *supra* note 13, at 66.

[88] *See e.g.*, Ambroise Descamps et al., *Algorithms and Competition*, 20 Competition L.J. 32, 36 (2021).

[89] *See e.g.*, Stephanie Assad et al., *Autonomous Algorithmic Collusion*, 37 Oxford Rev. Econ. Pol'y 459, 476 (2021).

[90] Ariel Ezrachi & Maurice E. Stucke, Virtual Competition 209 (2016).

[91] There are criticisms on a sector-specific regulation, including AI regulation, because its substantive provisions have certain features of "inflexible, one-size-fits-all" or even formalistic approaches that can impede innovation. *See* James B. Bailey & Diana W. Thomas, *Regulating away Competition: The Effect of Regulation on Entrepreneurship and Employment*, 52 J. Reg. Econ. 237 (2017); Alden Abbott, *Should the Federal Government Regulate Artificial Intelligence?*, Truth on the Market (May 29, 2024), https://truthonthemarket.com/2024/05/29/should-the-federal-government-regulate-artificial-intelligence/?.

The Recoupment Conundrum: Rethinking Predatory Pricing in the Age of Algorithms

JENNIFER PULLEN[*]
University of St. Gallen, Switzerland

Abstract

This chapter examines the challenge of predatory pricing in the context of artificial intelligence (AI), focusing on the assessment of recoupment. In US law, recoupment requires a high burden of proof, demonstrating a dangerous probability, while EU law generally presumes its possibility for dominant firms. The rise of AI in pricing strategies disrupts these frameworks by enabling dominant firms to target specific customer segments with below-cost prices, minimizing predation losses and facilitating recoupment. This challenges the US' stringent requirements and the EU's cost-based analysis. The chapter concludes that artificial intelligence necessitates a reevaluation of competition law concepts. As sophisticated artificial intelligence tools make recoupment easier to achieve, a potential convergence between US and EU approaches to recoupment might become likely. However, both jurisdictions may need to adapt their frameworks to effectively address the nuances of predatory pricing in the era of algorithmic pricing strategies.

---

[*] Jennifer Pullen, M.A. HSG in Law and Economics, is pursuing her Ph.D in Law at the University of St. Gallen (HSG), specializing in competition law in digital markets. Email: jennifer.pullen@unisg.ch.

The Recoupment Conundrum:
Rethinking Predatory Pricing in the Age of Algorithms

I. Introduction

557. According to William Landes, Ronald Coase allegedly complained that he was fed up with antitrust because "when the prices went up the judges said it was monopoly, when the prices went down they said it was predatory pricing, and when they stayed the same they said it was tacit collusion."[1] This quote highlights the inherent difficulty in conceptualizing anti-competitive pricing behavior. While, for example, low prices are generally considered beneficial for consumers, low prices as a result of predatory pricing raise antitrust concerns. The key to identifying predatory pricing lies in the firm's ability to recoup losses incurred during the initial predation period by raising prices later. This chapter explores the contrasting approaches used in the US and EU to assess this recoupment element in predatory pricing cases and examines how the growing use of artificial intelligence in pricing strategies challenges these existing legal frameworks.

558. While a substantial body of legal literature explores the potential for artificial intelligence to facilitate collusion, less attention has been paid to how pricing algorithms might constitute an abuse of a dominant position, particularly in the context of predatory pricing.[2] This chapter aims to address this gap by focusing on the concept of recoupment in predatory pricing strategies. For this, the following sections provide a short explanation of predatory pricing and recoupment (Section II), and then examine the influence of artificial intelligence on the recoupment process (Section III). The core change with artificial intelligence lies in its ability to minimize predation losses and facilitate recoupment. Artificial intelligence can analyze vast amounts of consumer data to identify customers most susceptible to targeted below-cost pricing. This allows dominant firms to focus their predation efforts on specific customer segments while maintaining competitive prices for their existing customer base. Additionally, artificial pricing models are often opaque, making it difficult for customers to detect price discrimination and hindering potential rivals from effectively reacting to predatory strategies. Furthermore, the dynamic nature of AI pricing blurs the lines between predation and recoupment phases. Prices can be adjusted continuously, making it challenging to pinpoint a distinct predation period followed by a separate recoupment phase. The chapter, thus, contends that the traditional approaches to assessing recoupment in predatory pricing cases in the US and EU might require some reconsideration in light of algorithmic pricing strategies.

---

1   Edmund W. Kitch, *The Fire of Truth: A Remembrance of Law and Economics at Chicago, 1932–1970*, 26 J.L. & Econ. 163, 193 (1981) citing a statement by William Landes.

2   *See* with same observation Christopher R. Leslie, *Predatory Pricing Algorithms*, 98 N.Y.U. L. Rev. 49, 49 and 52 (2023). For papers analyzing possible collusion concerns with pricing algorithms, *see,* for example, Lea Bernhardt & Ralf Dewenter, *Collusion by Code or Algorithmic Collusion? When Pricing Algorithms Take Over*, 16 Eur. Competition J. 312 (2020); Emilio Calvano, Giacomo Calzolari, Vincenzo Denicolò & Sergio Pastorello, *Artificial Intelligence, Algorithmic Pricing, and Collusion*, 110 Am. Econ. Rev. 3267 (2020); Steven Van Uytsel, *Artificial Intelligence and Collusion: A Literature Overview*, *in* Robotics, AI and Future of Law 155 (2018).

II. Predatory Pricing and Recoupment

559. While recoupment is just one part of predatory pricing strategies, it has become a highly contested criterion. This chapter delves into how US and EU courts handle the recoupment element within predatory pricing cases (Section II.2). First, however, a general overview of predatory pricing is provided (Section II.1).

1. The Concept of Predatory Pricing

560. Predatory pricing is an established anticompetitive abuse in both US[3] and EU law.[4] It involves a dominant firm strategically setting prices below cost in the short term, aiming to strengthen or maintain market power by foreclosing competitors from entering or staying on the market due to unsustainable pricing pressures.[5] Broadly summarized, theories of predatory pricing agree that to be predatory, the pricing strategy can only be considered potentially profitable for the firm if the impact on competitor behavior is factored in. Conversely, this condition can only be met if the below-cost pricing discourages new entrants or forces existing competitors to exit the market, ultimately granting the alleged predator increased market power. This then allows the firm to raise prices above competitive levels later, recouping any losses incurred during the predatory period.[6]

561. Thus, predatory pricing strategies manifest in two stages: the predation phase and the recoupment phase. While in the initial predation stage, the predatory company will engage in below-cost pricing, hoping to drive rivals out of the market, it will leverage its strengthened market position during the second phase to raise prices and recoup prior losses incurred during the predation phase. Predatory pricing strategies evidently cause inefficiencies throughout both phases. The predation phase, characterized by below-cost pricing, incentivizes consumers to acquire excessive quantities of the good, leading to a misallocation of resources away from potentially more beneficial uses.[7] The subsequent recoupment phase, characterized by the firms' introduction of higher prices, results in a reduction of output and consumption below optimal levels, thereby generating economic inefficiency, further harming consumers subjected to the higher price, especially those who were not beneficiaries of lower prices during the predation phase.[8]

---

[3] Section 2 of the Sherman Antitrust Act prohibits both the act of monopolization and attempts to monopolize, including predatory pricing practices.

[4] European Union competition law prohibits predatory pricing under Article 102 of the Treaty on the Functioning of the European Union (TFEU) as an abuse of a dominant market position.

[5] KENNETH G. ELZINGA & DAVID E. MILLS, *Predatory Pricing*, in 2 THE OXFORD HANDBOOK OF INTERNATIONAL ANTITRUST ECONOMICS 41, 41 (2015).

[6] David Spector critiques, though summarizes, the definition. *See* David Spector, *Definitions and Criteria of Predatory Pricing* 5 (MASS. INST. TECH., DEPT. ECON., Working Paper No. 01-10, 2001).

[7] Paul L. Joskow & Alvin K. Klevorick, *A Framework for Analyzing Predatory Pricing Policy*, 89 YALE L.J. 213, 224 (1979).

[8] *See* Christopher R. Leslie, *Predatory Pricing and Recoupment*, 113 COLUM. L. REV. 1695, 1742 (2013).

2. The Concept of Recoupment

562. This chapter delves into the potential implications of artificial intelligence on the doctrinal divergence between the US and EU courts' approaches to handling recoupment in predatory pricing cases. While US case law imposes a stringent *dangerous probability* standard for proving recoupment, European Union courts generally presume the possibility of recoupment in scenarios involving dominant firms and below-average-cost pricing, meaning agencies can make such behavior subject to antitrust scrutiny without having to prove a possibility of recoupment. To showcase the potential impact of artificial intelligence on this doctrinal divergence, the following sections first aim to provide a short overview of the contrasting US and EU approaches to recoupment in predatory pricing cases.

A. *Recoupment in the US*

563. The US courts have been notably influenced by Chicago School theories when evaluating potential cases of predatory pricing.[9] The scholarship questions the feasibility of recoupment, arguing that the endeavor becomes unattainable due to the substantial losses incurred during predation. Chicago School scholars believe that predatory conduct will hardly be possible as predatory behavior results in the predator absorbing losses that far exceed those imposed on their targeted competitors.[10] Moreover, with the subsequent increase in market share, the dominant company's relative losses would concurrently escalate.[11]

564. Furthermore, Chicago commentators contend that predatory pricing is implausible because defeated rivals are likely to temporarily exit the market and reenter when prices increase during the recoupment phase. John S. McGee posited that dominant firms engaging in below-cost pricing would witness rivals temporarily halting their operations while leaving their physical capacities intact for opportunistic reentry when prices rise.[12] Robert Bork further suggested that the ease of entering into a market corresponds to the ease of exiting, implying that rivals would reenter the market as easily as they left it.[13] Because these arguments assume reentry is inevitable and, thus, recoupment is practically impossible, predatory pricing strategies become an irrational business practice.[14] Therefore, preda-

---

9 Leslie, *supra* note 2, at 58, 61.

10 ROBERT H. BORK, THE ANTITRUST PARADOX 148 (1978); John S. McGee, *Predatory Price Cutting: The Standard Oil (N.J.) Case*, 1 J.L. & ECON. 137, 140 (1958).

11 Leslie, *supra* note 8, at 1965, 1732; John S. McGee, *Predatory Pricing Revisited*, 23 J.L. & ECON. 289, 297 (1980).

12 McGee, *supra* note 1, at 140; for more on the consequences of McGee's work, *see* ELZINGA & MILLS, *supra* note 5, at 42.

13 BORK, *supra* note 10, at 148.

14 Leslie, *supra* note 2, at 60 referring to Harry S. Gerla, *The Psychology of Predatory Pricing: Why Predatory Pricing Pays*, 39 SW. L.J. 755, 755–56 (1985). *See* also, for example, Jonathan B. Baker, *Predatory Pricing after Brooke Group: An Economic Perspective*, 62 ANTITRUST L.J. 585, 589 (1994): "Thus, in the Chicago zoological taxonomy, predatory pricing is a rare white tiger or a mythical unicorn." It should be noted that

tory pricing theories have faced persistent skepticism from Chicagoans, with scholars like Frank H. Easterbrook even suggesting they might be best forgotten as an antitrust offense altogether.[15]

565. The inclination to align with Chicago School views on predatory pricing (or price theory in general[16]) is evident in American legal precedent.[17] While the Supreme Court initially classified predatory pricing as anti-competitive conduct in *Standard Oil Co. v. United States*,[18] it asserted in *Matsushita Electrical Industrial Co. v. Zentih Radio Corp.*,[19] that predatory pricing schemes are seldom attempted and even less frequently successful.[20] In the landmark case of *Brooke Group Ltd. v. Brown & Williamson Tobacco Corp.*,[21] the Supreme Court opted for a novel framework, requiring claims of predatory pricing strategies to show evidence of below-cost pricing and establish proof of recoupment.[22] In particular, plaintiffs must demonstrate a likelihood that the alleged predatory pricing scheme would result in prices rising above competitive levels enough to offset the costs accrued during the predation period.[23] This entails establishing that the competitor had a *dangerous probability* of recuperating their investment in below-cost prices during predation.[24] Thus, the Supreme Court characterized predatory pricing as an investment strategy in low prices, necessitating plaintiffs to provide robust proof that such a strategy would almost certainly yield returns in the recoupment phase. However, meeting such a strict recoupment requirement is highly complex to achieve, rendering it nearly impossible to substantiate such claims in court.[25]

---

post-Chicago economics has challenged the notion that recoupment is nearly always unattainable (*see*, for example, PAUL MILGROM & JOHN ROBERTS, *New Theories of Predatory Pricing*, *in* INDUSTRIAL STRUCTURE IN THE NEW INDUSTRIAL ECONOMICS 112 (1990) or Patrick Bolton, Joseph F. Brodley & Michael H. Riordan, *Predatory Pricing: Response to Critique and Further Elaboration*, 89 GEO. L.J. 2495, (who criticize the Supreme Court's lack of adaption to newer theories).

15 Frank H. Easterbrook, *Predatory Strategies and Counterstrategies*, 48 U. CHI. L. REV. 263, 337 (1981).

16 Christopher S. Yoo, *The Post-Chicago Antitrust Revolution: A Retrospective*, 168 U. PA. L. REV. 2145, 2153 (2020).

17 *See* Bolton, Brodley & Riordan, *supra* note 14, at 2506. For an overview of how the Supreme Court did not incorporate post-Chicago economics into its decision-making in *Brooke Group*, *see* Baker, *supra* note 14, at 592.

18 Leslie, *supra* note 2, at 55 referring to Standard Oil C. v. United States, 221 U.S. 1 (1911); *see also* Christopher R. Leslie, *Revisiting the Revisionist History of Standard Oil*, 85 S. CAL. L. REV. 573 (2012) and Patrick Bolton, Joseph F. Brodley & Michael H. Riordan, *Predatory Pricing Theory and Legal Policy*, 82 (UNIV. TILBURG TILEC, Discussion Paper No. 1999-82, 1999).

19 Matsushita Elec. Indus. Co. v. Zenith Radio Corp., 475 U.S. 574 (1986)

20 *Id.* at 589.

21 Brooke Group Ltd. V. Brown & Williamson Tobacco Corp., 509 U.S. 209 (1993), *see* Steven Cernak, *The US Supreme Court establishes a test for proving the existence of predatory pricing (Brooke Group)*, E-COMPETITIONS June 1993, art. N° 98743.

22 The Supreme Court relied to some degree upon the Areeda-Turner Test as a foundational framework (Herbert Hovenkamp & Fiona Scott Morton, *Framing the Chicago School of Antitrust Analysis*, 168 U. PA. L. REV. 1841, 1844 (2020)). See also for the original article on the Areeda-Turner Test, Phillip Areeda & Donald F. Turner, *Predatory Pricing and Related Practices under Section 2 of the Sherman Act*, 88 HARV. L. REV. 1869 (1975).

23 *Brooke Group*, 509 U.S. at 210.

24 *Id.* at 224.

25 Sandeep Vaheesan, *Reconsidering Brooke Group: Predatory Pricing in Light of the Empirical Learning*, 12 BERKELEY BUS. L.J. 81, 82 (2015); B.A. PHILLIP E. AREEDA & HERBERT HOVENKAMP, ANTITRUST LAW § 726(2) (2), at 72 (3d ed. 2008); Leslie, *supra* note 8, at 1743.

B. Recoupment in the EU

566. European Union case law has established two different analytical methods for determining whether an undertaking has engaged in predatory pricing, with the specific approach depending on the cost level of the allegedly predatory prices. The seminal *Akzo*[26] decision provided a first clear verdict: if a dominant firm prices below its average variable costs, it is assumed to be inherently predatory as the loss incurred on each unit sold renders any justification beyond competitor exclusion improbable.[27] For pricing exceeding average variable costs but falling below average total costs, however, the Court of Justice in *Tetra Pak II*[28] held that predatory pricing can only be presumed if there is proof of predatory intent, which is not required if prices are below average variable costs.[29] Notably, establishing the potential for future recoupment was not a prerequisite for establishing predatory pricing conduct in either *Akzo* or *Tetra Pak II*. This approach was further solidified in *France Télécom*,[30] where the Court of Justice explicitly reinforced that demonstrating recoupment possibilities is not mandatory to qualify a specific pricing behavior as predatory.[31] However, the Court of Justice might have indicated a potential shift in more recent cases. For example, while prior case law focused on the dominant undertaking's intent and cost levels, *Post Danmark*[32] arguably introduced a requirement for demonstrating actual or likely exclusionary effects on competition, potentially impacting consumers. This has led some scholars, such as Jay M. Strader, to argue that the Court of Justice may have implicitly incorporated a recoupment analysis for pricing strategies exceeding average variable cost but falling below average total cost.[33]

567. The European Commission, however, maintains the cost-and-intent framework. In its Guidance Paper on Article 102 TFEU, the Commission acknowledges proof of recoupment in a footnote, highlighting that case law

---

26 Case C-62/86, AKZO Chemie BV v. Comm'n of the Eur. Cmtys., ECLI:EU:C:1991:286, *see* European Court of Justice, *The EU Court of Justice states that an undertaking with a market share of 50 per cent could be considered dominant unless the undertaking proves the opposite (AKZO Chemie)*, e-Competitions July 1991, art. No. 107401.

27 *Id.* at ¶ 71.

28 Case C-333/94 P, Tetra Pak International SA v. Eur. Comm'n (*Tetra Pak II*), ECLI:EU:C:1996:436.

29 *Id.* at ¶ 41 and 42.

30 Case T-340/03, France Télécom v. Comm'n of the Eur. Cmtys., ECLI:EU:T:2007:22, *see* Laurent Flochel, *Predatory pricing: The CFI confirms the fine imposed upon Wanadoo for predatory prices (Wanadoo-France telecom)*, Concurrences No. 2-2007, art. No. 13490.

31 *Id.* at ¶ 113. The Court of Justice thus relied on the Turner-Areeda Test when analyzing predatory pricing strategies (Raimundas Moisejeves, *Predatory Pricing: A Framework for Analysis*, 10 Baltic J. L. & Pol. 124, 131 (2017)).

32 Case C-209/10, Post Danmark A/S v Konkurrencerådet, ECLI:EU:C:2012:172, *see* Anne-Lise Sibony, *Selective rebates – Universal service obligations: The Court of Justice, Grand Chamber, rules that selective rebates targeting clients of a competitor are not abusive when prices are below incremental cost but cover marginal cost and when no intent to eliminate competitor has been established (Post Danmark)*, Concurrences No. 2-2012, art. No. 45619.

33 Jay Matthew Strader, *Post Danmark's Recoupment Test*, 1 Competition L. Rev. 205 (2014).

does not require such for finding predation.[34] Similarly, the Commission's 2005 discussion paper on exclusionary abuses also argues that dominance – a prerequisite for finding a violation under Article 102 TFEU – inherently suggests the possibility of recoupment due to high entry barriers.[35] This implies that the Commission considers the dominant undertaking's ability to recoup losses to be self-evident without differentiating between pre-predation and post-predation forms of dominance.[36] The current status quo, thus, suggests that recoupment is inherent in a dominant market. However, as the discussion paper highlights, this should not mean that the likelihood of recoupment is entirely irrelevant when analyzing predatory pricing cases.[37]

568. Nonetheless, the assumption that dominance, combined with a specific pricing structure, inherently guarantees future recoupment in predatory pricing cases has faced criticism.[38] For example, Liza Lovdahl Gormsen contends that a dominant position alone cannot establish the likelihood of recoupment, as a dominant undertaking's only defense would be to disprove dominance itself. This creates a paradox, as Article 102 TFEU only applies to dominant undertakings to begin with.[39] Similarly, Chiara Fumagalli and Massimo Motta highlight a potential paradox within the dominance requirement. The authors construe a theoretical model that presents a scenario where a dominant firm with a secure incumbency advantage engages in predatory pricing to stifle a smaller yet efficient competitor. For the predatory pricing strategy to be economically rational for the dominant firm, at least some portion of the market must remain contestable, allowing for potential future recoupment of losses. This requirement for contestability might, in turn, however, question the initial assumption of actual dominance.[40] Michael Funk and Christian Jaag argue, on the other hand, that relying on pre-predation dominance as proof of recoupment potential risks over-enforcement of competition law, as a dominant firm might well be engaging in legitimate price competition (resulting in lower prices[41]) without harmful

---

34 *Guidance on the Commission's enforcement priorities in applying Article 82 of the EC Treaty to abusive exclusionary conduct by dominant undertakings*, at n. 46, COM (2009) O.J. (C 45), at 7.

35 European Commission, *Discussion Paper on Abuse of Dominance*, 26 (2005), https://ec.europa.eu/commission/presscorner/detail/en/IP_05_1626.

36 Günter Knieps, Wettbewerbsökonomie: Regulierungstheorie, Industrieökonomie, Wettbewerbspolitik (2005).

37 European Commission, *supra* note 35, at ¶ 122.

38 *See*, for example, Org. Econ. Coop. & Dev., Predatory Foreclosure (2005), www.oecd.org/dataoecd/26/53/34646189.pdf.

39 Liza Lovdahl Gormsen, *How Well Does the European Legal test for Predation Go with an Economic Approach to Article 102 TFEU?*, 37 Legal Issues Econ. Integration 293, 301 (2010).

40 Chiara Fumagalli & Massimo Motta, *A Simple Theory of Predation*, 56 J.L. & Econ. 595 (2013).

41 After all, competitively determined low prices are a consequence of effective markets. Overly zealous enforcement actions could, therefore, inadvertently stifle competition (*see* Robert O'Donoghue & Jorge A. Padilla, The Law and Economics of Article 82 EC (2006)).

predation.⁴² The authors cite the study of de Miguel La Mano and Benoît Durant, which suggests that successful predation by a dominant firm may even be less detrimental to competition and consumers than predation by a non-dominant firm.⁴³ However, it is worth acknowledging that to do without the dominance hurdle entirely, would again lead to a heightened risk of stifling healthy competition, particularly in oligopolistic markets.⁴⁴ In sum, while some scholars might contest the approach, EU law generally presumes the possibility of recoupment based on a finding of dominance and certain pricing behavior. This contrasts with US antitrust enforcement, where the burden of proof rests on the plaintiff to demonstrate the likelihood of future recoupment by the dominant undertaking.

III. Artificial Intelligence and Recoupment

569. The previous chapters have elaborated on the contrasting approaches to recoupment analysis in the EU and US This chapter aims to explore the potential influence of artificial intelligence on this legal conundrum. This section provides an overview of artificial intelligence's characteristics (Section III.1). The following section then highlights how the growing sophistication of these tools influences pricing strategies (Section III. 2). Lastly, the chapter analyzes the potential impact of artificial intelligence on the assessment of recoupment in both jurisdictions (Section III. 3). The chapter concludes with an analysis of how artificial intelligence could influence both the EU's presumption of recoupment based on dominance as well as the US' requirement for specific proof and, ultimately, how these tools might affect the current divide between the two legal systems.

1. Artificial Intelligence – A Big Word

570. Defining artificial intelligence proves a tricky endeavor.⁴⁵ Nonetheless, to thoroughly analyze its impact on predatory pricing strategies, it is imperative to delineate the capabilities and functionalities of the diverse range of technologies encompassed within the term of artificial intelligence. Abstract attempts at definition, such as merely labeling it as *intelligent* technology, provide little assistance, especially within regulatory contexts.⁴⁶

---

42 Michael Funk & Christian Jaag, *The More Economic Approach to Predatory Pricing*, 14 J. COMPETITION L. & ECON. 292, 298 (2018). The authors propose a novel approach to analyzing predatory pricing by dominant firms, drawing upon the logic of merger control. In particular, they argue that the evaluation of recoupment potential should mirror the analysis of competitive effects in a hypothetical merger scenario between the dominant firm and the targeted competitor.

43 Miguel de la Mano & Benoît Durand, *A Three-Step Structured Rule of Reason to Assess Predation under Article 102* (OFF. CHIEF ECONOMIST, Discussion Paper, 2005).

44 MASSIMO MOTTA, COMPETITION POLICY: THEORY AND PRACTICE (2004); KNIEPS, *supra* note 36.

45 MARC SCHWABACHER & KAI GOEBEL, A SURVEY OF ARTIFICIAL INTELLIGENCE FOR PROGNOSTICS 2 (2007); Pei Wang, *On Defining Artificial Intelligence*, 10 J. A. GEN. I. 1 (2019).

46 Even John McCarthy, the founder of the term artificial intelligence, defined the phenomenon as the science and development of *intelligent* machines, particularly intelligent computer programs (John McCarthy, *What is Artificial Intelligence?*, STAN. UNIV. (Nov. 12, 2007), https://www-formal.stanford.edu/jmc/whatisai.pdf). However, using abstract terms like *intelligence* can result in circular definitions lacking clarity. If the termi-

571. The concept of artificial intelligence spans a broad array of technologies and applications.[47] From knowledge-based systems employing knowledge representation and argumentation techniques[48] to machine learning algorithms that construct models based on input data,[49] a wide spectrum of functionality exists, sometimes varying significantly. In essence, it can be simplified to say that while knowledge-based systems predominantly adhere to predetermined programmed rules, machine learning applications operate autonomously, recognizing patterns and determining their independent actions accordingly. Surden, thus, convincingly divides artificial intelligence for regulatory purposes into two categories: (1) machine learning, and (2) logical rules and knowledge representation.[50] Such a differentiation appears sensible as significant disparities exist in implementing these technological subsets. Machine learning, for example, demonstrates a capacity for autonomous pattern recognition, resulting in a higher degree of independence compared to knowledge-based systems, which are engineered to conform to pre-established rules.[51] Therefore, employing technologies from the former category may yield different consequences and challenges compared to those from the latter group. Additionally, it is essential to acknowledge the ever-changing nature of the term, given the continuous technological advancements that give rise to novel intelligent systems.[52] As elaborated in the following section, the evolution of pricing algorithms shows the significant transformations technologies can undergo over time.

---

nology utilized in formulating the definition necessitates further explication itself, the initial definition becomes unclear as well. *See* Miriam C. Buiten, *Towards Intelligent Regulation of Artificial Intelligence*, 10 EUR. J. RISK REGUL. 41, 43, 47 (2019); Philipp Hacker, *Europäische und nationale Regulierung von Künstlicher Intelligenz*, N.J.W. 2142, 2142 (2020). I drew the same conclusion in a previous paper while citing the aforementioned sources in JENNIFER PULLEN, *Künstliche Intelligenz als Erkennungsinstrument für Killer-Akquisitionen – Chancen und Herausforderungen*, in KARTELLRECHT UND ZUKUNFTSTECHNOLOGIEN 81, 82 (2024).

47 Consider as an overview STUART RUSSELL & PETER NORVIG, ARTIFICIAL INTELLIGENCE – A MODERN APPROACH 226 (4th ed. 2022) or PETER STONE ET AL., ARTIFICIAL INTELLIGENCE AND LIFE IN 2030 – ONE HUNDRED YEAR STUDY ON ARTIFICIAL INTELLIGENCE, REPORT OF THE 2015–2016 STUDY PANEL 1, 11 (2016), https://ai100.stanford.edu/sites/g/files/sbiybj18871/files/media/file/ai100report10032016fnl_singles.pdf. According to Russell and Norvig, artificial intelligence encompasses knowledge-based systems, machine learning, natural language processing, robotics, and computer vision, whereby each of these technological areas comprises a multitude of specific implementation options and are not mutually exclusive.

48 Knowledge-based systems refer to systems that utilize knowledge representation and reasoning techniques to solve problems and make decisions. These technologies fall back on an existing knowledge base, through which the application can conclude, for example, using logic (RUSSELL & NORVIG, *supra* note 47, at 226).

49 With machine learning, algorithms create a model based on observed data, which is then utilized both as a hypothesis about the world and as software capable of solving problems. Such a learning process can occur through supervised, unsupervised, or reinforcement learning techniques. (*Id.* at 669).

50 Harry Surden, *Artificial Intelligence and Law: An Overview*, GA. ST. U. L. REV. 1305, 1310 (2019) citing Rene Buest, *Artificial Intelligence Is About Machine Reasoning – or When Machine Learning Is Just a Fancy Plugin*, CIO (Nov. 3, 2017), https://www.cio.com/article/230943/artificial-intelligence-is-about-machine-reasoning-or-when-machine-learning-is-just-a-fancy-plugin.html. With regards to the different categories mentioned earlier *supra* note 47, the distinction also makes sense. Natural language processing, robotics, and computer vision all use machine learning and knowledge-based systems as applications (PULLEN, *supra* note 46, at 86).

51 REINHARD HEIL, *Künstliche Intelligenz/Maschinelles Lernen*, in HANDBUCH TECHNIKETHIK 242 (2021).

52 NILS J. NILSSON, THE QUEST FOR ARTIFICIAL INTELLIGENCE – A HISTORY OF IDEAS AND ACHIEVEMENTS 433 (2010).

2. Artificial Intelligence, Pricing and Recoupment

A. *Artificial Intelligence-Driven Pricing*

572. The digital age has ushered in a new era of pricing strategies, with businesses increasingly turning to artificial intelligence tools for support. Pricing algorithms have revolutionized how companies set prices, increasingly shifting them away from traditional, rule-based methods and towards a more data-driven and dynamic approach focused on profit maximization.[53] Artificial intelligence models, including programs some may rather consider simple software, automate price settings by either instructing algorithms to follow predefined rules and parameters established by humans or by utilizing more sophisticated machine learning algorithms that leverage real-time data to make dynamic price adjustments.[54] As detailed above, artificial intelligence encompasses a broad spectrum of technological approaches with distinct functionalities. While basic algorithms offer a rule-based approach,[55] adhering to pre-determined conditions, learning algorithms exhibit a level of autonomy.[56] More advanced learning algorithms might employ techniques that obscure the specific data points used to set a particular price, even from the people building or overlooking the model. Consequently, a company employing such an artificial intelligence tool may be unable to determine whether increased profits stem from potentially anti-competitive conduct by the algorithm or legitimate business optimization.[57] Therefore, the risks associated with machine learning differ substantially from those linked to knowledge-based systems.

B. *The Impact of Artificial Intelligence on Pricing Strategies*

573. Pricing by artificial intelligence offers businesses three key advantages.[58] First, algorithms facilitate dynamic pricing, allowing for real-time adjustments based on current fluctuations in supply and demand.[59] Second,

---

53 *See* ARIEL EZRACHI & MAURICE E. STUCKE, VIRTUAL COMPETITION: THE PROMISE AND PERILS OF THE ALGORITHM-DRIVEN ECONOMY 100 (2016).

54 ORG. ECON. COOP. & DEV., ALGORITHMS AND COLLUSION: COMPETITION POLICY IN THE DIGITAL AGE 16 (2017), https://www.oecd-ilibrary.org/finance-and-investment/algorithms-and-collusion-competition-policy-in-the-digital-age_258dcb14-en; Davide Proserpio et al., *Soul and Machine (Learning)*, 31 MARK. LETTERS 393 (2020).

55 *See*, for example, Michal S. Gal & Niva Elkin-Koren, *Algorithmic Consumers*, 30 HARV. J.L. & TECH. 309, 344 (2017) or Zach Y. Brown & Alexander MacKay, *Competition in Pricing Algorithms* 7 (Harv. Bus. Sch., Working Paper No. 20-067, 2021).

56 Leslie, *supra* note 2, at 65.

57 Competition and Markets Authority, *Pricing algorithms: Economic Working Paper on the Use of Algorithms to Facilitate Collusion and Personalized Pricing*, ¶ 2.10 (Oct. 2018), https://assets.publishing.service.gov.uk/government/uploads/system/uploads/attachment_data/file/746353/Algorithms_econ_report.pdf.

58 Diego Aparicio & Kanishka Misra, *Artificial Intelligence and Pricing*, 20 REV. MARK. 103, 105 (2023).

59 *Id.* at 105. Ride-hailing platforms like Uber often use *surge pricing algorithms*, a form of dynamic pricing (*see*, for example, Jonathan Hall, Cory Kendrick & Christ Nosko, *The Effects of Uber's Surge Pricing: A Case Study*, (2015)).

respective technologies empower businesses to put in place personalized pricing strategies.[60] And lastly, artificial intelligence enables price experimentation, where firms can utilize algorithms to test different pricing structures and gauge customer responses.[61]

574. While such pricing tactics existed before the rise of pricing algorithms, their widespread use has heavily influenced the extent to which price patterns are analyzed. Algorithms can assess and adjust prices for individuals and specific products within milliseconds.[62] The analytical capabilities of artificial intelligence far surpass the human skillset, as algorithms can process significantly larger volumes of data in a much shorter time, allowing for faster reaction times and better capitalization of dynamic market changes.[63] The burgeoning power of algorithms, the ubiquity of the internet and increasingly sophisticated data mining techniques all indicate a pronounced shift in pricing authority from human decision-makers towards artificial intelligence.[64]

575. Notably, introducing artificial intelligence as a pricing tool leads to a significant change in price fluctuations. Whereas traditional pricing models might exhibit variations over the years (e.g., supermarket detergent prices[65]), pricing driven by artificial intelligence can demonstrate substantial price changes within days or even hours (e.g., Uber ride fares[66]). Furthermore, artificial intelligence tools offer an alternative to uniform price increases. Algorithmic targeting allows firms to concentrate price hikes on customers with lower price sensitivity (inframarginal customers), thus allowing them to leverage their willingness to pay.[67] Finally, pricing

---

60   Aparicio & Misra, *supra* note 58, at 110. Today, pricing algorithms can offer highly personalized prices. Dubé and Misra find that personalization improves expected profits by an additional 19 percent and by 86 percent relative to the nonoptimized price. While total consumer surplus declines under personalized pricing, over 60 percent of consumers benefit from personalization (Jean-Pierre Dubé & Sanjog Misra, *Personalized Pricing and Consumer Welfare*, 131 J. POL. ECON. 131 (2023)). *See also* Qian Li, Niels Philipsen & Caroline Cauffman, *AI-enabled Price Discrimination as an Abuse of Dominance: A Law and Economics Analysis*, 9 CHINA-EU L.J. 51 (2023).

61   Aparicio & Misra, *supra* note 58, at 113. Aparicio, Metzman & Rigobon show online grocers' exploration of the price grid indicates the use of price algorithms for frequent, small-scale price experimentation (Diego Aparicio, Zachary Metzman & Roberto Rigobon, *The Pricing Strategies of Online Grocery Retailers* (NAT'L BUREAU OF ECON. RSCH., Working Paper No. 28639, 2021)). Similarly, firms can utilize randomized price changes to estimate price elasticities (Marshall Fisher, Santiago Gallino & Jun Li, *Competition-based Dynamic Pricing in Online Retailing: A Methodology Validated with Field Experiments*, 64 MGMT. SCI. 2496 (2018)).

62   EZRACHI & STUCKE, *supra* note 53.

63   Alexander MacKay & Samuel N. Weinstein, *Dynamic Pricing Algorithms, Consumer Harm, and Regulatory Response*, 100 WASH. U. L. REV. 111, 114 (2022). *See also*, Brown & MacKay, *supra* note 55, at 7

64   Salil K. Mehra, *Antitrust and the Robo-Seller: Competition in the Time of Algorithms*, 100 MINN. L. REV. 1323, 1324 (2016).

65   Aparicio & Misra, *supra* note 58, at 105 referring to Günter J. Hitsch, Ali Hortaçsu & Xiliang Lin, *Prices and Promotions in US Retail Markets*, 19 QUANTITATIVE MKTG. & ECON. 19, 289 (2021). Using weekly scanner data from offline stores, the authors found that Tide laundry detergent's price fluctuated between $5 and $8 from 2008 to 2012

66   Aparicio & Misra, *supra* note 58, at 105 referring to Diego Aparicio & Roberto Rigobon, *Quantum Prices* (NAT'L BUREAU OF ECON. RSCH., Working Paper No. 26646, 2021). The authors show that identical rides provided by UberX from Boston's Museum of Fine Arts to Boston's Celtics Stadium were priced significantly differently over the course of seven days. The prices ranged from $8.23 to $14.38.

67   Competition and Markets Authority, *supra* note 57, at 9. Aparicio, Metzman & Rigobon, *supra* note 61, for example, provide evidence that Amazon and Walmart, especially, personalize prices at the delivery ZIP code.

algorithms will tend to be less cautious in their pricing and prioritize short-term gains.[68] Algorithms tasked with maximizing market share could autonomously pursue tactics designed to drive out competitors, such as predatory pricing strategies.[69] Moreover, the *black box* nature[70] of certain machine learning models can exacerbate the issue. The model's decision-making might be so opaque that even the companies implementing the tools might neglect to realize a problematic strategy like predatory pricing behaviors.

3. The Impact of Artificial Intelligence on Recoupment

576. This subsequent Section shifts the focus towards the central question of this article: how does artificial intelligence influence the recoupment requirement in predatory pricing cases, a concept treated differently in the US and the EU? To answer this, it is first necessary to examine artificial intelligence's influence on a firm's ability to recoup. In short, and as shall be elaborated in the ensuing discussion, algorithmic targeting empowers a dominant company to minimize predation losses and maximize recoupment with fewer predation risks of market entry by competitors.[71]

577. Easier recoupment starts with artificial intelligence's ability to monitor and analyze individual consumer behavior and past purchases to identify consumers susceptible to targeted below-cost pricing.[72] The various technologies mentioned above can contribute in distinct ways: rule-based systems could conduct targeted pricing by following a predetermined set of parameters. Supervised models, on the other hand, could be trained to optimize pricing strategies based on historical data. Additionally, unsupervised machine learning could uncover new patterns, revealing novel possibilities for predatory pricing. The continuous advancements of algorithmic tools enable increasingly effective behavioral targeting, moving beyond customer segmentation to relying on algorithms to determine the optimal price to display to consumers.[73] However, it might be worth highlighting that personalized pricing strategies may not always involve explicit price differentiation. For example, firms might display a uniform online price while employing targeted coupons for

---

68 Leslie, *supra* note 2, at 66 citing Sonia K. Katyal, *Private Accountability in the Age of Artificial Intelligence*, 66 UCLA L. Rev. 54, 96 (2019), who in turn uses the example Amazon's pricing algorithm once setting the price of the book The Making of a Fly at $23,698,655.93. *See, also* Ezrachi & Stucke, *supra* note 53, at 77.

69 Ezrachi & Stucke, *supra* note 53.

70 *See*, for example, Frank Pasquale, The Black Box Society: The Secret Algorithms That Control Money and Information (2015).

71 Thomas K. Cheng & Julian Nowag, *Algorithmic Predation and Exclusion* 14 (Lund Univ. Legal Rsch., Paper Series No. 1, 2022).

72 Leslie, *supra* note 2, at 70, highlights that artificial intelligence might not necessarily be able to estimate willingness to pay but offers the capability to collect, manage, and analyze data far more quickly and efficiently.

73 Pascale Chapdelaine, *Algorithmic Personalized Pricing*, 17 N.Y.U. J.L. & Bus. 1, 18 (2020); Alan M. Sears, *The Limits of Online Price Discrimination in Europe*, 21 Colum. Sci. & Tech. L. Rev. 1, 7 (2019).

specific customer segments, allowing for a less detectable form of price discrimination.[74]

578. Progress in personalized pricing technologies makes predatory pricing strategies more viable. This is particularly the case when a dominant firm can leverage artificial intelligence to target below-cost pricing to its rivals' customers while maintaining competitive (or even supra-competitive) pricing for its existing customer base.[75] This approach allows cross-subsidization, where inframarginal customers effectively subsidize marginal customers. Below-cost pricing, consequently, can be implemented without incurring substantial losses.[76] Further, the cost of predation is reduced as the predatory price is offered only to a select group of customers and only for specific transactions.[77] Therefore, the losses that need to be recouped are considerably lower (if any actual loss occurs due to cross-subsidization).[78]

579. The ability of computational tools to personalize pricing is further amplified by the relative ease with which online sellers can disguise price discrimination.[79] As Thomas K. Cheng and Julian Nowag point out, online pricing is often opaque, leaving most customers unaware of their classification as inframarginal and unable to react effectively.[80] This strengthens the dominant firm's position during recoupment, as unsuspecting consumers are less likely to switch to competitors. Nontransparent algorithmic targeting allows dominant firms to selectively offer below-cost prices, effectively reducing the pool of potential customers accessible to new entrants. This deprives potential entrants of the economies of scale crucial for viable market entry.[81] Artificial intelligence tools further bolster recoupment strategies by enabling personalized pricing right from the outset. In particular, they blur the lines between predation and recoupment by allowing a dominant firm to charge below-cost prices to targeted consumers (predating) while charging higher prices to others (recouping).[82] As these two phases happen simultaneously, the recoupment process naturally accelerates. Furthermore, improved pricing algorithms allow the dominant

---

74   *See* Axel Gautier, Ashwin Ittoo & Pieter Van Cleynenbreugel, *AI Algorithms, Price Discrimination and Collusion: A Technological, Economic and Legal Perspective*, 50 Eur. J.L. & Econ. 405, 409 (2020).

75   Leslie, *supra* note 2, at 90.

76   Cheng & Nowag, *supra* note 71, at 9; Leslie, *supra* note 2, at 92.

77   Cheng & Nowag, *supra* note 71, at 9.

78   *Id.* at 9; *See* Ignacio Herrera Anchustegui & Julian Nowag, *Buyer Power in the Big Data and Algorithm Driven World: The Uber & Lyft Example*, CPI Antitrust Chron. 31 (2017); Leslie, *supra* note 2, at 92.

79   Leslie, *supra* note 2, at 77 citing Chapdelaine, *supra* note 73, at 18; Terrell McSweeny & Brian O'Dea, *The Implications of Algorithmic Pricing for Coordinated Effects Analysis and Price Discrimination Markets in Antitrust Enforcement*, 32 Antitrust 75, 80 (2017); *see also* P.K. Kannan & Praveen K. Kopalle, *Dynamic Pricing on the Internet: Importance and Implications for Consumer Behavior*, 5 Int'l. J. Elec. Com. 63, 70 (2001).

80   Cheng & Nowag, *supra* note 71, at 9; Charles A. Miller, *Big Data and the Non-Horizontal Merger Guidelines*, 107 Calif. L. Rev. 309, 340 (2019).

81   Cheng & Nowag, *supra* note 71, at 13.

82   Leslie, *supra* note 2, at 92.

company to continue offering personalized prices after the competitor has exited the market. Unlike in traditional uniform price-setting scenarios, the dominant company could get away with charging some consumers even more than the monopoly price if their willingness to pay allows for this.[83]

580. Predatory pricing strategies driven by artificial intelligence can complicate rivals' possible attempts to reenter. For example, rule-based systems could automate responses to competitor price changes by following certain pricing rules, supervised machine learning could be trained to detect competitors' attempts at reentry to the market, and unsupervised machine learning might reveal new patterns that indicate potential reentry bids. In general, algorithmic systems can automatically detect reentry bids and respond by immediately lowering prices back to predatory levels. Such automated responses might discourage potential entrants, who are unlikely to continue incurring fixed costs, knowing their market entry will again be met with lower prices.[84] While some might argue that smaller competitors could also leverage algorithmic pricing strategies, leading to a *stalemate* of prices, this scenario might be questionable if a dominant firm is involved. Notably, the latter is likely to have superior data sets and more sophisticated algorithms, making it harder for smaller competitors to compete with their own algorithms.[85] Therefore, dominant incumbents are likely to prevail in such predatory pricing wars.[86]

581. In conclusion, artificial intelligence has significant potential to facilitate recoupment in predatory pricing strategies, posing a challenge to the traditional understanding of recoupment enshrined in US antitrust case law, which presumes a low probability of successful recoupment. As shown, this presumption[87] is significantly weakened by the rise of algorithmic pricing, indicating that recoupment may no longer be as impossible as the US Supreme Court previously assumed.[88] Furthermore, the dynamic nature of artificial pricing algorithms obfuscates the distinction between predation and recoupment. The continuous and simultaneous price optimization often inherent in pricing algorithms makes pinpointing a distinct predation phase followed by a separate recoupment phase challenging.[89] These developments suggest that a legal framework centered on a clear separation between recoupment and predation and the requirement of proving a dangerous probability of recoupment may no longer be appropriate when addressing algorithmic predatory pricing.

---

83   *Id.* at 93. Interestingly, this approach comes close to perfect price discrimination, resulting in a reduction of deadweight loss due to increased output. However, while total welfare may increase, it raises questions about the distribution of this welfare shift, which will likely favor producers.

84   *Id.* at 86.

85   *See*, for example, Maurice Stucke & Allen Grunes, Big Data and Competition Policy 89 (2016).

86   *See* Ezrachi & Stucke, *supra* note 53, at 238.; *see also* Leslie, *supra* note 2, at 98.

87   The validity of this notion is beyond the scope of this chapter and thus has not been questioned.

88   Cheng & Nowag, *supra* note 71, at 14; Leslie, *supra* note 2, at 84.

89   Leslie, *supra* note 2, with the same conclusion, at 91.

The EU enforcement practice that does not require strict proof of recoupment, however, could be more suitable to the realities of pricing driven by computational tools. Since such tools facilitate recoupment for the reasons described above, a focus on proving future recoupment might be indeed unnecessary. However, the EU's framework might encounter other difficulties concerning algorithmic predatory pricing cases. The heavy reliance on cost analysis might not be suitable in light of personalized and nontransparent pricing by algorithms. Additionally, proving predatory intent becomes complex when dealing with autonomous machine learning algorithms, which lack human-like motives but still act independently.

582. With regard to recoupment, developments in algorithmic pricing suggest a natural convergence between the US and EU approaches. As recoupment seemingly becomes easier with algorithms, particularly in connection with the growing access to consumer data, it might become a less relevant requirement to prove, rendering its different treatment under EU and US law less significant. In the era of algorithmic pricing online, proofing recoupment might even become unnecessary altogether, making the existing conundrum between the different approaches to recoupment obsolete. The need for novel analytical frameworks may supersede the longstanding debate surrounding the different methods. In this context, the specific technologies employed may significantly impact how they should be treated from a regulatory perspective. For instance, machine learning systems might require a different analytical approach than rule-based systems due to their intricate and often opaque decision-making process, with this complexity making it harder to detect whether the systems are involved in predatory pricing strategies. Some machine learning systems also make it more difficult to attribute responsibility if there is a lack of human control over an autonomous system.

583. Notably, algorithmic pricing also impacts the offline world. Brick-and-mortar stores can also utilize computational tools for pricing. However, given that prices in physical stores are visible to all customers, they lack the capacity for the obscure, personalized pricing often achieved by online stores. Consequently, price discrimination is more difficult to implement in brick-and-mortar settings. Additionally, rapid price fluctuations, as seen online, are more challenging to achieve in offline environments. Thus, the possibilities of attaining recoupment in the offline world differ from the possibilities online, although the line between off- and online might become increasingly blurred. Nevertheless, a different antitrust treatment of the requirement of recoupment off- and online, for the time being, might appear plausible. To conclude, as with many areas touched by artificial intelligence, competition law concepts require reconsideration. The times are a-changing, but one thing is clear: both the US and EU need to adapt their approaches to ensure effective competition in the age of algorithmic pricing.

Computational Methods in the Evaluation of Mergers and Acquisitions

VICTORIIA NOSKOVA[*] AND OLIVER BUDZINSKI[**]
Ilmenau University of Technology, Germany

Abstract

Computational antitrust is gaining popularity among competition authorities all around the globe. One of the areas of its application is merger review. Standard steps of merger review include: (1) selection of merger and acquisition cases for notification, (2) investigation phase, (3) court procedure, (4) final decision (with or without remedies), and (5) ex post control (if applicable). In this chapter, each sub-section addresses one of these steps and discusses the applicability and the existence of computational tools within the respective step. Furthermore, we focus on two research questions: (1) does the use of computational tools allow for a better selection of merger projects, and (2) are predictions of merger effects based on computational tools better than those based on typical types of evidence? The chapter is based on insights from computational research for merger review, modern economics, institutional economics, and political economics. Among other results, we conclude that each step of the merger review process offers space for computational tools, which, despite having considerable imperfections, can enhance merger control

---

[*] Post-doctoral researcher, Chair for Economic Theory, Institute of Economics, Ilmenau University of Technology, Germany, email: victoriia.noskova@tu-ilmenau.de; ORCiD: https://orcid.org/0000-0002-6932-4447. Main research areas include digital economics and competition policy.

[**] Professor of Economic Theory, Director of the Institute of Economics, Ilmenau University of Technology, Germany, email: oliver.budzinski@tu-ilmenau.de; ORCiD: https://orcid.org/0000-0003-4096-072X. Main research areas include competition economics, antitrust policy, media economics, institutional economics, and sports economics.

proceedings. However, some elements of the institutional framework of merger control require adaptations to specific characteristics of computational tools, which should contribute to their wider application.

I. Introduction

584. General statistics of the distribution of competition authorities' cases between three main areas – cartel prosecution, abuse control, and merger review – reveal that merger cases prevail.[1] Another statistics shows a growing trend of the application of computational antitrust tools by authorities all around the globe.[2] However, the distribution of the number of computational tools that are specific to mergers in *ex-ante* stage analysis is reported to still be the lowest.[3] This chapter gives insights helping to understand the current state of computational antitrust with respect to evaluating mergers and acquisitions. We do so by looking into two questions: (1) does the use of computational tools allow for a better selection of merger projects, and (2) are predictions of merger effects based on computational tools better than those based on typical types of evidence?

585. In general, computational tools[4] can be divided into structural tools, i.e., those computational tools which can be applied disregarding of the type of conduct analyzed by authorities (e.g., tools for data collection and aggregation, file submissions, internal management tools for electronic exchange of files), and merger-specific tools (with the prime example being merger simulation models). Thus, when we talk about computational merger tools, we address approaches used in merger evaluation procedures which use computational power, i.e., merger-specific tools. Even if the same tool is used, these tools may be further divided into ex ante and ex post assessments, because expectations of a tool's performance vary based on this criterion.[5]

---

1. *OECD Competition Trends 2024*, OECD (2024), https://www.oecd-ilibrary.org/finance-and-investment/oecd-competition-trends-2024_e69018f9-en.

2. Thibault Schrepel, *Computational Antitrust: An Introduction and Research Agenda*, 1 STAN. COMPUT. ANTITRUST 1 (2021); THIBAULT SCHREPEL & TEODORA GROZA, THE ADOPTION OF COMPUTATIONAL ANTITRUST BY AGENCIES: 2021 REPORT (2022); THIBAULT SCHREPEL & TEODORA GROZA, THE ADOPTION OF COMPUTATIONAL ANTITRUST BY AGENCIES: 2ND ANNUAL REPORT (2023); THIBAULT SCHREPEL & TEODORA GROZA, COMPUTATIONAL ANTITRUST WITHIN AGENCIES: 3RD ANNUAL REPORT (2024).

3. Here we talk about support during the assessments, not general structural tools like a system of registering mergers, *see* current examples from Czechia, Taiwan and UK in SCHREPEL & GROZA 2023, *supra* note 2, at 89, 149, 156.

4. The term computational tool is to be understood in a framework of application of computational law, i.e., as a "branch of legal informatics concerned with the mechanization of legal analysis (whether done by humans or machines)", MICHAEL GENESERETH, COMPUTATIONAL LAW, WHITE PAP. CODEX – STAN. CTR. LEGAL INFO. 1 (2015). Thus, it is not a completely new area and it focuses on advanced data-driven and computational-power-driven developments, thus an area with dynamic development and no closed list of tools.

5. *See* example of assessment of application of merger simulation models in *ex ante* and *ex post* scenarios in Oliver Budzinski & Victoriia Noskova, *Prospects and Limits of Merger Simulations as a Computational Antitrust Tool*, STAN. J. COMPUT. ANTITRUST 56 (2022).

586. Aggregated steps of a standard merger review can be listed as follows (with the possibility of the final decision occurring after any of them):
   - Selection of cases for notification and routine check in case of simplified procedures;
   - Different phases of investigation (in the EU there are Phase 1 and in-depth analysis in Phase 2);
   - In the courts, which are a common feature in the U.S. and can play a role through judicial review in the EU;
   - Final decision with or without remedies; and
   - *Ex post* control (if applicable).

587. The chapter is structured as follows: each sub-section addresses one step of a standard merger review, discusses the applicability and the existence of computational tools within this step, and offers a reflection on the abovementioned research questions. We focus on currently available instruments or those that are already on the horizon, and refrain from speculating about distant future tools.

II. Selection of Merger Cases

588. Usually, the notification of mergers is based on thresholds, which are defined by regulation. Selecting cases here means that competition authorities cannot review every notified merger for reasons of capacities. Moreover, many mergers are not problematic in terms of competition as they combine only minor and small market shares and do not restrict competition to a significant extent. Therefore, selecting the cases among the many notifications that require detailed scrutiny is an important and relevant task. In this framework, the selection step follows a reactive approach toward those transactions about which authorities were notified. However, technological and theoretical advancements may offer the possibility to move towards a proactive approach, e.g., predicting merger activities and notifying that they should be reviewed.[6] One of the ideas is the fitness-based merger review. It offers to apply new research on the dynamics of complex systems to analyze the competitiveness of firms.[7] The relevant market is seen as a network of agents (firms, suppliers, and customers) and their interrelations. The given firm's fitness is "the proportionality constant that governs the relationship between its growth rate and this firm's number of connections," i.e., the ability

---

[6] There is no unified system developed for authorities up to date, but theoretical research offers some tools, e.g., Sandro Claudio Lera, Alex Pentland & Didier Sornette, *Prediction and Prevention of Disproportionally Dominant Agents in Complex Networks*, 117 PROCEEDINGS NAT'L. ACAD. SCI. 27090 (2020); Lorenzo Arsini, Matteo Straccamore & Andrea Zaccaria, *Prediction and Visualization of Mergers and Acquisitions Using Economic Complexity*, 18 PLoS ONE e0283217 (2023), https://journals.plos.org/plosone/article?id=10.1371/journal.pone.0283217.

[7] Robert Mahari, Sandro Lera & Alex Pentland, *Time for a New Antitrust Era: Refocusing Antitrust Law to Invigorate Competition in the 21st Century*, STAN. J. COMPUT. ANTITRUST 52 (2021).

to translate connections into growth.[8] The competitive situation in the market can be assessed based on such calculations and early warnings can be introduced to avoid the downsizing of such firms after the merger. Mahari, Lera, and Pentland provide empirical support that currently high-fitness firms are not only more likely to be acquired, but also often such transactions will be under the notification threshold.[9] This has led to many of them taking place in recent decades and to the emergence of the term "killer-acquisition" to emphasize the phenomenon of a small potential competitor being acquired by a larger corporation instead of growing to its full potential to compete with it.[10] Such an approach could offer a solution to the important topic of innovation dynamics and killer acquisitions. If the startup is too small and goes under the radar of competition, it can be found by this tool.

589. Another more practical approach is less predictive (and less ambitious) in nature. It focuses on the search process among the announced mergers and facilitates the decision whether they should be investigated in-depth and subsequently may be challenged. This has been described in part by the CMA, the UK competition authority,[11] whose DaTA unit's developed automated tool applies machine learning to track merger activity.[12] If support-system algorithms go through the documents and pre-sort them based on probabilities of harmful outcomes (which should be quantifiable for this task[13]), this would help to re-allocate valuable resources of the authorities to the relatively more problematic cases due to a better identification of the problematic cases (in terms of competitive effects). However, this would require keeping the list of possible post-merger effects and theories of harm up to date, with a focus on: (i) novel developments in industries (like recently in digital ecosystems), (ii) (ongoing) new insights from modern economics and legal sciences, and (iii) an ability to operationalize and/or quantify them. The relatively

---

8   *Id.* at 57.

9   *Id.*

10 The relevance and regulatory urgency of the killer acquisition phenomenon is subject to controversial academic debate. *See*, inter alia, Marc Ivaldi, Nicolas Petit & Selcukhan Unekbas, *Killer Acquisitions: Evidence from EC Merger Cases in Digital Industries* (TSE, Working Paper No. 1420, 2023); Colleen Cunningham, Florian Ederer & Song Ma, *Killer Acquisitions*, 129 J. POL. ECON. 649 (2021); Axel Gautier & Joe Lamesch, *Mergers in the Digital Economy*, 54 INFO. ECON. POL'Y 100890 (2021); Massimo Motta & Martin Peitz, *Big Tech Mergers*, 54 INFO. ECON. POL'Y 100868 (2021); Michael L. Katz, *Big Tech Mergers*, 54 INFO. ECON. POL'Y 100883 (2021); JUSTUS HAUCAP & JOEL STIEBALE, NON-PRICE EFFECTS OF MERGERS AND ACQUISITIONS, REPORT FOR THE EUROPEAN COMMISSION (2023).

11 Note that UK regulation does not include commonly used obligation to notify mergers based on threshold. Thus, in this jurisdiction application of described tools is facilitated by the institutional framework and aims to identify all potentially problematic mergers.

12 "The tool, which is still in development, collects news article data from various APIs and RSS feeds and passes them through a natural language machine-learning model. It then outputs a prediction as to whether each article relates to a merger." SCHREPEL & GROZA 2023, *supra* note 2, at 156.

13 The most typical effects which can be measured before and after merger are connected with price, although there is still a loophole on existence of other factors influencing price outcomes. *See* example of analysis of existing studies in Annika Stöhr, *Price Effects of Horizontal Mergers: A Retrospective on Retrospectives*, 20 J. COMPETITION L. ECON. 155–179 (2024).

easy cases which now require formal control,[14] can be potentially delegated to the algorithmic check.

590. In a hypothetical situation, "early prediction" is a great promise which can re-shape the whole merger control system and significantly increase its efficiency. It offers to save the time that is otherwise spent on potentially unproblematic cases, i.e., yields a better allocation of the competition authority's resources. Currently, size in terms of turnover thresholds and value in terms of transaction volumes serve as first indicators to distinguish between mergers that need to be notified and those that benefit from a safe harbor (by virtue of being very small firms in terms of turnover and a low transaction value of the merger or acquisition deal). Quick screens are then used to select the cases among those notified that deserve more scrutiny because of antitrust concerns. However, the evolution of the modern economy – in particular but not only the emergence of complex digital systems of markets (digital ecosystems[15]) and the widespread implementation of data-driven business models[16] – implies that the detection of anticompetitive concerns becomes more difficult. Even acquisitions of small companies can have substantial effects on dynamic competition and non-horizontal effects – often not readily identifiable in quick screens – and are playing an increasingly important role. Computational tools can be implemented and employed as an alternative to thresholds for notification (which require a change of regulation, depending on the jurisdiction) or, within the current framework, inside the authority, where algorithms will sort out unproblematic cases. On the other hand, all these selections will happen based on existing assumptions, which should be audited by experts to keep them in line with current regulations and state of the art economic knowledge, so that there is an appropriate re-allocation of the authority's resources.

591. For the practical realization of such promises, a continuous data stream with parameters relevant for competition is required, as are formalization and measurability of potential harmful effects. For this reason, it is questionable if the sole use is better than the traditional approach; instead, a combination of them may be appropriate. One practical hint here would be that if a supportive merger review algorithm is developed, details should not be made public to companies. Otherwise, they could experiment with ways to hide their mergers from the investigation (as for example, already happened with cartel detection tools[17]). Moreover,

---

14 European Commission, Consolidated Jurisdictional Notice under Council Regulation 139/2004 on the control of concentrations between undertakings, 2008, O.J. (C 160) 1, https://eur-lex.europa.eu/legal-content/EN/TXT/?uri=CELEX:52023XC0505(01).

15 Viktoria H.S.E. Robertson, *Antitrust Market Definition for Digital Ecosystems*, in Concurrences 3 (2021).

16 Oliver Budzinski & Björn A. Kuchinke, *Industrial Organization of Media Markets and Competition Policy*, in 21 Mgmt. & Econ. Commc'n 36–39 (2020).

17 Competition and Markets Authority, *About the Cartel Screening Tool*, Gov.UK (withdrawn, 2020), https://www.gov.uk/government/publications/screening-for-cartels-tool-for-procurers/about-the-cartel-screening-tool.

if companies employ algorithms themselves in order to select acquisition targets or merging partners[18] that maximize profits (in pro- or anticompetitive ways) without raising the suspicion of the merger control authorities, the interplay of algorithms may produce interesting and novel effects. On the one hand, if these algorithms are very similar, the above-mentioned circumvention of detection could become a pressing issue. On the other hand, if algorithms that companies use are systematically superior to those the authority uses, the companies' algorithm may find new strategies of anticompetitive combinations that are not on the agenda of the controlling algorithm. It may not be too unrealistic that, in the race towards the best algorithms, companies can invest and spend more money than the taxpayer is willing/able to do.

III. Investigation Phase

592. For Phase 1 investigations and Phase 2 investigations, the main questions will be which tools can predict merger outcomes or at least assist authorities in this task. Apart from empirical tools applied for market delineation (usually based on econometric techniques), all tools which are part of merger simulations can be used separately or in available combinations. Further on, strategic assessments of text documents could complement the analysis.[19] There exist several approaches for simulations, but the core steps are commonly:

– A shape of demand and a form of competition is selected to construct a model reflecting both of them;

– For reasons of quality control, the accuracy of the assumptions from step 1 is tested during calibration based on past (pre-merger) data;

– Finally full simulation based on previous steps is performed.[20]

593. As an outcome, a new post-merger situation is predicted. Simulations usually aim to predict effects on market prices and quantities as well as on consumers' and producers' rents but can also include predictions of product variety and other relevant factors of competition.

594. One can argue that unilateral and coordinated assessments should be done based on learning from two other fields: (i) measurements of unilateral effects inspired by computational tools used for abuse of dominance cases, and (ii) coordinated effects analysis based on cartel detection experience. Thus, these instruments, to some degree, can be adapted for

---

[18] Due to various deficiencies in the managerial part of the merging process, the empirical record of traditionally organized mergers and acquisitions is notoriously bad. *See* Oliver Budzinski & Jurgen-Peter Kretschmer, *Implications of Unprofitable Horizontal Mergers: A Positive External Effect Does Not Suffice to Clear a Merger!*, 10 CONTEMP. ECON. 13 (2016). Therefore, using computational methods may improve the efficiency and profitability of target or partner choices.

[19] Budzinski & Noskova, *supra* note 5; Apostolos G. Katsafados et al., *Machine Learning in Bank Merger Prediction: A Text-Based Approach*, 312 EUR. J. OPERATIONAL RSCH. 783 (2024).

[20] *Id.* at 59–60.

merger review.[21] On top of that, the dissemination of algorithmic support of several companies' actions, conduct, and strategies is growing, so, the authorities may need to develop the similar competences to understand the market better (e.g., the skill of auditing algorithms may affect the level of proof in merger decisions).[22]

595. The application of econometric tools is not new for antitrust proceedings. Baker highlighted dependencies of the outcomes of econometric analysis on assumptions made for it, and wrote that it can be seen as beneficial to help with the identification of those scenarios which are of the most importance for market outcomes (the most sensitive ones should get more attention).[23] This is an intrinsic problem arising from all predictive tools, which also cannot be solved with technological advancements. For instance, AI as a machine learning algorithm will still have its own framework of parameters to predict results, some of which can be challenged. The more diverse and complicated companies' structures and markets are, the more options can be suggested to model them. Here, the different types of mergers will have an influence on predictive power, and the authorities will have more and clearer-cut experience in horizontal mergers in comparison to vertical and conglomerate ones. In addition, new business structures like platforms and huge interconnected digital ecosystems can pose another challenge.[24] Apart from that, an automation of merger review would require a clear selection between competing views of antitrust law.[25]

596. The quality of the predictions is a key element of an implementation decision based on the results drawn from computational tools. For merger review, these tools "compete" with traditional approaches like non-quantified analysis of testimonies from witnesses and experts. Here, computational power allows the use of elaborated models to match the pre-merger market development via calibration on data. Specifically, advancements in computational power allow monitoring and collection of more data relevant for the analysis and increase quality and speed of calibrations. The results obtained are supposed to be more objective in comparison with personal opinions and interests of traditional witnesses. It bears noting, however, that the degree of complexity of a real world market environment is still not captured (and probably never will be) by any merger tools, including computational modelling.

---

21 With respective difference in *ex-post* and *ex-ante* character, Jan Amthauer et al., *Ready or Not? A Systematic Review of Case Studies Using Data-Driven Approaches to Detect Real-World Antitrust Violations*, 49 COMPUT. L. & SEC. REV. 105807 (2023).

22 Michal S. Gal & Daniel L. Rubinfeld, *Algorithms, AI and Mergers* (N.Y.U. L. ECON., Research. Paper No. 23-36, 2023).

23 Jonathan B. Baker, *Contemporary Empirical Merger Analysis*, 5 GEO. MASON L. REV. 347 (1996).

24 *See* overview of typical models in use in Budzinski & Noskova, *supra* note 5.

25 Anthony J. Casey & Anthony Niblett, *Micro-Directives and Computational Merger Review*, STAN. COMPUT. ANTITRUST 132 (2021).

Typical models, for instance, assume that the structure of competition remains the same as it was before the merger, thus, in cases when this remains true, predictions will be of a high quality. For the same reason, models usually capture short-run effects better before any other changes come into play and start to influence the underlying assumptions (the accurateness of the extrapolation of pre-merger trends of the market environment and the market competition mechanisms). However, the same problem of the fundamental unpredictability of the future, especially if innovation dynamics and/or external shocks are involved, is typical for all *ex-ante* tools, be they computational or not. Therefore, no additional disadvantage should be given to computational tools. In terms of merger control, special problems for all predictive tools arise exactly from those mergers whose impact on the underlying markets is so strong that they cause structural breaks, and the pre-merger trends are not good predictors for post-merger developments and dynamics any longer. Those may be the most interesting mergers for merger control to look at, though.

597. For practical discussion, the acceptance of non-structural tools in competition policy enforcement institutions plays a crucial role. While structural tools serving organizational and workflow tasks means are unlikely to face serious acceptance problems, instruments actually influencing and altering decision-making will attract more scrutiny in this regard. Any hypothetical effectiveness does not bring much if it does not fit into the actual practice of merger review, for instance, if the results of computational calculations cannot be used as a strong argument by the authorities during the proceedings or in the decisions. A mismatch may originate from the interplay of computational instruments and legal standards, practices, and principles.[26] As of today, some elements are well accepted (e.g., the existence and necessity of calculating market shares), some less so (e.g., agent-based modeling,[27] sophisticated simulation models[28]). Usually, this corresponds with the degree of sophistication of a given tool, which reflects in cases of computational tools a selection bias towards an application of the most explainable rather than the best. *De facto*, the standard of proof applied towards computational predictive tools is dependent on their complexity. Although this happens as a side effect rather than a deliberate decision from the competition policy makers, several court decisions confirm this observation.[29]

---

26 We will discuss the special problems of innovation in merger control tools facing traditional law proceedings in the next section in more detail.

27 Thibault Schrepel & John Schuler, *The End of Average: Deploying Agent-Based Modeling to Antitrust* (AMSTERDAM L. TECH. INST., Working Paper Series, 2024).

28 *See* in Oliver Budzinski & Isabel Ruhmer, *Merger Simulation in Competition Policy: A Survey*, 6 J. COMPETITION L. ECON. 277 (2010); Budzinski & Noskova, *supra* note 5.

29 For instance, Oracle/PeopleSoft merger in 2003, Nuon/Reliant merger in 2003, and a recent case of the AT&T/Time Warner merger in 2018 (although the cased did not apply a full merger simulation model). *See* discussion in Budzinski & Noskova, *supra* note 5.

IV. In the Courts

598. There are no specific tools for merger court proceedings apart from structural tools supporting the filling in of application forms or the drafting of some documents. If a merger is challenged in front of the court, then all tools described in previous stages and contributing to argumentation towards specific decisions will be under scrutiny. Often, defendants find it easier to challenge the nature and accurateness of review tools than the results of their employment. So, the question of potential problems of integrating computational-economic evidence into legal proceedings (i.e., institutional frameworks) plays an important role. Two factors need to be discussed in this matter: first, the inherent complexity of computational instruments, as well as, second, the predictive nature of merger review assessments.[30] One should note that there exists no specialized court for merger reviews, which means that judges of general courts will need to deal with results of quantitative economics and computer science without specified training for that. Moreover, used to different types of evidence and under time constraints, judges may find it hard to fully understand and assess the quality of predictions coming from the computational tools.[31] Furthermore, they may find it difficult to accept evidence produced by novel instruments which have not been "tested" in former proceedings and court decisions (lack of precedents). The incentives to accept innovations in investigation tools and novel methods are usually low for judges for several reasons: (1) they may struggle to understand how the computational method works in detail and, thus, fail to build sufficient trust in the reliability of its results, (2) the risk that an appeals court will declare the judgement void increases with the novelty and innovativeness of the evidence and its underlying methods, so that career considerations (e.g., reputation effects) will lead judges to go for conservative solutions, (3) the side that carries the burden of proof experiences an increasing likelihood of losing cases because going conservative implies rejecting the novel methods, thus eroding the

---

30 *Id.* at 71–73.

31 Athey argues that, for instance, predictive machine learning tools may be easier for judges to accept since the predictive power could be demonstrated in court. While advanced econometric tools often raise controversial discussions among the experts of the parties about the right model specifications and small differences in model specifications may change the numbers of the model (though often not the signs/direction of the effects!), machine learning models self-select the "best" model specifications, thus internalizing the – often controversial – choice of the "right" model (specification). However, the criteria of fit according to which machine learning models self-select can be subject to scope, deliberation, and controversy again (as can be multiple other programming choices when setting up the machine learning model). In our opinion, Athey – emphasizing that she speaks about the future and not about in-court experiences – underestimates the power of the incentives that drives the creativity of the parties' experts (computer scientists similar to econometrists) to generate controversy and to blur evidence of the other side. While only future will tell us, we think that without appropriate institutional changes, machine learning tools will experience a similar fate to advanced econometrics in front of the judges. In the end, predicting counterfactuals or future developments necessarily dwells into territory where you naturally cannot demonstrate the predictive accuracy for a lack of existence of the counterfactual and/or the future. Thus, we remain skeptical that judges find these even more black-box character tools easier to accept. *But see* Susan Athey, *Keynote: "Computational Antitrust at the DOJ"*, YouTube (2024), https://www.youtube.com/watch?v=ckKzuzPlSCY.

evidence of the prosecuting side. Regarding merger control, these issues may lead to an unintended erosion of enforcement.[32]

599. Logically, in this situation the judges have incentives to turn and, at the end of the day, rely more on testimonies of experts in general and tend to dismiss evidence produced by novel, innovative, and sophisticated tools, for instance, because of concerns regarding the accuracy of modelling techniques (as each predictive tool will use some assumptions and full merger simulations are the most complex computational approach towards assessment of effects on demand and competition after the merger). This at the same time favors the emergence of battles of expert opinions. For merger simulation models this can include such details as: "technical aspects (as in Kimberley-Clark/Scott or Staples/Office Depot), the required threshold for the certainty of the predictions (e.g., Oracle/PeopleSoft), the acceptability of simplification of reality in course of the modelling (e.g., Volvo/Scania), the reliability of the predictive power of models (e.g., Oracle/PeopleSoft, Nuon/Reliant), as well as the choice of the right/best/adequate economic theory (virtually all of the mentioned cases)."[33] If identification of the most accurate model is problematic to judges, their incentive structure is to follow an established institution and avoid procedural mistakes by ruling against the side that carries the burden of proof. At the same time, this solution shifts responsibility for a wrong decision (from a competition policy perspective) to such a party. This is why the *de facto* standard of proof for computational tools may unintendedly rise to relatively high standards, eroding an effective protection of competition paradoxically, despite the emergence of better (computational) tools.

600. The second factor relates to the predictive character of computational merger tools applied for assessment of post-merger effects which give a numerical result of such forecasting. This form of presenting the results may intrinsically give a false sense of the precision and certainty of even the most sophisticated computational tools which conflicts in legal proceedings with other types of evidence. This comparison contributes to the asymmetry of the burden of proof, where a much lower evidentiary threshold is required to discredit simulation results than to confirm them.

V. Final Decision with or without Remedies

601. Competition authorities must reach a decision on the degree of automation, since it will influence workforce allocation. There can be several levels of autonomy: (1) no automation (as was the case prior to adoption of digitalization), (2) simple assistance automation, e.g., with excel spreadsheets for an overview of mergers, (3) advanced assistance automation, e.g., the

---

32 *See* Oliver Budzinski, *An Institutional Analysis of the Enforcement Problems in Merger Control*, 6 EUR. COMPETITION J. 445 (2010).

33 For further discussion and quotes *see* Oliver Budzinski, *Competing Merger Simulation Models in Antitrust Cases: Can the Best Be Identified?*, 6 J. MERGERS & ACQUISITIONS. 28–29 (2009).

introduction of machine learning, (4) semi-autonomous automation based on more advanced machine learning, (5) domain autonomy, e.g., restricted autonomous results only for simplified proceedings, (6) fully autonomous, and (7) superhuman autonomy (the latter two reach beyond human intelligence and are still far away from being achieved).[34] A review of survey results shows that competition authorities are mainly on levels 1 to 3, with some trying level 4 instruments.[35] Computational tools can support decision-making in the form of interface solutions presenting an overview of available information (structural computational tools), provide sorted and structured arguments from the previous stages, and provide access to all similar cases and brief summaries of relevant similarities and differences etc. However, the more decisions rely on computational methods and the less these decisions are accessible to human control, the more problems of a legal, economic, and political nature arise.

602. Computational instruments and methods can also be helpful in the selection of appropriate remedies (if applicable). The system can offer modifications in scenarios and predictions of results after implementation of remedies. In addition, monitoring tools can automatically check the fulfillment of remedies after a deal has closed and notify authorities and companies if violations occur, especially in the case of behavioral remedies. Even more importantly, the unconvincing track record of remedies may be more transparently and quickly revealed, helping to modify and adapt remedies. This would be particularly important if merger control were to move towards a bifurcated system combining *ex ante* and *ex post* elements to accommodate more dynamic aspects of competition.[36]

VI. *Ex Post* Control

603. However, *ex post* control can do more than being an innovative and novel element in a more dynamic system of merger control. Apart from monitoring post-merger developments and the effectiveness of remedies, computational tools applied to previous cases can become a rich source for policy learning, since they enable continual feedback after decisions. Policy learning refers to the proposition that knowledge about the actual effects of previous merger control decisions allows for insights regarding future cases. It is not aimed at revising former merger control decisions.

604. So far, *ex post* analysis of merger control decisions has mainly been done in academia, based on available sources.[37] Therefore, the selection of cases for *ex post* analysis is unsystematic and data availability issues are critical.

---

34 Lance B. Eliot, *Antitrust and Artificial Intelligence (AAI): Antitrust Vigilance Lifecycle and AI Legal Reasoning Autonomy*, COMPUT. SCI. (Dec. 23, 2020), https://arxiv.org/abs/2012.13016.

35 SCHREPEL & GROZA 2023, *supra* note 2.

36 Patrice Bougette, Oliver Budzinski & Frédéric M. Marty, *In the Light of Dynamic Competition: Should We Make Merger Remedies More Flexible?* (ILMENAU INST. ECON., Discussion Paper No. 28-181, 2023).

37 For a list of studies and an analysis of meta results *see* Stöhr, *supra* note 13.

Accordingly, it might be advantageous to have competition authorities do post-merger assessments. Obviously, however, this would require allocating resources to this task, and questions concerning the independence of the *ex-post* reviews could arise, given the self-interest of the authority. Establishing an independent body as a controlling agency for competition authorities – a bit like the German Monopolies Commission[38] – which could commission systematic expert studies based on data collection and employment powers by the agency, may be a superior solution. Still, it would require additional resources, financed by taxpayers.

605. However, by eliminating the predictive character of the task – now looking back on the accuracy of a previous merger control decision – a more systematic approach to *ex post* studies is likely to facilitate the use of computational tools and methods and offer a learning room for the adequate employment of these measures. This not only involves creating knowledge about the potentials and pitfalls, prospects and limits of these novel tools, it may also enhance trust in the accurateness of their results, which may then in turn contribute to alleviate some of the issues with using predictive computational tools in antitrust (see Section IV. above).

VII. Conclusion

606. Having provided a review of application areas and reflected upon the feasibility of prospects and limits and compared these with traditional tools to see the effects that computational tools can give; we conclude the following:
    – First, each step of the merger review process allows for the implementation of computational tools.
    – Second, data availability is a crucial aspect and can work as a driver for a wider application of computational tools for merger review purposes – but the lack of it can also slow the application down.
    – Third, computational tools have considerable potential (especially with employing *ex-post* reviews in addition to *ex-ante* uses) to increase the quality of policy development both in merger control and in other types of antitrust cases.
    – Fourth, the use of computational tools is beginning to influence the selection of merger projects, where they have a certain hypothetical potential, but up to date it is not possible to provide an applied assessment.
    – Fifth, predicting merger effects based on computational tools does have benefits in comparison to typical types of evidence, but acceptance by institutions is an initial limiting factor to overcome before we can see a wider application and cases unlocking the full potential of computational merger tools.

---

38   For more details *see* Monopolies Commission, *Mission*, MONOPOLKOMMISSION (2024), https://www.monopolkommission.de/en/monopolies-commission.html.

607. In general, our analysis has shown that despite existing imperfections and challenges, the use of computational tools in merger control is possible and can be beneficial to further develop merger control proceedings. However, this does not necessarily happen automatically. Instead, in some areas especially of the computational/economics/law-interfaces, institutions need to be modified to accommodate the specific advantages and promises of computational tools. This is particularly true for the institutional incentives faced by judges in law courts.

608. While computational tools addressing streamlining and improving issues of agency-internal organization or workflows are unlikely to raise serious institutional concerns and offer potential to reap organizational efficiencies, non-structural tools are more sensitive to their institutional embedment. If computational tools are employed more frequently and successfully in the actual assessment of a merger proposal, several institutional changes could prove to be helpful:[39]

- As we have argued in this chapter, the allocation and nature of the burden of proof plays a relevant role. An ambitious burden of proof, demanding almost certainty (beyond any reasonable doubt) cannot be adequately met with computational instruments trying to predict counterfactuals or future competition processes. Here, an adequate stand of proof needs to be two-sided: safeguarding that the evidence brought by the competition authority is of the highest possible quality but at the same time requiring that challenges must take place at the same level of quality – and not just raise doubts. Otherwise, the side with the burden of proof will lose out.

- Rebuttable presumptions can provide a way to deal with cases where exceptions are possible but improbable – and then shift the burden of proof to the defending side seeking an exception.

- Incentives for judges may be reconsidered to allow for more innovation and innovative instruments. Better economic and technological knowledge for the judges – perhaps through changes in their education or by implementing specialized judges instead of general ones – could represent a relevant step. Further institutional changes depend sensitively on the respective law system and (maybe unintended) legal side effects must also be carefully considered.

- Systematic *ex-post* analysis of cases is helpful.

609. These are just a few examples derived from the reasoning in this chapter. Our main take-away, however is that progress in computational tools in merger control needs to consider the institutional interface and cannot be driven by technologic possibilities alone.

---

39 *See* for more elaborate discussions, Budzinski, *supra* note 33; Budzinski & Ruhmer, *supra* note 28; Oliver Budzinski & Annika Stöhr, *Towards a Systematic Controlling of Antitrust Decisions*, 1 CPI ANTITRUST CHRON. 45 (2018); Budzinski & Noskova, *supra* note 5.

PART III
Policy Responses to the AI Boom

The Folly of AI Regulation

JOHN M. YUN[*]

Antonin Scalia Law School
George Mason University, Virginia

Abstract

The explosive growth of AI related technology has drawn the attention of government authorities around the globe. As these authorities consider various regulatory proposals, this chapter advocates a model similar to the one used when the internet first emerged, that is, a relatively restrained approach to regulation. This position is founded on several core tenets. First, there can be trade-offs between technological growth rates and addressing specific harms. Thus, even if a regulation is ultimately successful in addressing a specific harm, if it dampens the rate of innovation, then this could lead to a net welfare loss. Second, premature regulatory solutions can crowd out market-based solutions, which may offer more efficient solutions to emergent harms. Finally, premature regulations can have the consequence of entrenching incumbents and raising barriers to entry, which, perversely, harms the competitive process rather than promoting it. Importantly, this proposal is not a call to ignore the dangers that AI generated output can pose – nor is it a call for a "more permissive" treatment of AI under existing laws or existing regulatory schemes of general application.

---

[*] Associate Professor of Law, Antonin Scalia Law School, George Mason University. I thank the editors, Alden Abbott and Thibault Schrepel, for excellent comments and suggestions, and Carol Stephenson, for careful edits.

The Folly of AI Regulation

I. Introduction

610. The setting is 2016, and legendary professional Go player Lee Sae "The Strong Stone" Dol anticipates an easy victory over his opponent, Google's DeepMind artificial intelligence (AI) AlphaGo.[1] Yet, with a decisive 4–1 victory, AlphaGo handily defeated humanity's champion.[2] AI had arrived. Well, sort of. After Google's DeepMind's victory, even though AI continued to advance,[3] it did not hit mainstream consciousness until six years later. On November 30, 2022, OpenAI's ChatGPT went live.[4] The launch of ChatGPT is notable because, for the first time, it put the power of AI into the hands of "everyday people,"[5] and the output was reasonably "intelligent." AI had indeed arrived.[6]

611. The emergence of ChatGPT has changed the conversation around how we educate, work, and play using AI.[7] Fundamentally, AI technology represents an immensely powerful tool that can, in a sense, mimic human intelligence – including our ability to glean the key points from a legal document, play games of strategy, design more efficient production processes, and even drive cars.[8] In less than two years, there has been an explosion of AI products and developments – ranging from small startups to the largest technology companies in the world.[9]

612. However, like all disruptive technologies, AI raises some fundamental questions. These questions involve, for instance, intellectual property (IP) rights,

---

[1] Lee Sae Dol, *8 Years Later: A World Go Champion's Reflections on AlphaGo*, GOOGLE (Mar. 19, 2024), https://blog.google/around-the-globe/google-asia/8-years-later-a-world-go-champions-reflections-on-alphago/.

[2] Demis Hassabis, *AlphaGo's Ultimate Challenge: A Five-Game Match against the Legendary Lee Sedol*, GOOGLE (Mar. 8, 2016), https://blog.google/technology/ai/alphagos-ultimate-challenge/.

[3] *See, e.g.*, An Economic Framework for Understanding Artificial Intelligence, *in* 7 ECONOMIC REPORT OF THE PRESIDENT 247–48 (Mar. 2024) (demonstrating how AI systems have improved relative to human performance since the late 1990s and started overtaking human performance in the late 2010s).

[4] Kristi Hines, *History of ChatGPT: A Timeline of The Meteoric Rise of Generative AI Chatbots*, SEARCH ENGINE JOURNAL (June 4, 2023), https://www.searchenginejournal.com/history-of-chatgpt-timeline/488370/.

[5] Song by Sly and the Family Stone. *See also* Krystal Hu, *ChatGPT Sets Record for Fastest-Growing User Base – Analyst Note*, REUTERS (Feb. 2, 2023, 3:33 PM GMT), https://www.reuters.com/technology/chatgpt-sets-record-fastest-growing-user-base-analyst-note-2023-02-01/.

[6] *See, e.g.*, Ethan Mollick, *ChatGPT is a Tipping Point for AI*, HARV. BUS. REV. (Dec. 14, 2022), https://hbr.org/2022/12/chatgpt-is-a-tipping-point-for-ai.

[7] *See, e.g.*, McKinsey, *The State of AI of 2023: Generative AI's Breakout Year*, MCKINSEY.COM (Aug. 1, 2023), https://www.mckinsey.com/capabilities/quantumblack/our-insights/the-state-of-ai-in-2023-generative-ais-breakout-year; Ajay Agrawal, Joshua S. Gans & Avi Goldfarb, *Artificial Intelligence Adoption and System–Wide Change*, 33 J. ECON. & MGMT. STR. 327 (2024).

[8] *See* Harry Surden, *ChatGPT, AI Large Language Models, and Law*, 92 FORDHAM L. REV. 1939, 1942 (2024) (defining AI as "a computer system to automate tasks that, when humans do them, require higher-order, cognitive skills associated with 'intelligence' – such as driving a car, playing chess, reading books, writing emails, or creating art or music."). Thus, a record player would not be considered AI, but software that could compose a new song based on the mood of the listener would be considered AI.

[9] *See, e.g.*, Yang Jie, *Nvidia's Biggest Manufacturer Doubles Down on AI*, WALL ST. J. (Apr. 18, 2024, 9:59 AM ET), https://www.wsj.com/tech/nvidia-ai-chips-demand-tsmc-strategy-e2b8e7d0; Jason Tabeling, *Perplexity AI: Exploring AI-Powered Search Beyond Google*, SEARCH ENGINE LAND (Apr. 25, 2024, 10:03 AM), https://searchengineland.com/perplexity-ai-exploring-ai-powered-search-beyond-google-439879.

privacy, antitrust laws, and the extent of First Amendment rights.[10] As policy stances take shape, reactions to AI have broadly fallen into three categories (noting that these categories are neither mutually exclusive nor exhaustive).

613. First, there are the optimists who highlight the benefits and opportunities of AI. Some of the achievements and predictions are extraordinary.[11] Some are more mundane.[12] Yet, they share the conviction that AI will improve the human condition and economic outcomes.

614. Second, there are the pessimists. Perhaps influenced by dystopian cinema going as far back as the 1960s through the 1980s, with movies such as *2001: A Space Odyssey*, *WarGames*, and *The Terminator*, there is a sense that the risks related to AI are too great. Some even believe that computers will become "sentient" and overtake humans and create unacceptable risks. Policy recommendations include calls to pause or halt AI technology.[13]

615. Finally, there are the "controllers." The view is that AI, like other aspects of the economy, should be regulated, made "safe," or, at the very least, have guardrails to ensure the proper development of the technology and associated markets.[14] The justification is based on the belief that we can enjoy the benefits of AI, while mitigating the downsides of AI through proper controls.[15] Thus, the policy question is not whether to regulate, but how much.[16]

---

10 *See, e.g.*, Cass R. Sunstein, *Artificial Intelligence and the First Amendment*, HKS (Apr. 28, 2024), https://papers.ssrn.com/sol3/papers.cfm?abstract_id=4431251 (exploring the question whether AI has First Amendment rights).

11 For instance, Google's DeepMind has mapped the structure and sequence of the 200 million known human proteins – a process that used to take years for a single protein. *See AlphaFold*, Google, https://deepmind.google/technologies/alphafold/. *See also* Brad Smith, *How Do We Best Govern AI*, Microsoft (May 25, 2023), https://blogs.microsoft.com/on-the-issues/2023/05/25/how-do-we-best-govern-ai/ ("We've seen AI help save individuals' eyesight, make progress on new cures for cancer, generate new insights about proteins, and provide predictions to protect people from hazardous weather. Other innovations are fending off cyberattacks and helping to protect fundamental human rights, even in nations afflicted by foreign invasion or civil war."); Alexei Oreskovic, *Box CEO Aaron Levie Remembers the Moment He Realized ChatGPT was a Game Changer*, Fortune (Apr. 1, 2024, 7:28 PM GMT+1), https://fortune.com/2024/04/01/box-ceo-aaron-levie-openai-chatgpt-game-changer/ ("With LLMs, [Box CEO Aaron] Levie says, that problem is instantly resolved – it's almost as if all the world's language barriers were suddenly erased and everything anyone has ever said becomes instantly intelligible to you.").

12 *See, e.g.*, Ajay Agrawal, Joshua Gans & Avi Goldfarb, *From Prediction to Transformation*, Harv. Bus. Rev. (Nov.–Dec. 2022), https://hbr.org/2022/11/from-prediction-to-transformation (detailing how AI has improved financial fraud detection, language translators, sailboat design and operations).

13 *See, e.g.*, *Pause Giant AI Experiments: An Open Letter*, Future of Life (Mar. 22, 2023), https://futureoflife.org/open-letter/pause-giant-ai-experiments/ (including signatures from prominent names such as Elon Musk and Steve Wozniak); Eliezer Yudkowsky, *Pausing AI Developments Isn't Enough. We Need to Shut it All Down*, Time (Mar. 29, 2023, 6:01 PM EDT), https://time.com/6266923/ai-eliezer-yudkowsky-open-letter-not-enough/.

14 *See, e.g.*, Economic Report of the President, *supra* note 3, at 289 ("Sensible policies to encourage responsible innovation, protect consumers, empower workers, encourage competition, and help affected workers adjust are critical."); Francesco Filippucci et al., *The Impact of Artificial Intelligence on Productivity, Distribution and Growth: Key Mechanisms, Initial Evidence and Policy Challenges*, OECD Artificial Intelligence Papers 1 (Apr. 2024), https://doi.org/10.1787/8d900037-en 92 (calling "for a comprehensive policy approach to ensure AI's beneficial development and diffusion, including measures to promote competition, enhance accessibility, and address job displacement and inequality.").

15 *See, e.g.*, Economic Report of the President, *supra* note 3, at 251 ("Often, discrimination may only be observable through sophisticated analysis of AI methods and outputs. Regulatory measures to help identify discrimination in critical markets are necessary.").

16 *See, e.g.*, Smith, *supra* note 11 ("As technology moves forward, it's just as important to ensure proper control over AI as it is to pursue its benefits.").

616. This raises a central question: when and how should governments promulgate new regulations? It is an age-old question that conceptually has a straightforward answer: when the benefits of the regulation outweigh the costs.[17] Indeed, much of the economic research into regulations attempt to quantity these benefits and costs.

617. For instance, Sam Peltzman pioneered research into the "unintended consequences" from regulation.[18] While his focus was on automobiles, his insight can be applied to all sectors – including AI. Similarly, Harold Demsetz cautioned against the nirvana fallacy, which is, in essence, a call to consider the right counterfactual when considering regulatory proposals.[19] Specifically, policymakers should compare the costs *and benefits* of the prevailing, unregulated system with the likely benefits *and costs* of the regulatory proposal. While Demsetz's proposal may seem obvious, there can be a tendency to compare an idealized version of a regulatory proposal (that is, focusing on the promises of regulation) to the realized outcomes of an unregulated system (that is, focusing on the imperfections of markets) – hence, falling prey to the nirvana fallacy.[20]

618. More broadly, Demsetz's idea of a nirvana fallacy was also taken up by James Buchanan.[21] Ultimately, the point is that, in order to regulate, not only should there be a clear market failure, but also strong grounds for believing that government intervention will yield net welfare-superior results to non-intervention. Within this consideration, public choice and rent-seeking considerations should also be weighed.[22]

619. The specific calls to regulate aspects of AI are far reaching and ever growing.[23] There are proposals to regulate AI within IP law – including whether AI systems can patent an idea, whether AI generated content is copyrightable, or whether inputs used to train AI involve fair use or

---

17 *See, e.g.*, Thomas A. Lambert, How to Regulate: A Guide for Policymakers (2017).

18 Sam Peltzman, *The Effects of Automobile Safety Regulation*, 84 J. Pol. Econ. 677 (1975).

19 Harold Demsetz, *Information and Efficiency: Another Viewpoint*, 12 J.L & Econ. 1 (1969).

20 *Id.*

21 *See, e.g.*, 7 James M. Buchanan, The Limits of Liberty, Between Anarchy and Leviathan, The Collective Works of James M. Buchanan, at 216 (2000) ("Governmental correctives to presumed particular flaws in the operation of markets were considered piecemeal and independent one from the other. More important, these correctives were presumed to work ideally once they were introduced. Because there existed no principle or vision of the process of governmental operation, the naive presumption was made that intention was equivalent to result.").

22 *See generally* James M. Buchanan & Gordon Tullock, The Calculus of Consent: Logical Foundations of Constitutional Democracy (1962).

23 *See, e.g.*, The White House, *Executive Order on the Safe, Secure, and Trustworthy Development and Use of Artificial Intelligence*, (Oct. 30, 2023), https://www.whitehouse.gov/briefing-room/presidential-actions/2023/10/30/executive-order-on-the-safe-secure-and-trustworthy-development-and-use-of-artificial-intelligence/; Senator Ron Wyden, *Wyden, Booker and Clarke Introduce Bill to Regulate Use of Artificial Intelligence to Make Critical Decisions like Housing, Employment and Education*, (Sept. 21, 2023), https://www.wyden.senate.gov/news/press-releases/wyden-booker-and-clarke-introduce-bill-to-regulate-use-of-artificial-intelligence-to-make-critical-decisions-like-housing-employment-and-education.

copyright infringement.[24] There are also calls to regulate AI's use in pricing under the justification that AI can facilitate collusion or engage in price discrimination.[25] There are also calls to protect labor from AI.[26]

620. Yet, these regulatory calls have striking parallels with the last disruptive technology that emerged over 30 years ago: when, in 1991, the World Wide Web publicly launched.[27] Many considered the web to be too dangerous to be left loose in the wild. Yet, Congress exercised surprising restraint during that early period of growth with the first notable inclusion of the internet in a regulation coming with the Telecommunications Act of 1996.[28] This lighter touch with regulation is likely responsible for the tremendous innovation and growth that we are still experiencing today.[29] While the calls to regulate the web and internet have not subsided, history can serve as a guide to caution against stifling innovation in a nascent and dynamic area of technology.

621. This chapter proposes that AI policy should follow a similar regulatory path as that used when the web first emerged. The argument for regulatory restraint is based on several core tenets. First, when dynamic technologies are emerging, regulations that dampen the growth rate of a technology can lead to immense losses in social welfare – even if those regulations lead to some short-term reductions in specific harms, e.g., reducing the number of deepfakes over the coming year.[30] Second, premature regulatory solutions can crowd out market-based solutions, which can offer more efficient solutions to emergent harms compared to a regulatory body and the associated administrative costs. Finally, given that AI markets are nascent, and many markets have yet to form, having overly broad, blunt *ex ante* policy instruments in the name of providing guardrails to ensure, for

---

24 *See, e.g.*, James Love, *We Need Smart Intellectual Property Laws for Artificial Intelligence*, SCI. AM. (Aug. 7, 2023), https://www.scientificamerican.com/article/we-need-smart-intellectual-property-laws-for-artificial-intelligence/.

25 *See, e.g.*, Jason Potts, *Sources of Innovation in Generative AI*, NETWORK L. REV. (Feb. 12, 2024) ("[A]mong those concerned with economic and geopolitical consequences, the prime public policy issue is competition, and specifically the prospect of industrial concentration into powerful AI agglomerations that control entire industries or markets.").

26 *See, e.g.*, Filippucci et al., *supra* note 14, at 5 (noting the dangers to inequality and inclusion).

27 *See, e.g.*, *A Short History of the Web*, CERN, https://home.cern/science/computing/birth-web/short-history-web ("[T]he first Web server in the US came online in December 1991, once again in a particle physics laboratory: the Stanford Linear Accelerator Center (SLAC) in California.").

28 Specifically, Title V of the Act (a.k.a., Communications Decency Act of 1996), § 230 (adding some protections to online service providers from third-party content).

29 This is not an assertion that there was *no* regulation or government involvement with the web or internet more broadly. Rather, this early period avoided heavy-handed regulations. *See generally* ADAM THIERER, PERMISSIONLESS INNOVATION: THE CONTINUING CASE FOR COMPREHENSIVE TECHNOLOGICAL FREEDOM (2016) (advancing the idea of "permissionless innovation" rather than the "precautionary principle," which deems that innovators have the burden of showing their inventions will not cause harm to specific groups, norms, traditions, existing laws, etc.).

30 A "deepfake" is "an artificial image or video (a series of images) generated by a special kind of machine learning called 'deep' learning (hence the name)." *See* UVA Information Society, *What the Heck is a Deepfake?*, UNIV. VA. https://security.virginia.edu/deepfakes. An example would be the unauthorized use of a celebrity's image in an online advert.

instance, "safety and security" or "to promote competition" – when the roads have yet to be paved – is putting the cart before the horse. Specifically, regulations can have the consequence of entrenching incumbents and raising barriers to entry, which, perversely, harms the competitive process rather than promoting it.

622. A call to avoid premature regulatory interventions is not the same as advocating exempting AI from laws of general applicability, public (e.g., antitrust, criminal, tax) or private (e.g., contracts, torts, property). Moreover, AI should not be treated more gently or harshly under the law. For instance, a scheme among competitors to use the same AI-powered pricing algorithm – perhaps orchestrated by a ringleader – should be closely examined by antitrust authorities to assess whether there is a Sherman Act, Section 1 violation. Similarly, a joint venture or merger between two cloud computing firms that fuel AI technologies could raise both Clayton Act, Section 7 or Sherman Act, Section 2 issues.

II. Avoiding Dampened Innovation Incentives

623. The wealth of nations is built on innovation.[31] To that end, even small changes to the rate of innovation can lead to generational losses in wealth. The reverse is also true. Even incremental increases in growth can lead to immense gains over time.[32]

624. For example, consider the growth rate of a technology, where we assume the technology has some level of "output" (T) that translates into products and services that offer value to the market participants. Let us start, in period 0, with T = 100 and an annual growth of 30 percent (which is likely orders of magnitude less than recent growth rates in AI innovation).[33] Due to compounding, after 10 years, T will grow to nearly 14X the original size. Now, consider instead a marginally lower growth rate of 25 percent. After the same 10-year period, T will grow slightly over 9X the original size, which is clearly still quite good – but 5X lower than the counterfactual growth rate of 30 percent. The point is that even a "modest" reduction in the growth rate of an emerging technology, e.g., 5 percent (in terms of absolute levels),

---

31 Robert Solow won the Nobel Prize in economics for demonstrating that gains in wealth are due primarily to innovation – not to marginal improvements in the efficiency of what already exists. *See* Nobel Prize Press Release, *The Royal Swedish Academy of Sciences*, (Oct. 21, 1987), https://www.nobelprize.org/prizes/economic-sciences/1987/press-release/. *See also* Philippe Aghion & Peter Howitt, *Capital, Innovation, and Growth Accounting*, 23 OXFORD REV. ECON. 79, 80 (2007) (finding "results are consistent with the neoclassical model which implies that, in the long run, *all* of the growth in output per worker is caused by technological progress.").

32 A particularly illustrative example is that of British cycling. *See* James Clear, *This Coach Improved Every Tiny Thing by 1 Percent and Here's What Happened*, https://jamesclear.com/marginal-gains ("Brailsford had been hired to put British Cycling on a new trajectory. What made him different from previous coaches was his relentless commitment to a strategy that he referred to as 'the aggregation of marginal gains' which was the philosophy of searching for a tiny margin of improvement in everything you do... Just five years after Brailsford took over, the British Cycling team dominated the road and track cycling events...").

33 *See, e.g.*, Xuli Tang et al., *The Pace of Artificial Intelligence Innovations: Speed, Talent, and Trial-and-Error*, 14 J. INFORMETRICS (Nov. 2020).

can result in significant, long-term losses in social welfare. These losses are magnified over even longer horizons.[34]

625. Further, let us consider the potential harms from a technology. Again, suppose that a regulatory apparatus lowers the innovation growth rate by 5 percent (in absolute terms) but offers some mitigation in social harms that could result from a technology. For instance, while the internet has clearly transformed our societies and economies, there are undoubtedly downsides such as cybercrimes and social media additions. Suppose that for every unit of a technology, T, there is some resulting harm. For instance, for every T = 100, then this results in a social harm of, say, $10 while creating a social benefit of $20. Thus, as T grows, the benefits will remain greater than the harms, but the magnitude of both will increase. Suppose that regulation reduces the harms from a technology by 20 percent; thus, total harm falls from $10 each period to $8. Let us further assume that the regulation continues to negatively impact the growth of innovation but does not negatively impact the social benefit that emerges from each unit of innovation. Under this scenario, at the end of 10 years, the net gains each period, that is, the benefits minus the costs, under a regulatory environment will be 21 percent less than the unregulated environment. Indeed, to achieve equivalence, regulation would have to reduce harms by nearly 50 percent to match the net gains from an unregulated environment. Of course, the prior exercise is highly stylized, but it offers a thought experiment to illustrate that even regulations which are highly effective in reducing harms (e.g., 20 percent) resulting from a technology can ultimately lower the welfare of society, relative to the counterfactual, if the regulatory apparatus slows innovation – even on the margin.

626. Are there examples to illustrate the potential harms from regulating innovative markets? Perhaps the EU's recent experience with its General Data Protection Regulation (GDPR) can at least serve as a cautionary tale. GDPR went into effect in 2018 and requires organizations to guarantee user rights related to access, consent, erasure, and data portability.[35] At the time, the regulation was hailed as transforming online privacy and discipling online firms from imposing privacy harms.[36] The reality has been quite different. An examination of the empirical evidence in the aftermath of GDPR reveals a complex story.[37] First, there are some clear gains in terms

---

34 The same exercise over a 15-year period results in a 51X growth (under 30 percent) v. a 28X growth (under 25 percent), which means a 5 percent difference results in nearly half the level of innovation after 15 years.

35 *See, e.g.*, Ben Wolford, *What is GDPR, the EU's New Data Protection Law?*, GDPR.EU (Feb. 7, 2024, 8:55 PM GMT) https://perma.cc/R58J-ZPSN.

36 *See, e.g.*, Nitasha Tiku, *Europe's New Privacy Law Will Change the Web, and More*, WIRED (Mar. 19, 2018), https://perma.cc/BRX4-SFWG. *See also* Adam Satariano, *G.D.P.R., a New Privacy Law, Makes Europe World's Leading Tech Watchdog*, N.Y. TIMES (May 24, 2018), https://perma.cc/Y7L2-7KSM ("'If we can export this to the world, I will be happy,' said Vera Jourova, the European commissioner in charge of consumer protection and privacy who helped draft G.D.P.R.").

37 *See, e.g.*, Garrett A. Johnson, *Economic Research on Privacy Regulation: Lessons from the GDPR and Beyond* 1 (NAT'L BUREAU OF ECON. RSCH., Working Paper No. 30705, 2022), https://www.nber.org/papers/w30705; Avi Goldfarb & Verina F. Que, *The Economics of Digital Privacy*, 15 ANNU. REV. ECON. 267, 281 (2023); John M. Yun, *A Report Card on the Impact of Europe's Privacy Regulation (GDPR) on Digital Markets*, 31 GEO. MASON L. REV. F. 104 (2024).

of less tracking and collection of data from users who opt out. Yet, there are also consequences to markets and innovation including a consistent finding that the regulation tends to entrench incumbents, raise barriers to entry, and harm startups, entry, and smaller firms.[38] Part of the reason for this entrenchment includes the ability of large incumbents better able to absorb regulatory costs than smaller rivals.[39] Relatedly, this disparate ability to absorb regulatory costs may dissuade some potential entrants, which further dampens competition. Overall, there is evidence that GDPR has led to outcomes that clearly appear harmful to markets and innovation.

627. Turning more specifically to the development of AI, an examination of AI patenting activity indicates a growing divide between the US and the EU.[40] While patents are an imperfect proxy for innovation, they offer at least a window into the number of ideas being generated within a given field.[41] In the figure below, reproduced from an OECD publication,[42] the US and China are the global leaders in AI patent applications with Japan a distant third. In contrast, the EU countries, even combined, contribute relatively little. Interestingly, if one were to have predicted in 2010, what would happen over the next decade, it is not clear that many would have predicted the exponential growth of patenting activity in the US and China. While we cannot arrive at a sound conclusion as to why the EU has fallen dramatically behind[43] without a sound causal study, it is highly plausible that the gap will only grow – especially, given the passage of the EU's Artificial Intelligence Act in 2024.[44]

---

38 *See* Johnson, *supra* note 37, at 1 ("The economic literature on the GDPR to date has largely – though not universally – documented harms to firms. These harms include firm performance, innovation, competition, the web, and marketing. On the elusive consumer welfare side, the literature documents some objective privacy improvements as well as helpful survey evidence."); Goldfarb & Que, *supra* note 37, at 280 ("Overall, the conclusion from these papers is that the GDPR led to an immediate reduction in web visits and revenue… and a reduction in the efficiency of online search… It also appears to have reduced the firms' ability to target advertising and track consumers… Competition appears to have decreased in the online advertising market…, and there was a decline in new firms, venture capital investment, and new apps… In summary, the early evidence in the aftermath of the GDPR is that it worked, in the sense that firms were using less data in the year following the law's passing. This, however, had costs in terms of firm profits, the consumer online experience, innovation, and competition. There is some suggestive evidence that the impact has declined over time, with both less consumer protection and less impact on concentration…"); Yun, *supra* note 37, at 124 ("What appears to be clear is that user gains in privacy are coming at a substantial cost to compliance that disproportionately harms smaller firms. Further, the more important dynamic incentive effects on innovation and competition are factors that will have substantial long-term implications.").

39 *See, e.g.*, Heli Koski & Nelli Valmari, *Short-term Impacts of the GDPR on Firm Performance* (RSCH. INST. OF THE FINNISH ECON. (ETLA) Working Paper No. 77, 2020) (estimating substantial compliance costs related to GDPR and finding smaller and medium-sized European firms were the most disadvantaged); Chinchih Chen et al., *Privacy Regulation and Firm Performance: Estimating the GDPR Effect Globally*, (OXFORD MARTIN WORKING PAPER SERIES ON TECH. AND ECON. CHANGE, Working Paper No. 2022-1, 2022) ("Firms exposed to the GDP experienced an 8 percent decline in profits, and the decline in profits of small companies is almost double the average").

40 Filippucci et al., *supra* note 14, at 27.

41 Importantly, patents are more than merely ideas. They memorialize the methods that embody the ideas, which creates the potential for technology markets that are instrumental to introducing and commercializing innovation. *See* DANIEL SPULBER, THE CASE FOR PATENTS (2021).

42 Filippucci et al., *supra* note 14, at 27.

43 *Id.* (indicating how the figure "highlights the strong geographic concentration of AI patenting activity in the US and China, with Japan and the EU substantially lagging behind and other countries contributing very little.").

44 *See, The AI Act Explorer*, ARTIFICIAL INTELLIGENCE ACT.EU (2024), https://artificialintelligenceact.eu/ai-act-explorer/.

Figure 1. AI patenting increased dramatically, with high cross-country concentration[45]

Number of PCT patent applications in AI-related technologies

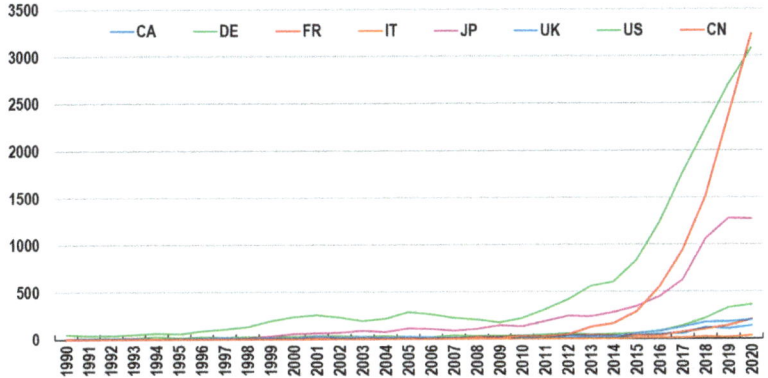

Note: Number of PCT patent applications in AI-related technologies. Data refer to patent applications filed under the Patent Co-operation Treaty (PCT), according to the filing date and the applicant's location, using fractional counts. AI-related patents are identified according to the methodology described in (Baruffaldi et al., 2020[83]).
Source: OECD, STI Micro-data Lab: Intellectual Property Database, http://oe.cd/ipstats, January 2024.

III. Premature Regulation Could Crowd Out Market-Based Solutions

628. When email first emerged, adoption grew rapidly, and the technology has indisputably created immense value.[46] Nonetheless, as with most technologies, there can be associated harms that result from widespread use. For email, a key concern is spam, which, by many accounts, was started in 1978 by Gary Thuerk, who was marketing a new computer model.[47] Spam and other types of unwanted emails are a form of asymmetric information, which can lead to market failures.[48] Yet, it was not until 2003, with the CAN-SPAM Act, that a major regulation governing email spam was passed.[49] The act is

---

45 Filippucci et al., *supra* note 14, at 27 ("Note: Number of PCT patent applications in AI-related technologies. Data refer to patent applications filed under the Patent Co-operation Treaty (PCT), according to the filing date and the applicant's location, using fractional counts. AI-related patents are identified according to the methodology described in (Baruffaldi et al., 2020). Source: OECD, STI Micro-data Lab: Intellectual Property Database, http://oe.cd/ipstats, January 2024.").

46 *See, e.g.*, Email uplers, *Theory of Email Evolution*, MAVLERS https://email.uplers.com/infographics/evolution-of-emails/ (last visited July 25, 2024) ("Today with an estimated 2.6 billion active email users present globally and revenues, generated through email marketing from different industries, expected to be $19.3 billion, email marketing gives tough competition to social media.").

47 *See, e.g.*, Microsoft, *Why is it Called Spam, Anyway? A Brief Inbox History*, MICROSOFT (May 12, 2023), https://www.microsoft.com/en-us/microsoft-365-life-hacks/privacy-and-safety/what-is-email-spam.

48 Specifically, with unwanted emails, it is difficult for a user to determine legitimate offers from those that can create a cybersecurity threat.

49 *See, e.g.*, Fed. Trade Comm'n, *Controlling the Assault of Non-Solicited Pornography and Marketing Act of 2003 (CAN-SPAM Act)*, FTC, https://www.ftc.gov/legal-library/browse/statutes/controlling-assault-non-solicited-pornography-marketing-act-2003-can-spam-act.

notable for its targeted and specific guidelines, e.g., opt-out requests must be honored within 10 days.[50] Yet, before and after the CAN-SPAM Act, market-based solutions emerged to manage spam.[51] Indeed, few individuals or organizations are likely hampered by spam to the degree that the technology represents a significant danger.

629. While AI will inevitably have different obstacles compared to email, asymmetric information will likely remain a major source of potential harms. Take for instance the concern over AI bias, which is generally defined as perpetuating existing human biases.[52] However, AI bias can be broader and be defined as asserting something that is simply untrue – irrespective of intent. Yet, "fake news" and false information in various forms are not unique to AI and have been with us for quite some time.[53] There are numerous mechanisms to deal with false information – whether AI generated or not – including laws dealing with the tort of fraud, deceit, defamation, and libel.

630. Importantly, if AI output is generally deemed unreliable due to misinformation, disinformation, hallucinations,[54] or bias, then the market will adjust to these information asymmetries. For instance, consider Google's misstep with its Gemini image generator.[55] Holding aside Google's intent or motivation, Gemini clearly created misrepresentations of historical figures.[56] The condemnation and mockery of Google was swift.[57]

631. This early AI episode illustrates a point made by Klein and Leffler.[58] Specifically, they explained that, even in the absence of formal laws, markets

---

50   *See, e.g.*, Denys Kontorskyy, *Legal Aspects of Email Marketing: Laws, Compliance, and Penalties*, MAILTRAP (Jan. 25, 2024), https://mailtrap.io/blog/email-marketing-laws/.

51   *See* Adam D. Thierer, *Whatever Happened to Leaving the Internet Unregulated?*, CATO INST. (Dec. 22, 2002), https://www.cato.org/commentary/whatever-happened-leaving-internet-unregulated ("As the market works to shift costs of commercial email back to the sender, we must be on guard against legislative confusion: How might the definition of spam expand beyond 'unsolicited' and 'commercial' email, and would such expansion be a good thing?").

52   *See, e.g.*, IBM Data and AI Team, *Shedding Light on AI Bias with Real World Examples*, IBM (Oct. 16, 2023), https://www.ibm.com/blog/shedding-light-on-ai-bias-with-real-world-examples/.

53   *See, e.g.*, Seth B. Sacher & John M. Yun, *Fake News is Not an Antitrust Problem*, CPI ANTITRUST CHRON. 1, 3 (Dec. 2017) ("[A]ccording to a Merriam-Webster article, the term fake news began to enjoy 'general use at the end of the 19th century.' The generation and distribution of intentionally false stories is not a new phenomenon even if it went under different names such as 'false news' or even outright 'lies.'").

54   Hallucinations occur when a large language model (LLM) "perceives patterns or objects that are nonexistent or imperceptible to human observers, creating outputs that are nonsensical or altogether inaccurate." *See* IBM, *What are AI Hallucinations?*, IBM, https://www.ibm.com/topics/ai-hallucinations.

55   *See, e.g.*, William Gavin, *Google's Gemini AI Generator Will Relaunch in a 'Few Weeks' after Spitting Out Inaccurate Images*, QUARTZ (Feb. 26, 2024), https://qz.com/google-gemini-ai-historically-inaccurate-1851287538.

56   *Id.* ("Users testing out the service last week had discovered that Gemini was having difficulties generating historical images of people accurately, usually by changing their race or sexuality.").

57   *See, e.g.*, David Gilbert, *Google's "Woke" Image Generator Shows the Limitations of AI*, WIRED (Feb. 22, 2024), https://www.wired.com/story/google-gemini-woke-ai-image-generation/.

58   Benjamin Klein & Keith B. Leffler, *The Role of Market Forces in Assuring Contractual Performance*, 89 J. POL. ECON. 615 (1981).

can operate based on both reputational effects and repeat purchases.[59] The idea is that firms will invest to establish brands and reputations, which act as a type of bond that firms do not want to lose by engaging in short-term opportunism that harms consumers – such as fraud. Further, repeat purchases can serve as a nongovernmental contractual enforcement mechanism to ensure performance on, for instance, hard to quantify features like the "taste" of a hamburger. Within this context, intentional bias and deepfakes are a type of fraud associated with AI. Like other types of fraud, the question is not whether we can reduce it to zero, but whether there are adequate punishments to efficiently deter the conduct. Punishments can come in the form of market-based mechanisms, such as those described by Klein and Leffler, or well-established legal remedies.

632. In sum, we simply do not know how markets will develop and deal with associated harms from a technology. Further, it is not a stretch to assert that, given the level of commerce and investments associated with AI, firms will not sit idly by if, for instance, bias and deepfakes start to derail their products.[60] Firms are highly motivated to prevent and address harms – given that consumers will be reticent to adopt AI technologies until there are assurances that the output is authentic and free from bias and falsehoods. Again, this is not to suggest that potential harms associated with AI technology should be exempt, or receive a lighter touch, from existing laws of general applicability. Perhaps, at some point, there will be a good justification for having AI-specific regulations to further disincentivize false information. If so, those regulations will likely be more successful if they target a specific harm or industry – rather than focusing on broader principles that are hard to define and remedy.[61]

IV. The Potential to Entrench Incumbents

633. While AI is still largely nascent in terms of products and associated markets, some have already sounded the alarm over the competitiveness of AI markets.[62] The concerns appear largely the same as those associated with social media, search engines, and operating systems. Specifically, the claim is

---

59   *Id.* at 616 ("[E]conomists also have long considered 'reputations' and brand names to be private devices which provide incentives that assure contract performance in the absence of any third-party enforcer."); *id.* ("Market arrangements such as the value of lost repeat purchases which motivate transactors to honor their promises may be the cheapest method of guaranteeing the guarantee.").

60   *See, e.g.*, Casey Newton, *Inside OpenAI's New Plan for Fighting Deepfakes*, PLATFORMER (May 7, 2024), https://www.platformer.news/open-ai-deepfakes-election-c2pa/. *See also* Chloe Aiello, *How This VC-Backed Startup is Fighting Deepfakes*, INC. (Feb. 16, 2024), https://www.inc.com/chloe-aiello/how-this-vc-backed-startup-is-fighting-deepfakes.html.

61   *See, e.g.*, Thibault Schrepel, *Decoding the AI Act: A Critical Guide for Competition Experts* (AMSTERDAM L. & TECH. INST., Working Paper No. 3-2023, 2023), (DYNAMIC COMP. INITIATIVE., Working Paper No. 4-2023, Oct. 23, 2023), at 16, https://papers.ssrn.com/sol3/papers.cfm?abstract_id=4609947("Despite the technical difficulties in ensuring compliance with the AI Act... and the vagueness of the language..., the fines for non-compliance are set high.").

62   Filippucci et al., *supra* note 14, at 33 ("As in markets for online platform services, market dominance by AI incumbents can worsen market outcomes (e.g., via lower output) but also generate abusive behaviour aimed at exploiting or foreclosing competitors in vertically (or horizontally) related markets when AI developers are also AI-powered suppliers in such markets.").

that AI markets have characteristics – e.g., first-mover advantages, network effects, and tipping, that will likely lead to greater market concentration.[63]

634. Yet, fears over first-mover advantages in AI – particularly from large platforms – have already been disproven. OpenAI's ChatGPT was not the first mover in the large language model product space. Meta's Galactica was launched several weeks before ChatGPT on November 15, 2022, and, only days later, Meta withdrew the product.[64] Despite Meta's success in social media and having a first-mover advantage, this did not translate into success in AI. Further, it was only recently that Meta relaunched its AI assistant under the name Llama.[65] Of course, perhaps there is a "second mover advantage" or a "once the market matures advantage" to incumbents, but such notions are largely indistinguishable from many other markets that tend to eventually settle on several, long-standing market leaders – including industries such as oil & gas, pharmaceuticals, automobiles, luxury goods, soda, toothpaste, etc.

635. As for network effects, it is important to consider both the strength and significance of these effects.[66] Network effects are also a market feature and not necessarily a market failure, and their presence does not ensure success or tipping towards one dominant firm.[67] Further, it is not clear the importance of network effects in AI related markets. Perhaps some will have more importance while others have very little. What should generally be avoided are presumptions based on assumed levels of network effects and, just as importantly, assumed harms associated from these effects.

636. Thus, rather than making premature presumptions about how AI markets will work and the nature of the competitive process, perhaps we should let the markets develop rather than imagining some idealized market structure and attempting to manipulate markets into that structure.[68] Adam Smith memorably warned against the hubris of market controllers.[69]

---

[63] *Id.* at 32 ("[P]otentially more than with other digital technologies, AI systems embed characteristics that can lead to both early mover advantage and market concentration, tipping and dominance, which could thwart competition and contestability in the provision of AI services.").

[64] *See* Will Douglas Heaven, *Why Meta's Latest Large Language Model Survived Only Three Days Online*, MIT TECH. REV. (Nov. 18, 2022), https://www.technologyreview.com/2022/11/18/1063487/meta-large-language-model-ai-only-survived-three-days-gpt-3-science/ (detailing how Meta's first mover advantage was not enough to overcome its failed initial attempt in AI).

[65] *See, e.g.*, Alex Heath, *Meta's Battle with ChatGPT Begins Now*, THE VERGE (Apr. 18, 2024, 4:59 PM GMT+1), https://www.theverge.com/2024/4/18/24133808/meta-ai-assistant-llama-3-chatgpt-openai-rival.

[66] *See, e.g.*, John M. Yun, *Does Antitrust Have Digital Blind Spots?*, 72 S.C. L. REV. 305, 314 (2020).

[67] *See, e.g.*, Catherine Tucker, *What Have We Learned in the Last Decade? Network Effects and Market Power*, 32 ANTITRUST 77, 77 Spring 2018, ("[N]etwork effects are not the guarantor of market dominance that antitrust analysts had initially feared."); Daniel F. Spulber, *Unlocking Technology: Antitrust and Innovation*, 4 J. COMP. L. & ECON. 915, 918 (2008) ("Despite being rarely observed, technology lock-in remains influential in competition policy.").

[68] *See, e.g.*, Hanno F. Kaiser, *Are "Closed Systems" an Antitrust Problem?*, 7 COMP. POL'Y INT'L 91 (2011) (detailing the futility in defining markets as "open" versus "closed" and, consequently, policies that favor one over the other).

[69] ADAM SMITH, THE WEALTH OF NATIONS (1776) ("The statesman, who should attempt to direct private people in what manner they ought to employ their capitals, would not only load himself with a most unnecessary attention, but assume an authority which could safely be trusted, not only to no single person, but to no council or senate whatever, and which would nowhere be so dangerous as in the hands of a man who had folly and presumption enough to fancy himself fit to exercise it.").

637. Further, regulating AI can lead to the consequence of entrenching incumbents and raising barriers to entry, which, perversely, harms the competitive process rather than promoting it.[70] As the experience of GDPR illustrates, incumbents may stand to benefit, on net, from regulation since, even if regulatory compliance is costly, incumbents are generally in the best position to weather those costs relative to smaller competitors and potential entrants. In a nutshell, barriers are likely to be unevenly felt.

638. Relatedly, there is the real potential for regulatory capture.[71] The idea is that, before a technology fully develops, interest groups and stakeholders will attempt to steer regulation in a manner that ultimately benefits their interests at the expense of consumers and other suppliers.[72] Leading AI companies have already publicly announced the need for AI regulation.[73] The public should be wary of those calls – particularly when those companies have a hand in shaping the regulation. Ultimately, regulatory capture is a rich area of economic research and a genuine concern.

V. Conclusion

639. While AI offers great promise, fundamentally, the technology is powered by people.[74] Since people are involved, AI can be used to generate immense value but also facilitate harms including perpetuating false information and deception. Within this context, this chapter has called for restraint on AI regulation, which is not a call to ignore the dangers that AI generated output can pose – nor is it a call for a disparate, "more permissive" treatment of AI under existing laws or existing regulatory schemes of general application. Rather it is a recognition that regulation itself creates costs on society including the potential to dampen the rate of innovation, crowd out market-based solutions, and entrench incumbents and larger technology companies.

---

70   *Cf.* Schrepel, *supra* note 61; Koski & Valmari, *supra* note 39; Chen et al., *supra* note 39.

71   *See* George J. Stigler, *The Theory of Economic Regulation*, 2 BELL J. ECON. & MGMT. SCI. 3, 3 (1971) ("A central thesis of this paper is that, as a rule, regulation is acquired by the industry and is designed and operated primarily for its benefit.").

72   *See, e.g.*, Thibault Schrepel, *The Fight for Open Source in Generative AI*, NETWORK L. REV. (2023), https://www.networklawreview.org/open-source-generative-ai/ ("The fact that incumbents feel threatened by open-source generative AI creates competition. That is a good thing. But it also creates an irresistible desire for regulatory capture. ... We are already seeing some large corporations pushing for regulations that require all companies to pay licences and royalties to content providers whose publications are used to train AI. Conveniently, large companies can afford these costs, while smaller and open-source competitors often cannot.").

73   *See, e.g.*, Daniel Howley, *Why Tech Companies Want the Government to Regulate AI*, YAHOO! FIN. (Oct. 4, 2023), https://finance.yahoo.com/news/why-tech-companies-want-the-government-to-regulate-ai-190732572.html.

74   *See, e.g.*, Agrawal, Gans & Goldfarb, *supra* note 12 ("[I]t is possible to imagine a fully automated solution that makes every decision. But that approach is surely a rarity. AI prediction provides information that improves decisions, which are made by people. Interestingly, with AI the difference is not so much whether machines do more but who the best people to make decisions are.").

The Case Against Preemptive Antitrust in the Generative Artificial Intelligence Ecosystem

Jonathan M. Barnett[*]
University of Southern California

Abstract

Since the launch of the ChatGPT application in late 2022, the generative artificial intelligence (genAI) ecosystem has elicited scrutiny from competition regulators in the United States, European Union, and other jurisdictions. Relying on the assumption that digital platform markets are prone to converge on entrenched monopoly outcomes, some commentators and regulators favor intervening preemptively in the genAI ecosystem. This contribution assesses whether there are reasonable antitrust grounds for taking such action. Available evidence suggests that technically and commercially competitive entrants can generally secure access to the inputs required to achieve entry, including funding, semiconductors, cloud-computing services, datasets, and foundation models. Consistent with this view, entry into the models and applications segments of the genAI is especially robust. Investments and alliances involving large technology platforms, venture-capital, and institutional investors, and model developers, which have elicited regulatory concern, currently appear to be efficient arrangements to aggregate the complementary assets required to produce genAI models and applications and face competition from other business models.

---

[*] Torrey H. Webb Professor of Law, Gould School of Law, University of Southern California. I am grateful for comments from the editors. Additional comments welcome at jbarnett@law.usc.edu.

The Case Against Preemptive Antitrust
in the Generative Artificial Intelligence Ecosystem

I. Introduction

640. The enforcement and interpretation of antitrust law currently stand at a historical juncture between two starkly divergent approaches. Following long-standing practice, antitrust enforcement has followed a fact-intensive approach that enforces antitrust law on a case-specific basis. Outside limited cases of explicit collusion, this approach generally involves a balancing analysis that weighs anticompetitive against procompetitive effects attributable to a contested business practice. In contrast, some regulators in the US, the EU, and other jurisdictions advocate a policy of preemptive enforcement toward large platforms in digital markets, which presumes that certain business practices and acquisitions are presumptively anticompetitive and places a significant burden on platforms that engage in those practices and acquisitions to show otherwise.

641. This approach toward antitrust enforcement is reflected most clearly in the EU's Digital Markets Act (DMA), which came into effect in late 2022 and largely bans certain practices when engaged in by the largest tech platforms,[1] and the Artificial Intelligence (AI) Act, which was enacted by the European Parliament in 2024 and empowers "national surveillance authorities" to access certain datasets and models of "AI systems" and then pass on certain information to EU and national competition regulators.[2] This anticipatory approach is also reflected in the revised Merger Guidelines adopted in late 2023 by the US Department of Justice and Federal Trade Commission (FTC), which contemplate merger challenges based on theories of potential entry, even in "markets that do not yet consist of commercial products."[3]

642. Concerning the generative artificial intelligence (genAI) market, the heads of the three most prominent competition agencies have endorsed, or are considering taking action under, a preemptive approach. FTC Chair Lina Khan has stated: "[W]e are still reeling from the concentration that resulted from Web 2.0, and we don't want to repeat the missteps of the past with AI."[4] Similarly, EU Competition Commissioner Margrethe Vestager has stated: "[I]f we don't act soon, we will find ourselves, once again, chasing solutions to problems we did not anticipate."[5] Sarah

---

1   Commission Regulation 2022/1925 of the European Parliament and of the Council of Sept. 14, 2022, on Contestable and Fair Markets in the Digital Sector and Amending Directives (EU) 2019/1937 and (EU) 2020/1828 (Digital Markets Act, hereinafter DMA), 2022, O.J. (L 265/1), at arts. 2(1) (definition of "gatekeeper") and (2) (definition of "core platform service") and art. 3 (setting forth rules for gatekeeper designation).

2   Thibault Schrepel, *Decoding the AI Act: A Critical Guide for Competition Experts* 4–5 (AMSTERDAM L. & TECH. INST., Working Paper No. 3-2023, 2023).

3   U.S. DEP'T OF JUST. AND FED. TRADE COMM'N, MERGER GUIDELINES 11 n.22 (Dec. 18, 2023).

4   Rana Foorhar, *The Great US-Europe Antitrust Divide*, FIN. TIMES, Feb. 5, 2024.

5   *Making A.I. Available to All – and to Avoid Big Tech's Monopoly – Vestager*, INSIGHT EU MONITORING (Feb. 20, 2024), https://ieu-monitoring.com/editorial/making-a-i-available-to-all-and-to-avoid-big-techs-monopoly-vestager/428826?utm_source=ieu-portal.

Cardell, CEO of the UK Competition & Markets Authority (CMA), has stated that "it is important to act now to ensure that a small number of firms with unprecedented market power don't end up in a position to control not just how the most powerful models are designed... but also how they are... used across all parts of our economy...".[6]

643. The preemptive approach toward antitrust enforcement relies on the assumption that digital markets are prone to converge on entrenched monopolies that are largely immune to competitive discipline. If it is sound to assume that digital markets generally pose an especially high risk of competitive harm, then it is sound to adopt an anticipatory regime that subjects leading platforms to per se prohibitions or strong presumptions of competitive harm that are difficult to rebut. If that assumption is unfounded, however, then rushing to condemn business practices undertaken by large digital platforms, or generally presuming that large platforms pose an elevated risk of competitive harm without thorough factual inquiry, could yield interventions that suppress innocuous or efficient business practices. Hence any rigorous case for anticipatory antitrust must deliver compelling evidence, rather than merely relying on theoretical assumptions, that specific digital markets are likely to converge toward a durable "winner-take-most" outcome that is shielded from competitive challenges.[7]

644. In this contribution, I use the rapidly growing market for genAI models and applications as a test case for assessing the merits of an anticipatory approach to antitrust enforcement. To undertake this exercise, I proceed in four steps.

645. In Section I, I describe the technological and economic characteristics of the genAI ecosystem. In Section II, I describe the principal concerns that have been raised by antitrust regulators concerning the genAI market. In Section III, I assess whether the genAI market currently exhibits competitive risks that persuasively favor anticipatory application of the antitrust laws, as distinguished from conventional *ex post* enforcement. I find that available evidence does not support taking anticipatory action in the genAI market; rather, it appears that competitive forces are generally responsive to actual or anticipated bottlenecks at each segment of the genAI supply chain. In Section IV, I apply a conventional balancing approach to assess the likely net competitive effects reasonably attributable to the transactions between large technology platforms and genAI model developers that have specifically elicited regulatory scrutiny. While these transactions may raise potential antitrust risks that merit regulatory attention, available

---

[6] Opening Remarks at the American Bar Association (ABA) Chair's Showcase on AI Foundation Models, Remarks by Sarah Cardwell, CEO of the CMA, delivered during the 72nd Antitrust Law Spring Meeting, CMA (Apr. 11, 2024).

[7] On evidence concerning the durability of winner-take-most outcomes in digital markets, see Jonathan M. Barnett, *Illusions of Dominance: Revisiting the Market Power Assumption in Platform Markets*, ANTITRUST L.J. (forthcoming 2024).

evidence favors the view that these transactions currently reflect an efficient adaptation to certain technological and economic characteristics of the genAI ecosystem.

1. Structure of the genAI Market

646. In this section, I review each segment of the genAI supply chain, including the major players, technologies, entry barriers, and other characteristics that may impact competitive conditions.

A. *Technology Overview*

647. A discussion of the structure of the genAI market requires familiarity with the technology that supports AI-enabled user-facing applications. For that purpose, this section provides an overview of the key elements of the genAI "technology stack," which in turn impacts the structure of the genAI supply chain.[8]

648. Broadly speaking, genAI-enabled applications rely on "foundation models" (FMs) that deploy machine-learning algorithms to identify and anticipate statistical relationships in the sequence of inputs provided by a curated dataset and, on that basis, to generate outputs in connection with various downstream tasks.[9] Inputs and outputs may comprise various modalities, including text, audio, music, images, video, or software code. For example, AI-enabled applications include ChatGPT and Gemini, which generate textual output, Dall-E and Midjourney, which generate visual output, Runway, which generates video output, GitHub Copilot, which generates code output, and "multimodal" applications that involve multiple input and output modalities. Assembling the datasets from which an FM "learns" is typically a costly undertaking that involves data collection and curation (known as "pretraining"). Refining the model to support a specific application then requires "fine-tuning" or "alignment" (sometimes involving specialized datasets or reinforcement learning based on data gathered from users), which is less costly than the pretraining process.[10]

649. To provide greater insight into the composition of a particular genAI market, the Table below presents a representative list of firms that supply "large-language models" ((LLMs), a subset of FMs for purposes of

---

[8] The discussion in this subsection is informed by COMPETITION & MARKETS AUTHORITY, AI FOUNDATION MODELS: INITIAL REPORT (2023) [hereinafter CMA REPORT]; Christophe Carugati, *The Generative AI Challenges for Competition Authorities*, 59 INTERECON. 14 (2024); Stefan Feuerrigel, Jochen Hartmann, Christian Janiesch & Patrick Zschech, *Generative AI*, 66 BUS. INFO. SYS. ENG'G 111 (2024); Matt Taddy, *The Technological Elements of Artificial Intelligence*, in THE ECONOMICS OF ARTIFICIAL INTELLIGENCE: AN AGENDA (Ajay Agarwal, Joshua Gans & Avi Goldfarb eds., 2019).

[9] There is no settled definition of a foundation model. On various definitions, see RISHI BOMMASANI ET AL., ON THE OPPORTUNITIES AND RISKS OF FOUNDATION MODELS, CENTER FOR RESEARCH ON FOUNDATION MODELS, STANFORD INSTITUTE FOR HUMAN-CENTERED ARTIFICIAL INTELLIGENCE, STAN. UNIV. 3–6, 48–49, 74–80 (2021).

[10] *Id.* at 6–9, 85–87.

natural-language processing (NLP) applications) and AI-enabled applications in the NLP market.[11]

Table 1. GenAI Natural-Language Processing and Answer Engine Market (excluding China)

Market segment	Models and Firms (selected, as of March 2024)
Models (LLMs)	GPT-4 (OpenAI), BERT (Google), LaMDA (Google), T5 (Google), PaLM (Google), LLama 3 (Meta), Jurassic-2 (AI21 Labs), Claude 3 (Anthropic), Contextual, Megatron-Turing-NLG (Nvidia, MS), BLOOM (Hugging Face), Cohere, Orca (MS), Luminous-Explore (Aleph Alpha), NeMo (Nvidia), Titan (Amazon), MM1 (Apple), Phi-2 (MS), Gemma (Google), Mistral 8x7b
Applications (chatbots)	ChatGPT (OpenAI), Gemini (Google), Azure AI (MS), Meta AI, Microsoft Copilot, Claude (Anthropic), Pi (Inflection AI), Perplexity AI, Jasper Chat, Grok (xAI), Amazon Q, HuggingChat (Hugging Face), Llama Chat (Meta)

Note: Name of developer entity appears in parentheses when that is not obvious from the name of the model or application. The most recent known version of each model is listed.

650. As can be seen, there are currently many firms active in each segment, which is expected, given the nascent stage of the technology and industry. Among these firms, OpenAI currently has a leadership position in the LLM segment of the genAI market, followed by Microsoft (which offers the Azure AI platform). According to one widely referenced report, OpenAI and Microsoft each held an estimated 39 percent and 30 percent share, respectively, of the "models and platforms" segment of the genAI market as of year-end 2023,[12] although it should be noted that there is not yet a settled methodology for defining relevant markets and measuring market share in the genAI context.

B. *GenAI Supply Chain: Make/Buy Choices*

651. GenAI-enabled applications represent the endpoint of a supply chain that relies on multiple inputs. The principal inputs include: (1) specialized semiconductors (specifically, graphic processing units or GPUs), (2) computing and data-storage services (typically known as "compute"), (3) data, (4) FMs and LLMs (including pretraining services), and (5) fine-tuning services (including a secondary market in fine-tuning tools). Like any market, a firm that seeks to provide a particular genAI application faces a "make/buy" decision concerning each of these required inputs, which in turn determines the firm's level of vertical integration. For ease of reference, these inputs are set forth in Figure 1 below.

---

11 For a fuller list, see CMA Report, *supra* note 8, at 21–24.

12 Philipp Wegner, *The Leading Generative AI Companies*, IOT Analytics (Dec. 14, 2023), https://iot-analytics.com/leading-generative-ai-companies/. The report's estimate relies on a market defined as "foundational models and platforms," which encompasses FMs and "software that enables the management of generative AI-related activities outside of foundational models."

The Case Against Preemptive Antitrust
in the Generative Artificial Intelligence Ecosystem

Figure 1. GenAI Market Supply Chain (NLP and Answer Engine Segment)

Semi-conductors (chips)	"Compute" services (cloud)	Data samples	LLM (incl. pretraining services)	Fine-tuning services and tools	NLP applications	→	End-users

653. At the 100 percent "make" end of the organizational spectrum, a firm that chooses to construct a vertically integrated supply chain can hypothetically secure funds and expertise to construct an FM, assemble a dataset through public-domain sources (or "scrape" proprietary content available online), engineer and train an LLM on the dataset, undertake fine-tuning to optimize the LLM's performance, and produce an application for distribution to the target user market. At almost the 100 percent "buy" end of the spectrum, a firm can choose only to produce a user-facing genAI-enabled application and secure all other inputs by contractual agreements with cloud-computing services and LLM providers, supplemented by specialized datasets (obtained by license or constructed independently from public-domain sources) and refined internally through fine-tuning for purposes of the application.

654. In real-world genAI markets, a rich variety of make/buy choices and resulting organizational structures can be observed. The highest levels of vertical integration characterize some incumbent platforms, such as Google and Microsoft, that are already active in the cloud computing market, have access to internal or external datasets, and, as shown in Table 1 above, have produced LLMs and end-user applications. Intermediate levels of vertical integration are exhibited by other major LLM providers, such as AI21 Labs, Anthropic, Cohere, Mistral AI, and OpenAI, which typically secure compute services by contracting with cloud-computing providers. The lowest levels of vertical integration are exhibited by independent genAI application providers in the NLP and answer engine market segments, which produce genAI-enabled applications for end-users but contract with cloud-computing services and LLM providers (and, in some cases, providers of specialized datasets or fine-tuning tools) for all other required inputs.

655. In general, the capital, talent, and expertise requirements for entry are highest at the top of the genAI supply chain and decrease (but remain significant) as the supply chain moves toward the point of market release. Currently the GPUs required to train and fine-tune genAI models are principally produced by Nvidia. However, Nvidia is not immune to competitive threats. Intel and AMD offer, or are reportedly developing, competing GPUs or processors for AI-specific PC devices. Alphabet, Amazon, IBM, Meta, and Microsoft have or reportedly are developing capacities to produce semiconductors for AI purposes,[13] and other firms are reportedly developing or already offer specialized "edge AI" chips

---

[13] CMA Report, *supra* note 8, at 12–13, 34; Carugati, *supra* note 8, at 15; Emilie David, *Chip Race: Microsoft, Meta, Google, and Nvidia Battle it Out for AI Chip Supremacy*, The Verge (Mar. 25, 2024), https://www.theverge.com/2024/2/1/24058186/ai-chips-meta-microsoft-google-nvidia/archives/2; Cade Netz, Karen Weise & Mike Isaac, *Nvidia's Big Tech Rivals Put Their Own A.I. Chips on the Table*, N.Y. Times, Jan. 29, 2024.

that can be embedded in devices.[14] Establishing a free-standing cloud-computing infrastructure is a capital-intensive undertaking that requires the installation and maintenance of physical data centers and transmission systems and, at this stage in the development of the cloud-services market, can probably only be feasibly supplied by leading incumbent providers such as Amazon (AWS), Google (Google Cloud), and Microsoft (Azure), smaller cloud-services providers such as Oracle and IBM, or well-resourced potential "Big Tech" entrants. Constructing an FM/LLM model "from scratch" is a capital-intensive undertaking but, as the genAI market shows, can be executed by entrants backed by sufficient external funding from VCs and institutional investors, such as Anthropic, AI21 Labs, Cohere, and Mistral AI.

656. Entry costs fall significantly in the applications segment of the market, where firms can rely on open-source FMs/LLMs (subject to contractual limitations specified by the FM/LLM provider) or enter into licenses for application programming interfaces (APIs) to closed-source FMs/LLMs, can secure data through publicly accessible sources or enter into licenses for proprietary data sources, and purchase fine-tuning services or execute those services independently. Hence, in general, it is expected that, as the genAI market develops, concentration will be highest in the upstream levels of the supply chain (cloud-computing services and FMs/LLMs) and decline as the supply chain approaches the end-user market (fine-tuning services and applications).

II. Preemptive Antitrust in the GenAI Market

657. The adoption of genAI technology since the release by OpenAI in November 2022 of ChatGPT 3.5 has been remarkably rapid, reaching over an estimated 150 million users by year-end 2023,[15] and has spawned the release of hundreds of other genAI applications involving various modalities, including text, images, speech, and more recently, music and video. The applications market has bifurcated into "horizontal" and "vertical" segments. Those segments refer, respectively, to applications that can be used across a range of industries (such as ChatGPT) and applications that are designed for use in a specific industry (such as an NLP tool for generating contractual agreements). In the short time since the release of ChatGPT, the genAI market has attracted scrutiny from competition regulators in major economies. In this section, I describe the principal statements made, and actions taken, by competition regulators in the US, EU, and UK concerning the adoption of an anticipatory approach to addressing purported risks to competition in the genAI market. I then discuss whether this anticipatory approach is compatible

---

14 COMPETITION AND MARKETS AUTHORITY, AI FOUNDATION MODELS: TECHNICAL UPDATE REPORT 12, 17–18 (2024), at [hereinafter CMA TECHNICAL UPDATE].

15 David Curry, *ChatGPT Revenue and Usage Statistics*, BUS. OF APPS (Jan. 15, 2024, updated July 2, 2024), https://www.businessofapps.com/data/chatgpt-statistics/

1. Regulatory Activity in the Generative AI Market

658. In the EU, US, and UK, regulators have launched investigations into agreements between large technology platforms and smaller LLM providers. In early 2024, the European Commission announced an investigation into the Microsoft/OpenAI relationship and issued a call for public comments on competition issues in the genAI market.[16] At the same time, the FTC announced that Alphabet, Amazon, Anthropic, Microsoft and OpenAI had been issued orders to provide relevant information concerning relationships involving Microsoft and OpenAI, Amazon and OpenAI, and Google and Anthropic.[17] In the UK, the CMA had initiated in 2023 an investigation into Microsoft's investment in OpenAI[18] and released a report that assesses risks to competition in the FM market,[19] which was followed by an updated report in 2024.[20] Competition authorities in Portugal, India, Hungary, France, and other countries have also issued statements expressing concern over potential antitrust risks in genAI markets.[21] Additionally, the Digital Markets Act (DMA), which entered into force in the EU in 2023, may apply to certain FMs, LLMs, or genAI applications if those models or applications are designated by regulators as "core platform services" that are offered by entities designated as "gatekeepers."[22] Critically, if the DMA applies to any such model or application, it would trigger quasi-per se prohibitions of certain tying and related practices that do not permit the gatekeeper to raise an efficiency defense.[23] As I will argue subsequently, the net competitive effect of certain transactional arrangements in the genAI market (in particular, partnerships between incumbent platforms and independent model developers) cannot be reasonably assessed without taking into account both the anticompetitive risks and procompetitive efficiencies attributable to those arrangements.

---

16 European Commission Press Release IP/24/85, Commission Launches calls for Contributions on Competition in Virtual Worlds and Generative AI (Jan. 9, 2024).

17 FED. TRADE COMM'N, RESOLUTION DIRECTING THE USE OF COMPULSORY PROCESS REGARDING THE INVESTMENTS AND PARTNERSHIPS INVOLVING GENERATIVE AI COMPANIES (Jan. 24, 2024).

18 COMPETITION AND MARKETS AUTHORITY, CMA SEEKS VIEWS ON MICROSOFT'S PARTNERSHIP WITH OPENAI (Dec. 8, 2023).

19 CMA Report, *supra* note 8.

20 COMPETITION AND MARKETS AUTHORITY, AI FOUNDATION MODELS: UPDATE PAPER (Apr. 11, 2024) [hereinafter CMA UPDATE].

21 Carugati, *supa* note 8, at 14. For a full list of AI-related initiatives at competition agencies (as of Jan. 2024), see Thibault Schrepel, Abdullah Yerebakan & Nikoletta Baladima, *A Database of Antitrust Initiatives Targeting Generative AI*, NETWORK L. REV. (Winter 2024).

22 DMA, *supra* note 1.

23 DMA, *supra* note 1, at arts. 5(7) (prohibition of ties between a core platform service and an identification service, web browser engine, or payment service) and 5(8) (prohibition of ties between two or more core platform services).

659. The alacrity with which regulators have focused on the genAI market reflects the ascendance of a preemptive approach toward antitrust enforcement in digital markets, which favors taking early action to intervene based on the expectation that those markets are likely to move toward an entrenched level of concentration due to a combination of network effects, switching costs, and economies of scale. This presumptive characterization of competitive conditions across digital environments contrasts with the conventional fact-intensive approach to antitrust enforcement. That approach assesses competitive conditions in particular markets on a case-specific basis and typically refrains from making predictive assessments of a specific market's future competitive trajectory.

2. The Potential Error Costs of Preemptive Antitrust

660. Preemptive antitrust reflects a significant departure from conventional practice among competition regulators, which have generally undertaken enforcement actions only in cases in which the enforcer can identify sufficient evidence to infer that a particular practice is actually or likely causing harm to competition.[24] That practice reflects both rule-of-law principles and the error-costs framework for antitrust enforcement, which seek to minimize false-positive and false-negative error costs by adhering to an appropriately calibrated evidentiary threshold for taking enforcement action or reaching a judicial determination of an antitrust claim. In the clearest cases of anticompetitive conduct involving explicit collusion, US antitrust law seeks to minimize false-negative error costs through the adoption of per se rules of antitrust liability that impose low evidentiary requirements.[25] EU competition law follows approximately the same approach toward offenses that are illegal "by object."[26] In all other cases (in which competitive harm is harder to diagnose with confidence without factual inquiry), US antitrust law has sought to minimize false-positive error costs through the rule-of-reason or similarly demanding evidentiary standards that weigh the competitive gains against the competitive losses attributable to a particular business practice.[27] Some commentators have asserted that EU competition law has shifted toward

---

24 To facilitate a discussion of competition policy in the US, EU, and UK, this is intended as a generic description of the legal threshold that must be met to defend legal action under antitrust or competition laws against a firm engaging in a particular business practice. The precise legal standard will differ in each jurisdiction.

25 Nw. Wholesale Stationers, Inc. v. Pac. Stationery & Printing Co., 472 U.S. 284, 289 (1985) (observing that antitrust law reserves "per se" treatment for practices "that have proved to be predominately anticompetitive") and Broad. Music, Inc. v. Columbia Broad. Sys., Inc., 441 U.S. 1 (1979) (noting that "agreements among competitors to fix prices... are among those concerted activities that the Court has held to be within the *per se* category").

26 DAMIEN GERADIN, ANNE LAYNE-FARRAR & NICOLAS PETIT, EU COMPETITION LAW AND ECONOMICS §§ 3.114–120 (2012) (discussion of concept of restriction "by object").

27 Herbert J. Hovenkamp, *The Rule of Reason*, 70 FLA. L. REV. 81, 83 (2018) (noting that US antitrust law applies some form of the rule of reason to all practices except "'naked' price fixing and market division agreements, a small subset of boycotts... and – by a very thin thread – some tying arrangements").

a similar balancing approach (with certain important differences) toward practices that can only be deemed illegal "by effect."[28]

661. None of the practices in the genAI market that have been identified by antitrust regulators as potentially or likely problematic fall into the special category of practices that so clearly pose anticompetitive harm that false-negative risks predominate and therefore per se treatment is warranted. These practices are not yet fully understood, take place in a market at an early stage in its development, and therefore fall into the large pool of practices that raise a mix of offsetting competitive effects and, in any particular case, are generally assessed by courts and regulators under some form of a balancing analysis. In the following discussion, I will therefore apply this type of balancing analysis (akin to the rule of reason) to assess whether there are reasonable grounds to anticipate a future anticompetitive outcome in the genAI market at a sufficiently high level of confidence to support preemptive intervention at the current nascent stage in the market's development.

III. Assessing Risks to Competition in the GenAI Technology Stack

662. Concerns expressed by some regulators and commentators over risks to competitive conditions in the genAI market often refer to the market generally, without specifying which segment of that market is exposed to these risks. A rigorous assessment must assess risks to competition at each segment or "layer" of the genAI supply chain, each of which presents different combinations of technological and economic characteristics that may raise different types and degrees of antitrust concern.[29] In this section, I undertake that exercise, which permits a tentative assessment based on currently available evidence at this early stage in the development of the genAI ecosystem.[30]

1. Applications Layer

663. At the applications layer of the genAI ecosystem, there do not currently appear to be significant risks to competition. The applications layer is unconcentrated and, given abundant opportunities for product differentiation and relatively low capital requirements for entry, there is little reason to believe that would change in the foreseeable future, although

---

28 GERADIN ET AL., *supra* note 26, at §§ 3.122–127 (discussion of concept of restriction "by effect"). On the EU's movement toward effects-based enforcement concerning exclusionary practices, see Linsey McCollum, *A dynamic and workable effects-based approach to abuse of dominance*, COMP. POL'Y BRIEF Issue 1, Mar. 2023; Damien M.B. Gerard, *The Effects-Based Approach under Article 101 TFEU and Its Paradoxes: Modernization at War with Itself? in* TEN YEARS OF EFFECTS-BASED APPROACH IN EU COMPETITION LAW (Jacques Bourgeois and Denis Waelbroeck eds., 2013).

29 For other layer-by-layer analyses of the genAI market, see CMA REPORT, *supra* note 8; Ela Glowicka & Jan Malek, *Digital Empires Reinforced? Generative AI Value Chain, Dynamics of Generative AI*, NETWORK L. REV. (Winter 2023).

30 To clarify, this exercise does not purport to definitively define antitrust-relevant markets within the genAI ecosystem.

a particular firm may establish a leadership position in a particular vertical or horizontal application. Reflecting a market segment characterized by low entry barriers and rich differentiation opportunities, the release of the ChatGPT chatbot in late 2022 has been followed by the entry of multiple competitors from incumbent platforms and smaller entrants, including (among others) Google's Gemini service, xAI's Grok service, Meta's Llama Chat service, and Anthropic's Claude service.[31]

2. FM/LLM Layer

664. The FM/LLM layer presents potentially greater competition concerns due to higher capital requirements, potential barriers to accessing data, and elevated concentration levels.

A. *Capital Requirements and Pretraining Costs*

665. Entry barriers in the FM/LLM layer may arise due to significant capital requirements, especially due to pretraining costs.[32] A report released by the CMA in 2023 found that pretraining costs vary depending on the size of the model, dataset, hardware, and other factors, ranging from $4 million to $100 million.[33] However, capital requirements to fund pretraining and other development costs are mitigated to the extent that entrants can access external funding through private or public markets. In the genAI market, private funding appears to be plentiful. The largest independent LLM developers have successfully secured hundreds of millions of dollars (or, in at least one case, billions of dollars) in outside funding from VCs, corporate VC, and institutional investors: as of March 2024, $5 billion for Anthropic, $641 million for Aleph Alpha, $553 million for Mistral AI, $435 million for Cohere, and $321 million for AI21 Labs.[34] Moreover, there are indications that pretraining costs are falling as a result of technical advancements.[35] Mistral AI has reportedly reduced the cost of pretraining a model to less than $22 million, which contrasts with the approximately $100 million reportedly spent to pretrain OpenAI's GPT-4 model, which supports the ChatGPT chatbot.[36] It has also been

---

31 For a fuller list, see Table 1.

32 Many antitrust scholars would not necessarily treat capital requirements (or relatedly, scale economies) as an entry barrier. Following the work of George Stigler, those scholars limit entry barriers to costs that are uniquely borne by an entrant. Other scholars are partial to Stigler's definition but nonetheless recognize that capital requirements and scale economies can have anticompetitive implications from a consumer-welfare perspective if these costs impede entry (which would tend to arise if there are capital-market imperfections). For discussion, see Preston R. Fee, Hugo M. Mialon & Michael A. Williams, What is a Barrier to Entry?, 94 AMER. ECON. REV. 461 (2004).

33 CMA REPORT, *supra* note 8, at 13–14, 34.

34 *See*, https://dealroom.co/ (last visited July 20, 2024).

35 Thibault Schrepel & Alex Sandy Pentland, *Competition between AI Foundation Models: Dynamics and Policy Recommendations* 8–9 (AMSTERDAM L. & TECH. INST., Working Paper No. 1-2023, 2023).

36 Sam Schechner, *The 9-Month-Old AI Startup Challenging Silicon Valley's Giants*, WALL ST. J. (Feb. 26, 2024, 12:07 PM ET), https://www.wsj.com/tech/ai/the-9-month-old-ai-startup-challenging-silicon-valleys-giants-ee2e4c48.

reported that Nvidia has succeeded in reducing model training costs below $500,000.[37] Other firms are developing "small language models" that are designed to run on smaller datasets for targeted applications and therefore are less costly to assemble and train.[38]

B. *Data Inputs*

666. Some regulators and commentators assert that data is a scarce resource that the largest technology platforms will inevitably control, which would in turn impede entry by independent model developers.[39] Yet it is not clear that this assertion describes the typical case in the data segment of the genAI ecosystem.[40] Reports released in 2023 and 2024 by the CMA identify multiple avenues through which model developers can secure data.[41] Developers can obtain data through open-access and closed-access sources (subject in the latter case to payment of a licensing fee),[42] and entrants can choose to invest in constructing, purchasing, or licensing proprietary databases for particular applications. The successful entry of LLM startups such as Anthropic and Mistral AI in text-generation, Midjourney and Stable Diffusion in image-generation, and RunwayML in video-generation (among other examples) suggests that data availability in multiple modalities may not typically be an impediment to entry.[43] Moreover, the demand for data by LLM entrants appears to have elicited entry by secondary data providers to meet that demand. These sources of data include "data warehouses" (also known as "data lakes"[44]), synthetic dataset providers, and other specialized data providers that support LLM models or genAI applications developers.[45] Lastly, recent innovations enable models sometimes to train on smaller datasets or to use

---

37 Thibault Schrepel, *Toward a Working Theory of Ecosystems in Antitrust Law: The Role of Complexity Science, Dynamics of Generative AI*, NETWORK L. REV. (Winter 2023).

38 Carugati, *supra* note 8, at 16; Tom Dotan & Deepa Seetharaman, *For AI Giants, Smaller is Sometimes Better*, WALL ST. J. (July 6, 2024, 5:30 AM ET), https://www.wsj.com/tech/ai/for-ai-giants-smaller-is-sometimes-better-ef07eb98.

39 For a review of these arguments, see Geoffrey A. Manne & Dirk Auer, *From Data Myths to Data Reality: at Generative AI Can Tell Us About Competition Policy (and Vice Versa)*, CPI (Feb. 2024), https://www.pymnts.com/wp-content/uploads/2024/02/CPI-ANTITRUST-CHRONICLE-Economics-of-Data-February-2024.pdf.

40 *Id.*; Carugati, *supra* note 8, at 14.

41 CMA REPORT, *supra* note 8, at 11; CMA UPDATE, *supra* note 20, at 6.

42 Carugati, *supra* note 8, at 16–17.

43 Manne & Auer, *supra* note 39.

44 For leading providers in this segment, see Snowflake, Inc., Form 10-K (Mar. 26, 2024); Synthesis AI, *Synthesis AI Launches Enterprise Synthetic Dataset on Snowflake Marketplace*, PR NEWSWIRE (May 11, 2023, 09:00 AM ET), https://www.prnewswire.com/news-releases/synthesis-ai-launches-enterprise-synthetic-dataset-on-snowflake-marketplace-301821525.html.

45 CMA REPORT, *supra* note 8, at 30–33; Arvind Murali, *The Ins and Outs of a Data Marketplace*, FORBES (May 12, 2021, 08:30 AM EDT), https://www.forbes.com/sites/forbestechcouncil/2021/05/12/the-ins-and-outs-of-a-data-marketplace/#:~:text=Key%20features%20that%20enable%20a,well%2Ddefined%20policies%20and%20procedures.

synthetic data,[46] which can sometimes enables FMs/LLMs to achieve superior performance for certain tasks. This technological advance may sometimes mitigate competitive advantages attributable to having access to the largest datasets.

C. *Market Share: Concentration and Durability*

667. The LLM layer in the genAI market appears to exhibit significant concentration, based on the estimated 39 percent and 30 percent market shares attributed in one widely referenced report to OpenAI and Microsoft, respectively, as of year-end 2023.[47] Note that Microsoft is OpenAI's largest investor (although, as noted subsequently, it does not enjoy any governance rights). However, it is not clear that these shares are durable or otherwise provide a reliable insight into competitive conditions in the models layer of the genAI ecosystem. There are several reasons for this.

668. There is a strong likelihood that the models layer of the genAI ecosystem will devolve into multiple differentiated FM segments tailored for particular industries (vertical applications) or uses (horizontal applications), in which case no single FM would dominate the models layer (or the LLM segment of the models layer). FM markets may disaggregate based on output or input modalities or specific uses (for example, text preparation, financial analysis, or medical applications), in which case no single "super model" would likely dominate. This has already occurred to some extent, resulting in a distinction between general-purpose FMs and specialized FMs.[48] Some model developers already offer more tailored models that target specific downstream tasks[49] and others have developed small language models that reduce memory, data, and compute requirements.[50] Additionally, developers can elect to develop applications that do not rely on a single LLM model or model provider, which would mitigate the market power of any single model provider in any particular segment of the genAI market.[51]

669. Even significant concentration levels in any appropriately defined models market would not necessarily translate into pricing power in the face of a meaningful threat of competitive entry. OpenAI faces actual and potential entry threats posed by competing models from well-resourced

---

46 Anil Ananthaswamy, *In AI, Is Bigger Better?*, 615 NATURE 202–205 Mar. 9, 2023; Schrepel & Pentland, *supra* note 5, at 5., 8-9.

47 Wegner, *supra* note 12.

48 Schrepel & Pentland, *supra* note 5.

49 Wegner, *supra* note 12, at 16.

50 CMA UPDATE, *supra* note 20, at 10–11.

51 Multiple vendors offer "LLM-agnostic" genAI application development platforms, see Adrian Bridgewater, *It's Time to Believe in AI Agnosticism*, FORBES (Feb. 12, 2024, 10:01 AM EST), https://www.forbes.com/sites/adrianbridgwater/2024/02/12/why-its-time-to-believe-in-ai-agnosticism/; Lucy Mazalon, *AI Wars: How Salesforce's Agnostic LLM Approach Works*, SFBEN (Aug. 4, 2023), https://www.salesforceben.com/ai-wars-how-salesforces-agnostic-llm-approach-works/.

incumbent platforms, such as Google, Meta, Amazon, and X, and independent VC-backed LLM providers. Some providers have sought to overcome OpenAI's early lead by offering LLMs on an open-source (subject to certain constraints) and royalty-free basis (unlike OpenAI, which is a closed-source LLM), such as Anthropic, Cohere, Google, Hugging Face, Mistral AI, and xAI.[52] In particular, Meta has pursued an open-source-based strategy[53] to promote adoption of its LLaMA model, which may offer a competitive alternative to OpenAI, which relies on a closed-source, subscription-based business model. One commentator observes that "thousands of closed-source and open-source models compete on various parameters like task requirements, language specifications, and data use."[54] The Hugging Face website (which hosts genAI models), offered over 350,000 models as of January 2024.[55] While this figure does not reflect differences in model size, performance, or other relevant parameters, it does not appear that models are a scarce asset in the genAI supply chain.

670. Notwithstanding the current proliferation of models in various modalities (text, image, speech, music, video, and code), it could be reasonably asserted that the models layer (or appropriately defined portions of the models layer) will ultimately converge on a small number of LLMs, which may be protected to some extent from competitive discipline. That outcome would be consistent generally with the history of digital markets, which tend to support a large number of providers initially but then evolve toward a handful of providers (which, depending on particular circumstances, may face potential entry threats that continue to exert competitive pressure). There are three principal reasons for this tendency toward "winner-take-most" outcomes: (1) economies of scale (resulting from high fixed costs and low variable costs), which compels exit by all but the most efficient providers, (2) network effects, which lead users to favor providers that are most popular with other users, and (3) switching costs, which lead users to resist migrating to entrants.[56]

671. Yet the models layer of the genAI market does not clearly exhibit the constituent elements of this paradigm model of a digital market. The genAI market exhibits economies of scale to the extent that constructing an FM/LLM imposes high fixed costs and lower variable costs. However, in contrast to other software markets, variable costs in the genAI context

---

52 Belle Lin, *Open-Source Companies Are Sharing Their AI Free: Can They Crack OpenAI's Dominance?*, WALL ST. J. (Mar. 21, 2024, 2:10 PM ET), https://www.wsj.com/articles/open-source-companies-are-sharing-their-ai-free-can-they-crack-openais-dominance-26149e9c. Mistral initially offered its models on an open-source basis but is releasing more advanced models on a closed-source basis, see Schechner, *supra* note 36. On Google's and Cohere's open-source releases, see CMA UPDATE, *supra* note 20, at 5–6.

53 Meta offers its models on an open-source, royalty-free basis but requires consent once the licensee's product exceeds 700 million users, see Lin, *supra* note 54.

54 Carugati, *supra* note 8, at 14.

55 Jonathan Gillham, *HuggingFace Statistics*, ORIGINALITY.AI (Jan. 29, 2024), https://originality.ai/blog/huggingface-statistics.

56 Barnett, *supra* note 7.

are significant due to substantial costs in continuously evaluating and fine-tuning the model's performance and executing "inferences" to deliver responses to user queries.[57] genAI markets do not necessarily exhibit network effects because the value placed by a user on a genAI-enabled model or application does not depend on the number of other users of that same model or application. Network effects may nonetheless arise if a model's performance improves as a function of the number of users who provide more data from which the model can "learn" and deliver more accurate results. Yet network effects may not be especially salient in the case of FMs/LLMs that are designed for specialized uses, in which case performance may be more closely tied to the quality of the dataset than the number of users.[58] There may be switching costs for applications developers if they must incur learning costs when migrating from one FM/LLM to another. However, those costs would be mitigated to the extent that applications are developed through platforms or tools that are "LLM-agnostic" (which, as noted previously, are already offered by some vendors[59]).

D. *The Overlooked Virtues of Winner-Take-Most Outcomes*

672. Even supposing that the models layer of the genAI ecosystem, or an appropriately defined subset of the genAI ecosystem, *did* converge on a small number of LLM providers consistent with the pattern observed in other mature digital markets, this is not clearly an adverse outcome as a matter of competition policy, when compared to other commercially and technically feasible market structures. (As I have argued elsewhere, this point applies generally in the antitrust analysis of digital markets and is an application of Harold Demsetz's famous warning against "nirvana fallacies" in public policy analysis.[60]) To the contrary: this outcome may in some cases represent a maximally efficient market structure that is best structured to cultivate the economic gains generated by the scale economies and network effects that *can only* arise under high concentration levels.[61] Scale economies confer benefits on users by spreading fixed costs over a large volume of transactions, which results in cost-savings for users (the precise proportion being determined by competitive forces) and therefore promotes output and, in the case of a new technology, adoption. Network effects are a necessary precondition for generating the economic value that arises from the capacity

---

57 BOMMASANI ET AL., *supra* note 9, at 87–88, 91–96; CMA REPORT, *supra* note 8 at 13–14, 36–38; Carugati, *supra* note 8, at 16.

58 Schrepel & Pentland, *supra* note 35, at 11–13. Indirect network effects may also arise when a single FM/LLM, or a handful of FMs/LLMs, become the dominant sources of genAI-enabled "plug-ins" in developers' applications (for example, a ChatGPT chatbot embedded in a contract-drafting application), *see* CMA REPORT, *supra* note 8, at 48–50, 56–57, 68–69.

59 *See supra* note 55.

60 Barnett, *supra* note 7. For the Demsetz source, see Harold Demsetz, *Information and Efficiency: Another Viewpoint*, 12 J.L. & ECON. 1 (1969).

61 For a fuller explanation of the arguments in this paragraph, see Barnett, *supra* note 7.

of digital technologies to reduce the costs of matching transacting parties. Antitrust policies and commentary that view scale economies and network effects as unqualified "competitive bads" and lower market concentration as an unqualified "competitive good," without taking into account the offsetting efficiencies and inefficiencies attributable to those same characteristics, can yield enforcement actions that distort market structures and, as a result, raise prices, limit output, and degrade quality.

3. Compute Layer

673. The "compute" layer encompasses firms that supply cloud-computing and data-storage services (also known as the "infrastructure-as-a-service" segment of the cloud market), which are inputs required by industry sectors that rely on cloud-enabled digital applications. The compute layer is prone to concentration due to capital requirements, economies of scale, and switching costs.[62]

674. Entry into the compute layer requires significant capital resources, especially because supplying cloud-computing services involves physical plant such as data centers and transmission cables. Currently the leading providers (outside China) are Amazon (AWS), Microsoft (Azure), and Google (Google Cloud), with the remainder of the market comprising Oracle, IBM, and smaller providers.

675. Economies of scale arise because providers incur high costs to install the physical plant and smaller (but significant) costs to maintain and upgrade that infrastructure. Again, it is important to keep in mind that economies of scale can benefit users by reducing per-unit costs and yielding cost-savings that are shared with users to some extent depending on competitive pressures in the end-user market. Consistent with this possibility, data covering the period 2012–2022 shows that the cloud-computing market has exhibited declining prices since the entry of Microsoft Azure and Google Cloud to challenge AWS, the pioneer.[63]

676. Switching costs arise because it is often costly for users to move data from an existing cloud-computing provider. Yet it is not clear that switching costs raise significant competition concerns in the cloud-computing market. Cloud-services users are typically sophisticated businesses that can anticipate switching costs and have developed "multi-cloud" strategies (using multiple providers and on-premises data storage) to mitigate reliance on a single provider.[64] This same type of hedging strategy can be observed in the genAI market: Anthropic, one of the largest independent LLM providers,

---

62   As noted previously (*see supra* note 34), many antitrust scholars do not treat capital requirements and scale economies as antitrust-relevant entry barriers unless these constitute costs uniquely borne by entrants.

63   Rachel Stephens, *IaaS Pricing Patterns and Trends 2022*, REDMONK (Apr. 11, 2023), https://redmonk.com/rstephens/2023/04/11/iaaspricing2022//; Rachel Stephens, *IaaS Pricing Patterns and Trends 2021*, REDMONK (Dec. 17, 2021), https://redmonk.com/rstephens/2021/12/17/iaas-pricing-2021/; Caroline Donnelly, *Public Cloud Competition Prompts 66% Drop in Prices since 2013, Research Reveals*, COMPUTERWEEKLY Jan. 12, 2016.

64   Barnett, *supra* note 7.

has reportedly entered into agreements to use the cloud-computing services of all three major providers (Microsoft, Amazon, and Google),[65] which deters opportunism by any single provider. Moreover, repeat-play cloud-services providers that seek to attract more users and retain existing users (which can shift usage across providers and in-house data storage) have incentives to accrue reputational goodwill by refraining from opportunism. This reputational feedback mechanism may be especially powerful since a significant portion of the potential cloud-services market remains unserviced[66] (including firms that have not yet shifted data storage to cloud services or firms that have only done so partially), which implies that the long-term penalty from provider opportunism toward existing users would likely exceed the short-term gains attributable to such actions.

4. Summary

677. The foregoing analysis can be summarized as follows. In the applications layer of the genAI ecosystem, there is little concern over risks to competition given low entry barriers, low levels of concentration, and the expectation that these characteristics will persist. In the LLM/models layer, OpenAI enjoys a first-mover advantage that appears to be vulnerable to challenge by a combination of integrated and VC-backed independent LLM providers. It is not clear whether the FM/LLM layer will converge toward a small number of leading providers, differentiate across various competitively relevant parameters, or develop mechanisms in the applications layer to facilitate interoperability across model providers.[67] Even if economies of scale did ultimately favor a small number of the most efficient providers in the FM/LLM layer (or in an antitrust-relevant segment of the FM/LLM layer), this market structure may efficiently reflect the economic and technological characteristics of model development. While concerns over concentration may seem more well-grounded in the compute market, these concerns are mitigated by intense competition among well-resourced providers over a market that is far from saturation and users' ability to diversify usage across cloud-services providers and on-premises data storage.

IV. Investments and Other Relationships in the GenAI Market

678. Regulators' concerns over the genAI market have focused on a sequence of transactions between incumbent platforms, some of which have developed FMs or LLMs, and smaller independent model providers. In this section, I describe some of these relationships and assess whether, at this early stage in the development of the genAI market, these relationships

---

65 Joseph Pisani, *Amazon Invests $2.5 Billion More in AI Startup Anthropic*, WALL ST. J. (Mar. 27, 2024, 7:35 PM ET), https://www.wsj.com/tech/ai/amazon-invests-2-75-billion-in-ai-startup-anthropic-87bb869e.

66 Barnett, *supra* note 7.

67 For related discussion, see CMA REPORT, *supra* note 8, at 46–48.

provide a basis for regulatory intervention on antitrust grounds, taking into account the error costs of unnecessary interventions that may distort or suppress efficient business practices.

1. Investments and Acquisitions in the GenAI Market

679. Transactions involving large digital platforms principally take the form of investments and, in certain cases, partnerships or acquisitions. Table 2 below sets forth a representative list of investments by large digital platforms in independent providers of LLMs and other FMs, applications, and applications hubs or tools in the genAI market. Note that, in all cases, the digital platform is part of a larger group of investors that principally includes VC and institutional investors.

Table 2. Major Investments by Large Technology Platforms in genAI Models, Applications, and Applications Hubs and Tools Developers (as of March 2024)[68]

Investee	Platform Investors	Total funding (all investors)	Principal product (investee)
AI21	Alphabet, Intel, Nvidia	$327M	FM/LLM
Anthropic	Alphabet, Amazon, Salesforce, SAP	$5B	FM/LLM
Cohere	Salesforce, SAP	$435M	FM/LLM
Databricks	Amazon, Microsoft, Salesforce	$4B	AI development tool
Essential AI	Alphabet, Nvidia	$64.8M	FM/LLM
Hugging Face	Amazon, Intel, Microsoft, Salesforce, Nvidia	$396M	AI development hub
Inflection AI	Microsoft, Nvidia	$1.5B	FM/LLM
Mistral AI	Microsoft, Salesforce	$553M	FM/LLM
OpenAI	Amazon, Microsoft	$12.3B	FM/LLM
RunwayML	Alphabet, Nvidia, Salesforce	$237M	Horizontal AI (video)

Note: "Total funding (all investors)" includes investments by non-platform investors (principally, VC and institutional funds).

680. As Table 2 shows, all independent LLM providers have secured investments from at least two large technology firms, in addition to VC and institutional investors (not shown in the Table). These investor-firms include: (1) four of the largest firms in the global digital ecosystem (Alphabet, Amazon, Microsoft and Nvidia), (2) Intel, which is seeking to compete in the production of specialized chips for the AI market, (3) Salesforce, a leading provider of cloud-enabled applications for the enterprise sector, and (4) SAP, a leading enterprise software firm. These

---

68 *See*, https://dealroom.co/ (last visited Mar. 24, 2024).

large technology firms tend to invest in multiple FM/LLM providers, suggesting a hedging strategy, and in each case (to my knowledge), the investing firm hold a minority ownership stake.[69] It should also be observed that all independent LLM providers have secured investments from more than one large technology platform (in addition to VC and institutional investors), which suggests that LLM providers are not usually reliant on any particular platform investor. Aside from investments, large technology platforms have also engaged in acquisitions of smaller firms in the AI market or entered into partnerships with smaller firms. According to CB Insights, Apple, Google, Meta, Microsoft, and Amazon collectively acquired a total of 98 firms in the AI market during 2010–2023.[70]

681. Large technology firms have also entered into certain relationships with LLM providers that involve a level of cooperation that extends beyond a conventional minority equity investment. These partnerships have elicited the greatest concern from competition regulators. According to the CMA, there were 90 partnerships in existence as of April 2024 between Google, Amazon, Microsoft, Meta, Apple, or Nvidia on the one hand, and model developers on the other hand.[71] Three of the most widely discussed transactions are described below.

682. *Microsoft/OpenAI.* In 2019, Microsoft committed to investing up to $10 billion in OpenAI in exchange for a 49 percent stake in the entity, which is ultimately controlled by a nonprofit entity.[72] Investors in OpenAI invest under an unusual capped profit structure that limits investors' maximum returns, although Microsoft is entitled to 75 percent of OpenAI's profits until it recoups the principal amount of its investment. Investors have no voting or other governance rights in the nonprofit entity that governs OpenAI. Additionally, OpenAI reportedly agreed to use Microsoft Azure as its exclusive cloud-computing services provider and Microsoft has deployed OpenAI's applications in its products and services.

683. *Microsoft/Inflection AI.* In 2024, Microsoft hired Inflection AI's co-founders and most of its employees and concurrently agreed to pay Inflection AI $650 million to license its FMs (on a non-exclusive basis),[73] effectively acquiring access to AI's principal human and innovation capital assets.

---

69 CMA Technical Update, *supra* note 14, at 30 (stating that, out of 90 partnerships between platforms and model developers entered into since 2019, most did not involve acquisition of a majority stake).

70 *The Big Tech Company Leading in AI Acquisitions*, CB Insights (Sept. 27, 2023), https://www.cbinsights.com/research/big-tech-ai-acquisitions/.

71 CMA Update, *supra* note 20, at 17.

72 This paragraph is informed by Tim Bradshaw et al., *How Microsoft's Multibillion Dollar Alliance with OpenAI Really Works*, Fin. Times, Dec. 15, 2023.

73 Shirin Ghaffary & Rachel Metz, *Microsoft to Pay Inflection AI $650 Million After Scooping Up Most of Staff*, Bloomberg (Mar. 22, 2024, 00:20 GMT), https://www.bloomberg.com/news/articles/2024-03-21/microsoft-to-pay-inflection-ai-650-million-after-scooping-up-most-of-staff?embedded-checkout=true.

684. *Amazon/Anthropic; Google/Anthropic.* In 2024, Amazon completed a $4 billion investment in Anthropic.[74] In connection with that investment, Amazon deployed Anthropic's services in its AWS cloud-computing service and Anthropic agreed to make use of Amazon's AWS cloud service and custom chips for training and operating AI models. This transaction follows a previous investment by Google of $2 billion in Anthropic, in connection with which Anthropic reportedly agreed to purchase a certain amount of Google Cloud services.

2. Do Investments and Acquisitions in the GenAI Market Pose a Risk to Competition?

685. Apart from cases of explicit collusion, antitrust analysis typically addresses practices that have ambiguous competitive implications in the absence of further inquiry. Any rigorous antitrust analysis of investment and acquisition practices in the genAI market must start from this agnostic position. Moreover, any such analysis must keep in mind that minority equity investments by large incumbents in startups – commonly known as "corporate venture capital" (CVC) – are a standard practice throughout the tech ecosystem and, so long as the investor does not have a controlling interest, this does not generally raise competition concerns. An approach that reflexively condemns CVC investments would have potentially far-reaching *anti*competitive effects by discouraging CVC investments and significantly constraining startups' access to risk capital, given the fact that, based on data for the period 2016–2020, CVC investments annually represented 20 to 25 percent of all reported VC investments.[75]

686. As noted, regulators have expressed concerns over investment and acquisition practices by large digital platforms in the genAI market.[76] These concerns appear to reflect the view that arrangements between incumbent platforms and model developers, such as the Microsoft/OpenAI, Amazon/Anthropic, and Microsoft/Inflection AI relationships, may constitute horizontal acquisitions (or a functional equivalent) that remove a potential or actual challenger in the LLM segment of the genAI market. These concerns confront two obstacles. First, it is not clear that these transactions can be construed as acquisitions. Second, it is not clear that these transactions remove a potential challenger from the LLM layer of the genAI market to an extent that significantly reduces competition, taking into account (as a rule-of-reason or similar balancing analysis requires) operational or other efficiencies attributable to a particular acquisition.

---

74 This paragraph is informed by Pisani, *supra* note 67.

75 BOSTON CONSULTING GROUP & INSEAD, CORPORATE VENTURE CAPITAL, https://www.insead.edu/sites/insead/files/assets/dept/centres/gpei/docs/corporate-venture-capital.pdf?vid=347 (citing data from CVC Insights: The 2020 Global VC Report).

76 *See supra* notes 6, 16–19.

A. Do Platform/LLM Provider Transactions Constitute Acquisitions?

687. The MS/Open AI transaction does not plausibly appear to constitute an acquisition since, by the reported terms of the transaction, it does not confer governance rights on Microsoft, which was not entitled to a seat on the OpenAI board until it received a non-voting observer seat in November 2023 (which it subsequently relinquished in July 2024).[77] The European Commission declined in April 2024 to open a formal investigation into the transaction because it found that the transaction did not constitute a change of control that fell within the scope of merger review.[78] Based on available information, the same diagnosis would seem to apply to the Amazon/Anthropic transaction, which involves a non-controlling equity investment. It may be argued that the Microsoft /Inflection AI transaction is functionally equivalent to an acquisition since, as part of the transaction, Microsoft hired the target's leadership and, through a license, acquired the right to use Inflection's LLM assets, although that characterization is challenged by the license's non-exclusivity clause.[79] Moreover, even if construed as the functional equivalent of an acquisition, it is not clear that the transaction materially reduces competition in the genAI market. Even without Inflection AI in the market as a separate entity, there remain other LLM providers, including vertically integrated providers such as Google and Microsoft, Meta's and xAI's open-source LLMs, independent VC-backed (and CVC-backed) providers such as AI21 Labs, Anthropic, Cohere, and Mistral AI, and many others as described previously.[80]

B. Do Platform/LLM Provider Transactions Pose Material Risks to Competition?

688. It might be objected that vertically integrated entities with existing cloud-computing capacities will have a formidable competitive advantage over independent LLM providers. If that is the case, then the FM/LLM market is prone to converge ultimately on only Google, Microsoft, and Amazon, the three leading cloud providers. This view relies on the assumption that any firm with an upstream leadership position in the cloud segment of the genAI supply chain would elect to foreclose firms situated at downstream

---

77　Hayden Field & Kif Leswing, *Microsoft Secures Nonvoting Board Seat at OpenAI,* CNBC (Nov. 30, 2023, 8:16 AM EST), https://www.cnbc.com/2023/11/29/microsoft-secures-non-voting-board-seat-at-openai-.html.; Ina Fried, *Microsoft gives up observer seat on OpenAI board,* AXIOS, July 10, 2024, https://www.axios.com/2024/07/10/microsoft-openai-board-seat-observer

78　Samuel Stolton, *Microsoft's $13 Billion OpenAI Deal to Avoid Formal EU Probe,* BLOOMBERG (Apr. 17, 2024, 17:24 BST), https://www.bloomberg.com/news/articles/2024-04-17/microsoft-s-13-billion-openai-deal-to-avoid-formal-eu-probe.

79　I do not purport to address whether the Microsoft/Inflection AI transaction is subject to merger review under the antitrust laws of the US or other jurisdictions.

80　*See supra* notes 32–33 and accompanying discussion.

points on the supply chain from accessing required cloud services.[81] In this scenario, an upstream firm would deny or limit LLM developers' access to cloud services, or deny or limit application developers' access to a required LLM model (including, in each case, charging licensing or other fees that are functionally equivalent to a denial or limitation of access).

689. Yet it is not clear that integrated upstream entities in the genAI market would have rational incentives to adopt these preclusive strategies toward downstream model and applications providers. Any such strategy requires that the integrated upstream entity expects that: (1) the immediate loss of revenues from declining to provide cloud services to a downstream entity (a model or applications developer) exceeds; (2) anticipated incremental revenues attributable to increased market power if those downstream entities are then compelled to exit from the relevant model or applications market. It is not obvious that this condition would typically be satisfied. There are two reasons. First, a cloud-services provider that acts opportunistically toward existing users may lose new users to competing cloud-services providers that do not engage in such actions. In the current cloud-services market, there is robust competition on price and other parameters among AWS, Microsoft Azure, and Google Cloud and, to a somewhat lesser extent, Oracle and IBM.[82] Second, if the models market fragments into differentiated segments that specialize vertically (based on industry) or horizontally (based on use), then downstream models and applications would constitute a complement, rather than a substitute, for the FMs or LLMs provided by integrated model providers. Upstream cloud-services providers would have an incentive to facilitate, not suppress, the growth of this market in complementary models and applications.

690. Consistent with these expectations, large tech platforms do not currently pursue preclusive strategies in the genAI market. Rather, each of the large cloud services provides a "models hub" that offers a selection of models for developers and other users, as well as fine-tuning services and other tools to facilitate model deployment.[83] This is not surprising: each model provider competes for adoption by developers and therefore seeks to facilitate developers' ability to choose among a broad variety of models, complemented by tools that accelerate the development process. While overlooked in much of the antitrust literature, technology platforms often compete for user adoption by providing costly assurances against acting opportunistically in the future[84] – in this case, by enabling developers to select from multiple model providers through the platform infrastructure. Amazon offers

---

81 See, e.g., COMMENT OF US FEDERAL TRADE COMMISSION TO THE US COPYRIGHT OFFICE, ARTIFICIAL INTELLIGENCE AND COPYRIGHT, DOCKET No. 2023-6, at 4 (Oct. 30, 2023); *see also supra* notes 4–5 and accompanying discussion.

82 Barnett, *supra* note 7.

83 The remainder of this paragraph relies on CMA TECHNICAL UPDATE, *supra* note 14, at 21–22, 25–26, information gathered from each platform's website, and Table 2 above.

84 Jonathan M. Barnett, *The Host's Dilemma: Strategic Forfeiture in Platform Markets*, 124 HARV. L. REV. 1861 (2011).

its Titan FM, but through its Bedrock development platform, also offers access to FMs and LLMs from other providers, including Anthropic, AI21 Labs, Cohere, Mistral, Meta, and Stability AI. Microsoft offers its Azure AI service, but extensively funds OpenAI and Mistral AI, and through the Azure Machine Learning Studio, offers access to LLMs provided by Open AI (on an exclusive basis), Meta, Nvidia, and Mistral, and open-source models from Hugging Face. Google offers its Gemini, PaLM 2, and other models, but funds A121, Anthropic, and Essential AI, and, through the Vertex AI development platform, offers access to more than 130 models, including (among others) models from Meta and Anthropic.

691. Assertions by some regulators and commentators that the genAI market demands preemptive intervention reflect a blanket assumption that digital markets inevitably converge toward entrenched monopoly outcomes, rather than relying on case-specific factual analysis that those outcomes *may* arise in a particular market. As I have shown elsewhere, the history of digital markets does not support a dominant tendency toward entrenched winner-take-all outcomes; rather, mature digital markets tend to exhibit high concentration levels, which are then periodically disrupted by an innovative entrant or technological development.[85] In the genAI ecosystem, it appears that the market may be evolving toward a structure composed of a mix of vertically integrated AI model providers and hybrid multi-firm arrangements involving various contractual relationships between LLM providers and larger platforms that supply the financial capital and physical plant to facilitate the training and deployment of new models. Figure 2 below provides a visual representation of these relationships among a selected group of LLM providers, platforms, and other investors.

Figure 2. Emergent Structure of the GenAI Ecosystem (Selected Relationships, as of March 2024)[86]

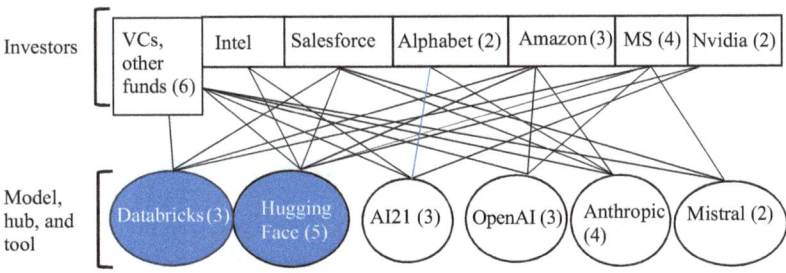

Note: Values in parentheses denote number of investees (for investors) or investors (for model, hub, and tool providers, excluding VCs and institutional investors). Solid lines denote investments. Circular shapes with blank background denote FM/LLM providers. Circular shapes with shaded background denote model hub or tool providers. Figure omits investment by SAP in Anthropic.

---

85 Barnett, *supra* note 7.

86 *See*, https://dealroom.co/ (last visited Mar. 24, 2024).

692. This market structure is hardly inconsistent with competitive conditions. At this stage in the ecosystem's development, there is robust competition between, on the one hand, stand-alone FM/LLMs maintained by some platforms, and, on the other hand, contractually governed alliances composed of large platforms, specialized LLM providers, and other firms that hold complementary assets. Both the alliance-based and vertically integrated forms of organization may support open-source or closed-source distribution strategies, including hybrid strategies in which only certain features of a model are available for licensing.

693. Those alliances (which include both horizontal and vertical components) are partially overlapping since (as described in Section IV.1) platforms and LLM providers have entered into various contractual arrangements involving capital investment, cloud-computing services, model development, and model deployment. These alliances compete with alternative business models being pursued by two leading and well-resourced platforms, Apple and Meta. In Apple's case, it entered into a partnership in June 2024 with OpenAI to integrate ChatGPT into Apple devices (but without making an investment in OpenAI), and has also invested extensively in internal development of various functionalities as part of its AI-enabled technology known as Apple Intelligence.[87] In Meta's case, it has adopted an open-source strategy in an apparent effort to seed adoption of its model in the genAI ecosystem, in a manner comparable to the open-source strategy pursued by Google when it released Android in 2008 and rapidly challenged seemingly entrenched incumbents such as Nokia and Blackberry in the mobile communications device market.

694. Regulators have focused on the alliances between incumbent platforms and LLM developers, which are interpreted presumptively as a preclusive strategy to "leverage" a platform's existing leadership position in a particular digital market to establish dominance in the genAI market.[88] Currently available evidence does not favor this presumption. These hybrid structures implement competing aggregations of the complementary inputs – chips, compute, data, financing, and talent – required to develop and promote adoption of differentiated "clusters" of FM/LLMs and genAI-enabled applications, which in turn efficiently fund technology development and promote technology adoption. At the same time, these alliances compete both with each other and stand-alone model providers, including platforms that are also members in one or more alliances, which in turn adopt a combination of open-source, closed-source, and hybrid distribution strategies. Moreover, there is little certainty that these

---

[87] Mark Gurman, *Apple Poised to Get OpenAI Board Observer Role as Part of AI Pact*, BLOOMBERG (July 2, 2024, 21:43 BST), https://www.bloomberg.com/news/articles/2024-07-02/apple-to-get-openai-board-observer-role-as-part-of-ai-agreement?.

[88] On regulators' views, see Tiffany Tsoi, Shona Ghosh & Bloomberg, *U.K.'s Antitrust Watchdog Sounds the Alarm over Big Tech AI Grip as It Uncovers an "Interconnected Web"*, FORTUNE (Apr. 12, 2024, 10:31 AM GMT), https://fortune.com/europe/2024/04/12/u-k-antitrust-watchdog-alarm-big-tech-ai-grip-interconnected-web-alphabet-openai-microsoft-mistral-cma/.

competing alliances and stand-alone providers will ultimately operate in a single genAI market for antitrust purposes; rather, the ecosystem may fragment into differentiated vertical or horizontal segments each populated by a different mix of alliances and stand-alone providers that offer a mix of closed-source, open-source, and hybrid models. It is unlikely that a single "super LLM" would then be in a position to exercise market power in any antitrust-relevant genAI market.

V. Conclusion

695. In a short period of time, competition regulators in the US, EU, UK, and other jurisdictions have shifted toward a preemptive approach that favors intervening in digital markets even in the absence of compelling evidence of actual or likely harm to competition. This approach rests on the view that digital markets are prone to converge on entrenched monopoly outcomes that are difficult to correct. This contribution assesses the wisdom of this approach in the generative AI ecosystem.

696. Based on currently available evidence, there is little compelling support for the propositions that: (1) the genAI ecosystem currently exhibits a sufficiently material risk to competition to warrant anticipatory intervention, or (2) that markets will be unable to foresee and mitigate any such risks through business-model and technological innovations. Concentration levels and entry barriers vary across potentially antitrust-relevant segments of the genAI ecosystem and widespread assumptions that critical inputs – specifically, chips, models, data, funding, and training services – are scarce as a general matter do not appear to be well-founded. Investments and other arrangements involving large tech platforms and model developers currently appear to be efficient aggregations of complementary assets to accelerate market release and technology adoption, which compete with each other and vertically integrated business models.

697. At this early stage in the genAI market's development, preemptive intervention on antitrust grounds appears to lack a reasonable justification and raises the risk of distorting and harming the market's future trajectory. To be clear, this is not a recommendation for a "do nothing" antitrust approach, especially given the propensity of digital markets to converge on winner-take-most outcomes that naturally raise anticompetitive concerns. At present, the most reasonable regulatory approach demands robust monitoring of various segments of the genAI ecosystem for potential anticompetitive practices that may inhibit entry or otherwise distort competition, while conditioning intervention on compelling evidence, rather than conjectural theories, of actual or likely antitrust harm.

Antimonopoly Tools for Regulating Artificial Intelligence

TEJAS N. NARECHANIA[*] AND GANESH SITARAMAN[**]
University of California, Berkeley
Vanderbilt University Law School, Nashville

Abstract

In this chapter, we make the case for an antimonopoly approach to governing artificial intelligence. We show that AI's industrial organization, rooted in AI's technological structure, suffers from market concentration at and across a number of layers. And we argue that an unregulated AI oligopoly has a range of undesirable consequences. In light of these conclusions, we show how antimonopoly market-shaping tools – the law of networks, platforms, and utilities, industrial policy, public options, and cooperative governance – can help facilitate competition. As policymakers debate governing AI early in its technological lifecycle, antimonopoly tools must be part of the conversation.

---

[*] Professor of Law, University of California, Berkeley, School of Law.
[**] New York Alumni Chancellor's Chair in Law, Vanderbilt University Law School.

I. Introduction

698. Since OpenAI released ChatGPT in November 2022, debates regarding regulating artificial intelligence (AI) among policymakers, technologists, and scholars have intensified. For all the interest in regulating AI, however, there has been comparatively little discussion of AI's industrial organization and market structure.[1] This is surprising because critical layers in the AI technology stack are already highly concentrated.[2] As in other areas, monopoly and oligopoly in these industries can not only distort markets, chill investment, and hamper innovation, but can also facilitate downstream harms to users, threaten national security and resilience, and help accumulate private power in relatively few hands.

699. In this chapter, we explain why and how policymakers ought to use antimonopoly tools to regulate the harms that come from concentration in the AI "technology stack." These include public utility tools, including structural separations, nondiscrimination requirements, and interoperability rules, public options for cloud computing and data resources, and consideration of competition when making industrial policy investments or engaging in procurement.

II. Understanding the AI Technology Stack

700. Understanding the AI technology stack is the foundation for identifying problems that arise from market power and concentration. Drawing on accounts from industry investors and analysts, AI's technology stack can be described in four basic layers: hardware, cloud computing, models, and applications.

   – *Hardware.* The hardware layer includes the production of microchips and processors – the horsepower behind AI's computations. This layer is extremely concentrated, with a few firms dominating important aspects of production. This is partly because as chip technologies have become more and more sophisticated, fewer firms are able to supply the needed technologies. Nvidia, which designs chips, has captured between 80 and 95 percent of the market of the GPU chip business used for AI.[3] Nvidia's chips are, in turn, manufactured

---

[1] To the extent there has been any discussion, it has largely focused on semiconductor manufacturing, and to a lesser extent, cloud infrastructure provision. But even then, these concerns have not generally been considered in the context of AI specifically. Among the rare works to examine competition aspects of AI are C. Scott Hemphill, *Disruptive Incumbents: Platform Competition in an Age of Machine Learning*, 119 COLUM. L. REV. 1973, 1975–81 (2019); Amba Kak & Sarah Myers West, *2023 Landscape: Confronting Technology Power*, AI NOW INST. (Apr. 11, 2023).

[2] *See infra*, Section II, for a discussion.

[3] Kif Leswing, *Meet the $10,000 Nvidia Chip Powering the Race for A.I.*, CNBC (Feb. 23, 2023), https://www.cnbc.com/2023/02/23/nvidias-a100-is-the-10000-chip-powering-the-race-for-ai-.html (noting that Nvidia has 95 percent market share for machine learning); *See also*, Zoe Corbyn & Ben Morris, *Nvidia: The Chip Maker that Became and AI Superpower*, BBC NEWS (May 30, 2023), https://www.bbc.co.uk/news/business-65675027; Asa Fitch & Jiyoung Sohn, *The Next Challengers Joining Nvidia in the AI Chip Revolution*, WALL ST. J. (July 11, 2023, 12:00 AM ET), https://www.wsj.com/tech/ai/the-next-challengers-joining-nvidia-

(i.e., fabricated) by Taiwan Semiconductor Manufacturing Corporation (TSMC),[4] which is far and away the world's dominant semiconductor manufacturer.[5] Only Samsung also fabricates the smallest, highest powered chips.[6] To make the smallest chips requires photolithography equipment, something only one company in the world, the Dutch firm ASML,[7] provides – and sells for between $150 and $200 million per machine.[8]

- *Cloud Computing.* The cloud computing layer consists of the computational infrastructure that is required to host the data, models, and applications that comprise AI's algorithmic outputs. This layer is also concentrated, with three firms (Amazon Web Services (AWS), Google Cloud Platform, and Microsoft Azure) dominating the market. It features several dynamics that tend toward concentration and make sustaining competition difficult, including extremely high capital costs and significant switching costs to move from one provider to another.

- *Models.* The model layer includes data, stored in unstructured "data lakes" or more structured "data warehouses", algorithmic models, which many think of as "AI", and modes of accessing these models, including model hubs (where developers can download and use publicly available models) and application programming interfaces (or APIs which allow developers to communicate with proprietary models that may not be publicly available). Some firms are fully integrated and offer all three products, which are then used to develop proprietary applications. Other firms only offer models and still others are more disaggregated, offering raw data or serving only as a model hub.

- *Applications.* Applications are the part of the sector with which the public interacts directly. When we enter a prompt into ChatGPT, for example, we use an application (ChatGPT). The application draws on all prior layers in the stack, it interacts with a model (GPT4), that model is stored in a cloud computing platform (Microsoft's Azure)

---

in-the-ai-chip-revolution-e0055485 (noting that Nvidia has more than 80 percent of the market); Wallace Witkowski, *Nvidia "Should Have At Least 90%" of AI Chip Market with AMD on its Heels,* MARKETWATCH (July 10, 2023, 12:50 PM ET), https://www.marketwatch.com/story/nvidia-should-have-at-least-90-of-ai-chip-market-with-amd-on-its-heels-13d00bff (projecting Nvidia will have 90 percent of the chip market).

[4] Arjun Kharpal, *Two of the World's Most Critical Chip Firms Rally After Nvidia's 26% Share Price Surge,* CNBC (May 25, 2023, 7:14 AM EDT), https://www.cnbc.com/2023/05/25/tsmc-asml-two-critical-chip-firms-rally-after-nvidias-earnings.html.

[5] TSMC's market share was estimated at 58.5 percent in 2022, with runner-up Samsung coming in at 15.8 percent. Peter Clarke, *TSMC, Globalfoundries Gained as Foundry Market Cooled,* EE NEWS (Mar. 13, 2023), https://www.eenewseurope.com/en/tsmc-globalfoundries-gained-as-foundry-market-cooled/.

[6] Saif M. Khan & Alexander Mann, *AI Chips: What They Are and Why They Matter,* CSET (Apr. 2020), https://cset.georgetown.edu/publication/ai-chips-what-they-are-and-why-they-matter/.

[7] *Id.*

[8] Kate Tarasov, *ASML is the Only Company Making the $200 Million Machines Needed to Print Every Advanced Microchip. Here's an Inside Look,* CNBC (Mar. 23, 2022, 1:00 PM EDT), https://www.cnbc.com/2022/03/23/inside-asml-the-company-advanced-chipmakers-use-for-euv-lithography.html.

and requires micro processing hardware (designed by Nvidia and fabricated by TSMC). While some firms in the application layer build their products on open-source – that is, free and publicly available – models, many others offer applications built upon existing proprietary foundation models. Yet others are vertically integrated, offering both the foundational model and applications built on them. Critically, though both types of firms compete in the applications market those that are not vertically integrated depend upon other firms for access to their models.

III. The Problems with an AI Oligopoly

701. Understanding the AI technology stack shows that significant portions of AI's industrial organization and market structure are likely to be, and already are, dominated by a small number of firms – and that these dominant firms are vertically integrating across the stack. This concentration – an AI oligopoly – is concerning for a variety of reasons, including abuses of power, national security and resilience risks, exacerbated economic inequality, and its effects on democracy.

1. Abuses of Power

702. Concentration across the AI stack creates opportunities and incentives for dominant firms to abuse their power, with consequences for competitors, would-be entrants, and the public. These abuses include, but are not limited to:

- *High Prices.* In hardware, dominant firms could demand monopoly and/or oligopoly prices for photolithography equipment, chip design, and chip manufacturing. Cloud computing firms might also charge monopoly or oligopoly prices. In the model layer, the high costs of obtaining good data and sufficient compute infrastructure constitute a steep barrier to entry, and foundation model providers might therefore be able to raise prices to downstream application developers for model access.

- *Self-Preferencing and Discriminatory Prices and Terms.* Monopoly or oligopoly firms at each layer in the stack may discriminate between downstream firms, offering better terms or prices to their vertically integrated businesses as opposed to competitors. Reports already indicate that Nvidia preferences some customers over others[9] and that TSMC prioritizes its relationship with Apple over other customers.[10] This poses substantial downstream risks to competition and innova-

---

9 Clay Pascal, *Nvidia H100 GPUs: Supply and Demand*, GPU UTILS (July 2023, updated Nov. 2023), https://gpus.llm-utils.org/nvidia-h100-gpus-supply-and-demand/.

10 Samuel Nyberg, *Apple Gets Special Treatment Amid Chip Shortage*, MACWORLD (Jun. 22, 2021, 1:01 PM PDT), https://www.macworld.com/article/677141/apple-gets-special-treatment-amid-chip-shortage.html.

tion. Model providers may also favor selected AI-based applications, including vertically integrated applications, through selectively exposed APIs.[11] In extreme cases, dominant firms could refuse to deal entirely with a customer who is also a competitor.[12] For example, Microsoft Azure could preference OpenAI, in which it has a financial stake, over other competitors, when offering access to its cloud computing infrastructure. Such self-preferencing has long precipitated competition concerns in network and platform industries.[13] Model providers might also favor their own applications over others by charging higher rates to third-party developers than their own in-house business lines, exclude some third-party applications from use of the model altogether, or they might prefer or advantage their own separate business lines. For example, if people ask Microsoft Bing what they should do this weekend, it might suggest playing the videogame Call of Duty – which Microsoft also owns.[14]

- *Lock-In Effects.* Cloud providers have taken steps to entrench their dominance by facilitating lock-in effects that raise the costs for consumers to switch providers through egress fees and multi-year contracts.[15] These effects exacerbate the already-high switching costs in compute, due to factors like personnel expertise in a particular platform.[16]

- *Copying.* In cloud, multiple firms have already complained that AWS has "strip mined" products developed atop AWS's platform, offering their own integrated version of the product, and thereby harming their company's value and future business.[17] Firms that copy applications from competitors and incorporate them into their own offerings deter competition to integrated offerings and even chill innovation. Venture capitalists describe this practice as creating a

---

[11] Philip J. Weiser, *The Internet, Innovation, and Intellectual Property Policy*, 103 COLUM. L. REV. 534, 579 (2003) ("In the government's antitrust case against Microsoft, for example, the government submitted evidence of a manager's statement that 'to control the APIs is to control the industry' and established that Microsoft's monopoly rested, in part, on its firm control of its APIs.").

[12] W. KIP VISCUSI, JOSEPH E. HARRINGTON, JR. & JOHN M. VERNON, ECONOMICS OF REGULATION AND ANTITRUST 82 (4th ed. 2005).

[13] *See, e.g.*, Tejas N. Narechania & Tim Wu, *Sender Side Transmission Rules for the Internet*, 66 Fed. Comm. L.J. 467, 470–78 (2015) (describing the history of competition concerns in network neutrality contexts); see also United States v. Microsoft, 253 F.3d 34 (D.C. Cir. 2001).

[14] We are indebted to Nick Garcia of Public Knowledge for this example.

[15] STAFF OF H. COMM. ON THE JUDICIARY, 117th CONG., INVESTIGATION OF COMPETITION IN DIGITAL MARKETS 98–99 (2020), https://www.govinfo.gov/content/pkg/CPRT-117HPRT47832/pdf/CPRT-117HPRT47832.pdf

[16] *Id.* at 96.

[17] Gerald Berk & Anna Lee Saxenian, *Rethinking Antitrust for the Cloud Era*, 51 POL. & SOC'Y 409, 415–17 (2023); see also Andrew Leonard, *Amazon Has Gone From Neutral Platform to Cutthroat Competitor, Say Open Source Developers*, ONEZERO (Apr. 24, 2019), https://onezero.medium.com/open-source-betrayed-industry-leaders-accuse-amazon-of-playing-a-rigged-game-with-aws-67177bc748b7; Jordan Novet, *Amazon Steps Up its Open-Source Game, and Elastic Stock Falls as a Result*, CNBC (Mar. 12, 2019, 1:42 PM EDT), https://www.cnbc.com/2019/03/12/aws-open-source-move-sends-elastic-stock-down.html; Jordan Novet, *Amazon's Cloud Business is Competing with its Customers*, CNBC (Nov. 30, 2018, 8:04 AM EST), https://www.cnbc.com/2018/11/30/aws-is-competing-with-its-customers.html.

"kill zone," wherein the likelihood of copying or acquisition by a dominant firm discourages investment in innovative companies and products.[18]

2. National Security and Resilience

703. Concentration at critical points in the AI technology stack raises significant concerns for national security and resilience. With very few chip companies, the possibility that one foundry could be shut down due to a pandemic, weather event, war, or other emergency is significant.[19] Such an event would cause supply chain challenges for both military and non-military critical infrastructure. Concentration in cloud computing raises risks too: an oligopoly of cloud providers, integrated up and down the AI stack and without interoperability between them, gives rise to substantial software supply-chain concerns. If a cloud provider is attacked in a cyberattack, or if a cloud provider's warehouse is affected by a severe weather event, or even if an employee makes a simple mistake, dozens of AI applications – and the operations, services, and websites that depend on them – could shut down for hours, days, or longer.[20] Just as concerning is that faulty foundation models, if offered by only one firm, can lead to widespread error that could be catastrophic in emergency situations.

3. Economic Inequality

704. Like concentrated power in other industries, an AI oligopoly is likely to further economic inequality. Concentration means that a small number of firms will capture the vast majority of the sector's profits, "showering wealth on the elite executives and engineers lucky enough to get in on the action."[21] Amplifying the concentration of income and wealth both arrests economic mobility[22] and is undesirable for those who seek a more egalitarian society. Moreover, concentration in AI can increase global inequality, as the dominant firms, located in a small number of industrialized and technologized countries, extract value from data that is harvested from other economies.[23]

---

18 Sai Krishna Kamepalli, Raghuram Rajan & Luigi Zingales, *Kill Zone*, (NAT'L BUREAU OF ECON. RSCH., Working Paper No. 27146, May 2020), https://www.nber.org/papers/w27146.

19 *See, e.g.*, KAREN M. SUTTER, JOHN F. SARGENT, JR. & MANPREET SINGH, SEMICONDUCTORS AND THE CHIPS ACT: THE GLOBAL CONTEXT, CONG. RES. SERV. 2 (May 18, 2023), https://crsreports.congress.gov/product/pdf/R/R47558.

20 *See, e.g.*, Nick Merrill & Tejas N. Narechania, *Inside the Internet*, DUKE L.J. (forthcoming 2023).

21 KAI-FU LEE, AI SUPERPOWERS: CHINA, SILICON VALLEY, AND THE NEW WORLD ORDER 171 (Mariner Books ed., 2021) (2018).

22 See Jared Bernstein & Ben Spielberg, *Inequality Matters*, THE ATLANTIC (June 5, 2015), https://www.theatlantic.com/business/archive/2015/06/what-matters-inequality-or-opportuniy/393272/.

23 Steven Weber & Gabriel Nicholas, *Data, Rivalry and Government Power: Machine Learning Is Changing Everything*, 14 GLOB. ASIA 23 (2019).

4. Democracy

705. An AI oligopoly can also contribute to democratic erosion. Concentration in AI may give a relatively small number of companies an outsized influence over the information ecosystem, complementing their outsized political influence through lobbying and other forms of political influence. Economic power also often translates into political power, as demonstrated by a voluminous literature in political science showing the extensive influence of the wealthy and interest groups on American politics.[24]

IV. Antimonopoly Public Utility Tools

706. Aspects of the AI sector share features with public utilities: they are essential inputs into downstream activities, are means to an end rather than ends in themselves, and feature high capital costs, network effects, and economies of scale. Indeed, machine learning itself, as one of us has shown elsewhere, has the characteristics of a natural monopoly, even under narrow definitions.[25]

707. Regulatory tools from the law of networks, platforms, and utilities (NPU) have long been applied to industries that feature monopoly or oligopoly characteristics.[26] These tools can be helpful for scaling enterprises, ensuring continuity of service, preventing monopoly and oligopoly abuses, avoiding destructive competition (i.e., races to the bottom), ensuring widespread access, promoting commercial development, and sustaining democracy.[27] We discuss here three of the most important tools from utilities law: structural separations, nondiscrimination requirements, and interoperability rules.

708. These regulations operate primarily *ex ante*, that is, by structuring the market to prevent likely, foreseeable harms *before* they arise. They therefore differ from traditional antitrust enforcement, which requires that harms take place and then for litigants to bring lawsuits on a case-by-case basis to challenge those harms. In structuring a more competitive and less abusive market, public utility tools ensure that an industry develops in a way that helps foster innovation and competition.[28]

---

24 Martin Gilens, Affluence and Influence: Economic Inequality and Political Power in America 97–123 (2012); Larry M. Bartels, Unequal Democracy 253–54 (2008); Martin Gilens & Benjamin I. Page, *Testing Theories of American Politics: Elites, Interest Groups, and Average Citizens*, 12 Persp. on Pol. 564, 573 (2014); Lee Drutman, Business of America is Lobbying (2015); Alyssa Katz, The Influence Machine: The U.S. Chamber of Commerce and the Corporate Capture of American Life (2015); Kay Lehman Schlozman, Sidney Verba & Henry E. Brady, The Unheavenly Chorus 404–11 (2012).

25 Tejas N. Narechania, *Machine Learning as Natural Monopoly*, 107 Iowa L. Rev. 1543 (2022).

26 Morgan Ricks, Ganesh Sitaraman, Shelley Welton & Lev Menand, Networks, Platforms, and Utilities: Law and Policy 8–10 (2022).

27 *Id.* at 11–21.

28 More precisely, these rules are designed to operate in contexts where policymakers determine that the benefits of the regulatory approach outweigh their costs, accounting for the possibilities that *ex post* enforcement will be too little, too late, as well as for the possibility that a regulatory approach may miss its mark. That is, accounting for the benefits and the costs, including error costs, they deem an *ex ante* regulatory approach to be superior. Moreover, if policymakers err, they remain democratically accountable. But errors that favor the market, and that have the effect of entrenching economic power – can be more difficult to correct. See, e.g., Narechania, *supra* note 25.

1. Structural Separations

709. Structural separations "limit the lines of business in which a firm can engage."[29] The central benefit of structural separations is that they prevent a business from self-preferencing or leveraging their power from one business-line into another, which, as noted, presents special competition concerns in networked and platform industries (as in the cloud computing and model layers). In addition to preventing conflicts of interest and leveraging profits to gain competitive advantage in another line of business, structural separations also limit the concentration of economic power and promote a diverse business ecosystem of users of the platform.[30] Perhaps most importantly, they can be more administrable than other policies, such as nondiscrimination rules (discussed in the next section). If a company is involved in the prohibited business line, it violates the rule. This is a far simpler approach than one that permits commingling and requires monitoring behavior.

710. With respect to AI, there are number of places where structural separations could be useful.[31] Perhaps most notably, structurally separating the cloud layer from higher layers in the stack could address the wide range of market dominance problems identified above. It would treat cloud computing platforms as utility providers of a service (namely, computational capacity) that is open for all kinds of uses and ensure that those providers cannot prioritize their own downstream business lines over their competitors'. Separation could also spur cloud providers to innovate on their cloud offerings, rather than on innovating through vertical integration.[32] This would, in turn, also facilitate innovation in downstream markets where cloud users could develop a range of products and services.

2. Nondiscrimination, Open Access, and Rate Requirements

711. One alternative to structural separation requirements are nondiscrimination and equal access rules, sometimes coupled with rate regulation.[33] Nondiscrimination rules allow a firm to operate two or more vertically-linked business lines, but require the firm to treat downstream businesses neutrally – including its own vertically-integrated business lines.[34] Equal pricing rules are an essential corollary to nondiscriminatory rules because firms could charge prohibitive prices as a workaround to

---

29 *Id.* at 28.

30 For a discussion of this example and others, including a theory of structural separations, see Lina M. Khan, *The Separation of Platforms and Commerce*, 119 COLUM. L. REV. 973 (2019).

31 *Cf.* William P. Rogerson & Howard Shelanski, *Antitrust Enforcement, Regulation, and Digital Platforms*, 168 U. PA. L. REV. 1911, 1934–36 (2020).

32 If regulators were to determine that integration of chips and cloud is desirable for effective service provision, then separating chips/cloud from higher levels in the stack would encourage innovation across both layers together – while preserving the innovative potential of competition further up the stack.

33 RICKS ET AL., *supra* note 26, at 24–26.

34 *Id.* at 24, 26, 29.

evade their open access obligation.[35] In some areas, regulators have also directly set the prices firms can charge. Rate setting "is usually directed toward preventing NPU enterprises from lowering output and raising prices," while simultaneously ensuring firms earn a reasonable return on invested capital.[36]

712. In the AI context, nondiscrimination and equal access rules could be adopted at multiple places in the stack. At the hardware level, given the scarcity of chips, fabricators and designers could be required to serve customers equally – at least until chip fabrication becomes more widely available. At the cloud level, cloud providers should be required to treat all downstream businesses in a nondiscriminatory fashion, be open to all comers, and offer transparent, uniform, publicly available prices. Open source and non-open source, but commercially available data warehouses and lakes could also be subject to nondiscrimination and equal access rules. This would enable many model developers to use the data to develop and train new models. Foundation models and APIs could also be subject to rules ensuring that app developers have reliable, nondiscriminatory access to these resources.

3. Interoperability Rules

713. Interoperability rules lower barriers to entry and thus stimulate competition by "allowing new competitors to share in existing investments" and "imposing sharing requirements on market participants."[37] In the telecommunications context, for example, rules that required a telephone provider to transfer a user's phone number to a competing provider (and thus require that the providers work together on an interoperable number portability system) facilitate competition by reducing switching costs for users. Those rules targeted a notable lock-in effect: it is quite cumbersome to let all your contacts know you have a new phone number (in part because your contacts cannot call you to ask for the new number).

714. Such requirements could be applied to parts of the AI stack, too. Policymakers might consider rules that improve interoperability among cloud platforms, easing transitions from one provider's system to another. As different providers of cloud computing services specialize – moving away from offering a pure commodity "compute" resource to more bespoke computing resources and incorporating specialized applications (or utilizing specialized hardware) – some applications developers have found it difficult to take advantage of specializations across different providers. A developer might wish, for example, to train a model on one cloud provider – but use a different one for inferential applications.

---

35 *Id.* at 24.

36 *Id.* at 25.

37 Narechania, *supra* note 25, at 1555.

Interoperability could facilitate that, potentially yielding better outcomes for participants in the downstream model and application layers, and ultimately for consumers.

715. Another type of interoperability rule is to mandate data sharing and shared learning through federated learning. Federated learning is a technical "approach to machine learning where a shared global model is trained across many participating clients that keep their training data locally."[38] Rules that require a federated learning approach among competitors may be attractive to policymakers seeking to induce competition while ensuring that no one application, vertically integrated with the underlying model, uniquely benefits from improvements made through continuous or reinforcement learning.[39] Instead, the model's improvements are derived from all the applications that use it – and are shared among all of them, too. Such forms of AI development may help to undermine the consolidation-driving network effects of the data sublayer.

4. Entry Restrictions and Licensing Requirements – The US

716. Congress has often established entry restrictions and licensing requirements for firms or individuals operating in many sectors of the economy. Such rules limit entry into a sector to firms that have registered with an appropriate regulator or otherwise have approval from the government (often in the form of a license or certificate).[40] These provisions are usually justified on one (or more) of three different grounds. First, entry restrictions can ensure safety and reliability. By placing conditions on entry into a sector, regulators can ensure that firms will operate safely and effectively. Airline pilots (and airlines themselves), for example, must be licensed. Likewise, nuclear power plants are licensed, in part, to ensure safe operation. Second, in some markets (particularly those typically characterized as natural monopolies or oligopolies) competition can lead to waste and ultimately deter capital investment.[41] In the railroad industry, for example, firms competed vigorously to build railroad tracks at a high cost – but fierce competition over price sent them into bankruptcy or merger. The result was wasted expense, abandoned rail lines, and eventual consolidation. Entry restriction can prevent these downsides, creating a stable environment for capital investment. And third, in sectors where universal service – i.e., ensuring that everyone can access the regulated service – is a

---

38 See, *TensorFlow Federated: Machine Learning on Decentralized Data*, Tensorflow, https://www.tensorflow.org/federated (describing one approach to federated learning) (last visited July 18, 2024).

39 This approach is distinct from the one adopted in Europe via Gaia-X, which predominantly regards federated data storage, for the purposes of complying data localization requirements (e.g., rules that certain personal data be housed in certain locales). By contrast, federated learning can describe an interoperable approach to training, in which multiple applications or users train a single, shared foundation model through an interoperable standard.

40 Ricks et al., *supra* note 26, at 29–30.

41 *Id.*

critical policy goal, regulators will often limit entry to the market.[42] This is because open competition often undermines universal service policy goals. Some services, like energy provision, have costs that vary across geographies: urban centers are typically cheaper to serve (and hence are more profitable), while rural areas can be more expensive. Without entry restrictions and related regulations, providers will tend to compete to serve the cheaper and more profitable customers (with those customers enjoying the benefits of competition), while neglecting the more expensive customer base. But entry restrictions coupled with duty-to-serve rules can ensure that everyone has access to the regulated service, often at regulated rates (typically regulated, in part, by averaging the high-cost customers with the low-cost ones).

717. Such requirements might be applied to the AI technology stack at various layers. First, entry restrictions might be deployed to ensure that certain foundation models and their associated applications are effective, and do not pose substantial risks to health and safety, or risks of bias. Indeed, the FDA's process for approving medical systems that incorporate AI resembles this approach. Similarly, licensing rules could oblige cloud providers to "know their customers," as in banking law, and ensure that entities in the model layer have checks in place to ensure non-discriminatory access, fair pricing, and safety. Applications could also be required to register with the model or cloud they use, to make it easier to identify and address dangerous or problematic behavior on a post hoc basis. Likewise, entry restrictions might help to address concerns about costly and wasteful investment – and the tendencies towards consolidation – in the model layer, which are characterized by high fixed costs, scale economies, and network effects.

V. A Public Option for AI

718. Public options are publicly provided goods or services that coexist with private market options and are offered at a set price.[43] While the term may be most familiar from debates over health care policy, public options are common in the United States. Public schools coexist with private schools. Public swimming pools with private ones. We have parcel post from the Postal Service, public golf courses, public libraries – the list goes on.

719. Public options come with a range of benefits. First, they can help ensure competition, as the public option disciplines private monopolists or oligopolists that might increase prices or reduce service quality.[44] Competition from private businesses, in return, also ensures that the public

---

42   *Id.*

43   GANESH SITARAMAN & ANNE ALSTOTT, THE PUBLIC OPTION 27 (2019).

44   *Id.* at 38–40.

option provides high quality service.[45] Relatedly, a public option adds to the number of providers of a good or service thereby diversifying and strengthening supply chain resilience and reliability. Second, public options expand access to goods or services that might be unaffordable or scarce in the private market. A public swimming pool or playground offers a convenient place for kids to play without residents needing to buy an expensive pool or swing-set for their backyard.

720. As we have seen, in the AI context, both of these features are important. The AI technology stack is already dominated by a small number of big technology companies, meaning that competition is limited within and across multiple layers. Moreover, the high cost and high demand for semiconductors and cloud infrastructure has led to scarcity at both of these layers – meaning that some kind of prioritization is likely. Firms might prioritize vertically integrated businesses and the most profitable activities over competitors or uses that serve the public.

1. A Public Option for Cloud Infrastructure

721. A public option for cloud infrastructure could serve as a helpful complement or alternative to tools from networks, platforms, and utilities (NPU) law – including structural separations or nondiscrimination requirements, and interoperability rules.[46] Because of high capital costs, network effects, and concerns from vertical integration, a public option for cloud could provide the cloud services that developers and end-users need – but without relying on oligopoly providers such as AWS, Microsoft Azure, and Google Cloud Platform. The public option could also ensure that cloud infrastructure is available at an affordable price to researchers and other users who might have different goals than private firms. A public option is also not unworkable; the United States has a long history of publicly run supercomputers, and Japan is in the process of building a public option supercomputer, which will make cloud services available to companies focusing on AI.[47]

722. While there are some existing proposals for public access to AI resources, it is unclear if these proposals are true public options or whether they will further entrench oligopoly firms at different layers in the AI stack. For example, the National Science Foundation's proposal to offer a National AI Research Resource (NAIRR) seeks to "democratize access to AI resources" and therefore "must primarily be sustained through

---

45 E.S. Savas, *An Empirical Study of Competition in Municipal Service Delivery*, 37 PUB. ADMIN. REV. 717 (1977); E.S. Savas, *Intracity Competition between Public and Private Service Delivery*, 41 PUB. ADMIN. REV. 46 (1981).

46 For a discussion, see our companion paper, Tejas Narechania & Ganesh Sitaraman, *An Antimonopoly Approach to Governing Artificial Intelligence*, YALE L. & POL'Y REV. (forthcoming 2025), https://papers.ssrn.com/sol3/papers.cfm?abstract_id=4597080.

47 Nikkei Staff Writers, *Japan's METI to Build New Supercomputer to Help Develop AI at Home*, NIKKEI ASIA (July 24, 2023, 15:02 PM JST), https://asia.nikkei.com/Business/Technology/Japan-s-METI-to-build-new-supercomputer-to-help-develop-AI-at-home.

Federal investment."⁴⁸ However, the NSF's proposal is unclear on whether NAIRR will be a public option or will simply contract with big technology companies for AI services. It suggests that NAIRR provide a mix of computational resources, including "commercial cloud" as an option.⁴⁹ It also recommends that NAIRR "include at least one large-scale machine-learning supercomputer," but then is unclear whether this would be a publicly run resource.⁵⁰ Recently introduced legislation to create a NAIRR states that the entity would offer "a mix of computational resources," including "on-premises, cloud-based, hybrid, and emergent resources," "public cloud providers providing access to popular computational and storage services," opensource software, and APIs.⁵¹ This structure appears to require some amount of non-oligopoly cloud provision, as the provision for an on-premises, cloud-based system is separate from the one that describes public cloud providers. But the draft legislation could be interpreted to lead only to contracts with existing cloud providers. Such a contract would further entrench their oligopoly, rather than increase competition. It might also place public access at risk, if prices increase or if the cloud service deprioritizes public uses. The NAIRR, if created and funded, should ensure there is a true public option, rather than merely a government contract for researchers to purchase compute and other resources from the biggest cloud providers.

2. Public Data Resources

723. Data are also foundational for AI applications and area resource that depend on extraordinary scale. Without considerable data – which must also be cleaned and processed – machine learning is not possible. If leading data sources are all proprietary, then the companies that control them could raise prices on downstream businesses or researchers who rely on that data for their models or applications or even deny them access, perhaps if they seek to develop a competitive product.

724. A public option for data could therefore "provide a pathway for start-ups and public-sector organizations to develop abilities and products."⁵² A public data resource could work in a few different ways. First, the government could ensure that when it makes public data available (e.g., weather

---

48 NATIONAL ARTIFICIAL INTELLIGENCE RESEARCH RESOURCE TASK FORCE, STRENGTHENING AND DEMOCRATIZING THE U.S. ARTIFICIAL INTELLIGENCE INNOVATION ECOSYSTEM: AN IMPLEMENTATION PLAN FOR A NATIONAL ARTIFICIAL INTELLIGENCE RESEARCH RESOURCE 22 (Jan. 2023).

49 *Id.* at 31.

50 *Id.* ("This could be made available by leveraging an existing supercomputer or newly procured through a competitive bid process managed by the Operating Entity in consultation with the Steering Committee and relevant advisory boards.")

51 CREATE AI Act of 2023, S. 2714, 118th Cong. § 5603(b) (2023). *See* Press Release, Representative Anna G. Eshoo, AI Caucus Leaders Introduce Bipartisan Bill to Expand Access to AI Research, (July 28, 2023), https://eshoo.house.gov/media/press-releases/ai-caucus-leaders-introduce-bipartisan-bill-expand-access-ai-research.

52 Ben Gansky, Michael Martin & Ganesh Sitaraman, *Artificial Intelligence is Too Important to Leave to Google and Facebook Alone*, N.Y. TIMES (Nov. 10, 2019).

data) that the data are not privatized. Public data would thus remain in public hands – and the government would develop public data warehouses that researchers and qualified businesses could use. Second, the government could develop a data resource akin to those that private companies have developed and that would compete with it. This would likely be difficult and costly, but it would offer another avenue for competition and access. Notably, the proposed NAIRR legislation also includes provisions for data access,[53] and the federal government already has several other initiatives under consideration that are aimed at releasing public datasets to support model development.[54]

725. With public options at these layers, technologists would be able to develop their own foundation models or applications, without relying on the AI oligopoly for cloud services or the underlying data. This would both improve competition with those firms and ensure that public-spirited uses could be pursued.

VI. Industrial Policy, Procurement and Competition

726. Policymakers should also consider competition policy when engaging in industrial policy spending and in their procurement decisions. One of the central questions for industrial policy in the AI sector is whether investment decisions will entrench dominant players or facilitate competition. Subsidies, loan guarantees, tax advantages, and procurement decisions directed toward dominant players may keep them in positions of leadership. In areas that have a tendency toward consolidation – due to economies of scale, network effects, high capital costs, and other factors – such policies could further extend their lead. But industrial policies could also be targeted at new, smaller, and innovative actors, in which case they can facilitate competition, rather than entrench market power.[55]

1. Industrial Policy and Semiconductors

727. In the hardware layer, scarcity and supply chain vulnerability are paramount concerns. To address these problems, the United States has already taken steps to incentivize the development of chip manufacturing within the US. The bipartisan Chips and Science Act of 2022[56] established a range of incentives to spur domestic production of cutting-edge chips. The Act committed $52.7 billion to the Departments of Commerce and Defense and the National Science Foundation to support US development

---

53   CREATE AI Act of 2023, S. 2714, 118th Cong. (2023).

54   *See, e.g.*, *AI Researchers Portal: Data Resources*, AI.GOV, https://ai.gov/research/ / (last visited July 18, 2023).

55   *See* Philippe Aghion, Jing Cai, Mathias Dewatripont, Luosha Du, Ann Harrison & Patrick Legros, *Industrial Policy and Competition*, 7 AM. ECON. J. MACROECON.1–32 (2015).

56   CHIPS Act of 2022, Pub. L. No. 117-167, Div. A, § 102, 136 Stat. 1372 (2022).

of semiconductor programs.⁵⁷ The Commerce Department's Chips for America program seeks to use federal funds to crowd-in private sector investment in order to develop at least two large-scale clusters for fabrication of chips.⁵⁸

728. As the Commerce Department develops its program, it should carefully assess whether federal funding will entrench power or increase competition. Government officials coordinating industrial policy efforts should consider market diversification and competition as a critical element in evaluating candidates for federal grants.⁵⁹

2. Procurement Decisions

729. Federal departments and agencies are also likely to make procurement decisions with private companies and consulting firms. These decisions could range from outsourcing development of AI applications to contracts with infrastructure providers. Here too, federal officials should consider the extent to which they can promote competition, rather than entrench dominance, when making these contracts. In particular, they may have flexibility to draft contracts in a way that allows them to build in pro-competition provisions. As just one example, departments and agencies could, where possible, require that public data be quarantined from privately held data, so that the dominant platforms do not leverage their public data resources to support their dominance in the private market. In addition, the AI Guide for Government, published by the AI Center of Excellence in the General Services Administration, should be updated to include best practices for federal agencies when considering competition as a factor in AI procurement decisions.⁶⁰

VII. Conclusion

730. An antimonopoly approach can help mitigate the harms of extreme concentration in the AI sector. By introducing public utility tools such as structural separations, nondiscrimination requirements, and interoperability rules public options, and pro-competition industrial and procurement policies, policymakers can shape how AI is developed, deployed, and used – and in the process, protect innovation, competition, national security, and the American people.

---

57 *Id.*

58 CHIPS FOR AMERICA, VISION FOR SUCCESS: COMMERCIAL FABRICATION FACILITIES 1, NAT'L INST. STANDARDS & TECH. (Feb. 28, 2023), https://www.nist.gov/system/files/documents/2023/02/28/Vision_for_Success-Commercial_Fabrication_Facilities.pdf.

59 Note that the Chips office does appear to want *two* clusters in the United States but does not commit to those being run by two independent firms. *See Id.*

60 *AI Guide for Government*, COE.GSA.GOV, https://coe.gsa.gov/coe/ai-guide-for-government/introduction/index.html (last visited Oct. 2, 2023).

Competition Policy after the Coming Wave of General Purpose Technologies

DANIEL A. CRANE[*]
University of Michigan

Abstract

A coming wave of general-purpose technologies centered on AI will fundamentally reshape the economic order, and disrupt the assumptions on which competition policy is built. While it may take years for this wave to fully arrive, it is not too early to begin planning for competition policy after the wave arrives and significantly reduces the possibility for competitive markets. Rather than mandating competition, post-wave competition policy is likely to involve direct regulation of AI to mimic the outcomes that competitive markets are supposed to obtain.

A coming technological wave, consisting of a variety of overlapping and mutually reinforcing general-purpose technologies including artificial intelligence (AI), robotics, quantum computing, synthetic biology, energy expansion, and nanotechnology,[1] is likely to fundamentally reshape the economic order, with profound consequences for competition policy. Exactly how and when remains the subject of speculation. This chapter considers the apparent paths of emerging technologies and attempts to work out their implications for competition law. The predictions are radical, but so are the technologies.

---

[*] Richard W. Pogue Professor of Law, University of Michigan. This chapter is adapted from my article *Antitrust After the Coming Wave*, (forthcoming NYU L. REV. 2024), with permission of the NYU L. REV.

[1] *See generally* MUSTAFA SULEYMAN, THE COMING WAVE: TECHNOLOGY, POWER, AND THE 21ST CENTURY'S GREATEST DILEMMA 92–102 (2023).

Competition Policy after the Coming Wave of General Purpose Technologies

Competition policy is premised on four assumptions or pillars that will likely buckle in the coming wave. These are: (1) competitive markets provide the best measure of information about consumer preference and producer efficiency, (2) competitive markets create incentive structures necessary to the maximization of human welfare, (3) consolidation of economic power in very large units of production or distribution is not inevitable, and (4) legal principles and their enforcement can meaningfully police anticompetitive conduct. All four assumptions are vulnerable to the coming technological wave. First, the information producing or discovery function of competitive markets will become subject to the challenge that AI and robotic systems may become far more adept at anticipating and fulfilling human wants and needs than market price signals. This will occur not only because AI and robotics systems will be able to forecast changing demand far more efficiently than markets, but also because AI systems augmented by synthetic biology and other technologies will restructure, reshape, and indeed begin to program the attributes of human demand. At the limit, consumer demand will no longer be exogenous to the system of production and distribution; it will be created by that system. Second, the competition paradigm on which antitrust law is based assumes that individual motivation is too multifaceted, ambiguous, unknowable, and variable to regulate directly, but that a competitive spur serves to direct human incentives toward beneficial outcomes by regulating competitive behavior. What changes with AI, is that operational commands and objective functions must be explicitly stated and coded, but processing steps are a black box. Thus, with humans, motivations are opaque, but processing steps tend to be clear. With AI and robots, motivations (directions) are clear, and processing steps are opaque. The prospect that key productive assets will have clearly knowable incentives, but unknowable processes will flip the entire antitrust paradigm on its head. Third, antitrust's competition paradigm assumes that it is possible for an economy to operate with multiple independent and rivalrous units. Already, the digital revolution and its associated scale economies and network effects have shifted scale dramatically toward the large and monopolistic. AI, robotic production, and the continuing shift in economic value from atoms to bits will likely turbocharge these effects, with an inevitable and perhaps unstoppable tendency toward monopolistic megafirms. Finally, antitrust's technologies for controlling anticompetitive behavior will run into a wall as far more powerful technologies for engaging in anticompetitive behavior emerge. While there may still remain some role for differentiation, specialization, and variety in the economy, the broad sweep of the coming wave of technologies will be to move toward extreme concentrations of economic power in a few hands.

How should policy planners be preparing for the coming wave? If the outcomes suggested by the technological trends materialize – even if they *half* materialize – attempting the prescribed continuation of a conventional competitive order is a fool's errand. Rather, any effective solutions will require moving beyond competition as the organizing economic concept and considering a different regulatory approach to engineering production and distribution to maximize human welfare, distribute power, and achieve equitable outcomes – an approach that harnesses the power of the coming wave technologies as both subject and means of regulation.

I. How the Four Pillars of Competition Policy Will Buckle in the Coming Wave

731. Like any legal or economic system, competition law has an underlying conceptual structure consisting of purposes and operations.[2] We can call these the "pillars of competition policy." The first two pillars – information and incentives – concern the justifications for committing our system of production and distribution to competitive markets as opposed to some other form of economic organization such as central planning or monopoly franchises. In short, competitive markets are thought to have significant performance advantages over alternative economic modalities because they are uniquely capable of generating necessary information about consumer demand and producer efficiency and because they give producers incentives to maximize consumer wellbeing. The second two pillars – structure and conduct – relate to the legal technologies available to the state to ensure that markets behave competitively. Anticompetitive structures – concentrated markets – are avoided through the implementation of merger policy and other branches of competition law concerned with market concentration. Anticompetitive conduct, such as cartels or abuses of dominance, is also prescribed. Putting it all together, competition law rests on the belief that competitive markets outperform other systems because they are better at solving information and incentive problems and that the law can enforce competition by policing both structure and conduct.[3]

732. The coming wave will not merely force companies to compete differently, as it is already doing (for example, by increasingly investing in AI-driven technologies).[4] It will also undermine the very reasons that they are required to compete and the state's techniques to ensure that they do. As the wave rolls, the four pillars on which the competition law system is predicated – information, incentives, scale and scope, and conduct control – will buckle.

---

2   I refer here generically to "competition law," glossing over the many important differences between the competition laws of different countries. At their core, however, most competition law systems – at least those cast in the model of the US or EU – share these commonalities.

3   Although the structuralist model of competition law – defined by the Structure-Conduct-Performance paradigm of the 1950s Harvard school – has largely been replaced by subsequent economic thinking, competition law continues to focus on anticompetitive market structures using such techniques as market concentration indices to predict a market's susceptibility to collusion or unilateral dominance.

4   *See* Marco Iansiti & Karim R. Lakhani, Competing in the Age of AI: Strategy and Leadership When Algorithms and Networks Run the World (2020).

Competition Policy after the Coming Wave
of General Purpose Technologies

1. Information

A. *Consumer Preference*

733. Friedrich Hayek famously argued that a single regulator or central planner never has sufficient information about individual wants and needs to be able to optimize economic allocation across all the complex and varied preferences of the individuals who comprise society.[5] But Hayek had in view a human mind, not a machine with a "God's-eye" view[6] of individual preferences. The coming technological wave will empower producers to anticipate and optimize consumer preferences without the need for price signals. Firms will have tremendously more information and predictive power about what customers value and how they would make tradeoff decisions given scarcity. As Ajay Agrawal, Joshua Gans, and Avi Goldfarb have written, AI systems are essentially "prediction machines" that vastly improve predictive power over consumer preferences.[7] They observe that at some point, a retailer's AI system will "cross[] a threshold where it becomes so good that the folks at Amazon could ask: 'If we're so good at predicting what our customers want, then why are we waiting for them to order it? Let's just ship it.'"[8] Amazon has already patented an "anticipatory shipping" technology.[9] Although producers and retailers have not yet reached the place where they are routinely initiating consumer transactions without waiting for an affirmative consumer signal, the trend toward AI-empowered consumer insight and information will accelerate precipitously in coming years, with no clear endpoint in sight.

734. We do not have to wait for future technological developments to see the revolutionary potential of AI systems in predicting or detecting what individuals want. AI can already be used to analyze consumer data to provide companies with markedly better predictions or real-time data regarding consumer preferences.[10] Deep convolutional neural networks can forecast retail sales with far better precision than human managers.[11] A 2023 literature review of sixty-four empirical papers on AI and consumer behavior found that early adopters of AI for marketing and

---

5   Friedrich von Hayek, *The Use of Knowledge in Society*, 35 AM. ECON. REV. 519 (1945).

6   *See* ARIEL EZRACHI & MAURICE E. STUCKE, VIRTUAL COMPETITION: THE PROMISE AND PERILS OF THE ALGORITHM-DRIVEN ECONOMY 72 (2016).

7   AJAY AGRAWAL, JOSHUA GANS & AVI GOLDFARB, PREDICTION MACHINES: THE SIMPLE ECONOMICS OF ARTIFICIAL INTELLIGENCE 2 (2022).

8   *Id.* at 37.

9   U.S. Patent No. US13/594 ,195 (filed Aug. 24, 2012) Method and system for anticipatory package shipping, https://patents.google.com/patent/US8615473B2/en.

10 Mohamed Zaki, Janet R. McColl-Kennedy & Andy Neely, *Using AI to Track How Customers Feel – in Real Time*, HARV. BUS. REV. (May 4, 2021), https://hbr.org/2021/05/using-ai-to-track-how-customers-feel-in-real-time.

11 Shaohui Ma & Robert Fildes, *Retail Sales Forecasting with Meta-Learning*, 288 EUR. J. OPERATIONAL RSCH 111 (2021); *see generally* Ming-Hui Huang & Roland T. Rust, *A Framework for Collaborative Artificial Intelligence in Marketing*, 98 J. RETAILING 209 (2022).

assessing consumer demand are using AI tools with increasing efficacy for a wide range of functions, including adding value to existing products and services, creating new products and services, and growing relationships with customers.[12] Use cases include, among many others, dynamic pricing, merchandise optimization, product information management, shelf optimization, and personalized content creation.[13] These techniques aim to replace a market's traditional discovery function by anticipating what consumers will want and how much they will be willing to pay for it, and, in many cases, planning for delivery of personalized goods or services to the consumer before they make their purchasing decision.

735. The coming technological wave will not only enable productive systems to detect consumer preferences with far greater precision, but it will also allow those systems to fulfill consumer preferences in a much more tailored fashion. This is already occurring on a widespread basis in the digital world, where predictive engines allow media and entertainment content providers to customize consumer offerings.[14] Increasingly, developments in AI and robotics will enable customization across wide swathes of the economy. A new generation of "smart factories" will allow manufacturing systems to perceive external environments, adapt to external needs, dynamically optimize operations, and deliver goods in small, customized batches.[15] AI-driven productive systems will not only detect consumer preferences at a highly granular level, but they will also increasingly be able to deliver differentiated, bespoke output matching individual consumers' preferences.[16]

736. That AI and related technologies will enable more highly accurate predictions and fulfillment of consumer preferences and hence consumer demand than any previous technology is a given, but the eventual effect of AI and related technologies is likely to go well beyond providing predictive information about consumer preferences. AI and other technologies of the coming wave are likely to go even further than merely influencing consumer preferences – they are beginning to directly program consumer preferences.

737. Consider the imminent prospect of human synthetic biology. The advent of CRISPR technology, now barely a decade old, is empowering the direct editing of gene sequencing to develop treatments for a wide variety of

---

12 Rajat Gera & Alok Kumar, *Artificial Intelligence in Consumer Behaviour: A Systematic Literature Review of Empirical Research Papers Published in Marketing Journals (2000–2021)*, 27 ACAD. MKTG. STUD. J. 1 (2023).

13 *Id.* at 2.

14 See Yashar Deldjoo, Markus Schedl & Peter Knees, *Content-driven Music Recommendation: Evolution, State of the Art, and Challenges*, 51 COMPUT. SCI. REV. 100618 (2024), https://www.sciencedirect.com/science/article/pii/S1574013724000029.

15 Jaifu Wan et al., *Artificial-Intelligence-Driven Customized Manufacturing: Key Technologies, Applications, and Challenges*, 109 PROC. IEEE 377 (2021).

16 Xingzhi Wang, Ang Liu & Sami Kara, *Machine Learning for Engineering Design Toward Smart Customization: A Systematic Review*, 65 J. MFG. SYS. 391 (2022).

human conditions.[17] DNA synthesizers, enhanced by advances in computational power and AI, will before long enable dramatic improvements in our ability to rewrite the code of life. Potential applications of these technologies include reversing the aging process, reconfiguring human genetics to enhance immune responses, and delivering medicines that are precisely tailored to a patient's biomarkers. Beyond genetic engineering, companies like Neuralink are working on brain interfacing technology or implants that connect the human brain directly to computer systems.[18] Reflecting on these rapidly scaling technologies, Suleyman asks, "[w]hat happens when a human mind has instantaneous access to computation and information on the scale of the internet and the cloud?"[19]

738. One answer to Suleyman's question is that the line between the will of the human agent and the technological systems that structure and program it will become blurred. Subject to ethical and political constraint – which will operate differently in different places – the coming technological wave may call into question the idea that consumers have demand functions that are separate from the systems that write a consumer's genetic code, program their brain, or curate, organize, and present the set of informational stimuli that shape preferences. As Eric Posner and Glen Weyl have written, once the computer "plans" the consumer, the comparative advantage of markets in discovering consumer preference dissipates.[20] Consumer demand is likely to shift from an exogenous fact that production and distribution systems seek to discover, to being shaped or even created by the production and distribution system itself. In such an increasingly likely scenario, it is hard to see why price signals are necessary to discover the consumer's preferences. The consumer's preferences will be both the input and output of the system.

739. Certainly, an economic system driven by technologies that directly extract or, in the extreme, directly program consumer preferences will impose costs. Even if algorithmic systems can improve consumer outcomes by giving consumers more of what they want, consumers may experience these systems as an intrusion on their autonomy.[21] The systems also might make mistakes by misidentifying consumer preferences. But it is important to remember that the case for market competition has never rested on perfection, but rather on comparative advantage. AI-driven systems may never – or not for a long time – provide *perfect* information necessary to the optimal allocation of social resources, but they only have to *outperform* markets to replace them. We are on a quick path to

---

17 SULEYMAN, *supra* note 1, at 82–83.

18 *Id.* at 91.

19 *Id.*

20 ERIC A. POSNER & E. GLEN WEYL, RADICAL MARKETS 288–93 (2018).

21 *See, e.g.*, Michael R. Hyman, Alena Kostyk & David Trafimow, *True Consumer Autonomy: A Formalization and Implications*, J. BUS. ETHICS 841 (2023); Quentin Andre et al., *Consumer Choice and Autonomy in the Age of Artificial Intelligence and Big Data*, 5 CUSTOMER NEEDS & SOLS. 28 (2017).

a point where AI-driven systems may provide much better information than competitive market price signals.

B. *Productive Efficiency*

740. The second discovery function of competitive markets is to identify the most efficient firms or producers and steer less efficient ones toward business activities in which they enjoy a greater comparative advantage. As with discovery of consumer preferences, the coming technological wave will dramatically increase the economic system's ability to observe comparative productive efficiency directly, without waiting for information revealed indirectly by competition and price signals. Employers are making extensive use of AI to track employee behavior, automate performance evaluations, recommend job improvements, supervise, and shift workloads.[22] Firms are also making increasing use of AI for evaluating the efficiency of suppliers through AI-enabled supplier scouting technologies.[23] As AI and machine learning grow exponentially in capacity and use cases, it will be increasingly possible to determine the comparative efficiency and performance of economic actors by observing them directly.

2. Incentives and Processes

741. The second pillar of competition policy is the assumption that market competition provides optimal incentives for firms to deliver value to consumers, especially by lowering prices, increasing output, innovating, and offering high-quality products and services. Since states of mind and motivations are difficult to detect or interpret, competition law instead scrutinizes firm behavior to determine whether the firm's conduct is consistent with the firm acting competitively or anticompetitively. Intentions are opaque, but processes are transparent – therefore, competition focuses on processes.

742. With AI-driven systems, the story is just the opposite: intentions are transparent, and processes are opaque. Of course, machines don't literally have intentions, but they do have objective functions: the mathematical functions that describe an optimization problem that machine learning is used to solve. For example, companies are increasingly delegating pricing decisions to algorithms.[24] A price-setting algorithm must be told what

---

[22] Prabhat Mittal, Rachna Bansal Jora, Kavneet Kaur Sodhi & Parul Saxena, *A Peer Review of the Role of Artificial Intelligence in Employee Engagement*, 9th International Conference on Advanced Computing and Communication Systems (2023); Siliang Tong, Nan Jia Xueming Luo & Zheng Fang, *The Janus Face of Artificial Intelligence Feedback: Deployment Versus Disclosure Effects on Employee Performance*, 42 STRATEGIC MGMT. J. 1600, 1600–01 (2020); Lionel P. Robert et al., *Designing Fair AI for Managing Employees in Organizations: A Review, Critique, and Design Agenda*, 35 HUM.-COMPUT. INTERACTION 544, 545 (2020).

[23] Michela Guida, Federico Caniato, Antonella Moretto & Stefano Ronchi, *Artificial Intelligence for Supplier Scouting: An Information Processing Theory Approach*, 20 INT'L J. PHYSICAL DISTRIB. & LOGISTICS MGMT. 387 (2023).

[24] John Asker, Chaim Fershtman & Ariel Pakes, *Artificial Intelligence and Pricing: The Impact of Algorithm Design* (NAT'L BUREAU OF ECON. RSCH., Working Paper Series No. 28535, 2021), https://www.nber.org/papers/w28535.

problem to solve, and what problem it is told to solve strictly determines what problem it solves. A pricing algorithm that is programmed to learn asynchronously (based solely on the returns from the actions it took) will tend to implement monopoly prices whereas algorithms programmed to update synchronously (based both on the returns from the actions it took and also on the returns from counterfactual actions it did not take) will tend to implement competitive prices.[25] Small changes in an algorithm's design – what it is told to do – can imply large changes in the prices it sets.[26] Unlike a human actor, an AI's "intentions" are perfectly clear and discernible once one has access to its objective function.

743. On the other hand, an AI-driven system's processes – the steps it takes to implement its objective function – are notoriously opaque. As Henry Kissinger, Eric Schmidt, and Daniel Huttenlocher write, AI platforms operate in ways that are "nonhuman, and, in many ways, inscrutable to humans."[27] Thus, Google engineers have found that AI-enabled search functions produce superior results than without AI, but they cannot explain the mechanism by which this occurs.[28] AI-driven operations thus involve a shift from human-mediated operations in which processes "could be paused, inspected, and repeated by human beings," to operations that produce outcomes whose operational steps are opaque.[29] Suleyman notes that "[i]n AI, the neural networks moving toward autonomy are, at present, not explainable."[30] Large language models such as Chat GPT and AlphaGo are "black boxes" with "outputs and decisions based on opaque and intricate chains of minute signals."[31] While a very general explanation of what the system has done may be possible, it is not possible to break down the system's actions into anything like the set of identifiable and understandable steps that are possible with respect to a human actor.

744. Human beings must write an algorithmic system's commands, and competitive stimuli might induce them to write commands that better suit social purposes, but the objective functions themselves can be directly scrutinized (through compulsory legal processes, if necessary) and have objectively determinable implications regardless of the subjective intentions of the programmers. Once a regulator or court has in view an AI's objective functions and the expertise to interpret them, it knows all it needs to know – and, given the opacity of the AI's operations, all that it may ever know – about whether or not the productive system will

---

25 *Id.*

26 J. Manuel Sanchez-Cartas & Engelos Katsamakas, *Artificial Intelligence, Algorithmic Competition and Market Structure*, 10 IEEE ACCESS 10575 (2022), https://ieeexplore.ieee.org/document/9684893.

27 HENRY A. KISSINGER, ERIC SCHMIDT & DANIEL HUTTENLOCHER, THE AGE OF AI AND OUR HUMAN FUTURE 107 (2021).

28 *Id.*

29 *Id.* at 107, 109, 185–06.

30 SULEYMAN, *supra* note 1, at 114.

31 *Id.*

behave "competitively." At that point, "competitively" loses saliency. If competition was deemed desirable because it induced firms to behave virtuously (e.g., by lowering prices and increasing quality), once it can be directly judged whether an AI has been instructed to behave virtuously or unvirtuously, whether or not the objective function is "competitive" becomes rather beside the point, and perhaps even unintelligible. The ultimate question is whether the algorithmic system has been programmed to produce socially desirable outcomes. At the limit, AI thus renders competition – the organizing principle of antitrust law – superfluous.

3. Scale and Scope

745. The second two pillars of competition law concern the operations of the competition law system. The first comprises the assumption that large-scale or monopolistic enterprise is avoidable through the vigilant application of competition law. The technologies of the coming wave will seriously challenge that assumption.

746. As with other digital technologies, AI-driven systems are subject to strong positive network effects, in which the utility of the platform to all users increases with the number of users.[32] But, unlike prior digital technologies,[33] it is hard to see a point at which returns to scale become negative. As AI and machine learning systems ingest increasing volumes of data, their algorithmic outcomes improve, which in turn allows their business outputs to improve, which in turn allows them to ingest more data, which in turn generates better business outputs, and so forth in a seemingly limitless virtuous cycle.[34] Thus, Marco Iansiti and Karim Lakhani argue that "[a]lgorithm-driven operating models are... almost infinitely scalable, as long as you can continue to add computing and storage capacity to the technology infrastructure" and that low-marginal-cost computing capacity and storage capacity are increasingly facilitated by the shift to cloud computing.[35] As network and learning effects accelerate, "the viability of competitive alternatives is diminished, and markets are driven toward concentration."[36] Other technologies of the coming wave will likely amplify these market-concentrating effects. Robotic or automated production will result in a continued shift toward high fixed cost, low marginal cost production, with the implication that dominant technologies will be highly scalable and displace traditional production based on human capital and

---

32 KISSINGER, SCHMIDT & HUTTENLOCHER, *supra* note 27, at 102–04.

33 *See* William J. Kolasky, *Network Effects: A Contrarian View*, 7 GEO. MASON L. REV. 577, 586–87 (1999); Alan Devlin, *Analyzing Monopoly Power Ex Ante*, 5 N.Y.U. J.L. & BUS. 153, 187 (2009); Michael S. Barr, *Banking the Poor*, 21 YALE. J. REG. 121, 202 (2004).

34 IANSITI & LAKHANI, *supra* note 4, at 97; *see also* Roxana Mihet & Thomas Philippon, *The Economics of Big Data and Artificial Intelligence*, 20 INT'L FIN. REV. 29, 30 (2019); Tejas N. Narechania, *Machine Learning as Natural Monopoly*, 107 IOWA L. REV. 1543, 1584–85 (2022).

35 IANISITI & LAKHANI, *supra* note 4, at 96.

36 *Id.* at 161.

labor inputs.[37] Firms with a comparative advantage brought about by quantum computing may obtain an insurmountable lead over firms that continue to rely on conventional computing and human intelligence. Firms that are able to engage in atomically precise manufacturing and create new synthetic compounds that far outstrip conventional production processes and materials in both cost and functionality will rapidly displace competitors using conventional production methods and materials.[38] In combination, the potential arises for economies of scale to stretch toward infinity, or at least far beyond the economic event horizon where the dominant firms' gravitational attraction collapses the entire market.[39]

747. The exact shape of AI's scale economy curve – and, particularly, the importance of big data in improving the performance of AI-driven productions systems – remains to be seen.[40] Returns to scale may flatten at some point, creating the necessary conditions for multiple competitive systems.[41] However, another economic feature of AI-driven systems – economies of scope – will also drive markets toward concentration. Just as AI-driven systems may operate with seemingly unbounded returns to scale for some period of time, they may also operate with seemingly unbounded increasing returns to scope.

748. A key feature of machine learning that differentiates it from prior technological innovations is its ubiquitous application to seemingly different problems by virtue of an AI system's ability to make predictions based on underlying patterns that were invisible to human agents. To a human production team, optimizing the design of an automobile and of a shoe may seem like very different problems. To an AI, they may be much more similar problems requiring similar predictive optimization techniques. This implies that an AI production system that achieves market dominance because it is very good at one task can be leveraged into many other fields with similar success.

749. How far AI's economies of scope will reach depends in large part on how far AI can progress from specialized intelligence to general intelligence. AI competitions and benchmarks are increasingly pushing AI systems to pursue all-purpose capability.[42] The open question is whether AI systems will perform less well as they become more general, or whether, to the contrary, systems "that are better for some tasks will also be better for other tasks."[43] Contrary to the operation of human intelligence, an

---

37   *See* Hamid Farooz, Zheng Lui & Yajie Wang, *Automation and the Rise of Superstar Firms*, (S.F. FED. RSRV. BANK, Working Paper No. 2022-05, Nov. 2023), https://www.frbsf.org/research-and-insights/publications/working-papers/2023/04/automation-and-the-rise-of-superstar-firms/.

38   K. ERIC DREXLER, RADICAL ABUNDANCE: HOW A REVOLUTION IN NANOTECHNOLOGY WILL CHANGE CIVILIZATION x (2013).

39   KATE CRAWFORD, ATLAS OF AI 211 (2022).

40   *See* Thibault Schrepel & Alex 'Sandy' Pentland, *Competition Between AI Foundation Models: Dynamics and Policy Recommendations* 9 (MIT CONNECTION SCI., Working Paper No. 1-2003, 2023).

41   *Id.* at 9.

42   José Hernádez-Orallo, *AI Generality and Spearman's Law of Diminishing Returns*, 64 J. A.I. RSCH. 529 (2019).

43   *Id.* at 530.

AI system that is the best at doing one task may also be the best at doing a number of other tasks – which implies very strong economies of scope.

750. As AI enters its general-purpose or "omni-use" phase, its applications will move out of discrete information-oriented silos to "permeate and power almost every aspect of daily life."[44] The same machine learning and robotic processes that already dominate large swathes of the economy will continue to spread to adjacent domains not typically thought of as the province of computers or automated production, with the effect of increasingly consolidating previously separate economic functions. Here again, AI's market-concentrating effects will be amplified by the other technologies of the coming wave. At the limit, the combination of AI, energy expansion, genetic engineering, quantum computing, and (in the distant but not unforeseeable future) nanotechnology empowering atomically precise manufacturing (APM)[45] could entail a transition of nearly all economic value from atoms to bits. The implication for market structure and competition from existing technologies and trends is the inexorable growth in scope of dominant technology platforms. Kissinger, Schmidt, and Huttenlocher forecast that the continuing development of AI will lead to a small number of international megafirms.[46] Suleyman forecasts that the coming technological wave will lead to the consolidation of economic power in the hands of a handful of "superstar" corporations with more scale and power than many nation-states.[47] The writing is on the wall for an economy characterized by many small, rivalrous producers, outside perhaps of fringe firms that customers patronize precisely because they want to find "quaint" alternatives to the dominant firm that controls the market.

4. Anticompetitive Conduct

751. The final pillar of the antitrust order that will buckle under the coming wave is the assumption that competition law is capable of preventing anticompetitive behavior. The first point to observe here is that the market-concentrating effects previously discussed will tend to make it either unnecessary or much easier for firms to engage in, or get away with, anticompetitive behavior. If a first mover in an AI-dominated market expands continuously in scale and scope, leaving potential competitors in the dust, it does not need to rely on exclusionary behavior to become or remain a monopolist. Or, if network effects and economies of scale result in a tight oligopoly with just a few firms, collusion among the firms becomes easier to undertake and more difficult to prove. Thus,

---

44 SULEYMAN, *supra* note 1, at 111.

45 *See* DREXLER, *supra* note 38, at ch 4; Sarita Kumari Yadav et al., *Impact of Nanotechnology on Socio-Economic Aspects: An Overview*, 2 REVS. NANOSCI. & NANOTECH. 127 (2013).

46 KISSINGER, SCHMIDT & HUTTENLOCHER, *supra* note 27, at 102–04.

47 SULEYMAN, *supra* note 1, at 188.

antitrust law's conduct rules, like its structure rules, will become either superfluous or much more difficult to enforce because of the structural changes precipitated by the coming wave.

752. Even apart from these structural changes, competition law's behavioral rules will become much more difficult, or eventually impossible, to enforce. In broad-brush terms, competition law's behavioral prohibitions fall into two categories (largely tracking Sections 1 and 2 of the Sherman Act or Articles 101 and 102 of the TFEU): collusion and exclusion. In the short term, AI and related technologies may have ambiguous effects on firms' ability to engage in those behaviors and on antitrust enforcers' ability to catch them. However, in the longer-term AI arms race between firms and enforcers, the firms have the decided advantage.

753. Starting with collusion: in order for conventional human-to-human price fixing to work, the cartelists must overcome a number of collective action problems, including coordinated output reduction and prevention of cheating or defection.[48] Existing AI technologies are already changing the nature of these problems, initially with ambiguous implications. AI-enabled improvements in demand forecasting may make initial collusion more feasible, but they may also increase the temptation to defection in periods of high predicted demand.[49] Conversely, while AI may facilitate cheating, it may also enable firms to better distinguish "rivals' cheating from unobserved negative shock demands,"[50] which in turn enables the cartel to mete out more effective punishment.

754. Those effects suggest ambiguity in the consequences of AI and algorithmic pricing when the new technologies are aids to what is still human driven decision-making on prices, agreements with competitors, and the punishment of defectors. But as pricing decisions are increasingly delegated to autonomous or semi-autonomous digital agents, the direction of the effects becomes less ambiguous, prices go up. For example, economic theory predicts that the adoption of algorithmic price-setting technologies that allows for more frequent price changes and automated price changes in response to price changes by rivals can increase price levels.[51] As markets become more concentrated, which could occur due to mergers or just the market-concentrating tendencies of the coming wave technologies discussed in the previous section, the upward pressures on prices intensify.[52] Similarly, the outsourcing of pricing decisions to a third-party

---

[48] Michael K. Vaska, *Conscious Parallelism and Price Fixing: Defining the Boundary*, 52 U. CHI. L. REV. 508, 512 (1985).

[49] Jeanine Miklos-Thal & Catherine Tucker, *Collusion by Algorithm: Does Better Demand Prediction Facilitate Coordination Between Sellers?*, 65 MGMT. SCI. 1552 (2019).

[50] Jason O'Connor & Nathan E. Wilson, *Reduced Demand Uncertainty and the Sustainability of Collusion: How AI Could Affect Competition* 3 (FED. TRADE COMM'N BUREAU OF ECON., Working Paper No. 341, 2019).

[51] Zach Y. Brown & Alexander Mackay, *Competition in Pricing Algorithms*, 15 AM. ECON. J.: MICROECONOMICS 109 (2023).

[52] *Id.*

pricing algorithm tends to make prices more sensitive to demand variation and hence leads to higher prices.[53] These effects are no longer the consequences of cartel facilitation or stabilization. They are consequences of a developing technology permitting firms to replicate cartel-like price structures without the need to participate in a cartel.

755. Legally and analytically, there are two dimensions to the problem of algorithmic price-setting. One is whether the antitrust enforcement agencies can continue to detect price-fixing once machines take over key pricing decisions. Cartel detection is already a tall order. In the pre-AI world, scholars estimated that fewer than 20 percent of cartel agreements are detected.[54] Although antitrust enforcers can develop their own AI tools to enhance cartel detection,[55] they will inevitably be in an arms race with firms developing their own AI technologies to avoid being caught.[56]

756. The other, and more important, dimension is the substantive question of what counts as agreement – a necessary predicate to finding illegality under the collusion or concerted practices provisions of most competition law regimes. The detection issue discussed in the previous paragraph assumes that there is some illegal behavior to catch. As AI and related technologies evolve from simply facilitating traditional cartel activity to replacing the need for explicit or implicit coordination among competitors on prices in order to achieve cartel-like price structures, it will be increasingly difficult for antitrust enforcers to make the case that the competitors have agreed to anything within the meaning of the antitrust laws. Ariel Ezrachi and Maurice Stucke have suggested four scenarios involving algorithmic collusion along a spectrum of increasing AI complexity.[57] The first two – a "messenger" scenario, where members of a cartel directly agree on an algorithm, and a "hub-and-spoke" scenario, where competitors separately outsource pricing decisions to a common algorithm – would likely be found to involve horizontal agreement under traditional antitrust principles.[58] A third "predictable agent" scenario, where each seller unilaterally creates its own algorithm knowing that it will likely facilitate parallel supra-competitive prices, could only be

---

53 Joseph E. Harrington & David Imhof, *Cartel Screening and Machine Learning*, 2 STAN. COMPUTATIONAL ANTITRUST 133 (2022).

54 *See* Emmanuel Combe, Constance Monnier & Renaud Legal, *Cartels: The Probability of Getting Caught in the European Union* 17 (BRUGES EUR. ECON. RSCH., Working Paper No. 12, 2008), https://www.coleurope.eu/sites/default/files/research-paper/beer12.pdf; Peter L. Ormosi, *A Tip of the Iceberg? The Probability of Catching Cartels*, 29 J. APPLIED ECON. 549 (2014); John M. Connor, *Cartel Detection and Duration Worldwide*, 2 COMPETITION POL'Y INT'L: ANTITRUST CHRON. (2011).

55 *See* Joseph E. Harrington & David Imhof, *Cartel Screening and Machine Learning*, 2 STAN. COMPUTATIONAL ANTITRUST 133 (2022); Martin Huber & David Imhof, *Flagging Cartel Participants with Deep Learning Based on Convolutional Neural Networks*, 89 J. INDUS. ORG. (2023); Thibault Schrepel & Teodora Groza, *The Adoption of Computational Antitrust by Agencies: 2nd Annual Report*, 3 STAN. COMPUTATIONAL ANTITRUST 55 (2023).

56 *See* Jason Furman & Robert Seamans, *AI and the Economy*, in INNOVATION POLICY AND THE ECONOMY VOL. 20 (Adam B. Jaffe, Josh Lerner & Scott Stern eds., 2019).

57 EZRACHI & STUCKE, *supra* note 6.

58 *Id.* at 39, 46.

captured with a considerable stretch in current antitrust doctrines.[59] In the final scenario – a "digital eye" with a "God's eye" view of the market – Ezrachi and Stucke argue that the "enforcement tool kit" may be "empty" because no human being ever makes a price-setting decision.[60]

757. The digital eye, an application of computer vision and something approaching AGI with vast computer vision capacity, is no longer a far-fetched idea. Advances in AI technology are quickly pushing firms from a "participative decision-making" model, in which human agents rely on AI-enhanced algorithmic tools to improve their own decision-making, to wholesale delegations of pricing and related decisions to autonomous AIs.[61] The machine is not told to collude; it is told to maximize profits. Through reinforcement learning and "autonomous trial-and-error experimentation," the machine learns to replicate cartel prices.[62] Without any agreement with a competitor, either explicit or even tacit, autonomous algorithmic pricing can succeed in doing what generations of competitors in smoke-filled rooms could not.

758. These observations as to the increasing difficulty of policing horizontal pricing decisions apply with equal force to the other principal branch of anticompetitive behavior – exclusionary strategies. As with collusion, the short-term effects of AI and related technologies may be to make traditional, human-directed exclusion strategies easier to implement, but with the offsetting potential for increases in counterstrategies by targeted competitors[63] and antitrust enforcers. As pricing decisions are increasingly delegated to digital agents with general profit-maximization instructions, it will be increasingly difficult for enforcers or courts to conclude that an AI's behavior is predatory. An AI does not have an "intention" to exclude competitors, nor would it have to directly consider the survival of competitors in making a pricing decision that would have the effect of excluding a competitor. For example, through machine learning, an AI might determine that when a new entrant shows up in the marketplace, the strategy that optimizes long-term profits is an immediate sharp price decrease. The AI does not even have to "know" that a new competitor has arrived to direct the price cut. It may simply be that when sales fall abruptly, the profit-maximizing solution is to cut prices aggressively. As with collusion, the ultimate answer to the question of why an AI sets a particular price may be no more granular than "because that's what was shown to make the most money in the long run, as the system was programmed to do."

---

59 *Id.* at 56.

60 *Id.* at 71.

61 *See* Cindy Candrian & Anne Scherer, *Rise of Machines: Delegating Decisions to Autonomous AI*, 134 COMPUT. HUM. BEHAV. 2 (2022).

62 Timo Klein, *Autonomous Algorithmic Collusion: Q-Learning Under Sequential Pricing*, 52 RAND J. ECON. 538, 538 (2021); Emilio Calvano, Giacomo Calzolari, Vincenzo Denicolò & Sergio Past, *Artificial Intelligence, Algorithmic Pricing & Collusion*, 110 AM. ECON. REV. 3267, 3267 (2020).

63 *See generally* Frank H. Easterbook, *Predatory Strategies and Counterstrategies*, 48 U. CHI. L. REV. 263 (1981).

759. The difficulties of preventing or catching predatory pricing after the coming wave will be equally true as to other exclusionary devices as well.[64] AI and machine learning-driven systems will work their ways to outcomes that eschew profit-depressing competition. Unlike human agents, the steps, strategies, and processes that underlie these outcomes will be opaque. Even assuming that it remains sensible to speak about AI and related technologies as having "exclusionary strategies" as opposed to the inexorable effect of consolidating economic power through the brute force of their internal logic, the conventional doctrines, tools, and techniques of antitrust law will be rendered largely obsolete by the economic changes implied by the coming wave.

II. Beyond Competition

760. If the coming wave is likely to kill off competition as the economy's organizing force, then what shape will be assumed by competition law's successor? The trite answer is regulation, which has long been assumed to stand in as the surrogate for competition in natural monopolies. Undoubtedly, in coming years, AI systems will be subject to increasing regulation along many dimensions, as they are already beginning to be, most notably under the EU's AI Act.[65]

761. But what form of regulation will replace competition? Traditional natural monopoly regulation focused on three elements: (1) controlling the monopolist's prices through a "cost plus" formula that required the regulated firm to submit its proposed prices for approval and allowed rates no higher than costs plus a reasonable profit, (2) guaranteeing universal access and prohibiting discrimination, and (3) prohibiting the regulated firm from leveraging its power in the monopoly market to adjacent competitive markets.[66] Only the second of these functions, to which much attention is already being paid,[67] makes much sense as applied to the coming wave's competition-eliminating tendencies. Conventional rate regulation was already challenging as to diversified entities that sold many different products drawing from a common cost pool, had relatively high fixed costs and low marginal costs, or sold both in regulated and unregulated

---

64 *See* Thomas K. Cheng & Julian Nowag, *Algorithmic Predation and Exclusion*, 42 U. PA. J. L. & BUS. 41 (2023).

65 Regulation 2024/1689 of the European Parliament and of the Council of 13 June 2024 laying down harmonised rules on artificial intelligence and amending Regulations (EC) 300/2008, (EU) 167/2013, (EU) 168/2013, (EU) 2018/858, (EU) 2018/1139 and (EU) 2019/2144 and Directives 2014/90/EU, (EU) 2016/797 and (EU) 2020/1828 (Artificial Intelligence Act), 2024, O.J. (L 2024), https://eur-lex.europa.eu/legal-content/EN/TXT/PDF/?uri=OJ:L_202401689.

66 *See* Jim Rossi & Morgan Ricks, *Forward to Revisiting the Public Utility*, 35 YALE J. REG. 711 (2018); WILLIAM K. JONES, REGULATED INDUSTRIES: CASES AND MATERIALS 1–3 (1975) (same); George L. Priest, *The Origins of Utility Regulation and the "Theories of Regulation" Debate*, 36 J.L. & ECON. 289, 295 (1993).

67 *See, e.g.*, Andrew D. Selbst & Solon Barocas, *Unfair Artificial Intelligence: How FTC Intervention Can Overcome the Limitations of Discrimination Law*, 171 U. PA. L. REV. 1023 (2023); Anya E.R. Prince & Daniel Schwarcz, *Proxy Discrimination in the Age of Artificial Intelligence*, 105 IOWA L. REV. 1257 (2020).

markets.⁶⁸ All of these things will be true of the post-wave megafirms, which will make a conventional price regulation model very difficult. As to the traditional leveraging problem,⁶⁹ in the post-wave economy, expansion into adjacent markets will be an embedded feature of massively scalable AI-driven enterprises. Strategic leveraging behavior of the kind employed by erstwhile regulated monopolists like AT&T⁷⁰ will not likely feature as a key regulatory concern. Regulation to prevent discriminatory behavior or other denials of universal service is likely to be part of the package of regulations focused on post-wave firms, but that will not address the tendency of a monopolist to maximize its profits by raising prices, reducing output, and degrading investments in quality and innovation.

762. The best candidate for a regulatory solution to this problem is harnessing the power of the AI system to regulate itself, subject to the public regulation of the AI's objective function.⁷¹ Here, we return again to the observation about the significant way in which AI systems invert the characteristics of human managers, whose intentions are opaque but whose processing steps are discernible. An AI's processing steps are opaque (the "black box" problem), but its intentions are precisely given by its objective function – the orders it is programmed to follow. AI programmers may thwart antitrust law's traditional prohibitions on collusion and exclusion by directing the AI to maximize profits without any strategic thought about competitors, at which point the AI would achieve that outcome without revealing any processing steps identifiable as collusive or exclusionary behavior. But what if an AI were programmed to do something different than maximize the firm's profits, for example, to achieve the highest total social surplus, and then allocate the surplus among the firm, its customers, and its workers according to some predetermined formula? A monopoly firm would not be inclined to specify that objective function on its own initiative, but regulation might.

763. Regulation of this kind would not restore competition, but it would aim to achieve the same ends as competition historically achieved, without requiring the messiness and waste of competitive markets. And it would do so by drawing on a benefit-sharing concept that is already deployed in some competition law systems. For example, under Article 101(3) of the TFEU, an agreement that restricts competition can be justified when it "contributes to improving the production or distribution of goods or to promoting technical or economic progress, while allowing consumers a fair share of the resulting benefit." Similarly, an AI might be directed

---

68 Jordan J. Hillman & Ronald R. Braeutigam, Price Level Regulation for Diversified Public Utilities 2–3 (1989); David Boies, Jr., *Experiment in Mercantilism: Minimum Rate Regulation by the Interstate Commerce Commission*, 68 Colum. L. Rev. 599, 647 (1968) (discussing ICCs with allocation of joint and common costs).

69 *See generally* Louis Kaplow, *Extension of Monopoly Power Through Leverage*, 85 Colum. L. Rev. 515 (1985).

70 Paul L. Joskow & Roger G. Noll, *The Bell Doctrine: Applications in Telecommunications, Electricity, and Other Network Industries*, 51 Stan. L. Rev. 1249, 1249–50 (1999).

71 Kai-Fu Lee & Chen Quifan, AI 2041: Ten Visions for Our Future 30 (2021).

to organize the firm's research and development, production, distribution, pricing, terms of service, and other attributes to maximize the well-being of specified categories of stakeholders (including profits for shareholders, wages or benefits for managers, wages or other terms of employment for employees, and low prices, high quality, innovation, and variety for customers), with specified criteria as to how surplus from the gains of trade should be allocated. Thus programmed, the AI might, for example, deploy automated processes to gather information on the costs of a particular medical condition and the benefits of a new treatment, invest firm resources in exploring a new pharmaceutical therapy, direct the production of the new drug, and then set its price and terms of distribution with an eye to sharing the surplus created between consumers and the firm.

764. Needless to say, all of these decisions would be ones of great complexity, and one might wonder how a regulator could possibly be up to the task of supervising the AI's programming to achieve these socially optimal outcomes. Without minimizing the challenges ahead, four concluding thoughts are presented on why this is nonetheless the likely path of post-wave, post-competition regulation.

765. First, there isn't much of a choice. For all of the reasons explored in this chapter, the coming wave is likely to destroy the assumptions and practices on which markets have traditionally been predicated and regulated. Markets will concentrate, competition will dwindle, and there is relatively little that any state or government can do to stop it. The big choices ahead are likely to be over ownership and regulation – who owns the megafirms (for example, do they become socialized and part of the state, or do they remain private or semi-private public utilities), and how should they be regulated? Whatever the ownership structures, the ultimate question will be what the AI that directs the deployment of resources and the fulfillment of human wants and needs is programmed to do. Someone will have to answer that question, and it seems unlikely that society will allow the answer to be given at the discretion of a handful of megafirm managers. Some democratically accountable oversight of the programmed instructions of an automated productive system affecting the lives of millions or billions of people seems desirable, and hopefully inevitable.

766. Second, a regulatory system that mandated some sharing of surplus rather than setting a firm's prices based on its costs would solve one of the longstanding problems with rate regulation: that the regulated firm, being guaranteed the same profit regardless of its effort, has little incentive to innovate or improve.[72] Traditional rate regulators did not have easily deployable tools to grant a regulated firm higher profits when the firm took actions like innovating that improved social welfare, but AI-driven systems may solve that problem by having far greater access to information about consumer needs and preferences and firm resources.

---

72 See Phillip Areeda, *Essential Facilities: An Epithet in Need of Limiting Principles*, 58 ANTITRUST L.J. 841 (1989).

767. Third, while there would be immense complexity involved in determining how to fulfill the objective function, that would be the AI's, not the regulator's problem. The regulator's task will be in understanding what the AI is programmed to do and requiring any adjustments necessary to give the AI a more socially minded mission. Further, that task would not need to be performed with standalone human intelligence – regulators will themselves need to rely on AI tools with comparable intelligence to those of the firms they regulate.

768. Finally, to say that a core regulatory function will be to supervise the programming of the AI's objective function is not to say that other kinds of regulatory supervision, whether automated or human, will not also be necessary or feasible. For example, the concept of "human-centered AI" (HCAI) focuses on deploying AI solutions that "amplify and augment rather than displace human abilities."[73] A regulatory HCAI approach might consist of initial pre-clearance on the deployment of a dominant firm's new AI system to ensure the sociability of its objective function, followed by ongoing monitoring of the firm's behavior and outputs by both AI and human regulators to ensure consistency with social, democratic, or economic values.

---

73 Werner Greyer et al., *What Is Human-Centered AI?*, IBM BLOG (Mar. 31, 2022), https://research.ibm.com/blog/what-is-human-centered-ai ("Human-Centered AI is an emerging discipline intent on creating AI systems that amplify and augment rather than displace human abilities. HCAI seeks to preserve human control in a way that ensures artificial intelligence meets our needs while also operating transparently, delivering equitable outcomes, and respecting privacy.").

The Tortuous Path to AI Act Compliance: A Competitive Burden for Companies

GODEFROY DE BOISCUILLÉ[*]

Côte d'Azur University and Paris Pantheon Assas University

Abstract

The AI Act will influence market competition in the EU, bringing significant challenges and obligations for businesses. The regulatory requirements could pose a considerable burden and constitute regulatory barriers to market entry. This chapter points out the tortuous path to AI Act compliance. It examines the main legal barriers that could therefore impact companies operating in the internal market, permanently preventing firms from entering a market or delaying the arrival of new companies.

---

[*] Associate Professor at Côte d'Azur University (EDEN). Co-Director of the Chair Competition & Digital Economics (CDE Chair) at Paris Panthéon Assas University. External expert at the European Institute of Public Administration EIPA Maastricht and Luxembourg.

The Tortuous Path to AI Act Compliance:
A Competitive Burden for Companies

I. Ex-Ante Regulation: Regulatory Barriers

769. The AI Act promotes an ex-ante regulation. To comply with the regulation, any EU company will have to follow five main steps: (i) be identified as a provider, (ii) then to classify their AI system as high-risk or low-risk, (iii) if it is high-risk, the AI provider or deployer must carry out a conformity assessment, (iv) it must ensure that stand-alone AI systems are registered in an EU database, (v) it must sign a declaration of conformity[1] and the AI systems should bear the CE marking to be placed on the market.[2] If significant changes occur in the AI system's lifecycle, the company must return to step (ii) and repeat the risk analysis. Compliance with any one of the standards is not sufficient. In other words, businesses must comply with all the obligations mentioned. Consequently, if a company carries out a compliance assessment on the use of an AI model that has not been properly registered, it could expose itself to legal risks. This European regulatory severity[3] creates regulatory barriers to market entry at many stages of this compliance journey. In our view, the main legal obstacles to market entry are to be found in points (ii) and (iii) mentioned above.

1. Overestimated Regulatory Obstacle

770. On the one hand, to understand how the AI Act impacts EU based companies, it is crucial to answer two fundamental questions: first, who is mainly targeted by the regulation? Second, are the definition of the key actors concerned so blurred that it is difficult for them to understand the scope of application?

771. Regarding the first question, the regulation gives a clear answer: it impacts key players within the AI value chain consisting of provider, deployer, product manufacturer,[4] importer,[5] distributor,[6] and authorized representative.[7] The most heavily regulated subjects under the AI Act are providers of AI

---

1 YourEurope, EUROPA.EU, (last visited Sept. 1, 2024: "An EU declaration of conformity (DoC) is a mandatory document that you as a manufacturer or your authorised representative need to sign to declare that your products comply with the EU requirements. By signing the DoC you take full responsibility for your product's compliance with the applicable EU law."), https://europa.eu/youreurope/business/product-requirements/compliance/technical-documentation-conformity/index_en.htm

2 YourEurope, EUROPA.EU, https://europa.eu/youreurope/business/product-requirements/labels-markings/ce-marking/index_en.htm (last visited Sept. 1, 2024). CE Marking: "Many products require CE marking before they can be sold in the EU. CE marking indicates that a product has been assessed by the manufacturer and deemed to meet EU safety, health and environmental protection requirements. It is required for products manufactured anywhere in the world that are then marketed in the EU."

3 VAGELIS PAPAKONSTANTINOU & PAUL DE HERT, THE REGULATION OF DIGITAL TECHNOLOGIES IN THE EU, ACT-IFICATION, GDPR MIMESIS AND EU LAW BRUTALITY AT PLAY 48–60 (2024), p. 56.

4 Persons that provide, distribute, or use AI systems in the EU with their products under their own name or trademark.

5 EU persons that release AI systems bearing non-EU based provider's name and mark.

6 Persons that make AI systems available in the EU Market.

7 EU persons appointed by a provider to perform obligations under the EU AI Act.

systems.[8] Concerning the second question, it has been said that AI law creates a blurred distinction between suppliers and deployers which makes it difficult to analyze legal risk. In this author's view, the debate seems overrated. The roles are clearly defined. A provider is a person or entity that develops an AI system and places it on the market.[9] A deployer is a person or entity that uses an AI system.[10] In short, companies will need to identify their precise role in order to determine their compliance obligations. Even when they don't fall into the provider "box", companies must remain vigilant. Extensive customization of an AI system by a company could lead to its reclassification as a provider.[11] In any case, the European regulation clearly defines the role of each party, and companies can refer to it to assess their obligations.

2. Underestimated Regulatory Obstacles

772. On the other hand, the most important legal barrier for companies relies in fact on the risk assessment of their AI system. All other heavy obligations depend on this qualification.[12] EU based companies must indeed analyze the risk of their AI system before implementing a conformity assessment, registering their AI system in the EU database and signing a declaration of conformity. In contrast to traditional legal approaches based on experience and a black-and-white approach shaped by interpretation, risk is a fuzzy concept that means "the combination of the probability of an occurrence of harm and the severity of that harm."[13]

II. Defining and Classifying Risks

773. First, the term "risk"[14] lacks a clear definition in all the versions of the AI Act.[15] It seems that the regulation confuses the sensitive sector in question with the nature of the AI system used. AI systems are indeed classified

---

8   Commission Regulation (EU) 2024/1689 of the European Parliament and of the Council of June 13, 2024 laying down harmonised rules on artificial intelligence and amending Regulations (EC) 300/2008, (EU) 167/2013, (EU) 168/2013, (EU) 2018/858, (EU) 2018/1139 and (EU) 2019/2144 and Directives 2014/90/EU, (EU) 2016/797 and (EU) 2020/1828 (Artificial Intelligence Act), art. 16.

9   Commission Regulation (EU) 2024/1689 (Artificial Intelligence Act), art. 3: "provider" means a natural or legal person, public authority, agency or other body that develops an AI system or a general-purpose AI model or that has an AI system or a general-purpose AI model developed and places it on the market or puts the AI system into service under its own name or trademark, whether for payment or free of charge.

10 Example: A bank that uses an AI system developed by a third party to make loan determinations is a deployer. Also, an organization can act as both a developer and a deployer. Example: a company develops AI software that monitors customer transactions and then uses it on its own platforms. The company is both a developer and a deployer.

11 Example: tailoring an AI system could tip companies into the provider category if the original AI system is high-risk and its tailoring results in an AI system different from the original (i.e., it is a "substantial modification"), but the system remains high-risk.

12 If it is high-risk, the provider or deployer must carry out a conformity assessment; it must ensure that stand-alone AI systems are registered in an EU database; and it must sign an EU declaration of conformity.

13 Commission Regulation (EU) 2024/1689 (Artificial Intelligence Act), art. 3(2).

14 *Id.* art. 3(2).

15 In fact, as mentioned, the latest version Commission Regulation (EU) 2024/1689 (Artificial Intelligence Act) defines the risk as "the combination of the probability of an occurrence of harm and the severity of that harm." It seems that it is less a definition of the risk than a method for quantitative estimate of risk probability in use–risk assessment.

high-risk because of the sectors in which they are applied:[16] biometrics-based systems, education, critical infrastructure, etc. The meaning of high-risk is debatable. AI systems should also be qualified as high-risk in terms of their autonomy, i.e., their ability to escape human control and prediction. The universal concern surrounding AI is the following: can it be smarter than humans? This is an obsessive question because intelligence has always been an instrument of domination in civilizations.[17] As a consequence, European institutions require IA to be transparent. The prejudice perceived is as follows: AI is not risky as long as humans can dominate it, i.e., control it, which leads to the next question. At what point does AI become high-risk? Following this reasoning, it should be so not just in terms of the risky factor, but in terms of its capacity to emancipate itself from human monitoring. On this last point, the dichotomy dividing the world of AI is helpful: on the one hand, there is the weak artificial intelligence known as the expert system, which implies the presence of a human expert. The reasoning of this AI system is perfectly transparent. The system makes simple syllogisms where it reasons by deduction. This type of AI is not risky, as it is perfectly predictable. On the other hand, advanced artificial intelligence such as machine learning, is based on reasoning by processing billions of pieces of data. This type of AI surpasses human intelligence in certain tasks. It can learn and generalize. It reasons not by deduction but by induction. Accordingly, the classification of high-risk AI systems is based on a caricatured view of risk. AI in the employment or education sector can be risk-free if the AI system is perfectly predictable and controlled. In contrast, a conversational agent (which seems to be classified as a low-risk AI system) may be high-risk if it starts conversing semi-autonomously and manipulating young people for example. Classification of AI systems regarding the sector is thus problematic. It could harm competition in certain sensitive markets where, in particular, small- and medium-sized enterprises (SMEs) would not have the financial capacity to meet all compliance obligations. For instance, a medium-sized company wishing to compete with Tesla in the autonomous car sector will be subject to a heavy regulatory burden, as an AI system developed in this field will be classified as high-risk AI.

1. Definition of AI system

774. Second, the risk is linked to AI systems which are not clearly defined either. Definition of "AI systems"[18] is too broad, which could complicate the risk assessment associated with the use of artificial intelligence. For

---

16 *Id.* Annex III.

17 Nuclear weapons and vaccines, for example, are inventions were made possible by human intelligence.

18 Commission Regulation (EU) 2024/1689 (Artificial Intelligence Act), "AI system means a machine-based system that is designed to operate with varying levels of autonomy and that may exhibit adaptiveness after deployment, and that, for explicit or implicit objectives, infers, from the input it receives, how to generate outputs such as predictions, content, recommendations, or decisions that can influence physical or virtual environments."

instance, stakeholders in the healthcare sector raised the point that the definition of AI is not consistent in the EU and potentially includes all medical technologies with software components that are not necessarily considered AI.[19] This observation could apply to many other technology sectors. Divergent interpretation of the definition of AI systems may lead to the fragmentation of the internal market and may decrease legal certainty for companies that develop AI systems, thus harming innovation and hence competition.

2. Amendment to the List of High-Risk Use Cases

775. Third, the risk of stand-alone AI systems can be assessed via a list that is potentially flawed and which can be amended by the Commission following the delegated act procedure.[20] The European Parliament's wording gives the Commission the possibility of frequently amending the list. Indeed, AI systems that pose a significant risk of harm *"to health and safety, or an adverse impact on fundamental rights"* could be added to the current AI Act Annex III.[21] In other words, a company that develops an AI system that is not listed in Annex III could still be subjected ex post to compliance obligations if the list is amended after the product is marketed. Immigration prediction tools is an example of a use case not listed in the Commission's AI proposal which was added in other versions of the text. On one hand, the delegated act procedure is necessary to anticipate rapid technological development which will quickly render the Commission's list obsolete. On the other hand, the apparent legal certainty for companies of referring to a list[22] is in fact contradicted by the Commission's ability to amend the list according to broad criteria such as the risk of harm to safety or fundamental rights. In other words, one can legitimately question the usefulness of this list, given that the Commission can easily modify it in the light of the broad criteria that empower it to do so. Broad delegation powers can create legal uncertainty for companies.

---

19 *Stakeholder Joint Statement on Access to Innovative Healthcare under the Artificial Intelligence Act (AI Act)*, MEDTECH EUROPE (June 15, 2023), https://www.medtecheurope.org/news-and-events/news/stakeholder-joint-statement-on-access-to-innovative-healthcare-under-the-artificial-intelligence-act-ai-act/.

20 Once a European law has been adopted, it may need to be updated to reflect developments in a particular sector or to ensure its correct implementation. The EU Parliament and the Council may authorize the Commission to adopt delegated or implementing acts. The Commission can adopt a delegated act on the basis of a delegation granted in the text of an EU law, in this case the AI Act. In the AI Act, the Commission can amend the list of high-risk AI systems, via delegated acts, to take into account the rapid technological development, as well as the potential changes in the use of AI systems, Commission Regulation (EU) 2024/1689 (Artificial Intelligence Act), § 52).The Commission's power to adopt delegated acts is subject to strict limits (for instance, the delegated act cannot change the essential elements of the law the legislative act, but the difficulty is often to determine precisely what are the essential elements of the legislative act). The Commission prepares and adopts delegated acts after consulting expert groups, composed of representatives from each EU country, which meet on a regular or occasional basis.

21 *Id.* art. 7: "The Commission is empowered to adopt delegated acts in accordance with Article 97 to amend Annex III by adding or modifying use-cases of high-risk AI systems where both of the following conditions are fulfilled: (a) the AI systems are intended to be used in any of the areas listed in Annex III; (b) the AI systems pose a risk of harm to health and safety, or an adverse impact on fundamental rights, and that risk is equivalent to, or greater than, the risk of harm or of adverse impact posed by the high-risk AI systems already referred to in Annex III."

22 The list is indeed useful to reassure companies that can refer to it to assess the risk of their AI system.

Companies would have to anticipate, at the time they innovate, whether the product in question could fall into the high-risk list in the near future. Several AI systems such as chatbots, while not on the list of high-risk AI systems, could pose a risk of harm or an adverse impact on fundamental rights. Consequently, the list of high-risk AI systems does not create legal certainty as the Commission claims. In contrast, the probable and frequent amendment of this list creates legal uncertainty. Uncertainty could have ambivalent effects on competition and market structure. According to Ashford, "although excessive regulatory uncertainty may cause industry inaction on the part of the industry too much certainty will stimulate only minimum compliance technology. Similarly, too frequent change of regulatory requirements may frustrate technological development."[23] This is precisely the risk at issue: too frequent changes to the list of high-risk AI systems may discourage new entrants into the market and a lack of stability in the regulatory framework could hinder innovation.

3. Risk of Harm

776. Fourth, AI systems are also in the high-risk list because of the risk of harm in several sectors.[24]. EU based companies must assess the risk in terms of: (i) severity, (ii) intensity, (iii) probability of occurrence, (iv) duration combined together, and (v) they must also determine whether the risk may affect an individual, a plurality of people or a particular group of people. Risk assessment forces companies to predict the future. It could also result in conflicting forecasts. The implementation of AI systems may for instance result in a high-risk severity but low probability of occurrence. It may also affect a group of people with low intensity over a long period. The problem is not the risk assessment itself. These types of assessments already exist in many sectors such as medical devices, food and drug administration. This works well in sectors that benefit from long-established practices and quantifiable results. Risk assessment for AI is clearly more challenging for EU based companies for a simple reason: the AI Act extends risk analysis beyond health and safety to assess impacts on fundamental rights. The regulation adopts a human-centric approach to protect fundamental rights and democracy. As a result, company developers will need to predict their AI system's impact on a wide range of factors based on very complex, fragmented, and evolving case law. The interpretation of fundamental rights is often known to be vague and controversial varying across legal systems within

---

23   Nicholas Ashford et al., *Using Regulation to Change the Market for Innovation*, 9 HARV. ENV'T L. REV. 419–466 (1985). – *See also,* Nicholas Ashford, George R. Heaton, Jr, *Regulation and Technological Innovation in the Chemical Industry*, LAW & CONTEMP. PROBS. 109–157 (1983).

24   AI Act, Regulation (EU) 2024/1689, Annex III. the Annex contains eight fixed areas. These are: (i) Biometric and biometrics-based systems; (ii) Management and operation of critical infrastructure; (iii) Education and vocational training; (iv) Employment, workers management and access to self-employment; (v) Access to and enjoyment of essential private services and public services and benefits; (vi) Law enforcement; (vii) Migration, asylum and border control management; and (viiii) Administration of justice and democratic processes.

the EU. The European Court of Human Rights (ECtHR) and the Court of Justice of the European Union (CJEU) are regularly criticized for delivering contradictory and insufficiently reasoned judgments. Decisions are often based on assertions that do not allow the litigant to understand the reasons that led the courts to make these choices.[25] In a nutshell, how can a company implement fundamental rights risk thresholds without the expertise or the authority to interpret legislation? The author wonders whether this assessment is reasonable for companies, especially for new entrants in a competitive market, who will not have the resources or expertise to master all these combinations and assess the damage.

4. Product Covered by Union Harmonization Legislation

777. Fifth, apart from the list in Annex III of the AI Act, some forms of AI are classed as high-risk where they are intended to be used: (i) as a product covered by Union harmonization legislation listed in Annex II, or (ii) are a safety component of a product.[26] This implies that EU based companies need to handle possible contradicting definitions between the definition of safety components in the AI Act and harmonization legislation. The meaning of "safety component" under the relevant harmonization legislation (machinery regulation for instance[27]) will sometimes differ from the meaning of "safety component" in the AI Act. One might point out that the relevant harmonization legislation prevails. Nonetheless, there is a clear issue of inconsistent interpretation resulting in a lack of legal certainty between AI law and other EU legislation.

5. Utopia of Compliance with Certain Rules

778. Sixth, high-risk AI systems activate compliance rules. Numerous obligations will be difficult to comply with,[28] among which are obligations relating to data governance; the subject of much criticism. Article 10 of the AI Act states that: "Training, validation and testing data sets shall be relevant, sufficiently representative, and to the best extent possible, free of errors and complete in view of the intended purpose[29] (…). Training, validation and testing data sets shall be subject to data governance and management

---

25 Hanneke C.K. Senden, Interpretation of fundamental rights in a multilevel legal system: an analysis of the European Court of Human Rights and the Court of Justice of the European Union, (Nov. 8, 2011) (Doctoral thesis, School of Human Rights Research Series, Intersentia, Antwerp), https://scholarlypublications.universiteitleiden.nl/handle/1887/18033.

26 The malfunction of the AI system embedded in a product could pose a danger to safety and health of persons.

27 Regulation (EU) 2023/1230 of the European Parliament and of the Council of 14 June 2023 on machinery and repealing Directive 2006/42/EC of the European Parliament and of the Council and Council Directive 73/361/EEC (Machinery Regulation), 2023, O.J. (L 165/1).

28 Quality of data sets used to train, validate and test the AI systems (Commission Regulation (EU) 2024/1689 (Artificial Intelligence Act), art. 10); technical documentation (Recital 46, art. 11, and Annex IV) ; Record-keeping in the form of automatic recording of events (art. 12); Transparency and the provision of information to users (Recital 47, and art. 13); Human oversight (Recital 48, and art. 14); Robustness, accuracy and cybersecurity (Recitals 49 to 51, and art. 15).

29 *Id.* art. 10(3).

practices appropriate for the intended purpose of the high-risk AI system."[30] These practices entail several necessary precautions, such as an assessment of the availability, quantity, and suitability of the data sets that are needed. Ebers et al have pointed out the impossibility of enforcing the regulation in practice as the data quantity requirement is not realistic from a technical perspective: it is not possible to precisely quantify the required amount of data for training an AI system.[31] Also, the obligation to maintain technical documentation is a standard that illustrates the paradox of the EU's digital single market. The core philosophy of the single market is to remove barriers that limit freedom of all kinds (services, goods, people, capital). The obligation to maintain technical documentation is a good example of an obstacle that can discourage innovation and the circulation of goods. This documentation seems extremely cumbersome[32] and could lead to a distortion of competition between companies that can afford to maintain such documentation and those that cannot.[33]

6. Overlap in Legislation

779. Seventh, the AI Act poses a more general problem of its relationship with other legislation. AI legislation intertwines and merges to the point of creating many rules imposing a disproportionate burden on providers to comply with a number of regulatory requirements. For instance, a large platform creating an AI system to generate all kinds of content could be simultaneously subject to the Digital Services Act, the Digital Markets Act, the AI Act, and the General Data Protection Regulation (GDPR). In the same way, a company that develops AI systems to provide new medical devices (MD) or in-vitro diagnostic medical devices (IVD) will have to comply with the GDPR, the AI Act, the Medical Devices Regulation (MDR), and the In-vitro Diagnostic Medical Devices Regulation (IVDR). But then again, the conformity assessment carried out by a company may have to be repeated several times by other companies in some sectors. For example, an AI system provider that enhances machine safety will have to implement a conformity analysis. This conformity assessment will have to be repeated by the machine manufacturer under the AI Act and the Machinery Regulation. To be concise, as soon as the AI Act calls for a conformity analysis of AI systems embedded in products, there will be

---

30   *Id.* art. 10(2).

31   Martin Ebers et al., *The European Commission's Proposal for an Artificial Intelligence Act – a Critical Assessment by Members of the Robotics and AI Law Society (RAILS)*, 4 MDPI 589–603 (2021), https://doi.org/10.3390/j4040403.

32   The technical documentation requires companies to detail the design specifications of the AI system (see Annex IV of the European Commission's AI Act) including the general logic of the algorithm and documented rationales and assumptions that were made during the design (*e.g.*, the main classification choices, what parameters it is optimized for, descriptions of outputs, etc.). The technical documentation must also assess whether the functioning of the AI system complies with several of EU fundamental rights covered in ch. 2 of the AI Act.

33   Thibault Schrepel, *Decoding the AI Act: A Critical Guide for Competition Experts* 4–5 (Amsterdam L. & Tech. Inst., Working Paper No. 3-2023, 2023).

a duplication of conformity assessments needed between the AI Act and all the regulations concerning the products at issue (AI Act vs Machinery Regulation, or AI Act vs Medical Devices Regulation, etc.).

780. Moreover, the AI act does not solve the issue of damages caused by AI systems. The interplay between the regulation and the AI Liability Directive is not clear. The regulation aims solely to prevent the risk of damage, without addressing the question of the civil liability regime applicable in the context of harm caused by AI systems. There is still legal uncertainty as to how the damage caused to specific businesses could be appropriately compensated. The AI Liability Directive[34] is still at the proposal stage and suffers from a number of shortcomings: (i) the proposal seems to favor the fault-based liability regime that is not suited to liability claims for damages caused by AI-enabled products,[35] (ii) the rebuttable presumption to facilitate proof of damage is admitted only under excessively strict conditions,[36] (iii) the relationship between the AI Liability Directive and the revised products liability directive creates overlaps and needs to be clarified.[37] One can question the consequences on competition of the lack of clarification on the civil liability model by underlining at least three potential effects. The first effect is to decrease the effectiveness of private enforcement of competition law related to damages caused by AI systems. In a stand-alone action, companies would have to prove the competitor is at fault. This would be very difficult to do given that AI is sometimes considered a "black box". Questions arise, such as: how can we prove what is opaque? How can we counter the "black box effect" of algorithms? Assuming this is possible, will victims have the means to do so? If so, at what

---

34 *Proposal for a Directive of the European Parliament and of the Council on adapting non-contractual civil liability rules to artificial intelligence (AI Liability Directive)*, COM (2022) 496 final (Sept. 28, 2022).

35 Indeed, how will victims be able to demonstrate the failure of the algorithm that caused the damage? The commission itself compares artificial intelligence to a black box. Proving the company is at fault (in other words, its negligence) by demonstrating that it is indeed responsible for the algorithm's failure implies proving the defect in the artificial intelligence system, which is a particularly difficult evidential burden to bear for the claimant. See, *Proposal for an AI Liability Directive*, COM (2022) 496 final, art. 4, "The fault of the defendant has to be proven by the claimant according to the applicable Union or national rules. Such fault can be established, for example, for non-compliance with a duty of care pursuant to the AI Act." *See also*, art. 1, indicating the subject matter and scope of this directive: "it applies to non-contractual civil law claims for damages caused by an AI system, where such claims are brought under fault-based liability regimes."

36 In principle, the AI Liability Directive will make it easier to prove a causal link between a relevant party's fault and the output of an AI system that causes the damage through rebuttable presumptions. In fact, "such a presumption should only apply when it can be considered reasonably likely, from the circumstances in which the damage occurred, that such fault has influenced the output produced by the AI system or the failure of the AI system to produce an output that gave rise to the damage." (*Proposal for an AI Liability Directive*, COM (2022) 496 final, § 25).

37 Indeed, as the proposals stand, it is highly likely that certain losses could give rise to an action both on the basis of the revised Defective Products Directive and on the basis of the AI Liability Directive. It would then be up to the claimant to make the right procedural choice, at the potential risk of seeing their action declared inadmissible because of the option chosen. COM(2022) 495 - Proposal for a directive of the European Parliament and of the Council on liability for defective products. Also, the European Parliament, on 12 March 2024, formally adopted the new Product Liability Directive. The Council of the EU still needs to formally adopt the directive, following which it will be published in the Official Journal of the EU and then enter into force 20 days after its publication. The new rules will apply to products placed on the market 24 months after entry into force.

cost? The second effect is to reduce access to the law. Indeed, the numerous reasons for AI systems to malfunction[38] added to the opacity of algorithms[39] will increase the cost of obtaining evidence. This will reduce access to the law due to the cost of civil litigation. Small- and medium-sized businesses will not have the financial resources to hire specialist firms to assess their losses. The final effect is that it will encourage large companies to cause damage to small businesses. The moment private enforcement of competition law becomes non-deterrent, the incentive to infringe market rules will be stronger.[40] The low risk for large companies developing AI of being exposed to liability and civil action from small- and medium-sized enterprises decreases their incentives to comply with the regulation.

III. Cost of Compliance

781. In more general terms, apart from the specific problems mentioned above, all the companies will face the same issue: the fragmentation of sources of EU law. This will have the consequence of significantly increasing the cost of compliance for EU based companies. The cost of compliance could be a strong economic barrier for all small- and medium-sized companies, reducing access to the single market.[41] The problem is not new, as similar criticisms have been raised regarding the GDPR.[42] AI startups are much more financially vulnerable than large companies.[43] According to the European Commission impact assessment of the AI Act, compliance costs are estimated to be between €193k to €330k.[44] The Commission has simply assessed the compliance cost for deploying high-risk AI systems. The cost of compliance due to the overlap with all the other legislation is not included in

---

38 Errors in data selection or labeling, choice of non-representative data, erroneous choice of algorithm, human bias, etc.

39 Jenna Burrell, *How the Machine "Thinks": Understanding Opacity in Machine Learning Algorithms*, 3 BIG DATA & SOC'Y 1–12 (2016).

40 Godefroy de Boiscuillé, *La faute lucrative en droit de la concurrence*, (Concurrences, Sept. 2022). *See also*, Godefroy de Boiscuillé, *Relevance and Shortcomings of Behavioral Economics in Antitrust*, 11 J.O EUROPEAN COMPETITION L. & PRAC. 228–237 (2020).

41 *Id.*

42 Regarding the PwC report, the cost of maintaining GDPR compliance amount to more than $1 million (about €900,000). The report mentions cases where this figure could be significantly higher. *See*, PwC REPORT, https://uk.insight.com/content/dam/insight/EMEA/blog/2017/06/GDPR-Infographic-design-final.pdf (last visited Sept. 1, 2024).

43 Weiyue Wu & Shoaoshan Liu, *Why Compliance Costs of AI Commercialization May Be Holding Start-Ups Back*, HARV. KENNEDY SCH. REV. (May 5, 2023) "Based on the OECD Regulatory Compliance Cost Assessment Guidance, we quantitatively compare the financial vulnerability of tech giants versus AI startups. We found that start-ups' operating margins are significantly impacted by compliance costs, in contrast to tech giants (...) When the fixed compliance cost increases by 200%, the operating margin of the startup changes from 13% to -7%, causing the firm to lose money. In contrast, such a change only causes a slight dip in the operating margin for tech giants."

44 EUROPEAN COMMISSION, STUDY SUPPORTING THE IMPACT ASSESSMENT OF THE AI REGULATION (Apr. 21, 2021), https://digital-strategy.ec.europa.eu/en/library/study-supporting-impact-assessment-ai-regulation.*See also*, *CECIMO Paper on Artificial Intelligence*, CECIMO (Nov. 1, 2023), https://www.cecimo.eu/wp-content/uploads/2022/10/CECIMO-Paper-on-the-Artificial-Intelligence-Act.pdf.

the European Commission's estimate. Also, dual or triple regulatory compliance (GDPR, AI Act, DMA, etc.) could lead to an accumulation of fines. Lack of legal clarity and the resulting compliance costs could permanently prevent companies from entering a market or delay the arrival of new companies.

782. However, it could be argued that the cost of compliance is not necessarily too high. It must only be proportionate to the objective pursued. The proportionality principle requires an assessment of whether EU measures are: (i) suitable for achieving the desired aim, (ii) are necessary to achieve the desired aim, and (iii) impose an excessive burden on individuals or companies, in relation to the objective to be achieved.[45] On the first two points, AI law is suitable for monitoring high-risk AI systems and necessary to protect citizens and fundamental rights. Nevertheless, on the third point, only the application of the AI regulation over time will answer the question of whether the regulatory burden on companies is too high. We could be predicting an unsolvable question. In principle, the burden imposed on companies is assessed regarding the specific objective to be achieved. But the AI Act follows many objectives.[46] Reading paragraph 1 of the AI regulation is dizzying, as are the numerous objectives.[47] The difficulty is therefore as follows: it is necessary to assess the conformity of the AI Act not only in relation to one objective (freedom of goods, for example), but also in relation to several objectives (freedom of goods, but also fundamental rights, security, ethical principles, etc.). The more objectives there are in the internal market, the more conflicts there will be between them. The greater the regulatory burden to protect fundamental rights, the greater the risk of limiting innovation and therefore the EU's economic prosperity, which is the general aim of the Common Market. Litigation in European law speaks for itself. The Charter of Fundamental Rights is full of principles that coexist and ultimately clash: the principle of equality versus the principle of freedom; the principle of freedom

---

45 The principle of proportionality is laid down in the Treaty on European Union (TFEU), art. 5(4).

46 The AI regulation appears to be in line with the objectives of the creation of a digital single market. The digital single market introduces a philosophy that is less liberal than the internal market. The main focus is on citizens' fundamental rights, security and respect for ethical principles, with freedom taking a back seat.

47 Commission Regulation (EU) 2024/1689 (Artificial Intelligence Act), § 1, "The purpose of this Regulation is to improve the functioning of the internal market by laying down a uniform legal framework in particular for the development, the placing on the market, the putting into service and the use of artificial intelligence systems (AI systems) in the Union, in accordance with Union values, to promote the uptake of human centric and trustworthy artificial intelligence (AI) while ensuring a high level of protection of:
    – health
    – safety
    – fundamental rights as enshrined in the Charter of Fundamental Rights of the European Union (the 'Charter') including
    – democracy
    – the rule of law
    – and environmental protection, to protect against the harmful effects of AI systems in the Union,
    – and to support innovation. This Regulation ensures the free movement, cross-border, of AI-based goods and services, thus preventing Member States from imposing restrictions on the development, marketing and use of AI systems, unless explicitly authorised by this Regulation."

versus security,[48] freedom of expression versus freedom of goods[49] or the right to privacy, freedom of movement versus the right to strike,[50] and so on. This leads to several observations that will increase the cost of regulation: first, the cohabitation of fundamental rights and the need to respect them in the AI Act will cause conflicts of rights. Second, by dint of enshrining multiple objectives in the name of diverse fundamental principles, the law will become less and less comprehensible, and the cost of regulation increasingly high. Third, the AI Act will cause an increasing overlap of fundamental rights in the case law of the CJEU. Fourth, as the regulation various fundamental rights, freedoms and principles put at the same level as essential objectives to be reached, the Court of Justice of the European Union will have to favor certain rights and freedoms to the detriment of others. Last, litigation before the CJUE will give rise to judgments where fundamental rights will prevail over fundamental freedoms. Indeed, if the AI regulation appears to be in line with the objectives of the Common Market, the digital single market introduces a philosophy that is less liberal than the internal market. The main focus is on citizens' fundamental rights, security and respect for ethical principles, with freedom taking a back seat.

IV. Conclusion

783. The AI Act increases the cost of compliance significantly for EU based companies. Legal obstacles to compliance with the regulation are numerous. This chapter points out the winding path to AI Act compliance by underlining at least seven legal barriers for businesses that increase the cost of compliance significantly: (i) the risk-based approach, which is subject to differing interpretations due to the broad and vague notion of risk, which is not a legal concept, (ii) the definition of AI, which is just as vague, raising the question of how to define risk on the basis of an ill-defined object, (iii) the legal uncertainty surrounding the list of high-risk AI systems that can be modified by the Commission under the delegated acts procedure, (iv) the risk assessment which involves predicting violations of fundamental rights on the basis of inconsistent and fragmented case law, (v) the lack of legal certainty due to possible conflicting definitions between the definition of safety components in the AI Act and that in harmonization legislation, (vi) unrealistic compliance

---

48 Examples from everyday life bear this out, from seatbelt laws to food safety regulations.

49 *See*, Case C-368/95, Vereinigte Familiapress Zeitungsverlags- und Vertriebs GmbH v. Bauer Verlag, ECLI:EU:C:1997:325. The case concerned Austria's ban on the sale of newspapers containing games of chance. The Court ruled that maintaining the diversity of the press and thus safeguarding the freedom of expression constituted "an overriding requirement justifying a restriction on the free movement of goods."

50 *See also*, Case C-112/00, Eugen Schmidberger Internationale Transporte Planzüge v. Austria, ECLI:EU:C:2003:333. This case illustrates a clash between the free movement of goods and the right of expression and the right of assembly.

rules that undermine incentives for innovation and distort competition in the internal market, and (vii) the duplication of conformity assessments between the AI Act and all product regulation and interaction with multiple legislation that creates uncertainty, overlap and collision. These defects can be corrected over time, but as always with the single market, it is a question of striking the right balance between protection for citizens and innovation for businesses.

Competition Policy over the Generative AI Waterfall

WILLIAM LEHR*
CSAIL/MIT, Cambridge, MA

VOLKER STOCKER**
Weizenbaum-Institute, Berlin

Abstract

The emergence of generative AI signals that we have reached a global waterfall moment – an inflection point that will result in profound and wide-ranging changes for society and the economy. In this chapter, we use the waterfall metaphor and focus attention on the policy challenges that need to be confronted as we go over the falls toward the uncertain future beyond. This focuses attention on the essential role that digital infrastructure plays in making (gen)AI possible, in establishing the effective bounds of what may be accomplished with it, and in shaping our digital future. This infrastructure is the ship that is carrying us over the falls, and its steerage capabilities will prove essential to enabling any hope for regulating AI, including enabling effective competition policy in the future digital economy. Of critical importance to those steerage capabilities will be the success in building a sound measurement ecosystem.

---

* William Lehr (wlehr@mit.edu) is an economist and research scientist in CSAIL/MIT and an associated researcher of the Weizenbaum-Institute in the research group Digital Economy, Internet Ecosystem, and Internet Policy.

** Volker Stocker (volker.stocker@weizenbaum-institut.de) is an economist and heads the research group Digital Economy, Internet Ecosystem, and Internet Policy at the Weizenbaum-Institute. Volker Stocker would like to acknowledge funding by the Federal Ministry of Education and Research of Germany (BMBF) under grant No. 16DII131 (Weizenbaum-Institut für die vernetzte Gesellschaft – Das Deutsche Internet-Institut).

Competition Policy over the Generative AI Waterfall

I. Introduction

784. We are in the advanced stages of a global digital economy transition. A continuous stream of innovation and investment in information and communications technology (ICT) over many decades has enabled this transition. More recently, the emergence and success of a class of machine learning (ML) technologies that are subsumed under the term generative AI (genAI) has been heralded as ICTs with transformative potential for widespread economic and societal changes.[1] Generative pre-trained transformers (GPTs) built on large language models (LLMs) have emerged as killer applications. The launch of OpenAI's ChatGPT in late 2022 served as a watershed event in raising public consciousness about AI and brought to the fore debates about the benefits and risks posed by AI. At the same time that investment in all things AI-related is surging, a number of prominent experts have expressed fears that AI poses an existential threat.[2] Taken together, these prognostications suggest we ought to take the potential that we are on the cusp of a waterfall moment seriously – an inflection point that will result in profound and wide-ranging changes for all aspects of society and the economy, including competition policy.

785. In this chapter, we redirect attention from what is new and novel about AI or genAI and the implications those novelties may have for legacy competition policy to focus instead on the challenges for policymaking during an inflection point – when we are heading *over* the waterfall. While it remains interesting and important to consider the details of what is novel about AI and its different incarnations, including genAI, and to debate what digital futures may (or should) look like, those are not the issues that concern us here. Instead, we focus on what is needed for evidence-based decision-making (EBDM) – the gold standard for good multistakeholder policymaking in complex domains

---

1   A number of scholars have argued AI has the potential to be a General-Purpose Technology (GPT). *See e.g.*, Timothy Bresnahan, *What innovation paths for AI to become a GPT?* Journal of Economics & Management Strategy, 33, 3015-316 (2024); Timothy Bresnahan & Manuel Trajtenberg, *General Purpose Technologies "Engines of Growth"?*, 65 J. Econ. 83–108 (1995); Flavio Calvino & Chiara Criscuolo, Generative AI And Productivity: Challenges, Opportunities and the Role of Policy, in Dynamics of Generative AI Symposium, Network L. Rev. (Thibault Schrepel & Volker Stocker eds., Winter 2023), https://www.networklawreview.org/calvino-criscuolo-generative-ai/; Merle Uhl, William Lehr & Volker Stocker, AI as a General Purpose Technology – Same but Different? 16th ITS Asia-Pacific 2023 Conference, Presentation, (Bangkok, Nov. 26–28, 2023), potentially on a scale sufficient to drive the Fourth Industrial Revolution, *see* Klaus Schwab, *The Fourth Industrial Revolution: Its Meaning and How to Respond*, World Econ. F. (Jan. 14, 2016), https://www.weforum.org/agenda/2016/01/the-fourth-industrial-revolution-what-it-means-and-how-to-respond/. Although this GPT shares an acronym with the "GPT" of ChatGPT, the latter refers to Generative Pre-Trained Transformer models, which is the type of large language model that the ChatGPT application runs on. *See* Gokul Yenduri et al., *GPT (Generative Pre-trained Transformer)–A Comprehensive Review on Enabling Technologies, Potential Applications, Emerging Challenges, and Future Directions*, 12 IEEE Access 54608-54649 (2023), for further discussion of such models.

2   Many prominent academics and industry leaders at the forefront of AI developments have warned that AI poses an existential risk to humanity. Those include Sam Altman, Bill Gates, Yuval Harari, Geoffrey Hinton, Bill McKibben, Elon Musk, Steve Wozniak, and many others. *See* Cade Metz & Gregory Schmidt, *Elon Musk and Others Call for Pause on A.I., Citing "Profound Risks to Society"*, N.Y. Times (Mar. 29, 2023), https://www.nytimes.com/2023/03/29/technology/ai-artificial-intelligence-musk-risks.html; Kevin Roose, *A.I. Poses "Risk of Extinction", Industry Leaders Warn*, N.Y. Times (May 30, 2023), https://www.nytimes.com/2023/05/30/technology/ai-threat-warning.html.

requiring multidisciplinary engagement.[3] In keeping with our waterfall metaphor, we focus attention on the critical role that digital infrastructure – i.e., the networked communications, computing, and storage resources – will play as the "ship" that will carry the digital economy over the falls.[4] Regulatory policies directed at ensuring that the digital infrastructure meets industrial policy goals for quality and reliability are important for ensuring the ship is appropriately "seaworthy" (or "economy worthy") and, hence, may prove critical in determining how we emerge from the falls.[5] Competition policy is an essential component of how we will manage digital infrastructure, but in this context, those goals need to be reconciled with industrial policies directed at ensuring universal access to reliable critical digital computing and communications infrastructure. To jointly achieve our competition and industrial policy goals, we will need a healthy measurement ecosystem to collect, share, and analyze the technical and economic performance data needed to inform and facilitate multi-stakeholder EBDM.

786. In Section II, we argue that AI's transformative potential is not so much new as a continuation of the process of digital automation. What is new is the way in which genAI in general, and ChatGPT as its most prominent example, brought the reality of digital automation to the general public's attention, rendering digital automation choices to appear as options that everyone needs (or soon will need) to confront in virtually every situation. Whether that perception is valid or not is not all that matters, and it would be foolish to presume that any single source of evidence or perspective would be accepted as trustworthy or probative in such a situation. The reality is that we are at a point where we see AI as confronting us with a digital Pandora's Box that genAI may open, and in so doing, potentially changing humanity's role in managing and controlling digital transformation. In the collective decision-making environment we confront, we do not know, nor can we be expected to easily agree on, what the best strategy should be, even if we might hope to agree on what the desired outcomes should be.[6]

---

3   *See* William Lehr & Volker Stocker, AI Regulation and 6G: The Measurement Ecosystem Challenge, in Dynamics of Generative AI Symposium, Network L. Rev. (Thibault Schrepel & Volker Stocker eds., Winter 2023), https://www.networklawreview.org/lehr-stocker-generative-ai/ for a discussion of EBDM. Regulating AI certainly requires substantial expertise about the technology, but also its economics and regulatory/legal governance options that need to be jointly considered.

4   By digital infrastructure, we intend the mix of public and private networks and computing resources commonly referred to as "the internet" or the internet ecosystem. This is a more inclusive set of industry participants, services, and ICT resources than may be appropriate in other contexts. *See* William Lehr, David Clark, Steve Bauer, Arthur Berger and Phillip Richter, *Whither the Public Internet?*, 9 J. Info. Pol'y vol. 9, 1–42 (2019).

5   Industrial policy determines what infrastructure is needed for what purposes. A key focus of ICT policies is to ensure that all citizens have access to basic infrastructure, which are facilities and services that are needed and used by virtually everyone and are necessary for the economy to function. As necessary preconditions, they are typically taken for granted in most contexts, and only noticed when problems arise. Not all of our digital infrastructure is basic infrastructure, but as the digital transition continues, more of it will be, and as we explain later, this may come to include AI-enabled applications.

6   For example, is it better to slow down or accelerate the pace of AI algorithm innovation and the scope and scale of AI application adoption? Assuming we could answer either of those questions, what policies would actually (best) accomplish those goals? Bad regulatory policymaking can result not only from failure to act but also from bad actions. Agreeing on the desired outcomes is a step in the right direction (i.e., that the digital economy should sustain and promote the same human values we hold dear in the pre-digital economy), but it is hardly sufficient.

787. In our trip over the falls, we certainly want to emerge better than before and with as few casualties as possible. Whether that calls for steady-at-the-helm steerage (or trust in our legacy policies) or nimble and flexible steerage to avoid breaking up on unforeseen rocks is, to a significant extent, unclear. What is obvious is that we cannot change our steerage capabilities, embodied in our legacy regulatory institutions and frameworks, overnight. It is also obvious that we need a healthy measurement ecosystem to provide the evidence needed for EBDM. Without such capabilities, hope for meaningful policy design and effective enforcement that reconcile digital infrastructure industrial policy and competition goals is limited. In Section III, we identify key features of the evidence-based measurement ecosystem that will be needed to characterize our approach for managing steerage as we approach our trip over the falls. Section IV concludes and cycles back to how a focus on digital infrastructure policy and the measurement ecosystem relates to the challenge of digital competition policy.

II. AI and a Waterfall Moment

788. AI comprises a range of technologies. A key economic impact of genAI, and ICTs more generally, is their contribution to the continuing process of automation.[7] Automation as a paradigm and driver of economic and societal change and progress has fueled fears of human replacement by machines long before AI emerged. Those fears go back to the first Industrial Revolution, which enabled mechanical power to augment and substitute for human power. With the invention of computers and modern telecommunication networks, the technological automation process changed to one of digital automation and transformation. Over time, this process has been steadily progressing from specialized silos of mainframe computing departments in the largest enterprises to networked computers on everyone's desktops to ICT intelligence being embedded into virtually everything we use and interact with in our professional and personal lives. AI is just the next step in adding new and ever-more capable tools to the ICT toolbox, making automation possible (and easier) in more ways and contexts.[8]

---

7 Automation is here defined as using technology, and more specifically digital technology, to augment or substitute for a human role in accomplishing a task. When digital technology is added to a system, it changes the roles of other (human and/or non-human) components in the system. In economic production function terms, it shifts the mix of factor inputs, their relative prices, and the marginal productivity of factor inputs. It most often results in restructuring the sub-tasks that comprise the system that characterizes the larger task. In computer software terms, the sub-tasks are akin to sub-routines, and the composability of sub-routines that may enable either general or task/system-specific functionality may be combined to support larger tasks. This composability and modularity of software systems contribute to their ability to be employed so widely and at so many different levels to enhance the productive capacity of so many economic activities. Indeed, if a task can be characterized as a Turing Machine, then it is amenable, at least in theory, to being automated. *See* JACK B. COPELAND, THE ESSENTIAL TURING (Oxford Univ. Press 2004).

8 Continued advances in those tools envision enabling ever more powerful and resource-intensive Augmented Reality/Virtual Reality (AR/VR) applications to provide real-time monitoring, recommendations, or automation for tasks such as the control of robots (e.g., in smart-factories), Unmanned Aerial Vehicles (UAVs)/Autonomous Vehicles (AVs), and enhanced Enterprise Resource Planning (ERP) decision-support and supply management systems for business enterprises.

789. *Smart-X* apps (i.e., augmenting any application or task, "X", with digital technology) existed before and without AI. What is potentially novel about genAI is its potential for automated self-improvement and for mining previously untappable sources of unstructured data. These are important innovations, but they have long been anticipated.[9] What has been obvious about the applications that genAI has spawned (e.g., ChatGPT) is the potential for such applications to expand universal access to digital tools to make them more accessible for anyone and everyone in every possible context.[10] While earlier ICT tools with sophisticated capabilities often required special training and expertise to implement and integrate, the new genAI tools and capabilities are accessible to everyone through their use of natural language and interactivity.[11] GenAI's interactivity capabilities can enable end-users to play a more direct role in innovation, enabling them to act as application developers and solvers of "local" problems.[12] In that sense, genAI has been hypothesized to activate and incentivize distributed innovation, thus releasing and harnessing previously untapped innovation potential.[13] It will take time to empirically assess the economic implications of AI. However, the perception of universal accessibility to digital automation tools that AI projects is now seen as a real-world possibility for many who previously had only considered the implications of automation as a distant prospect – if at all.[14]

790. When new technologies emerge, economists are often asked about their welfare implications. It should be obvious that it is premature to expect an authoritative answer to that question as it applies to AI or, more narrowly, to genAI. AI is a set of tools for automation that needs to be combined with other ICT tools, including those that comprise the digital infrastructure of

---

[9] Weizenbaum was an early AI pioneer and critic, arguing that "there are certain tasks which computers ought not be made to do, independent of whether computers can be made to do them." JOSEPH WEIZENBAUM, COMPUTER POWER AND HUMAN REASON: FROM JUDGMENT TO CALCULATION Preface, x (W.H. Freeman & Co. 1976).

[10] ChatGPT already has the appearance of enabling digital answers to questions on almost any topic more accessible to everyone. However, making it accessible in every context depends critically on the availability of capable digital infrastructure, and that means not just: (a) connectivity, but (b) computing resources, and (c) data (which, in computer infrastructure terms, just means storage).

[11] The goal of enhancing usability and accessibility to end-users without specialized digital skills is hardly new, but the interactive genAI applications raise those to a new level. For example, AI tools can enable end-users with no coding skills to code and can also enhance the productivity of those with skills by providing coding assistance, thus lowering adoption and innovation barriers. GenAI can, in conjunction with low code or no code environments, also release tremendous efficiency gains and speed up the development cycles of new applications (among other things).

[12] In this context, local means context-specific which can be an individual user in a specific task or a group of individuals in part of a "market." The end-users in such contexts have access to and can incorporate information and capabilities that are not directly observable or actionable upon by the providers of the technology.

[13] *See* Jason Potts, *Sources of Innovation in Generative AI*, in Dynamics of Generative AI (ed. Thibault Schrepel & Volker Stocker), Network Law Review, Spring 2024, available at https://www.networklawreview.org/jason-potts-generative-ai.

[14] In that sense, AI is a bit like climate change – its effects and implications may be anticipated fairly in advance by experts, but reacting to it may be postponed until folks are on the cusp of disaster (or not) and then forced to react to it in real-time as best they can. Make no mistake, though, even the mass-market consciousness implications of ICT automation effects have been forecast for a long time. For example, the folks interested in AI have been discussing the economic implications of digital automation and AI for a long time (*see e.g.*, WEIZENBAUM, *supra* note 9), and the economics of automation have provoked policy reactions since the dawn of the first industrial revolution (e.g., Luddite Movement).

networked computers, data centers, cloud platforms, and telecommunications networks that allow AI software to operate. The welfare implications of AI or any technology are not determinable without consideration of how the technology is used. While the technology and its uses are evolving as this chapter is written, we note that there is a polarized debate about whether they are likely to offer a tool for harm rather than for good.[15]

791. Examples of the potential for AI to deliver welfare-enhancing benefits include the potential for AI to enhance safe driving, make communication easier in more situations, and enable individuals to tackle tasks that otherwise would not be feasible (e.g., due to a lack of skills or labor intensity that render a cost-benefit trade-off negative). Practical examples may include facilitating commitments to coordinated climate change responses (e.g., making the Paris Accords relevant) or mining unstructured data for new insights to advance medical diagnoses. Similarly, genAI renders skills scalable and replicable, expanding access in contexts where needed skills are neither readily available nor affordable.[16]

792. Setting aside the existential risks posed by AI noted earlier, bad outcomes that AI may enable include creating opportunities for individuals to undertake new tasks prematurely, for example, when they lack the skills, literacy, or supporting resources to act safely. For example, smartphone GPS has led to under-prepared novices tackling backwoods trails with a false sense of security.[17] More abstractly, social media has expanded mass participation in collective decision-making, which seems superficially like a good thing for citizen engagement but has enabled today's mess of fake news and the

---

15 For varying perspectives on the potential benefits and harms of AI, *see*, for example, Daron Acemoglu, *Harms of AI* (NAT'L BUREAU OF ECON. RSCH., Working Paper No. 29247, Sept. 2021), https://www.nber.org/papers/w29247; Daniel Castro & Joshua New, *The Promise of Artificial Intelligence*, CTR DATA INNOVATION 32–35 (2016), https://www2.datainnovation.org/2016-promise-of-ai.pdf; Noam Chomsky, Ian Roberts & Jeffrey Watumull, *Noam Chomsky: The False Promise of ChatGPT*, N.Y. TIMES (Mar. 8, 2023), https://www.nytimes.com/2023/03/08/opinion/noam-chomsky-chatgpt-ai.html; William Lehr, *5G and AI Convergence, and the Challenge of Regulating Smart Contracts*, *in* EUROPE'S FUTURE CONNECTED: POLICIES AND CHALLENGES FOR 5G AND 6G NETWORKS 72–80 EUR. LIBERAL F. (ELF) ((2022), https://liberalforum.eu/publication/europes-future-connected-policies-and-challenges-for-5g-and-6g-networks/; Chris Stokel-Walker & Richard Van Noorden, *The Promise and Peril of Generative AI*, 614 NATURE 214–216 (2023); Ajay Agrawal, Joshua Gans, and Avi Goldfarb, *Do we want less automation?* Science, 381(6654), 155-158, 2023; and Erik Brynjolfsson, *The Turing Trap: The Promise & Peril of Human-Like Artificial Intelligence*, Daedalus, 151 (2): 272–287, 2022, available at doi: https://doi.org/10.1162/daed_a_01915.

16 For example, automated medical diagnostics can assist lower-skilled humans to achieve good medical care to poorer regions that are unserved or underserved. *See* Orly Lobel, *The Law of AI for Good* (SAN DIEGO LEGAL STUD., Paper No. 23-001, Jan. 26, 2023), https://papers.ssrn.com/sol3/papers.cfm?abstract_id=4338862, for more examples of "AI for good". Also, it is worth noting that AI (software-based) skills may be scaled more easily and quickly, or reskilled flexibly, than humans.

17 The false security is that smartphones will allow instant connection to help if needed, and the map navigation can allow sidestepping of orienteering skills that were previous prerequisites and byproducts of the process by which experienced hikers/mountaineers developed their expertise. Smartphones seemingly short-circuit that skills development process both directly and indirectly. Directly in that the smartphones enable partial replacement but require other skills (like knowing that a satellite phone may be needed when coverage of regular smartphone does not provide connectivity or knowing that backup power like solar cells may be needed when re-charging stations are not available) and capabilities (like being physically fit). Another analogy to the genAI debate is the potential loss in on-the-job training skills development that access to genAI systems may enable when young consultants use ChatGPT that proves less helpful to seasoned consultants. In short, the seemingly good of genAI may include seeds of future problems – like some processed foods being a quick cure for hunger but ultimately bad for one's long-term health.

threat to trust in non-digital institutions, communications, and processes on which all economic activity depends.[18] The new genAI tools raise concerns about the ways in which digital tools may be used to manipulate consumer choice to effect first-degree price discrimination (extract consumer surplus), promote fake news (to disrupt voting or otherwise lower the signal-to-noise ratio, potentially to flood content moderation systems), and create synthetic content (e.g., revenge porn or generally fake news, phishing, and other cyber-crime). Although genAI may make these digital threats to basic human rights more salient and dangerous, the abuse of digital technology to achieve such anti-social goals predates AI and, indeed, digital technology.[19]

793. While collective refinement of our understanding of the ways in which AI may be used for good or bad purposes will be critical for managing AI's economic implications, it is important to remember that AI's ability to have any impact depends on the state of our digital infrastructure. That infrastructure consists of the networked communications (i.e., connectivity, local and wide area networking), computing (i.e., servers and their capabilities, which depend on the chips and software that power them), and storage resources (or "data" in computer infrastructure terms).[20]

794. Digital infrastructure makes (gen)AI possible and establishes the effective bounds of what may be accomplished with it (in terms of use and innovation). It is, therefore, critical for shaping our digital future. Put simply, digital infrastructures determine AI capabilities, which determine AI use and innovation. AI use and innovation, in turn, feed back into AI technologies and the software applications that those AI technologies enable. Investment and innovation drivers operate in both directions. Given the essentiality of the digital infrastructure for the future of AI, the governance of this infrastructure – the resources that are required to sustain it, the ownership of its component parts, how those components are interconnected and co-dependent on each other, how the services those networked ICT infrastructure resources support, and how the overall system is controlled – will prove critical for regulating AI's uses.[21]

---

18   Daniel C. Dennett, *The Problem with Counterfeit People*, THE ATLANTIC (May 16, 2023), https://www.theatlantic.com/technology/archive/2023/05/problem-counterfeit-people/674075/.

19   Fake news to manipulate political outcomes or criminal abuses of technology have long and ancient histories.

20   See, for example, Lehr & Stocker, *supra* note 3 and Volker Stocker, Guenter Knieps & Christoph Dietzel, *The Rise and Evolution of Clouds and Private Networks – Internet Interconnection, Ecosystem Fragmentation*, Paper presented at the 49th Research Conference on Communication, Information and Internet Policy (TPRC) (Aug. 23, 2021, rev'd Oct. 8, 2021), https://papers.ssrn.com/sol3/papers.cfm?abstract_id=3910108. Moreover, the Google I/O '24 Keynote in May 2024 has emphasized this, *see* GOOGLE (May 2024), https://io.google/2024/explore/a6eb8619-5c2e-4671-84cb-b938c27103be/. In the Keynote, CEO Sundar Pichai mentioned that the company invests and develops across the entire stack from digital infrastructure (e.g., Google's customized AI Hypercomputer fueled by ever-more capable (and increasingly energy efficient) CPUs, TPUs, and GPUs and "more than 2 million miles of terrestrial and subsea fiber" which is "over 10 times the reach of the next leading cloud provider") to genAI to consumer-facing applications and services. Such infrastructure (capability) investment and innovation leads use and (uniquely) enables downstream innovation and investment, which in turn drives innovation and investment in infrastructure.

21   *See* Lehr & Stocker, *supra* note 3, William Lehr and Volker Stocker, *Comments to the European Commission's Call for Contributions on the Topic of "Competition in Virtual Worlds and Generative AI*, (Mar. 11, 2024), https://papers.ssrn.com/sol3/papers.cfm?abstract_id=4813002.

III. Navigating Waterfalls Requires a Measurement Ecosystem

795. Considering the current state of AI development and its potential to have transformative GPT-like effects for good or ill, there is significant room for confusion and hysteria to disrupt efforts to pursue sound EBDM. This potential has not been helped by the large number of prominent experts who have raised alarms that AI poses an existential threat.[22] If we are indeed at an inflection point in history, and following our metaphor, heading over the falls, then it is worthwhile considering what might best prepare us to survive our journey. In other words, when you approach a waterfall, the waterfall is not new. However, it certainly warrants new or more attention when you are about to go over the top than when it is still miles ahead.

1. Requirements for a Healthy Measurement Ecosystem

796. In such a situation, at least three aspects are critical: (i) real-time intelligence for context and situational awareness, (ii) the capability for a degree of automated responses ("muscle memory") to react to situational awareness, and (iii) planning for recovery and adjustments after emerging from the falls. All three of these aspects require a healthy measurement ecosystem.

797. Real-time intelligence and automated responses are critical for allowing effective helm control in the midst of the trip over the falls when processes/reactions too fast for human cognition may be necessary. The muscle memory is also built into the fabric and capabilities of our existing regulatory institutions, markets, and technologies. Whatever we confront in our path over the falls, much of our response will be the byproduct of our muscle memory and its automatic responses to real-time, situational intelligence to best address new challenges as they arise. We need to take stock of those capabilities and act quickly to shore-up weaknesses. One obvious issue we see is that there has not been sufficient investment in capacity-building to ensure the needed multi-stakeholder measurement ecosystem is prepared. If you anticipate needing to engage with situations like the one of the drunk looking for his keys under the streetlight, you might want to make sure you have better streetlights and maybe more help to find the keys, and when found, to make sure the drunk does not then drive.[23] More resources need to be directed to support multidisciplinary research on how the technical, economic, and regulatory policy forces are jointly driving the evolution of our digital infrastructure and how we need to reform the regulatory institutions most directly tasked with their management.[24]

---

22 *See, supra* note 4.

23 David H. Freedman, *Why Scientific Studies Are So Often Wrong: The Streetlight Effect*, Discover Mag. (Dec. 9, 2010, 12:00 AM), https://www.discovermagazine.com/the-sciences/why-scientific-studies-are-so-often-wrong-the-streetlight-effect.

24 William Lehr & D. Sicker, *Communications Act 2021*, 18 J. High Tech. L. 270–330 (2018), https://sites.suffolk.edu/jhtl/home/publications/volume-xviii-number-2/.

798. Moreover, planning for the future beyond the falls is needed both to help generate consensus during the trip over the falls and to prepare for possible future states to short-circuit (to the extent feasible) the post-fall adaptation time and costs. This requires measurements and data to inform design decisions that guide future planning. While some might use it as an argument against action, knowing that the future will be different in ways that cannot be fully anticipated does not invalidate the effort to prepare for anticipated future states, even when those are uncertain and, with some probability, unforeseen.[25] We therefore believe that a sound strategy is one that enables preparation for what is expected to be likely, but as well, both for what may be expected but unlikely and for the unexpected. Good EBDM needs to balance efforts across those tasks. Accomplishing that goal will depend on the state of knowledge (i.e., data/market-intelligence, risk assessments) and coordination/response capabilities. Without information, there is no evidence with which to make decisions. However, the existence of evidence does not mean it will be made use of appropriately (i.e., shared and acted upon). Also, aggregating the values of multiple decision-makers with different profiles (i.e., risk tolerance, expectations of outcomes) does not uniquely identify the best courses of action. This further highlights the need for more multidisciplinary research on the structure of the government and industry institutions and tools directed at managing digital infrastructure.[26]

799. The problems identified above are not unique to policymaking about genAI but arise whenever society is confronted with a technological waterfall moment, where the state of current institutions and shared knowledge lack clarity and capacity to be able to act decisively and from a foundation of legitimacy based on some notion of consensus. What constitutes adequate consensus and what that foundation should be has been an enduring challenge for public choice, with long roots in philosophy, political science, and economics. Perhaps of particular relevance is the challenge posed by Arrow's Impossibility Theorem, which denies hope for identifying any best-case algorithmic social choice solution for aggregating the preferences of individuals. As the work of numerous economists has argued, the right governance or social choice model is highly context dependent.[27] The legacy of institutions, laws,

---

25 This echoes Donald Rumsfeld's comment in 2002 regarding the Iraq war when he was the US Secretary of Defense: "As we know, there are known knowns; there are things we know. We also know there are known unknowns; that is to say we know there are some things we do not know. But there are also unknown unknowns – the ones we don't know we don't know." *See* Michael Shermer, *Rumsfeld's Wisdom*, 239 Sci. Am. 38 (Sept. 1, 2005), https://www.scientificamerican.com/article/rumsfelds-wisdom/.

26 The government institutions include existing legacy domain regulators such as those tasked with regulating telecommunications infrastructure. Industry institutions include standardization bodies and research collaborations.

27 The political economy literature on public choice theory is long and complex. It provides a rich foundation for assessing what economics may provide by way of insights regarding the foundations and challenges for EBDM. At the risk of grossly oversimplifying, what it highlights is how good EBDM depends critically on the nature of the information challenges confronting decision-makers and the structure and capabilities

and cultural norms all factor into determining what actions are possible and evaluating alternatives. That is certainly the case in today's global digital economy, where no single regulatory authority, nation state, or other group of stakeholders has the magic key to determine what should be done. Moreover, in the governance mélange we confront, strategic interests that often conflict need to be reconciled, so identifying the boundaries for what might constitute possible cooperative solutions is, in itself, a difficult challenge.[28] As noted earlier, the point of this chapter is that with respect to AI generally (and certainly that also applies to genAI), we are at a waterfall moment where the level of global interest is high, and the uncertainty regarding what to do and the potential for overlapping and potentially incommensurably conflicting policy goals are significant.

800. Nevertheless, although the general problems are not unique to genAI, it is possible to offer several speculations about what genAI may bring that is indeed new. Two features are that genAI has the potential to be self-learning and to access previously inaccessible unstructured data. That raises the potential for automated measurement and decision-making capabilities for EBDM that have previously not been feasible. For example, genAI tools may be used to monitor infrastructures (traditional infrastructures like electric power grids, and telecommunications networks, as well as institutional infrastructures like judicial decision-making or regulatory oversight[29]), and initiate alarms, data collection (surveillance)

---

of the institutions and markets that collect, process, and make use of that information to make decisions. That is fundamentally an asymmetric, imperfect information environment in which every element is fraught with and must contend with ignorance, strategic behavior, and incomplete contracting among the individual stakeholders that are engaged in the challenge of social choice decision-making. That applies to the collection of the evidence in the first instance and then its subsequent sharing with the process by which social choice decisions are made. For key milestones in this literature. See ERIC MASKIN & AMARTYA SEN, THE ARROW IMPOSSIBILITY THEOREM (Columb. Univ. Press 2014), https://doi.org/10.7312/mask15328 on Arrow's Impossibility Theorem which noted the challenges inherent in aggregating the preferences of individuals to make social decisions that meet sound criteria for accommodating potentially diverse preferences while still manifesting consensus. JAME M. BUCHANAN & GORDON TULLOCK, THE CALCULUS OF CONSENT: LOGICAL FOUNDATIONS OF CONSTITUTIONAL DEMOCRACY (Univ. Mich. Press 1965). addressed the challenge by noting the contractual nature of social governance and its contingent legitimacy on individual consent that may be context dependent. Other economists helped further develop the political economy/social choice literature by recognizing how the nature of the context (see, e.g., Vincent Ostrom & Elinor Ostrom, Public Goods and Public Choices, Paper presented at the Polycentricity and Local Public Economies in Readings from the Workshop in Political Theory and Policy Analysis, (Univ. Mich. Press 1999), https://spia.uga.edu/faculty_pages/tyler.scott/teaching/PADP6950_Fall2016_Thursday/readings/Ostrom_Ostrom_1977.pdf or the administrative and bureaucratic structure of governance institutions may impact outcomes, see Matthew D. McCubbins, Roger G. Noll & Barry R. Weingast, *Structure and Process, Politics and Policy: Administrative Arrangements and the Political Control of Agencies*, 75 VA. L. REV. 431–482 (1989).

28 This certainly applies to the question of whether more or less regulatory intervention is desirable in this waterfall moment, and in any case, that is a debate that goes beyond the more modest goals of this chapter.

29 For example, AI tools could be used to scrape legislative and court documents in search of evidence and insights into bias or noise. While such analyses have already been undertaken, the ability to potentially access previously uninvestigated data existing in paper and digital media is greatly expanded by the amalgam of AI Natural Language Processing/Computer Vision (NLP/CV) tools and the analytic capabilities of ML algorithms to collect and extract insights. Those same methods could be applied to any corpus of information or data of interest, including real-time measurements of the performance of physical infrastructure such as video scans of robots traversing pipelines to look for potential faults or scanning cyberspace to look for malicious code. What is especially novel is the generality of the capabilities of genAI to be applicable across many decision-making domains, calling for the automation of monitoring and enforcement actions.

or other actions that could supplement or substitute for today's non-AI based mechanisms. To the extent that such capabilities might be developed and employed, they would confront the same strategic manipulation challenges that confront today's efforts at oversight and monitoring of social choice governance efforts – but this time around, those tasks could be automated.[30]

801. In light of the uncertainty and confusion regarding the effects of AI, it is unclear what regulatory actions or controls may make steerage feasible or beneficial. A real possibility is that it is best to trust in existing regulatory institutions and adopt a "steady-at-the-helm" approach that eschews any significant regulatory action to address AI. That strategy might be promising if the ship had been properly built in advance in preparation for the trip over the falls. Indeed, some who believe no regulatory action is warranted may believe that either there are no falls or that existing institutions are adequate to deal with whatever challenges AI may bring. Maybe they are right, but the push for stronger AI regulatory remedies demonstrates that that view is not universal.

802. In any case, although any and all regulatory interventions are imperfect, and all current efforts are incomplete and leave many critical details indeterminate, it seems regulatory intervention is a foregone conclusion. If we were far from the falls, that indeterminacy would be less of an issue, but if we are in the midst of the transition, we may find ourselves locked into a bad trajectory by our failure to act sooner or by acting incorrectly too soon. Thus, we face significant policy risks no matter what we do. However we decide, it is unclear whether we have the requisite institutional capacity to render our policy initiatives effective.

2. Measurements and Policymaking over the Falls

803. Regardless of one's take on these issues, better measurement capabilities are needed to support better evidence-based decision-making. The measurement ecosystem needed to support decision-making in the current environment will necessarily need to embody certain key requirements, three of which are outlined below.

A. *Multi-Stakeholder Trusted Process*

804. Evidence-based decision-making must rely on a multi-stakeholder, multi-perspective context. There is no single point of view, and there should

---

[30] In the context of AI, this increases the challenges for rendering the methods and capabilities of the AI understandable and transparent to humans that are interacting with the AIs. The efforts to make AI understandable, especially as it applies to the multi-layer neural nets that power ML algorithms, has prompted an extensive literature. For further discussion of the challenges for enabling Explainable AI, *see* IBM (2024), https://www.ibm.com/topics/explainable-ai (last visited July 31, 2024); Jonathon Phillips, Carina Hahn, Peter Fontina, Amy Yates, Kristen Greene, David Broniatowski, and Mark Pryzybocki, *Four Principles of Explainable Artificial Intelligence*, NISTIR 8312 (Sept. 2021), https://nvlpubs.nist.gov/nistpubs/ir/2021/NIST.IR.8312.pdf, or Aorigelo Bao & Yi Zeng, *Understanding the Dilemma of Explainable Artificial Intelligence: A Proposal for a Ritual Dialog Framework*, 11 HUMANITIES & SOC. SCI. COMMC'NS 1–9 (2024), https://www.nature.com/articles/s41599-024-02759-2.pdf.

not be one. While we need cross-layer and end-to-end composability of measurements, we need an algebra to make sense of different measurements and means to facilitate a trusted exchange of information.[31] This conviction is based on the recognition that the ICT industry ecosystem is, in fact, and by desire (i.e., as a key goal of competition policy) a set of complex, competitive value chains with distributed economic control.[32] An aspect to consider is the extent to which downstream complementors depend on upstream inputs and how easily they can switch or effectively multi-home. Arguably, this depends on the application under consideration.

805. Sectoral regulators have the mandate to promote competition, and a lack of competition in infrastructure will fundamentally color competition in any using sector. Therefore, digital competition policy has to consider the reality and complexity of ICT value chains that shape the internet ecosystem as a complex, multi-stakeholder (global, not national) set of horizontally and vertically connected networks – scenarios of "platforms of platforms" and competition among ecosystems must be embraced.

806. In their battle for profits, the digital platforms and infrastructure providers will often disagree on the right way to assess technical or economic performance, what should be the market or product focus for a measurement question, and what inferences should be drawn from measurements. The policy process must recognize that there will be many reasonable ways to address a question, but also many more unreasonable and simply incorrect ways. While determining what is reasonable and what is not often involves gray areas where the benefit of the doubt should be granted to be more inclusive, there is also wide scope for all honest participants to agree on what is clearly inadmissible evidence based on its lack of verifiable provenance, factual errors, or flawed methods. For example, software errors in the implementation of an

---

31 Zoraida Frias et al., Measuring NextGen Mobile Broadband: Challenges and Research Agenda for Policymaking, 32nd European International Telecommunications Society Conference, EuroITS2023 (Madrid, June 19–20, 2023), https://www.econstor.eu/handle/10419/277959. Lehr & Stocker, *supra* note 3.

32 The ownership of relevant digital infrastructures, the control over relevant resources, and the extent of competition vary across the value chain. For example, the foundation models in genAI like GPT-4 (OpenAI/Microsoft), Gemini (Google), Llama2 (Meta), etc., are general-purpose models trained on massive amounts of data and require huge investments in computing and networking resources to build and operate. The market for such models may possibly constitute an oligopoly in light of the significant capital costs required to develop such models and the cost economies that result from this. However, it is unclear how much training data is really needed to support a competitive general-purpose foundation model or whether more specialized genAI models may be more useful. How the valuation for genAI technologies may evolve and how competitive it may be at different levels is uncertain (*see e.g.*, Ela Głowicka & Jan Málek, *Digital Empires Reinforced? Generative AI Value Chain*, in Dynamics of Generative AI Symposium, NETWORK L. REV. (Thibault Schrepel & Volker Stocker eds., Winter 2023), https://www.networklawreview.org/glowicka-malek-generative-ai/; Paul Seabright, Artificial Intelligence and Market Power, in Dynamics of Generative AI Symposium, NETWORK L. REV. (Thibault Schrepel & Volker Stocker eds., Winter 2023), https://www.networklawreview.org/seabright-artificial-intelligence/; Anouk Van der Veer & Friso Bostoen, Two Views on Regulating Competition in Generative AI, in Dynamics of Generative AI Symposium, NETWORK L. REV. (Thibault Schrepel & Volker Stocker eds., Winter 2023), https://www.networklawreview.org/veer-bostoen-generative-ai/.

algorithm,[33] bad input data,[34] or novel methods[35] that are not easily evaluated challenge processes for determining what evidence should be admissible. While many edge cases may prove challenging, there are lots of cases where experts can readily detect and identify errors, even when such identification may not be easy for non-experts. Using standardized test protocols with transparent data and methods enhances the trustworthiness of the measurements and the decision-making processes that make use of them. Unfortunately, not all questions or stakeholders may be able to avail themselves of similarly high-quality evidence. We believe that flexibility in the demands for rigor, transparency, and adherence to established measurement protocols or methods will be needed.

807. AI tools can assist in navigating these challenges in multiple ways. First, AI may facilitate human understanding and access to complex measurement data and analysis tools. This can level the playing field for expertise

---

33 For example, detecting coding errors is amenable to a wide array of techniques. One argument in favor of using open systems instead of systems based on proprietary code is the wider array of auditors open systems enable to check, detect, and correct coding errors. However, there are also good reasons for having proprietary and closed-code systems such as the need to protect intellectual property and exploit efficiencies that may be less easily available in open systems. Debating the benefits of open versus proprietary software code for genAI is beyond the scope of this chapter, but the choices as to what code is open and what closed will raise challenges for the evaluation of AI systems comprised of reusable and layered code. If lower-level systems are updated, updating practices for applications based on them may need to be audited. Addressing such software management challenges calls for ensuring standards for good software coding practice were employed in system development, which includes ensuring clear documentation. Although not due to a software coding error, a practical example of a measurement implementation problem occurred in measurements of broadband throughput performance in the US that factored into Federal Communications Commission (FCC) policy decision-making around 2010. A number of empirical testing protocols relied on single-threaded Transmission Control Protocol (TCP) sessions to measure the speed capability of broadband access services, resulting in measurements that suggested peak data rates that were well below those promised. One important cause for such measurements was that with single-threaded TCP measurements, the software was capping the data rate before the service supporting those measurements was reached. To accurately measure the speed of local broadband services, the measurement applications needed to move to multi-threaded techniques. For a discussion of this and other problems with measuring broadband speeds, *see* Steve Bauer, William Lehr & David Clark, *Gigabit Broadband Measurement Workshop Report*, 50 COMPUT. COMMC'NS REV. (Jan. 2020), https://dl.acm.org/doi/pdf/10.1145/3390251.3390259; Steve Bauer, William Lehr & David Clark, *Understanding Broadband Speed Measurements, Research Conference on Communication, Information and Internet Policy*, TPRC (Sept. 2010), https://papers.ssrn.com/sol3/papers.cfm?abstract_id=1988332.

34 If the input data used to train an AI or inform a decision is obviously biased or otherwise bad (e.g., made-up, full of errors) then "garbage in–garbage out" (GIGO) applies. There are many analytic techniques for detecting such problems, but sometimes knowing the provenance and source of the data can serve as a notice calling for special attention. For example, sponsored data or evidence that obviously supports the strategic position of the sponsor does not make such analysis or data "bad" but it should raise additional concerns that it might be bad. In such cases, seeking additional third-party independent data and investigating data from those with opposing positions may prove helpful. However, not all sources are equally capable or well-informed as noted in preceding notes. AI tools provide a powerful array of capabilities that supplement other non-AI data science tools that can be used to check data for evidence of fraud, bias, excessive noise, or anomalies that raise GIGO alarms.

35 For example, genAI is itself a novel method for discovering, analyzing, and presenting evidence. When decision makers are confronted with novel methods of analysis (relative to status quo best practices) that do not have established technical standards and where there is a lack of consensus and shared expertise among the community of experts called upon to vet evidence, those presenting the evidence confront an additional burden of proof to explain the methods and attest to the validity of those methods. This is certainly a transparency issue and will require prior work among those promoting the technology and methods to establish suitable Application Programming Interfaces (APIs) as well as Key Performance Indicators (KPIs) to enable independent experts to look under the hood of complex systems. AI and genAI increase this burden but can also serve as tools to help address it. If a genAI produces a result that is not understandable or at variance with the results produced by alternative methods, then that does not mean the result is wrong, but it should render the result more suspect. The rush to put in place strong AI regulations at this waterfall moment may be seen as an example of this problem. While we certainly suspect genAI is quite important – potentially a GPT (*see*, *supra* note 3) – we are quite uncertain as to what the right action should be.

across stakeholders.[36] Second, genAI may allow users to extract insights from unstructured data, or data that was previously not readily accessible as evidence for decision-making.[37] Third, AI can help facilitate real-time monitoring for situational awareness.[38] Fundamentally, AI provides a powerful toolset to augment an already rich toolset of digital measurement and data analysis and processing tools to support evidence-based decision-making.

808. In making use of these tools, it is important to remember that not every question needs to be or can be answered, and not everyone needs to be or is capable of being involved in every question. That leads us to consider a second key point.

B. *Balancing Asymmetric Information Requirements*

809. The acquisition and use of data are fundamentally asymmetric; the observation of data and its use are typically separable. Data sharing is essential for collective decision-making, but what and how data is shared is also important.[39] Competition policy has to balance the natural tensions between

---

[36] For example, AI-powered user interfaces (exploiting NLP and CV technologies) can assist in summarizing complex data in ways that are easier for humans to grasp. This can include dynamic visualizations, automated annotations, and interactive dashboard tools that can help users explore the data. An extreme version of such AI-powered technologies could be a full-blown digital twin model of the real-world phenomenon under investigation. Such models already exist for analyzing the behavior of IT sub-systems and components within specific and circumscribed contexts, and in many cases, make use of pre-AI data science methods and technologies. However, applications like ChatGPT expand the accessibility to such methods (e.g., allow users to quickly recover access to knowledge about formulas, definitions, and standard uses of data science results produced using non-AI methods like econometrics). Additionally, when ML techniques are used within the model itself, alternative strategies and designs may be employed to help make the results more intelligible by the relevant audience. For example, transparency regarding the ingested training data as well as the ML algorithms and foundation model, which may have been used to process the data, can prove useful in evaluating the risk of hallucinations and the trustworthiness of the evidence. ML models can produce high levels of confidence in their estimates of probabilities that may prove completely inapplicable if the training data used for the ML is not appropriate. Whether used as a turbo-powered user-interface (to fill gaps in different stakeholders' knowledge or skill sets) to address more complex EBDM contexts (what to measure? What inferences to draw? What decisions to make? etc.) or as part of the analytic engine providing the answers, genAI technologies can expand the range of stakeholders that can consider a wider array of complex evidence.

[37] For example, genAI techniques have been used to investigate video from surgical procedures and from cyberattack records to uncover new diagnostic insights to enable earlier detection of medical and cybersecurity problems. For use in medical diagnosis, *see* Luis Pinto-Coelho, *How Artificial Intelligence Is Shaping Medical Imaging Technology: A Survey of Innovations and Applications*, 10 Bio Eng'g 1435 (2023), https://www.mdpi.com/2306-5354/10/12/1435; Mohamad Koohi-Moghadam & Kyongtae Ty Bae, *Generative AI in Medical Imaging: Applications, Challenges, and Ethics*, 47 J. Med. Sys. 94 (2023), https://doi.org/10.1007/s10916-023-01987-4, or Adam Conner-Simons & Rachel Gordon, *Using AI to Predict Breast Cancer and Personalize Care*, MIT News (May 7, 2019), https://news.mit.edu/2019/using-ai-predict-breast-cancer-and-personalize-care-0507. For use in cybersecurity, *see* Yagmur Yigit et al., *Review of Generative AI Methods in Cybersecurity*, (Mar. 19, 2024), https://arxiv.org/abs/2403.08701; Venkata Ramana Saddi et al., Examine the Role of Generative AI in Enhancing Threat Intelligence and Cyber Security Measures, *2nd International Conference on Disruptive Technologies, ICDT* 537–542 (Greater Noida, India, Mar. 15–16, 2024), 10.1109/ICDT61202.2024.10489766, or Anna Fitzgerald, *How Can Generative AI Be Used in Cybersecurity? 10 Real-World Examples*, Secureframe (May 15, 2024), https://secureframe.com/blog/generative-ai-cybersecurity.

[38] The ability of AI to facilitate the automation of monitoring and real-time data analysis is already well-documented. One obvious example is facial recognition systems, but the examples cited in the preceding notes are also illustrative.

[39] A core function and capability of digital tools is to enable the sharing of information, and there are a plethora of pre-AI and AI tools that facilitate that function. The evolution of the network infrastructure from the silo networks of legacy telephony networks to today's 5G+ networks have vastly expanded the capabilities of our digital networks to collect, process, and make use of information. See William Lehr, Fabian Queder, and Justus

the desire for transparency to ensure the efficient use of information and promoting best practices based on the best available information and intellectual property protections, with the latter being necessary for innovation and investment incentives. Transparency and disclosure requirements are key tools for competition policy. They are utilized to share information (data) to facilitate the realization of positive network externalities, to reduce entry barriers for new firms that need complementary assets (e.g., foundation models, cloud resources, GPUs, etc.), and to facilitate regulatory enforcement directly by regulators and indirectly by market forces.[40] Intellectual property protection (e.g., trade secrets, patents, copyright as legal tools) is provided to ensure incentives to innovate and invest accordingly. Intellectual property rights may enable the creation of market power that can allow innovators to recover their *ex-ante* investments.[41] There are concerns that genAI poses a threat to existing regimes since genAI may make it easier to "reverse engineer" the creative process and thereby bypass the protections that would otherwise be provided by copyright or patents.[42]

810. It is also worth noting that intellectual property rights, by design, are a tool for creating monopoly power. This should remind us that the

---

Haucap, *5G: A new future for Mobile Network Operators, or not?*, Telecommunications Policy, 45(3) 2021, doi:https://doi.org/10.1016/j.telpol.2020.102486. AI and genAI expand the frontiers for how such capabilities may be automated to provide anywhere, on-demand, information collection, processing, and actionable decision-making that may take place in contexts where bounded human intelligence and responsiveness are unable to act (e.g., facilitating the real-time control of UAVs or managing the complex interactions between the applications and digital resources needed to support seamless connectivity). One of the attractions of genAI applications like ChatGPT is the ways in which it renders information accessibility easier for more individuals, even when prior digital tools like pre-LLM search engine tools also rendered the information accessible. A key feature of the future is that the advances in the underlying 5G+ infrastructure are enabling much richer and more granular ways to collect and provide measurement results on-demand, but since the networking elements that enable such measurement are distributed across a more complex collection of industry players (ISPs, cloud and edge service providers, etc.), there are significant strategic challenges that need to be addressed to ensure the right information is collected and shared. That is highly context dependent which means that blanket transparency or disclosure rules or calls for all software to be based on open systems are crude and ill-advised regulatory approaches. Although such rules definitely have an important role to play, designing transparency and disclosure requirements and the proper integration of open and closed systems are complex challenges. For more on how the measurement ecosystem is changing for 5G+ networks, *see* Frias et al., *supra* note 31.

40 That is, regulators and competitors rely on the data that market participants are required to disclose to focus their regulatory or strategic actions.

41 To be clear, if economists thought that the only things that intellectual property rights like patents accomplish is to create monopoly power that then threatens competition, the intellectual property regime would not have been created. The core idea behind patents is to facilitate a market for innovation results, that requires both disclosure (sharing) of the innovative results while also creating a legal framework for innovators to realize some of the benefits of their innovations – ideally, sufficient to provide the dynamic incentives for future innovators to invest in productive R&D. Moreover, it is the abuse of monopoly or market power that poses the threat to competition that justifies regulatory interventions, not typically the mere existence of it (although the creation of market power is a factor in merger review). For a discussion of the challenges and need for considering the complex relationship between intellectual property, innovation incentives, competition, and social welfare, *see* Alden F. Abbott & Daniel F. Spulber, *Antitrust Merger Policy and Innovation Competition*, 19 J. Bus. & Tech. L. (2023); *see also*, Daniel F. Spulber, The Case for Patents (World Sci. Publ'g Co., 2021).

42 A genAI that creates "new" works that are otherwise indistinguishable from other works by an artist might be near-perfect substitutes that would erode the artist's ability to realize economic value from their returns. If the genAI was trained using the artist's earlier work and those uses were not approved (e.g., did not fall under fair-use exceptions), then that might constitute a copyright violation. A genAI program may also be useful in reverse-engineering a patent to find ways to deliver the same functionality without infringing on an existing patent. Whether such applications might constitute patent or copyright infringement is not clear-cut. That lack of clarity, in itself, renders the economic efficacy of existing intellectual property laws less effective. It is worth noting, however, that genAI may also help in copyright and patent enforcement. Crowdsourcing has already proved very effective in establishing the existence of prior art in challenges for patent validity. Similar genAI methods offer additional ways to augment or support searches for prior art.

focus of competition policy actions is not to eliminate market power that may arise as a byproduct of industry economics but to protect against its socially harmful abuses. One way in which intellectual property and competition policy overlap is with standard essential patents. The ICT industry relies heavily on the creation of industry standards that help define industry structure (e.g., where interfaces are and how they should behave), and those standards often have to make use of patented technologies. Standard-setting processes often commit patent-holders to adopt fair licensing practices if their technologies are to be part of the standard, but the enforcement of those commitments has raised significant competition policy challenges. Thus, intellectual property and competition policy are closely related.

811. More abstractly, just as security and privacy or competition and innovation are inextricably bound, so too are transparency (disclosure) and intellectual property policy. AI will prove to be a tool for policymakers and for others both for enforcing *and* circumventing transparency (disclosure) and intellectual property rules.

C. *EBDM for shared infrastructure*

812. The asymmetric deployment of digital infrastructure and distribution of digital skills/literacy will give rise to asymmetric adjustment cost and equity issues. "Road mapping" or scenario-based planning will be required to engage discussions and plan for such contingencies.[43]

813. Moreover, the future of digital infrastructure will require the sharing of resources at all levels (from edge to core, from radio frequency to computing, and from real estate to software platforms). Sharing resources is a fundamental design principle for computer networks.[44] More capable digital infrastructure makes mass-market customization feasible without foregoing the scale and scope economies of shared infrastructure.[45] Enabling such sharing will require customized mapping from resources

---

43 That is, figuring out what plans are needed, designing those plans, and prioritizing among the plans will all require consensus building exercises. The evaluation and real-time-tuning of plans-in-action will require collecting evidence before, during, and after policies are implemented.

44 That is, network resources are provisioned with capacity for aggregate peak demand, taking advantage of the multiplexing opportunities to share capacity among many users and uses when those individual demands can be separated along some dimension – space, time, or context – so that they do not coincide at the peak.

45 ICTs have increasingly facilitated the more flexible customization of production processes to enable the joint realization of scale and scope economies without sacrificing end-user demand for customization. The push to deliver "market-of-one" mass customization goes back to the 1990s. *See* B. JOSEPH PINE II, MASS CUSTOMIZATION: THE NEW FRONTIER IN BUSINESS COMPETITION (Harv. Bus. Sch. Press 1992); Ashok Kumar, *From Mass Customization to Mass Personalization: a Strategic Transformation*, 19 INT'L J. FLEXIBLE MFG. SYS. 533–547 (2007), 10.1007/s10696-008-9048-6 and depends on smart-supply chains and production processes to support just-in-time production processes, a concept that goes back to the 1970s, Damodar Y. Golhar & Carol Lee Stamm, *The Just-In-Time Philosophy: A Literature Review*, 29 INT'L J. PROD. RSCH. 657–676 (1991), https://doi.org/10.1080/00207549108930094. There are many ways that AI, or more narrowly, genAI can contribute to advancing these capabilities Sukhpal Singh Gill et al., *AI for Next Generation Computing: Emerging Trends and Future Directions*, 19 INTERNET OF THINGS 100514 (Aug. 2022), https://doi.org/https:// doi.org/10.1016/j.iot.2022.100514; Romeo Giuliano, The Next Generation Network in 2030: Applications, Services, and Enabling Technologies, 8th International Conference on Electrical Engineering, Computer Science and Informatics, EECSI 294-298 (Semarang Indonesia, 2021), https://ieeexplore.ieee.org/

to consumption at increasingly granular levels, which in turn, will require significantly more granular performance measurement capabilities.

814. Reconciling social welfare, industrial, and competition policies confronts policymakers with a complex mix of challenges. While social welfare and industrial policy goals call for ensuring universal accessibility to essential infrastructures, competition policies seek to ensure that these challenges are met via competitive markets – in so far as that is possible. However, dynamic economic viability requires that investment costs either be recoverable, or a steady stream of subsidies be provided to make up the deficit. The recovery of shared costs presents a fundamental economic problem since no user wants to pay more than they are required to pay.[46] Moreover, while dynamic efficiency requires that usage covers its incremental costs, economics does not identify a *best way* to allocate the burden for the recovery of shared costs. Ramsey pricing offers a second-best, market-based method, but that requires significant information about user demand and costs, and Ramsey pricing is not easily implemented in decentralized markets.[47]

815. Universal service programs are a generalization of public policy initiatives that seek to redistribute economic costs and benefits of market outcomes. The redistribution of economic costs often involves the design of subsidy programs that have to identify who and how to collect the funds to redistribute (tax policies) and then how to direct those subsidies to address equity disparities and promote optimal investment. In the digital infrastructure context, industrial policies call for ensuring affordable access to essential digital infrastructure is available to all citizens. In the telecommunications infrastructure context, the focus in recent years has been on ensuring adequate broadband network connectivity for everyone, with the threshold data rate for what constitutes acceptable and socially desirable broadband service increasing.[48] However, to support ambitious Smart-X applications, more than just faster connectivity will be required. That will mean more edge computing to meet the needs of more demanding

---

document/9624241; ALEXANDROS KALOXYLOS ET AL., AI AND ML–ENABLERS FOR BEYOND 5G NETWORKS, WHITE PAPER 5G PPP TECHNOLOGY BOARD, version 1.0 (May 11, 2021), https://zenodo.org/records/4299895; Xuemin Shen et al., *AI-Assisted Network-Slicing Based Next-Generation Wireless Networks*, 1 IEEE OPEN J. VEHICULAR TECH. 45–66 (2020), https://ieeexplore.ieee.org/document/8954683.

46 Free-rider problems are common.

47 *See* MICHAEL A. CREW & JOSEPH C. SCHUH, MARKETS, PRICING, AND DEREGULATION OF UTILITIES 13, 213–216 (Kluwer Academic 2002) for a discussion of Ramsey pricing and some of the reasons why it is a second-best solution.

48 In 2000, the FCC used 200kbps in both directions (downstream/upstream) as the threshold speed for identifying the availability of broadband services. By 2015, the threshold had been increased to 25Mbps/3Mbps. Today's broadband targets are much higher in many regions of the world. For example, in the US, the Infrastructure Investment and Jobs Act has established the so-called Broadband Equity, Access and Deployment (BEAD) program that considers regions underserved where broadband access is below the targets of 100Mbps/20Mbps BROADBAND USA, https://broadbandusa.ntia.doc.gov/funding-programs/broadband-equity-access-and-deployment-bead-program (last visited July 31, 2024). In the EU, the Digital Decade policy program has established gigabit connectivity for all households in the EU, *see* European Commission, *Europe's Digital Decade* (2024), https://digital-strategy.ec.europa.eu/en/library/digital-decade-policy-programme-2030.

Smart-X applications (not only regarding performance but also data residency, privacy, security, etc.).[49]

816. In the future, the notion of what should be recognized as essential digital infrastructure needs to be expanded beyond connectivity to include computation and storage resources. Moreover, as AI-powered Smart-X apps evolve and grow in importance, some of those may need to be encompassed in our notion of what constitutes essential digital infrastructure. For it to be feasible for any such Smart-X apps to develop to the level where such a consideration is relevant, the digital infrastructure needs to be upgraded. Many of the anticipated genAI app developments that are expected to hold economic promise are reasonable candidates for being part of a set of socially desirable services that may shape the scope and design of future universal service obligations and other industrial policy initiatives for our digital future. Examples may include genAI financial advisors to make good financial advice available to everyone, genAI doctors to expand access to healthcare services, or genAI BuyingBots to enable consumers to identify better deals for themselves on digital markets. It is noteworthy that each of these examples shows that ensuring that the requisite digital infrastructure is available, affordable, and usable in the relevant context is a necessary requirement – not sufficient, but necessary.

817. The road mapping exercises and the creation of the capabilities to collect and share more granular measurements of how infrastructure and services are being deployed and used – and what effects may be observed from such use (on end-user behavior, options for behaving differently, and outcomes of relevance to society) – will be needed to sustain informed discussion and consideration of the many questions that will need to be addressed by EBDM processes.

818. Much of the relevant data to inform our understanding of the technical and economic performance of the digital infrastructure will be derived from the operations of the various networks involved.[50] The measures of traffic and its end-to-end performance in terms of throughput, latency, reliability, and technical and economic Quality of Service (QoS) in

---

49 Over time, computing and storage resources have become increasingly deployed closer to end-users, or into edge networks, complicating challenges of where the boundary should be between service provider and end-user networks, William Lehr et al., Edge Computing: Digital Infrastructure Beyond Broadband Connectivity, Annual Research Conference on Communications, Information and Internet Policy, TPRC (Sept. 22–23, 2023), https://papers.ssrn.com/sol3/papers.cfm?abstract_id=4522089. For example, what computing functions should be undertaken in which servers located (and controlled) by whom? Increasingly, as end-user and cloud service providers locate networked computing and storage resources closer to (and on the property and on the personal devices of) where end-users are using those digital technologies, expanded thinking of how to regulate edge-networks is needed. *See* William Lehr & Volker Stocker, *Next-Generation Networks: Necessity of Edge Sharing*, 5 Frontiers Comput. Sci. 1099582 (2023), https://www.frontiersin.org/journals/computer-science/articles/10.3389/fcomp.2023.1099582/full.

50 The relevant networks are not limited to those owned and operated by legacy telecommunications network operators, which traditionally have been responsible for providing transmission services. The computing, data storage, and higher-level software service platforms (e.g., cloud computing, content delivery networks, etc.) are also key parts of the digital infrastructure needed to support Smart-X applications. Digital technology makes it increasingly feasible to shift where digital functionality is provided and the reconfiguration of organizational structures. Consequently, regulatory frameworks premised on legacy classifications of service

different usage contexts (by market, by provider, by application, by end-user, etc.) will depend on a mix of in-network and outside measurement infrastructures (vantage points, active and passive probes, and networks to aggregate and process those measurements).

819. Fortunately, it appears likely that network operators will develop the measurement capabilities of the digital infrastructure because these are required for them individually to provide the sorts of dynamically customizable, scalable, and resilient services they are investing in as part of their next-generation network and future digital economy plans. Those plans are premised on the hope and expectation that ever-more-capable AI-driven Smart-X applications will need those capabilities.

820. Although the measurement capabilities are evolving, we cannot assume that the necessary information will be collected and appropriately shared among the relevant industry participants. The collection of the information is costly, has strategic value, and raises privacy considerations that make the design of the measurement ecosystem challenging.[51] Not all possible measurements are worth making, not all measurements that are made should be shared, and not everyone needs to know all the information that is shared. If the entire digital infrastructure of networked computing and communications resources were controlled by a single company, the infrastructure and competition policy challenge would be one of regulating a monopoly. In that case, policymakers would rely on the monopolist for realizing the industrial policy goals of ensuring universal service of reliable, high-quality digital infrastructure were met. To do that, the details for how information and decisions were shared among the constituent components would be left to the monopolist to manage, subject to regulatory oversight over the outcomes and performance goals. Instead, competition policymakers could focus on metrics related to the consumer experience and whether the monopolist was abusing its market power. However, that is not the case.

821. Instead, what we have is digital infrastructure that is comprised of a complex fabric of interconnected digital ecosystems, each of which is associated with multiple layered platforms with complex horizontal and vertical business and technical interactions. For example, Apple, Amazon, Microsoft, and Google are each associated with digital ecosystems with digital network and service platforms that, at different levels, support a host of other ICT-focused firms that sometimes are concentrated within a single ecosystem and sometimes operate across multiple ecosystems. Moreover, these players are hardly unique globally in offering products that compete either at select levels (e.g., search, cloud computing, social

---

providers into ISPs, edge/application, or digital platform providers need to be re-examined. See William Lehr, David Clark, Steve Bauer, and KC Claffy, *Regulation when platforms are layered,* TPRC47: Research Conference on Communications, Information, and Internet Policy, Washington DC, 2019, available at: https://papers.ssrn.com/sol3/papers.cfm?abstract_id=3427499.

51 Frias et al., *supra* note 31; William Lehr, *The Changing Context for Broadband Evaluation,* in TRANSFORMING EVERYTHING?: EVALUATING BROADBAND'S IMPACTS ACROSS POLICY AREAS, edited by K. Mossberger, E. Welch, and Y. Wu, Oxford University Press, 2022.

media, devices) or across multiple levels. The consumers and end-users of all of these services are also hardly homogeneous across or within markets however those markets are defined (i.e., by service, geographic location, or demographics). This is fundamentally a multi-stakeholder environment that has to embrace and reconcile diverse social and economic policy goals that will require and provide scope for consideration of multiple perspectives on evidence comprised from diverse measurement inputs.

822. The measurement infrastructure will necessarily mirror the complexity of the industry fabric and will depend on measurement elements (probes, sensors, sources of data) that are controlled, owned, and operated by both private and public stakeholders. That diversity of measurement sources and infrastructure is important but will raise additional challenges of how (best) to protect the ability to consider diverse but reasonable measurements while excluding fake or simply incorrect data.[52] Obviously, there will be gray areas – especially when it comes to evaluating subjective opinions or experiences – where such a determination may not be feasible. However, not all areas are gray. In technical matters, science may make it easier to apply theory and bring to bear the force of repeatable/reproducible, robust experimental methods to validate data. That is much harder in situations involving economic data, especially when that relates to making inferences about incentives or private valuations.

823. However, as digital infrastructure is increasingly integrated ever more deeply into our lives, we cannot avoid joint consideration of the technical/economic measurement challenges. That requires us to improve our economic data measurement capabilities while respecting the privacy and strategic challenges that such collection will entail. The same capabilities that will allow informed decision-making in a multi-stakeholder environment at a more fine-grained (local) level will also provide the technology that could enable the surveillance economy and the risk of dictatorial control.[53] To navigate those waters, and mixing metaphors, we need to improve our streetlighting. That is, the infrastructure "police" (i.e., the governance institutions tasked with enforcing industrial and competition policies) need better techno-economic "streetlight" capabilities to find the keys that the drunk will likely drop, and to help protect everyone so that the drunk can get where he needs to go without driving drunk. That will require efforts along multiple

---

52 A fuller investigation of the challenges and options for EBDM in this context is needed. One "challenge" area is to reform industry standard setting and self-regulatory governance institutions to provide more guidance regarding the appropriate definition of metrics and measurement procedures and disclosure/transparency policies. In note 31, we described the challenges of measuring broadband throughput (speed) performance. Deciding how best to measure and disclose those measurements to consumers has been assigned as a task for industry self-regulation.

53 As noted earlier, although the potential problems of a surveillance economy do not depend on AI or genAI, AI technologies render those capabilities more powerful and usable. NLP and CV enable automated facial-recognition that can take advantage of expanding IoT infrastructure supporting the collection of video and other media data (e.g., via CCTV systems). AI-powered Graphical User Interfaces (GUIs) and dashboards can significantly enhance the accessibility and use of surveillance media that previously was more difficult to integrate and make use of. Moreover, such systems can be integrated with real-world systems like the locks and security devices we use to secure buildings, etc.

fronts, but one important direction is to invest more in multidisciplinary capacity building.[54] That will require institutional reforms to existing sectoral regulators and expanded capabilities and resources to enable both better long-range planning as well as fast-response teams to address issues as they arise.

3. Competition and Infrastructure Policy over the Falls

824. In this chapter, we have indirectly shown that to craft good genAI competition policy, it will be important to focus on digital infrastructure policy, with special attention o building a healthy measurement ecosystem to support EBDM. This is due to multiple reasons. First, the digital infrastructure provided by the amalgam of entities that comprise the ICT sector is the nursery and foundation that makes genAI possible and where the direction of innovation and its intersection with real-world effects may be best first observed. Second, the tools to collect the data needed to make sound evidence-based policy will be embedded in and dependent on how the digital infrastructure evolves, constraining the possibility space of asymmetric information challenges for digital competition policy. And third, because in the uncertain future we are embarking on as we go over the falls, we need institutional multidisciplinary capacity to jointly address the technical and economic issues that will continuously arise and challenge pre-digital (legacy) approaches to competition policy.

825. In managing the digital infrastructure, the evolution of which will dictate the competition policies confronted within the ICT sectors responsible for delivering that infrastructure as well as the sectors that will make use of the digital infrastructure, we can identify three core competition issues. These will need to be addressed and require sound multidisciplinary research perspectives.[55]

826. First, competition policy will need to ensure that shared access is possible for all bottlenecks that may arise in the fabric of digital infrastructure.[56] A bottleneck is a resource that is necessary and cannot be bypassed. In digital infrastructure, a bottleneck facility may be the result of a natural monopoly (which might require regulatory protection to be sustainable) or winner-takes-all economics. Ideally, having at least a duopoly or more competitive choices, or at least contestability for every resource, may

---

54 The capacity building certainly requires making sure the policymakers and agencies on the front lines of providing security and emergency responses have access to AI technology expertise since it is certainly the case that those who will be posing security threats, the malicious actors, and other sources of threats to society (e.g., accidental problems) will be utilizing AI tools. However, as we have tried to make clear here, the challenges are not unique to genAI or even AI but are a byproduct of the increasing digitalization of all aspects of our economic and social lives. The multidisciplinary capacity needed requires more holistic and active integration of expert engagement among engineers, lawyers, and social scientists – including economists and others – who are broadly familiar with digital technologies. Too narrow a focus on genAI or AI expertise – although some of that is needed – will likely balkanize efforts to promote good EBDM.

55 The following discussion draws on Lehr & Stocker, *supra* note 21.

56 William Lehr & Douglas Sicker, *Telecom Déjà Vu: A Model for Sharing in the Broadband Internet*, 28 INFO. & COMMC'NS TECH. L. 1–36 (2019), https://www.tandfonline.com/doi/abs/10.1080/13600834.2019.1653546.

render active competition policy intervention unnecessary, and in many cases, the threat of regulatory action may be sufficient to deter any bottleneck monopolist from abusing control of the relevant resource. Identifying where bottlenecks may exist, and which ought to be regulated is not easy. In the context of legacy telecommunications regulation, it is clear that last-mile internet service providers (ISPs) are not the only players with market power in the internet ecosystem, and legacy regulations are in need of updating in light of changes in how edge networks are currently designed and provisioned.[57] AI tools have the potential to enable more dynamic and flexible multi-homing and network resource configuration options that may shift the balance of power in the fabric of digital infrastructure and the locus of where (or if) bottlenecks exist.[58] Of course, AI capabilities might also be used to reinforce and sustain the market power of entities with bottleneck control over particular resources.[59]

827. Second, competition policy will need to enable and ensure that ecosystem switching is possible. That will require a level of interoperability. However, that may not be feasible or desirable at all levels. Final consumers (end-users) ought to have choices for the digital infrastructure they depend on, and those choices should not be overly constrained by artificial barriers imposing switching costs. In telephony, number portability improved retail-level competition by enabling users to avoid having to incur the costs of changing to a new telephone number when they switched to a different network provider. AI technologies can facilitate the economic management of switching costs in ways that can both increase and decrease those costs for consumers. For example, AI translators can make multi-homing easier, while impediments to the portability of data used to personalize an AI may make it more difficult to switch across ecosystem platforms. For example, an appeal to users of sticking with a single ecosystem (e.g., Apple or Android) is the convenience (and reduced learning/adaptation costs) of being able to seamlessly access services across multiple devices. Ensuring that that is feasible while protecting end-user privacy and security is far from a simple technical challenge. By comparison, enabling number portability for fixed and later mobile telephone services was easy.

828. Moreover, another impediment to consumer switching among ecosystems or sub-components in digital infrastructures (i.e., mix-and-match selection of applications, devices, services, and service providers) depends on consumers having adequate access to the information needed to make

---

57 For example, the (efficient) management of next generation networks will require network slicing technologies that will pose a risk of harmful discrimination if abused.

58 For example, *see* Michael Gal & Amit Zac, Is Generative AI the Algorithmic Consumer We Are Waiting For?, in Dynamics of Generative AI Symposium, Network L. Rev. (Thibault Schrepel & Volker Stocker eds., Winter 2023), https://www.networklawreview.org/gal-zac-generative-ai/ who note the potential for genAI to support enhanced consumer bots, but with ambiguous welfare implications. *See* Schrepel & Potts, *supra* note 13 for a discussion of how the degree of openness of an AI system can help foster an information commons which enhances the likelihood that innovative results will be shared more widely and helps mitigate competition concerns.

59 Głowicka & Málek, *supra* note 32.

informed decisions. That may depend on disclosure and transparency policies. Once again, AI can be a tool to aid or hamper the ability to facilitate access to market participants to make informed decisions. In line with our explanations of EBDM above, perhaps the best hope for promoting the existence of what is needed is to ensure a robust market of third-party providers and users of network performance relevant data. Direct government data collection efforts (e.g., via test labs, sponsored research by academics, etc.) should be part of, but not the sole source of such efforts.

829. Finally, there is the real potential that a needed component of digital infrastructure does not exist (or is not available at the required quality, capability, or usability[60]). That may be more a matter for industrial policy than for competition policy, but the reason the needed component may not exist might be a competition failure that may be directly related to the risk of market power or to some other coordination failure, irrespective of any market power issues (e.g., incomplete markets, asymmetric information, or "Prisoner's Dilemma" type issues).[61] The threat of market power may deter investment because of a lack of trust by investors in what may be a winner-take-all competition that any winner will be able to earn risk-adjusted returns if it succeeds (to offset the expected losses should the investor fail). One reason for public provision of infrastructure is because it is indeed more efficient for there to be a monopoly provider that will provide the facility and then share access on a non-discriminatory, cost-based basis, or equivalently, provide the needed digital infrastructure component as a public utility service.[62] While relying on provision by competitive markets is the preferred alternative, figuring out when public participation (from provision of the service to subsidies) and management of the resource may be better in the evolving digital infrastructure future is not easy. Once again, AI can be both a source for solutions and problems.

---

60   Availability is not sufficient for usability. That depends on the resource being affordable to those who are expected to use it productively, and that those users have the requisite complementary resources in terms of skills and ancillary tools (devices, applications, data, etc.) to make use of the resource. The point is that these are never simple EBDM challenges.

61   Markets may fail to be efficient even if there is no risk of market power for numerous reasons. For example, participants in the market may fail because of incomplete markets that may prevent the negotiation of enforceable contracts. Moral hazard and adverse selection are examples of problems that arise because of incomplete contracts, asymmetric information, and externalities, see BERNARD SALANIÉ, THE ECONOMICS OF CONTRACTS: A PRIMER (MIT Press 2005). In the Prisoner's Dilemma, each player has a dominant strategy that precludes cooperation even though the failure to cooperate is worse for both (see Anatol Rapoport, *Prisoner's Dilemma*, in THE NEW PALGRAVE DICTIONARY OF ECONOMICS (Palgrave Macmillan 2018), https://doi.org/10.1057/978-1-349-95189-5_1850).

62   The level of public sector engagement and control in the provision of infrastructure can vary. Historically, the US favored a privately owned public utility model, wherein the AT&T Bell System was granted a monopoly franchise that was protected and supported by public subsidies. Many other nations favored government-owned public utilities (PTTs). The appropriate level of government engagement in the provision of basic infrastructure varies by context and type of infrastructure. For example, there is limited experience with providing private road networks, whereas electric power, natural gas, and water infrastructure often have a mix of private and publicly owned providers. With telecommunications infrastructure, components like outside structures (poles, conduits, rights of way) may be publicly owned, whereas the active components (servers, radios, etc.) may be privately owned.

830. In our fast-moving, fluid world of digital technologies, the future – which is inherently uncertain – arrives fast. "Unknown unknowns" as well as "known unknowns" intermingle with the asymmetric dispersion of insights. This is why it is so essential to build a measurement ecosystem that allows good data to be collected and shared to build toward consensus visions of both our areas for agreement and disagreement. The risk of bad data and noise, whether malicious or as a byproduct of collective ignorance, poses a serious threat to a sound dialog.

831. As markets/technology move faster and ICT control capabilities become more agile (in part because of further advances in AI and genAI), it becomes increasingly important to be able to rapidly decide how to adapt and act on those decisions.[63] However, as path dependency is unavoidable, forecasting into the uncertain future will also become more important.

---

63 This imperative is accentuated by the increasing returns that characterize many information technology investments, including genAI. *See* Thibault Schrepel & Alex Pentland, Competition between AI Foundation Models: Dynamics and Policy Recommendations (MIT CONNECTION SCI., Working Paper No. 1-2003, June 28, 2023), https://papers.ssrn.com/sol3/papers.cfm?abstract_id=4493900.

AI, IP, and Competition Policy: Adjusting Policy Levers to a New GPT

QUENTIN B. SCHÄFER[*]

University of Strathclyde

Abstract

This chapter contributes to the emerging debate on the intersection between IP as a set of legal norms and AI as a novel technology. The issue of the regulation of IP as applied to AI is highly topical in light of the rapid progress and increasing impact of AI as a technology on our lives and the numerous debates surrounding IP protection for AI tools and outputs. The chapter examines the suitability of different policy levers to improve and maintain incentives to invent and invest in AI. It argues that the fundamental technological and economic uncertainty surrounding AI renders anticipatory doctrinal adjustments in IP law, such as AI inventorship, undesirable. Rather, scholarship should focus on the design of policy levers capable of flexibly accommodating a variety of different technological and market outcomes, in particular on the IP-Competition Interface. It also proposes a research agenda for the IP-Competition Interface in relation to AI, focusing on obligations to share access to closed systems and IP rights.

---

[*] Lecturer (Assistant Professor) in Competition and Private Law, University of Strathclyde. Correspondence: quentin.schafer@strath.ac.uk. I would like to thank the editors, as well as Pablo Ibáñez Colomo, Bowman Heiden, Ioannis Lianos, and Malin Petrén for helpful comments. All errors are my own. The author has no conflict of interest to disclose.

AI, IP, and Competition Policy: Adjusting Policy Levers to a New GPT

I. Introduction

832. Artificial Intelligence (AI) is likely to be more than merely another arrow in humanity's ever-expanding quiver of tools. Rather, and this is without attempting to be sensationalist, we are most likely to usher in a world imprinted by a new general purpose technology (GPT),[1] a technology rivaled in its impact on human welfare, the economy, and civilization only by agriculture, metal working, the printing press, the steam engine, semi-conductors, and the internet.[2] Given its predicted impact, AI has been seen as forming part of a forthcoming "fourth industrial revolution."[3]

833. Continual technological human progress since the industrial revolution has arguably led us towards exponential progress.[4] The most ambitious view of this is the singularity, of autonomous self-replicating and self-improving AI.[5] But AI need not reach such lofty heights to fundamentally transform products and production processes. Facial recognition, self-driving cars, and chatbots are merely the first apparent, consumer-facing emanations of AI in our new economy. The incorporation of AI into production processes offers significant benefits to the economy at large by improving productivity and increasing innovation.

834. Given AI's potential economic impact, it is not surprising that AI, IP, and competition law have started to collide. Numerous lawsuits at the intersection of AI and IP have been filed implicating issues varying as widely as the ability of AI-generated inventions to be patented[6] and the scope of copyright exemptions in training AI foundation models.[7] Similarly, the impact of AI on competition policy has become much discussed; for example, advances in computation may offer firms the opportunity for algorithmic collusion[8] but also grant enforcers the ability to deploy computational techniques.[9]

---

1 Timothy Bresnahan & Manuel Trajtenberg, *General Purpose Technologies: "Engines of Growth"?* (NAT'L BUREAU OF ECON. RSCH., Working Paper No. 4148, 1992), https://www.nber.org/papers/w4148.

2 Manuel Trajtenberg, *Artificial Intelligence as the Next GPT: A Political-Economy Perspective*, in THE ECONOMICS OF ARTIFICIAL INTELLIGENCE: AN AGENDA 175 (Ajay Agrawal, Joshua Gans & Avi Goldfarb eds., 2019); Iain M. Cockburn, Rebecca Henderson & Scott Stern, *The Impact of Artificial Intelligence on Innovation: An Exploratory Analysis*, in THE ECONOMICS OF ARTIFICIAL INTELLIGENCE: AN AGENDA 115 (Ajay Agrawal, Joshua Gans & Avi Goldfarb eds., 2019).

3 Klaus Schwab, *The Fourth Industrial Revolution*, FOREIGN AFFAIRS, (Dec. 12, 2015), https://www.foreignaffairs.com/world/fourth-industrial-revolution.

4 AZEEM AZHAR, EXPONENTIAL: HOW ACCELERATING TECHNOLOGY IS LEAVING US BEHIND AND WHAT TO DO ABOUT IT (2021).

5 *See e.g.*, Vernor Vinge, *The Coming Technological Singularity: How to Survive in the Post-Human Era*, in VISION-21: INTERDISCIPLINARY SCIENCE AND ENGINEERING IN THE ERA OF CYBERSPACE 11 CP-10129 (1993), https://ntrs.nasa.gov/api/citations/19940022855/downloads/19940022855.pdf.

6 Thaler v. Comptroller-Gen. Pat., Designs and Trade Marks [2023] UKSC 49.

7 Getty v. Stability AI [2023] EWHC (Ch) 3090.

8 *See e.g.*, Emilio Calvano et al., *Artificial Intelligence, Algorithmic Pricing and Collusion*, SSRN (Apr. 1, 2019), https://papers.ssrn.com/sol3/papers.cfm?abstract_id=3304991&utm_source=pocket_reader.

9 *See e.g.*, Thibault Schrepel, *Computational Antitrust: An Introduction and Research Agenda*, 1 STAN. COMPUT. ANTITRUST 1 (2021).

835. If AI is as crucial a technology as commonly thought, law ought to promote its development by maximizing the incentives to innovate and invest in AI.[10] Yet, we would do well to recognize the limited capacity of our brains to foresee future technological and economic developments before we (may) transition from neurons to silicon. As of 2024, our ability to effectively foresee the exact way that AI as a technology and as an economic force will shape markets is more limited than the numerous suggestions for changes to legal norms in the recent academic literature on AI, IP, and AI and competition law may make it seem.

836. The main argument of this chapter is that the complexity of the technology stack for AI, compared to previous GPTs, as well as fundamental technological and economic unknowns make accurate predictions as to the correct scope of IP entitlements difficult. As a result, we ought to look to self-executing policy levers rather than speculative anticipatory readjustments of areas of law when we are looking for legal norms that promote invention and investment in AI innovation. In this chapter, I examine European IP and competition law for such levers, finding examples i.a., in the PHOSITA standard in patent law and the IP-Competition Interface.

837. Section II examines the general difficulties of anticipatory changes to IP doctrine in light of AI; I argue that the complexity of the AI tech stack as well as the differential application of IP law to different parts of the stack make *ex ante* adjustments to IP doctrine difficult. Section III examines the idea of different types of policy levers that allow us to calibrate responses to technological and economic change in the context of AI; policy levers are preferable to inflexible changes in doctrine considering the pervasive uncertainty surrounding the technological and economic consequences of AI. Section IV takes a look at different opportunities for policy levers on the IP-Competition Interface, setting out a research agenda to examine this insufficiently explored topic.

II. AI, Law and Technology, and New GPTs

838. When I speak of AI, I mean deep learning, i.e., statistical techniques of learning from data using artificial neural networks.[11] New technologies are commonplace in our modern economy. What is particularly noteworthy about AI is that it may go beyond a mere technology. AI

---

10 *See e.g.*, Ryan Abbott, *Intellectual Property and Artificial Intelligence: An Introduction*, in Research Handbook on Intellectual Property and Artificial Intelligence (Ryan Abbott ed., 2022); Peter Georg Picht & Florent Thouvenin, *AI and IP: Theory to Policy and Back Again – Policy and Research Recommendations at the Intersection of Artificial Intelligence and Intellectual Property*, 54 I.I.C. 916 (2023); Axel Voss, *What New Legal Rules Could Foster Competition and Innovation Dynamics In The Generative AI Ecosystem?*, Network L. Rev. (Feb. 14, 2024), https://www.networklawreview.org/voss-generative-ai/.

11 As described e.g., in François Fleuret, The Little Book of Deep Learning (2023), https://fleuret.org/public/lbdl.pdf; compare e.g., Ray Kurzweil, How to Create a Mind: The Secret of Human Thought Revealed (2012) for a different view of how to achieve (super)human-level AI.

is likely to be a general-purpose innovation in the method of invention (IMI).[12] Put simply, AI may help us make further scientific and technological advances.

1. Layers of the AI Tech Stack

839. AI is not a "market" in the economic sense or as one would define a market in competition law. AI as a technology is an input to technology and product markets. Accordingly, an entire stack of markets which are distinguished by their different technological operation, their different economics, and the different degrees of protection afforded by IP rights, is shaped by AI. We can list several clearly separate layers of the technology required for AI and its outputs. First, AI requires computational hardware. Second, AI models are trained on a substantial amount of data. Third, foundation models can be trained to apply to a variety of use cases. Fourth, applications may utilize foundation models to produce more specialized outputs. Fifth, the AI outputs produced by these applications, for example pictures created by Dall-E or ChatGPT's response to a query, form a final layer.[13] These different levels of the markets do not correspond well to market definition in competition and antitrust law; generally, the further downstream the market, the more market definition diverges from the scheme set out above.

2. Adjusting to New GPTs in the Past: The Role of Intellectual Property

840. AI is not the first technological development that law had to keep pace with. Neither is it the first GPT that intellectual property law has had to accommodate. I briefly detail the responses of intellectual property law to a few GPTs. Overall, IP law has evolved in response to technological change, rather than in anticipation of it.

841. In the common law system, modern intellectual property law entered the statute book with the English Statute of Monopolies of 1624[14] and the Statute of Anne of 1710,[15] establishing entitlements to patents[16] and copyright, even if they resembled today's bureaucratic, and democratized system only to an extent. These statutes predate the Industrial

---

12 Cockburn, Henderson & Stern, *supra* note 2.

13 For a similar disaggregation of the different layers of markets related to AI, *see* Thibault Schrepel & Alex Pentland, Competition between AI Foundation Models: Dynamics and Policy Recommendations (MIT CONNECTION SCI., Working Paper No. 1-2003, June 28, 2023), https://papers.ssrn.com/sol3/papers.cfm?abstract_id=4493900.

14 For a history of the earlier common law, *see* TERRELL ON THE LAW OF PATENTS ¶¶ 1-08-1-16 (Douglas Campbell et al. eds., 20th ed. 2024).

15 For a history of the earlier English law, *see* 1 COPINGER AND SKONE JAMES ON COPYRIGHT ¶¶ 2-46-2-54 (Gillian Davies et al. eds., 18th ed. 2021).

16 The first instances of patent grants with exclusionary rights for inventions occurred in middle age Venice, *see e.g.*, Ted Sichelman & Sean O'Connor, *Patents as Promoters of Competition: The Guild Origins of Patent Law in the Venetian Republic*, 49 SAN DIEGO L. REV. 1267 (2012).

Revolution, commonly agreed to having started around the middle of the eighteenth century.[17] Bently and Sherman trace the development of modern intellectual property rather to the middle of the nineteenth century.[18] Even on the more charitable view, the IP system was not a response to the age of consistent economic and technological progress that began with the industrial revolution (and has continued since). The IP law that existed seems to have had a minor effect on the accelerating the technological development of the industrial revolution in any event. Mokyr argues that the patent law in England, where the industrial revolution began, did not give Britain a head start in the industrial revolution, even if it gave inventors the hope to "try their luck."[19]

842. So too, in modern times, the history of intellectual property is one of *ex post* adaptation to technological changes rather than of *ex ante* promotion of technological change through IP. Controversies surrounding the patentability of organisms,[20] DNA,[21] computer software, or the implications of hyperlinking[22] and APIs[23] in copyright came about because firms seeking legal protection for inventions and creations sought to retrofit existing standards to new technology. They did not result from legislatures proactively creating standards in IP law that promoted the increased or decreased ability to obtain IP rights for technologies of a new type.

843. Accordingly, IP protection has not historically been used to promote the development of novel technologies at the frontier of IP law. As a further example, consider the steam engine. Boldrin and Levine have argued that patents on Watt's steam engine significantly delayed the industrial revolution.[24] While this account has been vigorously disputed by other scholars,[25] it remains true that the positive effect of IP protection on incentives to create the steam engine was minimal.

844. It is obvious that intellectual property law was not designed with AI in mind. Most major intellectual property statutes are a product of the twentieth century.[26] The two foundational international treaties on IP date

---

17   JOEL MOKYR, THE ENLIGHTENED ECONOMY: AN ECONOMIC HISTORY OF BRITAIN 1700–1850 4 (2009).

18   BRAD SHERMAN & LIONEL BENTLY, THE MAKING OF MODERN INTELLECTUAL PROPERTY LAW: THE BRITISH EXPERIENCE 1760–1911, 208–210 (1999).

19   MOKYR, *supra* note 17, at 410.

20   Diamond v. Chakrabarty, 447 U.S. 303 (1980), https://supreme.justia.com/cases/federal/us/447/303/.

21   Ass'n for Molecular Pathology, et al. v. Myriad Genetics, Inc., et al., 569 U.S. 576 (2013).

22   *See e.g.*, Case C-610/15, Stichting Brein v. Ziggo (Pirate Bay), ECLI:EU:C:2017:456.

23   Google v. Oracle, 593 U.S. ___ (2021).

24   MICHELE BOLDRIN & DAVID K. LEVINE, AGAINST INTELLECTUAL MONOPOLY (2008), http://www.dklevine.com/general/intellectual/againstfinal.htm.

25   George Selgin & John Turner, *James Watt as Intellectual Monopolist: Comment on Boldrin and Levine*, 47 INT'L ECON. REV. 1341 (2006); George Selgin & John L. Turner, *Watt, Again? Boldrin and Levine Still Exaggerate the Adverse Effect of Patents on the Progress of Steam Power*, 5 REV. L. & ECON. 1101 (2009).

26   For illustration, note the UK's Registered Designs Act 1949; Patents Act 1977 and Copyright, Designs and Patents Act 1988.

back to the nineteenth century.[27] Current controversies surrounding the patentability of AI-created inventions[28] and the application of copyright exemptions to data used for the training of AI models[29] mirror the past trend that adjustments of IP law postdate the emergence of new technological paradigms.

3. General Difficulties in Adjusting IP Law to AI

845. As Mokyr notes, macro inventions are largely serendipitous and difficult to predict.[30] In the case of AI, the current AI "spring" became possible as a result of the increased accessibility of computational power, rather than as a result of transforming discrete R&D into AI. It is inherently difficult to prepare for new GPTs. We have a limited degree of foresight as to which will be the next GPT. Given the AI "winter" of the late twentieth century, before the advances in machine learning taken in the 2010s, few technologists and fewer scholars had made their bets on AI.[31] Yet, even when we can predict what the next GPT will be with substantial confidence, as we can today with AI, it is an even more complex endeavor to predict optimal adaptations of IP law to AI. I offer four reasons for this.

846. First, AI exhibits substantial connected, fundamental technological and economic uncertainty. Second, the complexity of the AI tech stack coupled with the variety of markets impacted by AI as a GPT makes singular doctrinal solutions difficult to find; in other words, error rates are likely high for wholesale adjustments of legal doctrine. Third, given the anticipated economic impact of AI[32] and the potential distortion from inadequate IP doctrine, error costs are likely to be high. Fourth, IP law exhibits considerable doctrinal, economic, and political "stickiness" i.e., resistance to change.

A. *Fundamental Technological and Economic Uncertainty*

847. It comes as no surprise to any IP or competition lawyer that there is still no general consistent theory of the interaction between law and innovation. The academic literature has admittedly made progress

---

27 Paris Convention for the Protection of Industrial Property 1883; Berne Convention for the Protection of Literary and Artistic Works 1886.

28 *Thaler, supra* note 6.

29 In the US, databases normally do not attract copyright unless they can be protected as a compilation, *see* Feist Publ's, Inc. v. Rural Tel. Servs Co., 499 U.S. 340 (1991); In the EU, there is a sui-generis right protecting databases, *see* Directive 96/9/EC of the European Parliament and of the Council of Mar. 11, 1996 on the legal protection of databases. Individual Member States may offer protection through national copyright law, 1996, O.J. (L 77) 20. For example, § 4 the German Copyright Law protects databases as the "author's own intellectual creation."

30 Joel Mokyr, The Lever of Riches: Technological Creativity and Economic Progress 295 (1990).

31 For a history of AI, including the period(s) of the "AI winter", *see e.g.*, Stuart J Russell & Peter Norvig, Artificial Intelligence: A Modern Approach 17–27 (2021).

32 Philippe Aghion, Benjamin F. Jones & Charles I. Jones, *Artificial Intelligence and Growth, in* The Economics of Artificial Intelligence: An Agenda 237 (Ajay Agrawal, Joshua Gans & Avi Goldfarb eds., 2019).

since Schumpeter, Arrow, and Machlup's famous paradox,[33] but not to the extent of being able to causally connect intellectual property protection and innovation on a consistent theoretical basis. This is exacerbated by the coupled technological and economic uncertainty surrounding AI.

848. One core problem facing the regulation of AI is that there is no certainty that the current trend of returns from increasing the amount of compute thrown at models (known as "scaling") will continue. At present, there is evidence that there is some scope for the optimization of current model sizes.[34] But, if the trend does not continue, it is possible that the progress in the field may continue using other architectures or that it may stagnate.[35] Similarly, we may run into issues of limited compute or limited energy at some point if further returns require increases in the scale of foundation models.[36] This fundamental technological uncertainty generates a substantial degree of economic uncertainty. The shape of the markets producing or using AI in the future is unknown. For example, new computational techniques decreasing the need for further scale would significantly decrease the importance of the hardware layer and are likely to have substantial implications for the ability of smaller players or even consumers to apply AI to tasks.[37]

849. Further, the general trajectory of AI as an economic force is unknown. Aghion, Jones, and Jones show that economic growth depends on a number of different technological factors.[38] At the extreme, AI may run into physical limits or, counterintuitively, the speed of successive AI-generated innovation may reduce the incentives for the implementation of innovation for the benefits of humans.[39] Another point of uncertainty is the extent to which AI allows for the automation of tasks. Illustratively, contrary to expectations that the first tasks to be automated, would be tasks of low-skilled workers, recent advancements in generative AI seem to automate "nonroutine, cognitive tasks performed by high-skill workers."[40]

---

33 FRITZ MACHLUP, AN ECONOMIC REVIEW OF THE PATENT SYSTEM 80 (1958) ("[i]f we did not have a patent system, it would be irresponsible, on the basis of our present knowledge of its economic consequences, to recommend instituting one. But since we have had a patent system for a long time, it would be irresponsible, on the basis of our present knowledge, to recommend abolishing it."); *see also* Robert P. Merges, *Updating Machlup: The (Still Uncertain) Case for Patents*, THE MEDIA INST. (Aug. 10, 2015), https://www.mediainstitute.org/2015/08/10/updating-machlup-the-still-uncertain-case-for-patents/.

34 *See e.g.*, Marah Abdin et al., *Phi-3 Technical Report: A Highly Capable Language Model Locally on Your Phone*, COMPUT. SCI. (May 23, 2024, 22:42 UTC), https://arxiv.org/abs/2404.14219.

35 Azeem Azhar, *The Bedlam of AI*, EXPONENTIAL VIEW (Apr. 25, 2024), https://www.exponentialview.co/p/the-bedlam-of-ai?publication_id=2252&utm_medium=email&utm_campaign=email-share&triggerShare=true&r=qovd.

36 Dwarkesh Patel, *Mark Zuckerberg – Llama 3, Open Sourcing $10b Models, & Caesar Augustus*, DWARKESH (Apr. 18, 2024), https://www.dwarkeshpatel.com/p/mark-zuckerberg (last visited Apr 26, 2024).

37 *See* Schrepel & Pentland, *supra* note 13.

38 Aghion, Jones & Jones, *supra* note 32.

39 *Id.* at 259–261.

40 *Id.* at 239.

If automation remains possible only to a limited extent, economic growth may be hamstrung by scarce labor.[41]

850. Similarly, the optimal market structure for the markets involved in AI is unknown. At this time, it is far from straightforward to assert, with any degree of confidence, that the AI tech stack will (in the absence of market-shaping regulation) organize itself in a vertically integrated oligopoly or in an interconnected "AI value chain." In any event, the impact of AI on the market structure of other product markets is likely to be sector-dependent and therefore complex.

B. *The Complexity of the AI Tech Stack and the Variety of Impacted Markets*

851. The number of levels of the AI stack is characteristic of a continuing increase in complexity of successive GPTs. Fire was a natural phenomenon. Agriculture was more complex, involving e.g., techniques of irrigation and crop rotation. Ironworking was more complex than bronzeworking, requiring more advanced tools to shape the metal. The steam engine was based on (for the time) advanced mechanical, chemical, and thermodynamic principles. Starting with computing and communications, complexity increased even more significantly. AI as a GPT is has significantly more "moving parts" than previous GPTs; a natural result of technological progress.

852. In broad terms, there are four layers to the AI tech stack: (1) infrastructure/hardware, (2) data, (3) AI models, (4) applications.[42] When we are concerned with IP protection, we may add a fifth layer: (5) AI outputs.[43] AI may produce inventions that are implemented in physical media (e.g., the food container allegedly created by DABUS)[44] or it may produce creative works that have physical representations (e.g., an AI-designed piece of art painted by robotic hands).

853. Doctrinal questions of IP protection attach to virtually every single layer of this tech stack (aside from the hardware level, which follows established principles of intellectual property law). AI tools, including foundation models, may or may not be patentable, depending on their "technical contribution" and exclusions to patentability.[45] Applications are located on the same precipitous edge between patent law and copyright law as

---

[41] *Id.* at 256; *see also* Daron Acemoglu, *The Simple Macroeconomics of AI* (NAT'L BUREAU OF ECON. RSCH., Working Paper No. 32487, 2024), https://www.nber.org/system/files/working_papers/w32487/w32487.pdf?utm_source=substack&utm_medium=email.

[42] *See* Schrepel & Pentland, *supra* note 13.

[43] Reto M. Hilty, Jörg Hoffmann & Stefan Scheuerer, *Intellectual Property Justification for Artificial Intelligence*, *in* ARTIFICIAL INTELLIGENCE AND INTELLECTUAL PROPERTY (Jyh-An Lee, Reto M. Hilty & Kung-Chung Liu eds., 2021).

[44] *See Thaler*, *supra* note 6.

[45] MATT HERVEY & MATTHEW LAVY, THE LAW OF ARTIFICIAL INTELLIGENCE ¶¶ 8-004–8-064 (2021), *see also* ¶¶ 8-134–8-146 on copyright subsistence.

other computer software.[46] Many more substantial questions attach to the protectability of AI outputs. When it comes to AI, there are a number of different technologies involved, all of which contribute to AI as a GPT and which, in our balkanized IP system, are capable of attracting different types of IP rights and different strengths of protection.

854. Alongside changes in the economics of markets wrought by AI, justifications for IP protections may be affected by AI. Hilty, Hoffmann, and Scheuerer have argued that, in light of AI, deontological justifications decrease in value due to reduced human interaction.[47] As such, the economic justifications for the protection for AI tools are weak as there is little chance misappropriation.[48] There may, however, be better justification for the IP protection of (at least some) AI outputs.[49]

855. Similarly, the uncertainty surrounding the future prevalence of open source models compared closed source models is liable to substantially affect the dynamics of the markets comprising the AI stack.[50] Open source models present fewer concerns in competition law[51] and voluntarily forgo the exclusivity promised by intellectual property protection. Intellectual property laws can, depending on their configuration, promote or retard open source.[52]

856. It is worth noting that, despite the legal complexity and uncertainty surrounding the application of IP law to AI, considerable investment in the technology is taking place[53] at an accelerating pace.[54]

C. *Error Costs*

857. The potential impact of AI on economic growth and human welfare is significant.[55] Accordingly, adjusting institutions to promote AI's development and deployment has tremendous value.

---

46 Peter R. Slowinski, *Rethinking Software Protection*, in ARTIFICIAL INTELLIGENCE AND INTELLECTUAL PROPERTY 341 (Jyh-An Lee, Reto M. Hilty & Kung-Chung Liu eds., 2021); *see also* Hao-Yun Chen, *Copyright Protection for Software 2.0?*, in ARTIFICIAL INTELLIGENCE AND INTELLECTUAL PROPERTY 323 (Jyh-An Lee, Reto M. Hilty, & Kung-Chung Liu eds., 2021) (on copyright protection for partially AI-generated software).

47 Hilty, Hoffmann & Scheuerer, *supra* note 43.

48 *Id.*

49 *Id.*

50 For an overview, *see* Thibault Schrepel & Jason Potts, *Measuring the Openness of AI Foundation Models: Competition and Policy Implications* (SCI. PO DIGIT. GOVERNANCE & SOVEREIGNTY CHAIR, Working Paper, May 14, 2024), https://papers.ssrn.com/sol3/papers.cfm?abstract_id=4827358.

51 *Id.* at 27–28.

52 *Id.* at 23–25.

53 European Parliament Research Service, *AI Investment: EU and Global Indicators*, EPRS (2024), https://www.europarl.europa.eu/RegData/etudes/ATAG/2024/760392/EPRS_ATA(2024)760392_EN.pdf.

54 *AI Startup Funding More Than Doubles in Q2, Crunchbase Data Shows*, REUTERS, (July 9, 2024, 1:24 PM GMT+1), https://www.reuters.com/technology/artificial-intelligence/ai-startup-funding-more-than-doubles-q2-crunchbase-data-shows-2024-07-09/.

55 Aghion, Jones & Jones, *supra* note 32; *see also* Commission Regulation (EU) 2024/1689 of the European Parliament and of the Council of June 13, 2024 laying down harmonised rules on artificial intelligence and amending Regulations (EC) 300/2008, (EU) 167/2013, (EU) 168/2013, (EU) 2018/858, (EU) 2018/1139 and (EU) 2019/2144 and Directives 2014/90/EU, (EU) 2016/797 and (EU) 2020/1828 (Artificial Intelligence Act), 2024, O.J (L Recital 4).

858. At the same time, it has become a mainstream position that AI must be trustworthy and accountable.[56] The deployment of AI clearly incurs some risks, in particular in relation to fundamental rights.[57] The EU's AI Act prohibits AI in certain applications involving i.a. biometrics and manipulative techniques[58] and imposes other obligations in "high-risk" scenarios, i.a. to avoid replicating existing biases[59] as well as in relation to general purpose AI models.[60] These concerns are now finding their way into the first regulatory instruments, first and foremost the EU's AI Act.

859. Regardless of one's view of the likelihood of such adverse consequences, the error costs of incorrect IP doctrine surrounding the development and deployment of AI are substantial. Several cautionary tales in the past illustrate the difficulty of getting doctrine "right." Where IP sought to anticipate future developments and incentivize technologies outside the existing paradigm, it has generally not been tremendously successful. Going back to the example of the steam engine, Lemley argues that:

> [T]he subsequent development of steam engines was arguably driven by the Boulton-Watt patents, but not in the way we normally expect patents to work. Instead, the lockup effected by Watt's broad patent fights drove subsequent inventors to seek different approaches to the steam engine. It was one of those different approaches, designed to avoid the Boulton-Watt patents, that actually succeeded in making steam engines practical.[61]

860. In recent decades, arguably, patent protection granted for business methods and software in the US has done more harm than good.[62]

861. On a more technical level, adjustments in doctrine can result in unanticipated consequences. The incentive to generic drug makers of market exclusivity after filing patent challenges granted by the Hatch-Waxman Act resulted in the lock-up of markets under the pay for delay paradigm,[63] subsequently addressed through competition law.[64]

---

56 European Commission, *Ethics Guidelines for Trustworthy AI*, (Apr. 8, 2019), https://digital-strategy.ec.europa.eu/en/library/ethics-guidelines-trustworthy-ai; Council of Europe, *Framework Convention on Artificial Intelligence and Human Rights, Democracy and the Rule of Law*, (2024), https://rm.coe.int/1680afae3c; Commission Regulation (EU) 2024/1689, Recitals 1 and 2.

57 *See* Commission Regulation (EU) 2024/1689, Recital 5.

58 *Id.* at art. 5.

59 *Id.* at arts. 6–49.

60 *Id.* at arts. 51–56.

61 Mark A. Lemley, *The Myth of the Sole Inventor*, 110 MICH L. REV. 709, 717 (2012).

62 *See e.g.*, Bronwyn H. Hall, *Business Method Patents, Innovation, and Policy* (U.C. BERKELEY COMPETITION POL'Y CTR., Working Paper No. CPC03-39, 2003); James Bessen & Robert M. Hunt, *An Empirical Look at Software Patents*, 16 ECON. MGMT. STRATEGY 157 (2007); *but see* Daniel F. Spulber, *Should Business Method Inventions Be Patentable?*, 3 J. LEG. ANAL. 265 (2011).

63 C. Scott Hemphill, *Paying for Delay: Pharmaceutical Patent Settlement as a Regulatory Design Problem*, 81 N.Y.U. L. REV. 1553 (2006).

64 Fed. Trade Comm'n v. Actavis, Inc., 570 U.S. 136 (2013), *see* Bradley Graveline & Jennifer M. Driscoll, *The US Supreme Court establishes a rule that blurs the lines between antitrust and patent law (Actavis)*, E-COMPETITIONS June 2013, art. No. 66794.

862. These examples are intended to show that many changes in IP law intended to address issues arising from frontier technologies or that seek to shape market outcomes have high error rates and error costs where the requisite experience has not been gained. The technical and economic complexity of the AI tech stack makes such errors more rather than less likely.

D. "Stickiness" of IP doctrine

863. Lastly, IP law exhibits a high degree of "stickiness." This is for four interconnected reasons. First, IP law is politically charged and polar. Second, IP doctrine takes time to unfold. Third, international law limits flexibility. Fourth, IP and its existing structure are culturally embedded.

864. First, the political nature of IP makes existing law "sticky." Proposed substantial changes in IP doctrine often face substantial special interest opposition and manifestly little political resolve in light of the long-term amorphous benefits flowing from any change.[65] In particular, previous beneficiaries of IP protection often retain substantial influence on the future of IP doctrine. Meanwhile, the drawbacks of excessive IP protection lie in the future, can be difficult to analyze, and often have no strong constituency being harmed and capable of organizing around the benefits of weak IP protection.

865. Second, IP doctrine requires substantial time to demonstrate an effect. The crucial effects of IP protection are on future investment in innovation, which takes years if not decades to materialize.

866. Third, the existing political and legal inflexibility of much of international IP law contributes to "stickiness." International IP law permits flexibility in relation to limitations and exceptions through the "three step test."[66] There is less scope when it comes to altering the fundamentals of existing categories of law. Nevertheless, it is worth noting that this has not prevented the emergence of new sui generis rights, such as database rights or semiconductor designs, as well as regulation that impacts substantially on the use of related subject matter, such as data privacy regulation.

867. Fourth, our conception of IP and its existing structure is culturally embedded and therefore difficult to change. And, regardless of academic debates surrounding IP justifications, it remains culturally, instinctively correct to reward inventors and creators and protect their works against infringement by others.

---

65 *See e.g.*, HERBERT J. HOVENKAMP, *United States Antitrust Policy in an Age of IP Expansion* (U. IOWA LEGAL STUD., Research Paper No. 04-03, 2004).

66 World Intellectual Property Organization, Berne Convention for the Protection of Literary and Artistic Works, 1979 (as amended Sept. 28, 1979), art. 9(2); Agreement on Trade-Related Aspects of Intellectual Property Rights, 1994, art. 13 (copyright), art. 30 (patents).

868. In the course of the advent of modern IP law, entitlements to intangibles became organized in closed categories.[67] These categories have become (to an extent) fixed through law, culture, and institutional inertia. Yet, doctrinally retrofitting existing categories on the basis of limited legislative changes or judicial interpretation is liable to make a limited amount of sense in the context of AI. For example, when we are concerned with AI as an IMI, our main concern with the IP system may shift from the provision of sufficient incentives to invent to the concern that insufficient material remains in the public domain. Recent years have seen continual additions to the classical categories of IP law, but such additions have been adopted after certain types of innovations became mainstream rather than in furtherance of bringing them about.

869. In addition, given that IP rights are "property" rights in the sense that they are conceived of as being less amenable of subsequent limitation or deprivation than non-property rights,[68] it is not trivial to limit the scope of IP rights *ex post*. I take a look at the role of competition law in this context in Section IV.

III. IP Policy Levers in the AI-Context

870. Institutions are critical to economic growth and technological progress. It should be little surprise that our present-day institutions require some adjustments to continue functioning in the novel technological and economic context of AI.

1. What is a Policy Lever?

871. The idea of a "policy lever" originates from a 2003 article by Lemley and Burk in which they argued in favor of sector-specific patent law on the basis that patents are crucial to some sectors (e.g., pharmaceuticals) but counterproductive in others (e.g., computer software).[69] By "policy lever", Lemley and Burk mean a flexible legal standard that allows courts to keep up with evolving technologies.[70] Some of the policy levers exhibited by Lemley and Burk, such as utility and non-obviousness, in patent law are *ex ante* policy levers. They apply at the time that an IP right comes into existence, either through automatic subsistence (e.g., copyright) or through a grant procedure (e.g., patents). Some, such as experimental use, apply *ex post*, at the time of determining infringement.

872. We can contrast a "policy lever" with a mere doctrinal change, whether effected by the legislature or judicial interpretation; both types of change may occur as a result of changing circumstances, but a policy lever is a

---

67 SHERMAN & BENTLY, *supra* note 18.

68 *See e.g.*, Charter of Fundamental Rights of the European Union, Dec. 7, 2000, art. 17, 2000, O.J. (C 364) 1.

69 Mark A. Lemley & Dan L. Burk, *Policy Levers in Patent Law*, 89 VA. L. REV. 1575 (2003).

70 *Id.* at 1579.

flexible instrument taking account of both the old and the new, whereas a doctrinal change discards the old principle, rendering cases in the old paradigm subject to the new rule.

873. The idea of a policy lever then translates straightforwardly to our instant problem. We are not (only) concerned with different sectors, we are also concerned with different levels of the AI tech stack which have vastly different abilities to benefit from AI, vastly different inherent protections against spillover effects, and vastly different ways in which firms may monetize their technology.

2. Prioritizing Policy Levers

874. As noted above, in IP law, major doctrinal changes have substantial error costs, as well as high error rates when they are made in anticipation of uncertain future developments. One example of such a manual policy change would be allowing AI authorship and inventorship for all types of IP rights. Abbott has argued that recognizing AI inventorship and creatorship would promote incentives to invent.[71] Yet, the implications of such a change are hardly foreseeable and it is utterly unclear whether the increased incentives to invent would exceed, e.g., concerns raised by patent thickets[72] or harms to follow-on innovation. This is particularly so because the question of AI authorship/inventorship crosses over many different IP rights and many different sectors of industry.

875. Designing IP doctrine to respond flexibly to economic and technical developments is preferable to "manual" adjustments in the context of dynamic technologies and markets. In the dynamic context, policy levers incur lower error rates and error costs. IP theory suggests clearly that IP protection is not necessary where inventors may appropriate the returns from an invention even without IP protection. Granting IP protection where unnecessary results merely in the deadweight loss of monopoly, which is commensurate to the market power granted, i.e., the breadth and width of an IP right. Given the distinct nature of information protected by IP rights at each layer of AI technology, differential IP protection is a policy lever in itself.

876. For example, in the realm of patents for AI outputs, the much-discussed PHOSITA (person having ordinary skill in the art) standard could operate as an existing policy lever. If an invention is obvious to a PHOSITA, it fails to surmount the "obviousness" hurdle and is therefore not patentable. The PHOSITA standard may come to include the use of AI tools in the course of invention, thereby raising the non-obviousness standard,

---

71 Ryan Abbott, *Artificial Intelligence, Big Data and Intellectual Property: Protecting Computer Generated Works in the United Kingdom*, in RESEARCH HANDBOOK ON INTELLECTUAL PROPERTY AND DIGITAL TECHNOLOGIES (Tanya Aplin ed., 2020); Abbott, *supra* note 10.

72 Raphael Zingg, *Foundational Patents in Artificial Intelligence*, in ARTIFICIAL INTELLIGENCE AND INTELLECTUAL PROPERTY (Jyh-An Lee, Reto M. Hilty & Kung-Chung Liu eds., 2021).

constraining the ambit of patentable subject matter, and expanding the public domain.[73] If the market continues to exhibit returns from scale, PHOSITA will likely include AI on the basis that virtually every person skilled in the art would have found it obvious to use AI in the process of invention. If it does not, development slows and inventing using AI will not be commonplace, thus the PHOSITA will presumably not include AI. The PHOSITA standard is therefore sensitive to the future development of the industry. It is also worth noting that, at the same time, the present design of the PHOSITA standard may have unintended consequences. It is conceivable that the patentability bar may be raised so high that only well-resourced inventors (with advanced AI) will be able to obtain patents.[74] This would be an example of a poorly constructed policy lever.

877. Along the same lines, Zingg has argued that patentability, disclosure, non-obviousness, and novelty can act as policy levers to prevent the emergence of AI patent thickets.[75] Such flexible standards are ideal for our current stage of moving into a novel industrial paradigm, whose economic and technological effects are uncertain.

IV. AI and the IP-Competition Interface as a Policy Lever

878. When uncertainty regarding future market outcomes is substantial, *ex post* policy levers that are sensitive to the *ex post* economic context may be more appropriate than *ex ante* policy levers. In particular, the law has already evolved a policy lever to regulate IP overprotection through the IP-Competition Interface. This addresses the concern that the overprotection of IP doctrine may cause an excessive monopoly deadweight loss as well as harms to follow-on innovation.

879. Here, I focus on the IP-Competition Interface. It is worth noting that there have been other suggestions for flexible e*x-post* policy levers outside the realm of IP and competition law. For example, Pendleton has suggested to replace IP with a tort of unfair competition in light of the uncertainty of litigation outcomes and AI.[76] Scheuerer has suggested reliance on a flexible doctrine of unfair competition law before market failures are clearly apparent.[77] I do not discuss *ex-post* policy levers in IP law, e.g., experimental use, as suggested by Burk and Lemley.

---

73 *See e.g.*, HERVEY & LAVY, *supra* note 45, at ¶¶ 8-089–8-096].

74 *Cf.* Picht and Thouvenin, *supra* note 10, at 921.

75 Zingg, *supra* note 72.

76 Michael D. Pendleton, *An Abject Failure of Intelligence: Intellectual Property and Artificial Intelligence*, in RESEARCH HANDBOOK ON INTELLECTUAL PROPERTY AND ARTIFICIAL INTELLIGENCE (Ryan Abbott ed., 2022).

77 Stefan Scheuerer, *Artificial Intelligence and Unfair Competition – Unveiling an Underestimated Building Block of the AI Regulation Landscape*, 70 GRUR INT'L 834 (2021).

880. Given that, perhaps to the surprise of some readers, the principles of the IP-Competition Interface are not well worked out in any jurisdiction, this section presents more as a research agenda than as a list of policy recommendations.

1. The State of the IP-Competition Interface

881. The IP-Competition Interface is a policy lever that deserves more attention than it has been given in the past two decades. While considerable scholarly attention has been applied to the Patent-Antitrust Intersection,[78] in particular, the intersection between non-patent IP rights and competition law has remained significantly under theorized and under analyzed.[79] Given that competition law restricts market power *ex post* whereas IP rights grant some degree of market power *ex ante*, it follows that competition law can be used to curtail the market power granted by IP rights. The extent to which this is legitimate has been a question debated for over a century.[80] Jurisprudence at present adheres strongly to a view of the "complementarity" of IP and competition law.[81] Complementarity implies a lack of deference to IP rights in the course of competition enforcement. Beyond this, the content of "complementarity" remains unclear to a substantial extent.

882. The competition laws do not apply in the same manner to every type of right.[82] It is obvious from the case law, at least in the EU, that application of the competition laws routinely restrains the ambit of IP holders' exercise of their (intellectual) property rights. Property rights including IP rights are subject to some degree of deference in the case law, even if such deference tends to be implicit. While property rights may have been suggested by the CJEU to be irrelevant in the context of Article 101 TFEU,[83] it is clear, that access to essential facilities will not be subject

---

[78] *See* i.a. William F. Baxter, *Legal Restrictions on Exploitation of the Patent Monopoly: An Economic Analysis*, 76 YALE L.J. 267 (1966); WARD S. BOWMAN, PATENT AND ANTITRUST LAW: A LEGAL AND ECONOMIC APPRAISAL (1973); Louis Kaplow, *The Patent-Antitrust Intersection: A Reappraisal*, 97 HARV. L. REV. 1813 (1984); Michael A. Carrier, *Unraveling the Patent-Antitrust Paradox*, 153 U. PA. L. REV. 761 (2002).

[79] Daryl Lim & Peter K. Yu, *The Antitrust-Copyright Interface in the Age of Generative Artificial Intelligence*, 74 EMORY L.J. (forthcoming 2024).

[80] At least since Bement v. Nat'l Harrow Co., 186 U.S. 70 (1902), https://supreme.justia.com/cases/federal/us/186/70/; *see also* Amelia Smith Rinehart, *E. Bement & Sons v. National Harrow Company: The First Skirmish Between Patent Law and the Sherman Act*, 68 SYRACUSE L. REV. 81 (2018).

[81] U.S. Dep't of Just. & Fed. Trade Comm'n, *1995 Antitrust Guidelines for the Licensing of Intellectual Property*, (1995), https://www.justice.gov/atr/archived-1995-antitrust-guidelines-licensing-intellectual-property; Case T-172/21, Valve Corp. v. Eur. Comm'n, ECLI:EU:T:2023:587; *see* Quentin B. Schäfer, *Case T-172/21 Valve v Commission – Revisiting the Territorial Character and Probabilistic Nature of Intellectual Property Rights in Competition Enforcement*, 45 EUR. COMPETITION L. REV. 233 (2024); *see also* Faziel Abdul & Stef Geelen, *The EU General Court upholds fines for a video games developer and holds that the existence of IP rights does not preclude the application of a cartel ban (Valve)*, E-COMPETITIONS Sept. 2023, art. No. 115906.

[82] Compare Pierre Régibeau & Katharine Rockett, *The Relationship between Intellectual Property Law and Competition Law: An Economic Approach, in* THE INTERFACE BETWEEN INTELLECTUAL PROPERTY RIGHTS AND COMPETITION POLICY 505 (Steven D. Anderman ed., 2007).

[83] Case 56/64, Consten and Grundig v. Eur. Comm'n, ECLI:EU:C:1966:41; Case T-65/98, Van den Bergh Foods v. Eur. Comm'n, ECLI:EU:T:2003:281, *see* Nicolas Petit, *The EU General Court rejects the application for annulment dismissing the argument that the reluctance of retailers to sell products of other ice-cream manufacturers could not be attributed to the exclusivity clause (Van den Bergh Foods/Commission)*, E-COMPETITIONS Oct. 2003, art. No. 57069.

AI, IP, and Competition Policy: Adjusting Policy Levers to a New GPT

to duties of access without a clear abuse under Article 102 TFEU, with standards for IP rights being higher than for other property rights.[84]

883. While the exact implications of "complementarity" remains in doubt, at a minimum, complementarity implies a recognition that both IP and competition law are parts of an integrated system of regulation promoting innovation. In markets shaped by emerging technologies, the *ex-post*, economics-sensitive view of competition law is of particular importance. Competition law can play two roles: (1) It can promote openness throughout the AI value chain, and (2) it can restrain existing IPRs for AI tools and AI outputs where they prove undesirable.[85] If we take seriously the idea that competition law can restrain IP law where IP grants are excessive, it may offer a convenient device that is explicitly sensitive to issues of market power, collusion, and abuse. Such terms are natural to competition lawyers but not to IP lawyers.

2. Hardware, Talent, Data, and Orthodoxy

884. Before we get to where the IP-Competition Interface has some novel utility, it is worth taking a look at the "hardware" layer i.e., in markets for the hardware needed to produce and run AI models. At present, it seems that IP rights offering strong protection are largely addressed at the hardware level of the AI tech stack. As noted above, there are no significant debates on the need for increased or decreased IP protection in relation to hardware.

885. We have already seen that the value and applicability of IP differs significantly throughout the AI value chain. The hardware side offers little in the way that departs from ordinary principles of competition law. The present scarcity of compute, even if it does not abate, is unlikely to be problematic per se. Yet, enforcers must be aware of the possibility of anti-competitive foreclosure when compute is scarce, but such considerations must be balanced against the possible returns from further scale. The low degree of antitrust concern is consistent with the strength of IP and trade secrecy in this context. Another potential issue may be presented by constrained markets for talent, but outside merger control, competition law is ill-equipped to address such issues. Accordingly, there is also no particular need to adjust the principles of the IP-Competition Interface in this regard. In the absence of property rights to data (outside

---

84 Cases C-241/91 and C-242/91 P, RTE v. Eur. Comm'n (Magill), ECLI:EU:C:1995:98; Case C-7/97, Oscar Bronner v. Mediaprint, ECLI:EU:C:1998:569; Case T-612/17, Google v. Eur. Comm'n, ECLI:EU:T:2021:763, *see* Frédéric Pradelles & Mary Hecht, *The EU General Court confirms the Commission's decision to fine a Big Tech company for abusing its dominant position in online search by discriminating against comparison shopping services to favour its own offering (Google Shopping)*, E-COMPETITIONS Nov. 2021, art. No. 106676.

85 Suggestions in this vein have been made in the more general context of the IP system, *see* Michael W. Carroll, *One for All: The Problem of Uniformity Cost in Intellectual Property Law*, 55 AM. U. L. REV. 845 (2006); Michael W. Carroll, *Tailoring Intellectual Property Rights to Reduce Uniformity Costs*, 1 *in* RESEARCH HANDBOOK ON THE ECONOMICS OF INTELLECTUAL PROPERTY LAW 377 (Ben Depoorter & Peter S. Menell eds., 2019); *see also* PABLO IBÁÑEZ COLOMO, THE NEW EU COMPETITION LAW (2023) (arguing that the Commission has been engaged in proactive "market-shaping" in the context of IP rights).

the largely defunct database right) or stronger data protection legislation, we are unlikely to see significant issues on the IP-Competition Interface surrounding this layer of the AI tech stack.

3. Promoting Openness throughout the Value Chain

886. We may draw more interesting conclusions on the remaining layers of an AI value chain. The strength of competition law is its ability to take the economic context of conduct into consideration when acting as a policy lever. This includes the presence of IP rights.

887. In the context of data used to train models, the role of copyright exceptions is currently subject to extensive litigation. Yet, where copyright exemptions are held not to be applicable and copyright allows firms to prevent the training of AI models on their data, suits enforcing such intellectual property may come to be seen as anti-competitive in certain circumstances. Concerns regarding anti-competitive effects may apply in particular to proprietors of particularly crucial data sets, such as scientific publishers, newspapers, or social media platforms. Case law surrounding the copyright misuse doctrine in the US,[86] or the case law surrounding the enforcement of IPRs[87] in the EU, could support this view, but significant further research is required. So far, misuse-type doctrines have not been integrated well on a theoretical or practical basis into the larger scheme of market regulation.[88] An alternative to misuse-type doctrines which displace the ability of the IP holder to enforce its rights may be the grant of effective compulsory licenses in competition law (discussed in Section IV.4 below).

888. In the context of foundation models as well as applications and ecosystems, we see highly competitive dynamics with trends towards vertical integration through partnerships[89] and acquihires.[90] Nonetheless, the CMA noted concerns that foundation models are tilting towards incumbents with better ability to attract hardware, talent, and strategic partnerships.[91] Similarly, some scholars have argued that the market mirrors digital markets in the rise of platforms and winner takes all dynamics.[92] Others have

---

86 Lasercomb Am., Inc. v. Reynolds, 911 F.2d 970 (1990).

87 Case 53/87, CICRA and Maxicar v. Renault, ECLI:EU:C:1988:472; Case 238/87, AB Volvo v. Erik Veng (UK) Ltd., ECLI:EU:C:1988:477; Case T-111/96, ITT Promedia NV v. Eur. Comm'n, ECLI:EU:T:1998:183.

88 *See generally* DARYL LIM, PATENT MISUSE AND ANTITRUST LAW: EMPIRICAL, DOCTRINAL AND POLICY PERSPECTIVES (2013).

89 *See e.g.*, Emilia David, *Microsoft's Mistral Deal Beefs up Azure without Spurning OpenAI*, THE VERGE (Mar. 4, 2024, 5:36 PM GMT), https://www.theverge.com/24087008/microsoft-mistral-openai-azure-europe.

90 Tabby Kinder, *Microsoft Hires DeepMind Co-Founder Mustafa Suleyman to Run New Consumer AI Unit*, FIN. TIMES (Mar. 19, 2024), https://www.ft.com/content/5feedf3a-ff7a-4c89-9b1d-f9b48834ff4c; Tim Bradshaw, *UK Regulator Examines Microsoft and Amazon's AI Dealmaking*, FIN. TIMES (2024), https://www.ft.com/content/0597a834-c101-4d01-893d-87c2eca122c9.

91 COMPETITION & MARKETS AUTHORITY, AI FOUNDATION MODELS: TECHNICAL UPDATE REPORT (2024).

92 Austan Goolsbee, *Public Policy in an AI Economy*, *in* THE ECONOMICS OF ARTIFICIAL INTELLIGENCE: AN AGENDA 309, (Ajay Agrawal, Joshua Gans & Avi Goldfarb eds., 2019).

taken the opposite view.⁹³ Where IP law or technical protection forecloses the market for foundation models unduly, competition law may step in by requiring access to the model to be granted on non-discriminatory terms.⁹⁴ At the time of writing, it is unclear what role FRAND (fair, reasonable, and non-discriminatory) obligations will play in the AI value chain. Yet, what is certain is that FRAND obligations can be an additional policy lever. The concept of FRAND is sensitive to: (a) the essentiality of a market demanding interoperability, i.e., a standard, as well as (b) the essentiality of a particular IP right. The AI Act foresees a right of complaint by downstream providers against upstream undertakings in the AI Value Chain but stops short of imposing access obligations⁹⁵ outside a duty to provide interoperability information by providers of general-purpose AI models.⁹⁶ In addition, it is worth noting that the French Autorité de la Concurrence recently fined Google for abusive conduct involving AI data mining and the breach of previous commitments in relation to copyright and related rights.⁹⁷

889. Where access is provided to part of a platform, it seems that EU competition law is capable of guaranteeing that access to any part of the platform will not be provided on discriminatory terms.⁹⁸ The emphasis of the recent judgment of the General Court in *Google Shopping* on discrimination makes this apparent.⁹⁹

890. Further research and case law is necessary to ascertain the principles of FRAND licensing under competition law and its potential application to AI. It remains unclear as a matter of law how FRAND rates are set

---

93 Stephen Dnes et al., *DCI Submission to the European Commission (EC) on Generative Artificial Intelligence* (Mar. 11, 2024), https://papers.ssrn.com/sol3/papers.cfm?abstract_id=4756739.

94 *RTE (Magill)*, *supra* note 84; *Google Shopping*, *supra* note 84; *see also* Pradelles & Hecht, *supra* note 84.

95 Commission Regulation (EU) 2024/1689, art. 89(2).

96 *Id.* art. 53(b); *see also* Quentin B. Schäfer, *Economic Regulation and the AI Act: Perspectives from Competition Law, Data Protection, and Intellectual Property*, in AI ACT AND AI LIABILITY DIRECTIVE (Nikolaus Forgo, Ceyhun N. Pehlivan & Peggy Valcke eds., 1st ed. 2024); Quentin B. Schäfer, *Article 89*, in AI ACT AND AI LIABILITY DIRECTIVE (Nikolaus Forgo, Ceyhun N. Pehlivan & Peggy Valcke eds., 1st ed. 2024).

97 Autorité de la Concurrence, *Decision 22-D-13 of 21 June 2022 Regarding Practices Implemented in the Press Sector*, (2022), https://www.autoritedelaconcurrence.fr/sites/default/files/commitments/2022-10/Decision%2022D13%20V%20EN.pdf, *see* Frédéric Marty, *Commitment procedure: The French Competition Authority makes binding by decision the commitments proposed by a major platform leading to the compliance of its practices with the law protecting related rights (Google)*, CONCURRENCES No. 4-2022, art. No. 109571; Autorité de la Concurrence, *Decision 24-D-03 of 15 March 2024 Regarding Compliance with the Commitments in Decision 22-D-13 of 21 June 2022 of the Autorité de La Concurrence Regarding Practices Implemented by Google in the Press Sector*, (Mar. 20, 2024), https://www.autoritedelaconcurrence.fr/fr/decision/relative-au-respect-des-engagements-figurant-dans-la-decision-de-lautorite-de-la-0.

98 *Google Shopping*, *supra* note 84, *see* Pradelles & Hecht, *supra* note 84.

99 *Id.* Lena Hornkohl, *Article 102 TFEU, Equal Treatment and Discrimination after Google Shopping*, 13 J. EUR. COMPETITION L. & PRAC. 99 (2022). US antitrust law provides for weaker access and essential facility obligations, especially after Verizon Commc'n Inc. v. L. Offs. of Curtis v. Trinko, LLP, 540 U.S. 398 (2004), *see* Frédérique Daudret-John & François Souty, *United States of America: The US Supreme Court rules on the current relationship between antitrust law and sectoral regulation in the telecommunications sector (Trinko)*, CONCURRENCES No. 1-2004, art. No. 1603; There are no cases in which an IPR has been held to constitute an essential facility in US antitrust law, *see* HERBERT HOVENKAMP ET AL., IP AND ANTITRUST: AN ANALYSIS OF ANTITRUST PRINCIPLES APPLIED TO INTELLECTUAL PROPERTY LAW ¶¶ 13–17 (3rd ed., 2020 Supplement ed. 2017).

under prevailing case law. Similarly, it remains unclear when FRAND licenses enforced by competition law are appropriate outside the SEP-specific scenario addressed in *Huawei*, and on what legal basis FRAND rates are enforced.[100]

891. FRAND licenses offer an attractive policy lever especially where the minimum efficient scale of foundation models is substantial but there is little vertical integration. In such a case, obligations to provide access on FRAND terms are capable of addressing a whole host of problems arising from IP protection as well as technological protection. The imposition of FRAND terms is backed by the prohibition on anti-competitive abuses in Article 102 TFEU.

4. Compulsory Licenses and Access to Essential Facilities

892. Promoting openness prevents anticompetitive behavior[101] and is particularly important if future markets, e.g., in foundation models, result in few models on the basis that scale trumps specialization but resist vertical integration. A different view of the future of AI is one where models are highly specialized and require less scale or integrate vertically to a substantial degree. The response of competition law may be of a different character in such a future.

893. EU competition law may provide *ex-post* access to facilities essential to a market's operation where a firm has gained dominance.[102] After *Google Shopping*,[103] technical facilities are accessible under the *Bronner* doctrine,[104] whereas IP rights are accessible under the *Magill* case law.[105] Such restrictions are dependent on the capability and occurrence of anti-competitive agreements and abuses in a particular economic context. This capability varies depending on the opportunity of IP rights to foreclose competition outside of the right's core market, as well as the how "essential" access to the IP right is to competition. I have argued at length elsewhere

---

100 *See* Haris Ćatović, Refusal to License Intellectual Property Rights as Abuse of Dominant Position in EU Competition Law: The Implications of the Huawei Judgment (2015) (Thesis, University of Stockholm); Izarne Marko Goikoetxea, *Huawei v. ZTE Should Have Been Treated as a Refusal to Contract – to Grant SEP Licences – and Not as a New Category of Abuse*, 40 EUR. COMPETITION L. REV. 67 (2019), *see also* James Killick, Katarzyna Czapracka & Daniel Hoppe-Jänisch, *The EU Court of Justice sets out specific requirements with which an SEP holder needs to comply to be able to seek an injunction without abusing its dominant position (Huawei / ZTE)*, E-COMPETITIONS July 2015, art. No. 74861.

101 Schrepel & Potts, *supra* note 50.

102 US antitrust law provides for weaker access and essential facility obligations, *see generally*, *supra* note 99.

103 *Google Shopping*, *supra* note 84; *RTE (Magill)*, *supra* note 84; *see also* Pradelles & Hecht, *supra* note 84.

104 *Bronner*, *supra* note 84.

105 *Maxicar*, *supra* note 87; *AB Volvo*, *supra* note 87; *RTE (Magill)*, *supra* note 84; Case C-418/01, IMS Health GmbH & Co. OHG v. NDC Health GmbH & Co. KG, ECLI:EU:C:2004:257, https://eur-lex.europa.eu/legal-content/EN/TXT/PDF/?uri=CELEX:62001CJ0418&qid=1544200303106&from=EN; *see* James Killick, *The EU Court of Justice clarifies the applicable legal standard for compulsory licensing (IMS Health)*, E-COMPETITIONS Apr. 2004, art. No. 37125; Case T-201/04, Microsoft v. Eur. Comm'n, ECLI:EU:T:2007:289, *see* Catherine Prieto, *The Microsoft Saga: The Court of First Instance upholds the Commission's finding that Microsoft had abused its dominant position in the PC operating system market (Microsoft)*, CONCURRENCES No. 4-2007, art. No. 14478.

that competition law is unique in allowing for the compelling of not just a compulsory license but also the transfer of sufficient non-protected know-how to operationalize the relevant invention, and providing access to technologically closed systems.[106] Where right holders seek to enforce excessive IP rights, the discretion to deny injunctive relief or to impose compulsory licensing is capable of restricting the ambit of such rights in the public interest.[107] Duties to deal and license may counteract trends of vertical integration and the potential harms of foreclosure.

894. Yet, there remain substantial unknowns in this regard. At minimum in relation to patent rights, it seems doubtful as to how far the monopoly protection of patents can be restricted *ex post* by competition law, given that the case law pertains largely to non-patent IP rights. The case law surrounding SEPs offers an additional pointer, suggesting that at least those rights that firms have committed to licensing in a standard must be licensed on FRAND terms, and that such obligations are subject to control under Article 102 TFEU.[108]

895. Other scholars have suggested using competition law to provide access to "blocking patents."[109] In the context of a technology as crucial as AI, we may wish to do so if severe blocking AI patents arise. This need not be through competition law but could be through the buyout of IP rights by the government.[110] Along this line, in the EU, the protections for private property in Article 345 TFEU and Article 17 of the Charter of Fundamental Rights, as well as on the Member State level should be examined in the course of further scholarship.

V. Conclusion: Towards an Integrated Innovation Policy

896. As Aghion, Jones, and Jones point out, "[i]t would be interesting to explore how AI affects [the] complementarity between the two policies [of patent protection and competition policy]."[111] This work must continue not just in economics but also in law.

897. The general point of this chapter is that the ideal shape of IP law *ex post* is highly dependent on uncertain technical developments. To think we have AI regulation figured out, even remotely, is scholarly hubris. But we can say some things about the *ex-ante* case of how to promote the development of *any* technical developments that go in this direction, as

---

106 Quentin B. Schäfer, *Reconsidering the Limits of EU Competition Law on the IP-Competition Interface*, 15 J. Eur. Competition L. & Prac. 188 (2024).

107 *Id.*

108 Case C-170/13, Huawei v. ZTE, ECLI:EU:C:2015:477, *see* Killick, Czapracka & Hoppe-Jänisch, *supra* note 100.

109 Andreas Heinemann, *Blocking Patents and the Process of Innovation*, *in* New Developments in Competition Law and Economics 149 (Klaus Mathis & Avishalom Tor eds., 2019); Angelika S. Murer, Blocking Patents in European Competition Law: The Implications of the Concept of Abuse (2022).

110 Michael Kremer, *Patent Buyouts: A Mechanism for Encouraging Innovation*, 113 Q.J. Econ. 1137 (1998).

111 Aghion, Jones & Jones, *supra* note 32, at 263.

long as we constrain ourselves to designing policy levers that take account of different market and technology outcomes.

898. I then recommend the following: relying on manual doctrinal policy changes in IP law to accommodate the technological and economic effects of AI is generally not ideal because the risk of errors is high, error costs are substantial, and norms of substantive IP law exhibit a high degree of "stickiness." Scholarship should focus on the analysis policy levers in doctrinal analysis to accommodate AI in all its potential future guises.

899. Arguably, the complexity of IP law as well as its interaction with competition law are assets that allow the setting of complex policy levers within a pre-existing doctrinal framework. Yet, an integrated approach of policy levers for AI goes beyond IP and competition law, clearly implicating corporate law, tort, finance[112] etc. The real world confirms this assessment. Tax credits have already been successful in stimulating investment into hardware in the context of the US Chips Act.[113] Along these lines, the availability of venture capital seems to have been a significant driver of technologies at the frontier.[114]

900. Another issue worth highlighting is that IP and competition law (and regulation as a whole) form parts of systemic competition against other regulatory and economic models.[115] This applies to both AI tools and AI outputs, and includes regulatory competition which increasingly takes effect beyond the regulator's borders.[116] The EU's AI Act includes obligations related to copyright that apply extraterritorially.[117] Jurisdictions that fall short in this systemic competition are likely to be cast down in their productive capacities in the future. Scholars in competition and IP law can contribute to avoiding such a future.

---

112 *See* CARLOTA PEREZ, TECHNOLOGICAL REVOLUTIONS AND FINANCIAL CAPITAL: THE DYNAMICS OF BUBBLES AND GOLDEN AGES (2002).

113 Chris Miller, *The Chips Act Has Been Surprisingly Successful So Far*, FIN. TIMES (2024), https://www.ft.com/content/26756186-99e5-448f-a451-f5e307b13723.

114 Josh Lerner & Ramana Nanda, *Venture Capital's Role in Financing Innovation: What We Know and How Much We Still Need to Learn*, 34 J. ECON. PERSPS. 237 (2020); *see also* Anu Bradford, *The False Choice Between Digital Regulation and Innovation*, 118 Nw. U. L. REV. (forthcoming 2024).

115 *See* ANU BRADFORD, DIGITAL EMPIRES: THE GLOBAL BATTLE TO REGULATE TECHNOLOGY (2023).

116 *See* ANU BRADFORD, THE BRUSSELS EFFECT: HOW THE EUROPEAN UNION RULES THE WORLD (2020); BRADFORD, *supra* note 115.

117 Commission Regulation (EU) 2024/1689, art. 53(1)(c)–(d); Recital 106.

Biographies

Alden Abbott

Alden F. Abbott is Senior Research Fellow in the Mercatus Center at George Mason University. A widely published scholar and lecturer, he oversees the Center's research on antitrust and competition policy. He also is the regular antitrust columnist for Forbes online and a Leader of the American Bar Association's Antitrust Seciton. Prior to joining Mercatus, he served as General Counsel of the U.S. Federal Trade Commission. Mr. Abbott is co-author (with Shanker Singham) of Trade, Competition, and Domestic Regulatory Policy (Routledge 2023), a unique study of the global interactions among antitrust, regulation, trade, and intellectual property laws. His current research focuses on artificial intelligence, innovation, and government-sponsored anticompetitive market distortions that undermine the global economy. He also is a member of the Growth Commission, a special research body that issues reports on national economic policies that have slowed economic growth.

Cora Allen

Cora Allen is an associate at Wilson Sonsini Goodrich & Rosati, where her practice focuses on a broad range of antitrust issues, including mergers and acquisitions, litigation, and criminal investigations. Cora obtained her law degree at the Washington University School of Law, where she graduated magna cum laude and was an Executive Editor of the Washington University Law Review. While in law school, Cora also served as a law clerk for the antitrust and competition law practice at Kim & Chang in Seoul, Korea.

Jonathan M. Barnett

Jonathan M. Barnett is the Torrey H. Webb Professor of Law at the University of Southern California Gould School of Law, where he directs the law school's Media, Entertainment, and Technology Law Program. He is the author of The Big Steal: Ideology, Interest, and the Undoing of Intellectual Property (Oxford 2024) and Innovators, Firms, and Markets: The Organizational Logic of Intellectual Property (Oxford 2021). He specializes in antitrust and intellectual property law, with a focus on innovation policy and strategy in technology markets. He has published widely in scholarly journals in law and economics and regularly comments on antitrust and innovation policy issues. He is a graduate of Yale Law School.

Christian Bergqvist

Christian Bergqvist specializes in EU Competition Law, particularly in its application to deregulated and network-tied sectors (telecom, energy, post, and transport) and platforms, Big Tech, and the companies collectively referred to as FAANG (Facebook, Amazon, Apple, Netflix, and Google). The latter presents some of the same issues as the former. Christian Bergqvist has extensive experience in competition law as an academic and practitioner, often bridging the two professions, and has served as NGA to the Danish NCA and DG COMP. Before becoming a full-time academic, Christian Bergqvist was a lawyer with Danish Tier-1 law firms.

Oliver Budzinski

Oliver Budzinski is Professor of Economic Theory and Director of the Institute of Economics at Ilmenau University of Technology, Germany. He obtained his Ph.D. from the University of Hanover and held post-doc positions at Philipps-University of Marburg, New York University, and University of Duisburg-Essen, and served as a full professor at the University of Southern Denmark, Campus Esbjerg. On topics of competition policy, industrial and institutional economics as well as media, cultural and sports economics, Oliver Budzinski published 3 books, more than 70 articles in peer-reviewed academic journals, and more than 60 chapters in conference proceedings and handbooks.

Yo Sop Choi

Yo Sop Choi is Professor of Law at the Graduate School of International and Area Studies, Hankuk University of Foreign Studies in Seoul. He specializes in competition law and EU law. His research interests mainly focus on comparative studies of competition law and digital policies related to data protection, artificial intelligence, and consumer protection. Professor Choi has been as a visiting scholar at various universities, including Keio University, European University Institute, University of Zurich, IE Law School, Heinrich Heine University Dusseldorf, Waseda University, National Taiwan University, Chiang Mai University, Osaka Metropolitan University and Wuerzburg University. He is a member of the Academic Society of Competition Law (ASCOLA) and the Korean Competition Law Association (KCLA). His articles have been published in the peer review journals of competition law and intellectual property rights.

Diane Coyle

Diane Coyle is Bennett Professor of Public Policy at the University of Cambridge, and an academic advisor to the UK's Competition and Markets Authority and Office for National Statistics. She was a member of the Furman Review, 'Unlocking Digital Competition', and has previously held a number of public service roles including serving as a member of the UK Competition Commission 2001-2009, and vice chair of the BBC Trust 2006-2014. Her research focuses on the digital economy, productivity and economic measurement. Her most recent book is Cogs and Monsters: what economics is and what it should be.

Daniel Crane

Daniel Crane is the Richard W. Rogue Professor of Law at the University of Michigan and of counsel to Paul, Weiss, Rifkind, Wharton & Garrison LLP. He is the author of many books and articles on competition law, including The Institutional Structure of Antitrust Law (Oxford University Press, 2011).

Hayane Dahmen

Hayane Dahmen is a postdoctoral research fellow at the University of Toronto, where she investigates how AI will change competition policy. She has previously worked at the World Bank, at The Alan Turing Institute, and in private practice as an antitrust lawyer. Hayane holds a PhD investigating how technological and geopolitical changes will impact antitrust (Cambridge; Salje Medal for most outstanding PhD in the Arts and Humanities, Clare Hall); a Master in Public Administration (Harvard); a Master of Laws (Georgetown), a JD (British Columbia) and a BA (McGill).

Stephen Dnes

Stephen Dnes is a lecturer in law at Royal Holloway, University of London and Partner in Dnes & Felver PLLC. After graduating from the University of Cambridge and the University of Virginia School of Law, Stephen started his career working on EU competition law, including the early stages of the long-running Google antitrust matters. He later worked on antitrust cases in the USA including matters before the US FTC, DOJ, and the Federal Maritime Commission. Stephen has been engaged for expert advice in several high-profile antitrust cases. In 2024, he authored several of the key amendments to the UK's new competition law.

Fausto Gernone

Fausto Gernone is a competition economist currently visiting Haas School of Business, U.C. Berkeley, and a PhD candidate at the UCL Institute for Innovation and Public Purpose. His work stands at the crossroad of the fields of industrial organisation, information theory and strategy of the firm. In particular, his research explores the dynamics of competition between complementary products in the context of digital ecosystems, investigating the role of coordination devices like industry standards, APIs, and open-access initiatives. Fausto has years of experience in the field of competition policy, working on mergers and antitrust, as well as Standard Essential Patents and State aid issues.

Teodora Groza

Teodora Groza is a Ph.D. student and lecturer at Sciences Po Law School. Her research focuses on three axes: 1) the impact of new modes of organizing economic activity on antitrust, 2) the interaction between regulatory interventions and innovation, and 3) data governance. She is the Editor-in-Chief of the Stanford Computational Antitrust Journal.

Aaron Hoag

Aaron Hoag is the Chief of the Technology & Digital Platforms Section (TDP) of the Antitrust Division at the U.S. Department of Justice. Since joining the Antitrust Division in 1997, he has played a key role in many of the Division's technology-related matters. He served as the Special Counsel for Microsoft Decree Enforcement from 2005 to 2011. He was selected as Assistant Chief of TDP in 2013 and became Chief of the section in 2016. He has joined in supervising all the section's matters since then, including the Division's successful litigation challenges to Bazaarvoice's acquisition of PowerReviews and EnergySolutions' proposed acquisition of Waste Control Specialists, and the ongoing search case against Google.

William Lehr

Dr. William Lehr is a telecommunications and Internet industry economist and consultant with over thirty years of experience. He regularly advises senior industry executives and policymakers in the U.S. and abroad on the market, industry, and policy implications of events relevant to the Internet ecosystem. He is a research associate in the Computer Science and Artificial Intelligence Laboratory (CSAIL) at the Massachusetts Institute of Technology and at the Weizenbaum Institute in Berlin. Dr. Lehr's research focuses on the economics and regulatory policy of the Internet infrastructure industries and the implications of IT technologies such as AI, Smart Contracts, and Wireless.

Mariateresa Maggiolino

Mariateresa Maggiolino is an expert in Competition Law and Regulation, serving as a Professor at Bocconi University, where she is also the director of the five-year Master's program in Law. With three monographs to her name and numerous publications in top international and national journals, her work spans the critical intersections of competition law, intellectual property, banking law, and data law, making her an important voice in these dynamic fields.

Godefroy de Boiscuillé

Dr. Godefroy de Boiscuillé is currently an associate professor and researcher at the joint laboratory GREDEG/CNRS at Côte d'Azur University (Nice Sophia Antipolis). Godefroy de Boiscuillé is also the co-director of the "Competition & Digital Economics" Chair at University of Paris Pantheon Assas. In other respects, he has been appointed external expert at the European Institute of Public Administration (EIPA Maastricht & Luxembourg). He has published a book which has been awarded the Concurrences Best PhD Award and the international Jacques Lassier Prize. His research focuses mainly on European Law, Competition Law, and Digital Law.

Tejas N. Narechania

Tejas N. Narechania is a Professor of Law at the University of California, Berkeley, School of Law, where he focuses on a range of technology law matters. He has a J.D. from Columbia Law School, and a B.S. (Electrical Engineering and Computer Science) and a B.A. (Political Science) from the University of California, Berkeley. His research projects have appeared in interdisciplinary venues, from law reviews to computer science proceedings, and they have been cited by policymakers and in the press.

Mark Niefer

Mark J. Niefer is a lawyer and Ph.D. economist specializing in competition law and policy. He is an Adjunct Professor of Law at Antonin Scalia Law School – George Mason University, where he teaches antitrust; he also is a consultant to competition agencies abroad. He served the United States Department of Justice Antitrust Division for more than twenty-five years (1997-2024) in key roles that included International Advisor for digital markets and Deputy Chief Legal Advisor. He received the John Marshall Award, the Department of Justice's highest award for attorneys.

Victoriia Noskova

Victoriia Noskova is a Postdoctoral Researcher in Economic Theory Group at Ilmenau University of Technology, Germany. She has a double Master's degree in Media Economics from TU Ilmenau and Business Economics from Saint Petersburg State University, Russia. She obtained her Ph.D. from the TU Ilmenau. Her primary research interests lie in competition economics, institutional economics and media economics. Her Ph.D. projects were focused on the interplay between competition and data with focus on voice assistants, social networking and video advertising markets. Victoriia Noskova has published five peer-reviewed papers addressing various questions within her core field.

Maureen Ohlhausen

Maureen Ohlhausen co-leads the antitrust and competition practice at Wilson Sonsini Goodrich & Rosati in Washington, D.C., after previously chairing the global antitrust and competition practice at another top tier firm. She represents clients in technology, healthcare, life sciences, and retail industries in civil merger and non-merger cases before government agencies. A former acting chairman and commissioner of the Federal Trade Commission, Ohlhausen focuses on antitrust and consumer protection issues, particularly involving privacy and technology, including AI. Earlier in her career, she led the FTC's Internet Access Task Force, which issued a significant broadband competition report. A thought leader, Ohlhausen has published widely, testified before Congress, and received both the FTC's Robert Pitofsky Lifetime Achievement Award and the Internet Freedom Award.

Taylor Owings

Taylor Owings is a partner at Wilson Sonsini Goodrich & Rosati in New York, where she represents clients in civil merger and non-merger matters both before government agencies and in private litigation, with an emphasis on issues arising in technology-focused industries. Previously, Taylor served in the U.S. Department of Justice Antitrust Division Front Office, where she served as counsel to the Assistant Attorney General and chief of staff. In this role, she investigated and litigated antitrust cases involving technology companies, digital economy business models, and the exercise of intellectual property rights.

Jennifer Pullen

Jennifer Pullen, M.A. in Law and Economics HSG, is a PhD student at the University of St. Gallen in Switzerland. Her research strongly focuses on the intersection of legal frameworks and economic principles, particularly in the context of digital economies and artificial intelligence. Her PhD thesis analyzes potential adaptations to the substantive assessment in merger control for Big Tech acquisitions. In addition to her research, Jennifer serves as a research and teaching assistant at the University of St. Gallen, where she teaches undergraduate courses and supports research initiatives under the guidance of Prof. Dr. Miriam Buiten at the Institute for Law & Economics. She has previously worked as a Digital Economic Policy Analyst at Digital Policy Alert.

Lazar Radic

Lazar is Assistant Professor of Law at IE University and Senior Scholar for Competition Policy at the International Center for Law & Economics. Lazar is a qualified lawyer and has worked in band one law firms in Spain and Serbia, both in the competition law department. He holds degrees in Law and Political Science from the University of Madrid, LL.Ms from the University of Amsterdam and the EUI, and a PhD in law from the EUI.

Camilla Ringeling

Camila Ringeling is an associate at Hausfeld LLP DC, her practice focuses on antitrust, competition, and complex civil and international human rights litigation. Camila has over ten years of experience advising on antitrust and competition matters. Before her JD, she worked as a consultant at the FTC Office of International Affairs and the World Bank. She also worked in the public and private sectors in the European Union and Chile. Academically, she contributed to George Mason's Global Antitrust Institute as a research assistant and is currently a senior fellow at The George Washington University Competition and Innovation Lab.

Quentin B. Schäfer

Quentin B. Schäfer holds the position of Lecturer (Assistant Professor) in Competition Law and Private Law (permanent/tenure-track) at the University of Strathclyde, Glasgow. He is also currently a PhD candidate at Darwin College and the Faculty of Law, University of Cambridge. He holds an LLM from the University of Cambridge and an LLB from King's College London. His main research interests lie in the areas of competition law, intellectual property, and the regulation of technology and innovation.

Thibault Schrepel

Dr. Thibault Schrepel, LL.M., is an Associate Professor of Law at the Vrije Universiteit Amsterdam University (Amsterdam Law & Technology Institute), and a Faculty Affiliate at Stanford University (CodeX Center), where he founded the "Computational Antitrust" project that brings together over 65 antitrust agencies.

Thibault is also the founder of the Network Law Review, and the host of the "Scaling Theory" podcast. In recent years, Thibault has focused most of his research on blockchain antitrust, computational antitrust, and complexity theory. He has written the world's most downloaded antitrust articles on SSRN in 2018 ("The Blockchain Antitrust Paradox"), 2019 ("Collusion by Blockchain and Smart Contracts"), 2020 ("Blockchain Code as Antitrust"), 2021 ("Computational Antitrust: An Introduction and Research Agenda"), and 2022 ("Complexity-Minded Antitrust").

Ganesh Sitaraman

Ganesh Sitaraman holds the New York Alumni Chancellor's Chair in Law at Vanderbilt University Law School and is the Director of the Vanderbilt Policy Accelerator for Political Economy and Regulation. He is the author of numerous books and articles, including Networks, Platforms, and Utilities: Law and Policy (2022) (with Morgan Ricks, Shelley Welton & Lev Menand).

Daniel F. Spulber

Daniel F. Spulber is the Elinor Hobbs Distinguished Professor of International Business and Professor of Strategy at the Kellogg School of Management, Northwestern University. He is also Professor of Law (Courtesy) at the Pritzker School of Law, Northwestern University. Spulber received his Ph.D. in economics from Northwestern University. Previously, Spulber taught at Brown University, the University of Southern California, and Cal Tech. Spulber has served as an expert witness in antitrust, regulation, digital platforms, and intellectual property. Spulber is the founding editor of the Journal of Economics & Management Strategy. He has published fourteen books including The Case for Patents.

Volker Stocker

Volker Stocker is an economist specializing in the digital economy and the Internet ecosystem. He leads the research group "Digital Economy, Internet Ecosystem, and Internet Policy" at the Weizenbaum Institute in Berlin. Additionally, Volker is a post-doctoral researcher in the Internet Architecture and Management Group at TU Berlin and an associated researcher at the Max Planck Institute for Informatics in Saarbrücken. He is an affiliated scholar at the Dynamic Competition Initiative and a Vice Chair of the International Telecommunications Society (ITS) Europe. His primary research interests include the economics, evolution, and regulation of digital infrastructures, platforms, and technologies, as well as the Internet ecosystem.

Kristian Stout

Kristian Stout is the Director of Innovation Policy at the International Center for Law & Economics. A former technology entrepreneur and computer science lecturer at Rutgers University, Kristian has been a fellow at the Internet Law & Policy Foundry and the Eagleton Institute of Politics. He served on the FCC's Broadband Deployment Advisory Committee and chaired the Asset Forfeiture Working Group for the NJ State Advisory Committee to the U.S. Commission on Civil Rights.

David J. Teece
David J. Teece is a professor in the Graduate School at UC Berkeley. He is a scholar-entrepreneur who has shown how to join theory and practice. He advises governments and corporations around the world and founded the Berkeley Research Group which has 1,500 professionals in over 40 offices worldwide. He has contributed to the competition policy literature primarily through his development of the dynamic competition paradigm which emphasizes the central role that innovation plays in driving competition.

Aleksandra Wierzbicka
Aleksandra Wierzbicka is an associate at Cleary Gottlieb Steen & Hamilton, based in Brussels. She holds an LLM in European Law and Economic Analysis (ELEA) from the College of Europe, a Master in Global Business Law and Governance from Sciences Po Law School and a Master of Law from the Jagiellonian University. She received the Global Competition Law Centre's (GCLC) prize for her Master thesis at the College of Europe. Aleksandra is Academic Outreach Chair at Stanford Computational Antitrust project hosted by the Stanford University Codex Center.

John M. Yun
John M. Yun is an Associate Professor of Law and an economist who specializes in research at the nexus of antitrust, intellectual property rights, data, and privacy. Prior to joining the faculty at GMU, he was an Acting Deputy Assistant Director in the Bureau of Economics, Antitrust Division, at the U.S. Federal Trade Commission (FTC). He has also taught economics at Georgetown University, Emory University, and Georgia Tech. He holds a BA in economics from the University of California, Los Angeles and a PhD in economics from Emory University.

Laura Zoboli
Laura Zoboli is an Assistant Professor at the University of Brescia and the Scientific Coordinator of the Centre for Antitrust and Regulatory Studies at the University of Warsaw, where she started her academic tenure. She serves as the Managing Editor of the Yearbook of Antitrust and Regulatory Studies and is the Co-Director of the ASCOLA Center Europe Chapter. Laura earned her PhD from Bocconi University, where she currently teaches Data Law. Among her research experiences, she has been a research fellow at the Berkman Klein Center at Harvard University and the Max Planck Institute for Innovation and Competition in Munich. Laura also clerked at the Competition and IP Court in Milan and is admitted to the Italian bar.

Table of Contents

Contributors I
Overview III
Foreword V
    Cani Fernández
Introduction VII
    Alden Abbott and Thibault Schrepel
        Setting the stage VII
        Roadmap VII
        Market Dynamics, Mergers, and Partnerships in AI VIII
        AI Challenges for Competition Law IX
        Policy Responses to the AI Boom XI
Biographies XIII

PART I
Market Dynamics, Mergers, and Partnerships in AI

Open-Source Generative AI from a Competition Policy Perspective 3
    Diane Coyle and Hayane Dahmen
        I. Introduction 4
        II. The Meaning of Open Source 5
        III. Competitive Effects of Open-Source Models in GenAI 8
        IV. The AI Ecosystem: Business Models and Incentives 11
        V. Conclusions 15

Competing in the Age of AI: Firm Capabilities and Antitrust Considerations 17
    Fausto Gernone and David J. Teece
        I. The Economics of AI 18
            1. What is AI? 18
            2. Constituents of an AI Model 18
            3. AI as a Platform 20
            4. Competitive Landscape 21
        II. Organizational Benefits of AI 22
            1. Ordinary Capabilities 23
            2. Superordinary Capabilities 24
            3. Dynamic Capabilities 25
            4. Outsourced versus In-House AI 27

III. Potential Competition Issues	28
1. Access to Data	29
2. Access to Skilled Labor	30
3. Access to Compute	31
IV. Conclusions	34

Antitrust and Innovation Competition: Investments and Partnerships in Artificial Intelligence ... 35
Daniel F. Spulber

I. Introduction	36
II. Investments and Partnerships in Artificial Intelligence	38
1. Microsoft and OpenAI	38
2. Amazon and Anthropic	40
3. Google and Anthropic	41
III. Implications for Antitrust Policy Toward Innovation Competition	42
IV. Conclusion	45

Mergers by Other Means? AI Partnerships and the Frontiers of (Post-)Industrial Organization ... 47
Teodora Groza and Aleksandra Wierzbicka

I. Introduction	48
II. If You Can't Acquire Them, Team Up with Them: AI Partnerships	51
III. Post-Industrial Organization	56
IV. Catch Them If You Can: Competition Law	59
V. Conclusion	63

Preserving Competition in Generative AI: Addressing the Merger Conundrum ... 65
Mariateresa Maggiolino and Laura Zoboli

I. Introduction	66
II. What Are the FTC, the CMA, and the EC Afraid Of?	67
III. The Possible Legal Rules	70
IV. The Possible Theories of Harm	73
V. The Acquisition of the Best Kid on the Block	78
VI. Well Begun is Half Done	80

Artificial Intelligence, Uncertainty, and Merger Review ... 83
Mark J. Niefer and Aaron D. Hoag

I. Introduction	84
II. Artificial Intelligence: A Brief Overview	86
1. AI and Foundation Models	86

    2. Foundation Model Development and Deployment 87
    3. Uncertainty About the Evolution of the FM
       Ecosystem 89
 III. Merger Review Involving AI 92
    1. Assessing Effects 92
       A. *Merger of Unintegrated Foundation Model Developers*... 93
       B. *Merger of Integrated Foundation Model Developers* 94
       C. *Other Sources of Uncertainty in Assessing Effect* 95
    2. Defense Arguments 96
 IV. Accounting for Uncertainty in Mergers Involving AI 99
  V. Conclusion 104

PART II
AI Challenges for Competition Law

What is the Relevant Product Market in AI? 107
Lazar Radic and Kristian Stout
   I. Introduction 108
  II. What Makes AI Markets Internally Heterogeneous 114
 III. What Makes AI Similar to Non-AI Products and Services 117
 IV. Tentative Principles for Market Definition in AI 123
    1. Who Are the Consumers? What is the Product
       or Service? 125
    2. Does AI Fundamentally Transform the Product
       or Service? 128
  V. Conclusion 130

Defining AI Markets: Who is Afraid of Digital Ghosts? 151
Stephen Dnes
   I. Introduction 134
  II. (How) Should Market Definition Approach Innovation
     Markets? 135
    1. Are Relevant Markets Relevant to AI? 135
    2. Market Definition in Innovation Markets 136
       A. *From Paperclips to Hal 9000: What AI Analysis*
          Can Learn from a 1997 Case about Legal Pads 137
       B. *Innovation Analysis since Staples (1997)* 138
 III. Application to Emerging Markets Including AI 139
    1. Which Ghosts Did You Have in Mind? 140
    2. Guidelines position 141
       A. *Relationship with Consumer Analysis* 142

IV. Instances of Agency Intervention	144
A. *Facebook/Giphy*	144
B. *Meta/Within*	145
C. *Microsoft/OpenAI; Microsoft/Mistral*	146
D. *Amazon/Anthropic*	147
E. UK CMA Foundation Models Paper	147
F. US FTC AI-related Information Request	148
V. Conclusion	148

Finding the *Ghost in the Shell*: EU and US Antitrust Enforcement of AI Collusion 151
Christian Bergqvist and Camila Ringeling

I. Does AI-Supported Decision-Making Raise Concerns?	153
1. What Does AI-Supported Decision-Making Do?	154
2. Enforcers Have Addressed Coordination through AI	155
A. The EU Has Successfully Prosecuted AI Monitoring and Facilitating Conduct	156
B. Enforcement of AI Monitoring and Facilitating Conduct Has Also Been Successful in the US	157
3. Potential Concerns Are Limited to the Boundaries of Concerted Practices	158
II. The Notion of an Understanding in EU and US Antitrust	159
1. Conspiracy under EU Competition Law	160
A. *Concerted Practice*	160
(1) Contact between undertakings	161
a) Indirect contacts using an intermediary	162
b) Indirect contact through hub-and-spoke agreements	163
(2) Directed at coordinating	164
a) Unilateral declarations through public media	165
(3) Parallel conduct	166
B. *Vertical Concerted Practice*	166
2. Can AI Collusion be Effectively Prosecuted under Article 101?	167
A. *Possible Options to Close the Gaps*	167
3. Conspiracy under US Antitrust	168
A. *Concerted Action*	169
(1) Parallel conduct and plus factors	171
a) Invitations to collude	172
b) Information exchanges	173
c) Indirect contact through hub-and-spoke agreements	175
B. *Vertical Conspiracies*	178

	4. Can AI Collusion Be Effectively Prosecuted in the US under Section 1 US under the *Per Se* Standard?............	179
	A. What does US Case Law Tell us about AI Collusion?...............................	182
	B. Possible Options to Close the Gaps........................	183
III.	The *Ex-Ante* Approach	184
IV.	Finding the Ghost in the Shell	185

What About Bob? Revisiting the Intersection of Antitrust Law and Algorithmic Pricing in 2024 187

Maureen Ohlhausen, Taylor Owings and Cora Allen

I.	Introduction...	188
II.	First Factor: Does the User Exercise Independent Discretion Over the Output of the Algorithm?	188
	1. In re: RealPage, Inc. Rental Software Antitrust Litigation (No. II)...	190
	2. Duffy v. Yardi Systems, Inc.................................	192
III.	Factor Two: Does the Algorithm Use Non-Public Data to Determine a Price?..	195
IV.	Factor Three: Does the Algorithm Manipulate Market Conditions On Its Own?..	199
V.	Conclusion	200

Korean Competition Rules on Algorithmic Discrimination 201

Yo Sop Choi

I.	Algorithms and Competition Law in Korea.............................	202
II.	Substantive Provisions of Korean Competition Law	204
III.	Korean Competition Policy on Self-Preferencing and Algorithmic Pricing: The Guidelines	207
IV.	Korean Competition Law on Algorithmic Discrimination: Fairness in Online Rankings	209
V.	A Suggestion for Competition Policy on Algorithmic Discrimination	212

The Recoupment Conundrum: Rethinking Predatory Pricing in the Age of Algorithms.. 215

Jennifer Pullen

I.	Introduction...	216
II.	Predatory Pricing and Recoupment	217
	1. The Concept of Predatory Pricing	217
	2. The Concept of Recoupment............................	218
	A. Recoupment in the US	218
	B. Recoupment in the EU	220

III. Artificial Intelligence and Recoupment	222
1. Artificial Intelligence – A Big Word	222
2. Artificial Intelligence, Pricing and Recoupment	224
A. Artificial Intelligence-Driven Pricing	224
B. The Impact of Artificial Intelligence on Pricing Strategies	224
3. The Impact of Artificial Intelligence on Recoupment	226

Computational Methods in the Evaluation of Mergers and Acquisitions 231
Victoriia Noskova and Oliver Budzinski

I. Introduction	232
II. Selection of Merger Cases	233
III. Investigation Phase	236
IV. In the Courts	239
V. Final Decision with or without Remedies	240
VI. *Ex Post* Control	241
VII. Conclusion	242

PART III
Policy Responses to the AI Boom

The Folly of AI Regulation 247
John M. Yun

I. Introduction	248
II. Avoiding Dampened Innovation Incentives	252
III. Premature Regulation Could Crowd Out Market-Based Solutions	255
IV. The Potential to Entrench Incumbents	257
V. Conclusion	259

The Case Against Preemptive Antitrust in the Generative Artificial Intelligence Ecosystem 261
Jonathan M. Barnett

I. Introduction	262
1. Structure of the GenAI Market	264
A. Technology Overview	264
B. GenAI Supply Chain: Make/Buy Choices	265
II. Preemptive Antitrust in the GenAI Market	267
1. Regulatory Activity in the Generative AI Market	268
2. The Potential Error Costs of Preemptive Antitrust	269

III. Assessing Risks to Competition in the genAI Technology Stack... 270
   1. Applications Layer 270
   2. FM/LLM Layer.......... 271
      A. Capital Requirements and Pretraining Costs.......... 271
      B. Data Inputs 272
      C. Market Share: Concentration and Durability.......... 273
      D. The Overlooked Virtues of Winner-Take-Most Outcomes 275
   3. Compute Layer.......... 276
   4. Summary.......... 277
IV. Investments and Other Relationships in the genAI Market....... 277
   1. Investments and Acquisitions in the genAI Market.......... 278
   2. Do Investments and Acquisitions in the genAI Market Pose a Risk to Competition?.......... 280
      A. Do Platform/LLM Provider Transactions Constitute Acquisitions? 281
      B. Do Platform/LLM Provider Transactions Pose Material Risks to Competition? 281
V. Conclusion 285

Antimonopoly Tools for Regulating Artificial Intelligence............. 287
Tejas N. Narechania and Ganesh Sitaraman
  I. Introduction.......... 288
  II. Understanding the AI Technology Stack 288
  III. The Problems with an AI Oligopoly 290
     1. Abuses of Power.......... 290
     2. National Security and Resilience 292
     3. Economic Inequality 292
     4. Democracy.......... 293
  IV. Antimonopoly Public Utility Tools 293
     1. Structural Separations.......... 294
     2. Nondiscrimination, Open Access, and Rate Requirements 294
     3. Interoperability Rules.......... 295
     4. Entry Restrictions and Licensing Requirements – The US.... 296
  V. A Public Option for AI.......... 297
     1. A Public Option for Cloud Infrastructure.......... 298
     2. Public Data Resources.......... 299
  VI. Industrial Policy, Procurement and Competition 300
     1. Industrial Policy and Semiconductors.......... 300
     2. Procurement Decisions 301
  VII. Conclusion 301

Table of Contents

Competition Policy after the Coming Wave of General Purpose Technologies 303
   Daniel A. Crane
     I. How the Four Pillars of Competition Policy Will Buckle in the Coming Wave 305
        1. Information 306
           A. Consumer Preference 306
           B. Productive Efficiency 309
        2. Incentives and Processes 309
        3. Scale and Scope 311
        4. Anticompetitive Conduct 313
     II. Beyond Competition 317

The Tortuous Path to AI Act Compliance: A Competitive Burden for Companies 321
   Godefroy de Boiscuillé
     I. Ex-Ante Regulation: Regulatory Barriers 322
        1. Overestimated Regulatory Obstacle 322
        2. Underestimated Regulatory Obstacles 323
     II. Defining and Classifying Risks 323
        1. Definition of AI system 324
        2. Amendment to the List of High-Risk Use Cases 325
        3. Risk of Harm 326
        4. Product Covered by Union Harmonization Legislation 327
        5. Utopia of Compliance with Certain Rules 327
        6. Overlap in Legislation 328
     III. Cost of Compliance 330
     IV. Conclusion 332

Competition Policy over the Generative AI Waterfall 335
   William Lehr and Volker Stocker
     I. Introduction 336
     II. AI and a Waterfall Moment 338
     III. Navigating Waterfalls Requires a Measurement Ecosystem 342
        1. Requirements for a Healthy Measurement Ecosystem 342
        2. Measurements and Policymaking over the Falls 345
           A. Multi-Stakeholder Trusted Process 345
           B. Balancing Asymmetric Information Requirements 348
           C. EBDM for shared infrastructure 350
        3. Competition and Infrastructure Policy over the Falls 355

AI, IP, and Competition Policy: Adjusting Policy Levers to a New GPT 359
Quentin B. Schäfer
    I. Introduction 360
    II. AI, Law and Technology, and New GPTs 361
        1. Layers of the AI Tech Stack 362
        2. Adjusting to New GPTs in the Past: The Role
           of Intellectual Property 362
        3. General Difficulties in Adjusting IP Law to AI 364
           A. Fundamental Technological and Economic Uncertainty 364
           B. The Complexity of the AI Tech Stack and the Variety
              of Impacted Markets 366
           C. Error Costs 367
           D. "Stickiness" of IP doctrine 369
    III. IP Policy Levers in the AI-Context 370
        1. What is a Policy Lever? 370
        2. Prioritizing Policy Levers 371
    IV. AI and the IP-Competition Interface as a Policy Lever 372
        1. The State of the IP-Competition Interface 373
         2. Hardware, Talent, Data, and Orthodoxy 374
        3. Promoting Openness throughout the Value Chain 375
        4. Compulsory Licenses and Access to Essential Facilities ... 377
    V. Conclusion: Towards an Integrated Innovation Policy 378

Concurrences
Antitrust Publications & Events

20 YEARS ANNIVERSARY

The Institute of Competition Law

The Institute of Competition Law is a publishing company, founded in 2004 by Dr. Nicolas Charbit, based in Paris, London and NewYork. The Institute cultivates scholarship and discussion about antitrust issues though publications and conferences. Each publication and event is supervised by editorial boards and scientific or steering committees to ensure independence, objectivity, and academic rigor. Thanks to this management, the Institute has become one of the few think tanks in Europe to have significant influence on antitrust policies.

AIM

The Institute focuses government, business and academic attention on a broad range of subjects which concern competition laws, regulations and related economics.

BOARDS

To maintain its unique focus, the Institute relies upon highly distinguished editors, all leading experts in national or international antitrust: Bill Kovacic, Mario Monti, Eleanor Fox, Laurence Idot, Frédéric Jenny, Ioannis Lianos, Richard Whish, etc.

AUTHORS

4,000 authors, from 85 jurisdictions.

PARTNERS

- Universities: University College London, King's College London, Queen Mary University, Paris Sorbonne Panthéon-Assas, etc.

- Law firms: Allen & Overy, Cleary Gottlieb Steen & Hamilton, Baker McKenzie, Hogan Lovells, Jones Day, Norton Rose Fulbright, Skadden Arps, White & Case, etc.

EVENTS

Brussels, Dusseldorf, Hong Kong, London, Milan, New York, Oslo, Paris, Singapore, Warsaw and Washington DC.

ONLINE VERSION

Concurrences website provides all articles published since its inception.

PUBLICATIONS

The Institute publishes Concurrences Review, a print and online quarterly peer-reviewed journal dedicated to EU and national competitions laws. e-Competitions is a bi-monthly antitrust news bulletin covering 85 countries. The e-Competitions database contains over 26,000 case summaries from 4,000 authors.

Concurrences
Competition Laws Review

Concurrences Review

Concurrences is a print and online quarterly peer reviewed journal dedicated to EU and national competitions laws. It has been launched in 2004 as the flagship of the Institute of Competition Law in order to provide a forum for academics, practitioners and enforcers. Concurrences' influence and expertise has garnered contributions or interviews with such figures as Christine Lagarde, Bill Kovacic, Emmanuel Macron, Antonin Scalia and Magrethe Vestager.

CONTENTS

More than 15,000 articles, print and/or online. Quarterly issues provide current coverage with contributions from the EU or national or foreign countries thanks to more than 2,500 authors in Europe and abroad.

FORMAT

In order to balance academic contributions with opinions or legal practice notes, Concurrences provides its insight and analysis in a number of formats:
- Forewords: Opinions by leading academics or enforcers
- Interviews: Interviews of antitrust experts
- On-Topics: 4 to 6 short papers on hot issues
- Law & Economics: Short papers written by economists for a legal audience
- Articles: Long academic papers
- Case Summaries: Case commentary on EU and French case law
- Legal Practice: Short papers for in-house counsels
- International: Medium size papers on international policies
- Books Review: Summaries of recent antitrust books
- Articles Review: Summaries of leading articles published in 45 antitrust journals

BOARDS

The Scientific Committee is headed by Laurence Idot, Professor at Panthéon Assas University. The International Committee is headed by Frederic Jenny, OECD Competition Comitteee Chairman. Boards members include Douglas Ginsburg, Benoît Cœuré, Howard Shelanski, Richard Whish, Wouter Wils, Joshua Wright, etc.

ONLINE VERSION

Concurrences website provides all articles published since its inception, in addition to selected articles published online only in the electronic supplement.

WRITE FOR CONCURRENCES

Concurrences welcome spontaneous contributions. Except in rare circumstances, the journal accepts only unpublished articles, whatever the form and nature of the contribution. The Editorial Board checks the form of the proposals, and then submits these to the Scientific Committee. Selection of the papers is conditional to a peer review by at least two members of the Committee. Within a month, the Committee assesses whether the draft article can be published and notifies the author.

e-Competitions
Antitrust Case Laws e-Bulletin

e-Competitions Bulletin

Case law database

e-Competitions is the only online resource that provides consistent coverage of antitrust cases from 85 jurisdictions, organized into a searchable database structure. e-Competitions concentrates on cases summaries taking into account that in the context of a continuing growing number of sources there is a need for factual information, i.e., case law.

- 26,000 case summaries
- 4,000 authors
- 85 countries covered
- 30,000 subscribers

Sophisticated editorial and IT enrichment

e-Competitions is structured as a database. The editors make a sophisticated technical and legal work on all articles by tagging these with key words, drafting abstracts and writing html code to increase Google ranking. There is a team of antitrust lawyers – PhD and judges clerks – and a team of IT experts. e-Competitions makes comparative law possible. Thanks to this expert editorial work, it is possible to search and compare cases by jurisdiction, legal topics or business sectors.

Prestigious Boards

e-Competitions draws upon highly distinguished editors, all leading experts in national or international antitrust. Advisory Board Members include: Sir Christopher Bellamy, Ioanis Lianos (UCL), Eleanor Fox (NYU), Frédéric Jenny (OECD), Jacqueline Riffault-Silk (Cour de cassation), Wouter Wils (King's College London), etc.

Leading Partners

- Association of European Competition Law Judges: The AECLJ is a forum for judges of national Courts specializing in antitrust case law. Members timely feed e-Competitions with just released cases.

- Academics partners: Antitrust research centres from leading universities write regularly in e-Competitions: University College London, King's College London, Queen Mary University, etc.

- Law firms: Global law firms and antitrust niche firms write detailed cases summaries specifically for e-Competitions: Allen & Overy, Baker McKenzie, Cleary Gottlieb Steen & Hamilton, Jones Day, Norton Rose Fulbright, Skadden, White & Case, etc.

Concurrences +
THE COMPETITION LAW PORTAL

20 years of archives
40,000 articles

4 DATABASES

Concurrences
Access to latest issue and archives

- 15,000 articles from 2004 to the present
- European and national doctrine and case law

e-Competitions
Access to latest issue and archives

- 25,000 case summaries from 1911 to the present
- Case law of 85 jurisdictions

Books
Access to all Concurrences books

- 70 e-Books available
- PDF version

Conferences
Access to the documentation of all Concurrences events

- 600 conferences (Brussels, Hong Kong, London, New York, Paris, Singapore and Washington, DC)
- 350 PowerPoint presentations, proceedings and syntheses
- 550 videos
- Verbatim reports

NEW

New search engine
Optimized results to save time

- Search results sorted by date, jurisdiction, keyword, economic sector, author, etc.

New modes of access
IP address recognition

- No need to enter codes: immediate access
- No need to change codes when your team changes: offers increased security and saves time

Mobility

- Responsive design: site optimized for tablets and smartphones

Printed in the USA
CPSIA information can be obtained
at www.ICGtesting.com
JSHW060839061024
70889JS00008B/5